D1130442

Studies in Economic Theory

Volume 31

More information about this series at http://www.springer.com/series/2584

Jean-Michel Grandmont

Kazuo Nishimura · Alain Venditti
Nicholas C. Yannelis
Editors

Sunspots and Non-Linear Dynamics

Essays in Honor of Jean-Michel Grandmont

 Springer

Editors
Kazuo Nishimura
Research Institute for Economics
  and Business Administration
Kobe University
Kobe
Japan

Nicholas C. Yannelis
Department of Economics
University of Iowa
Iowa City, Iowa
USA

Alain Venditti
Aix-Marseille University (Aix-Marseille
  School of Economics), CNRS-EHESS
  and EDHEC Business School
Marseille
France

ISSN 1431-8849   ISSN 2196-9930   (electronic)
Studies in Economic Theory
ISBN 978-3-319-44074-3   ISBN 978-3-319-44076-7   (eBook)
DOI 10.1007/978-3-319-44076-7

Library of Congress Control Number: 2016947482

Printed on acid-free paper

This Springer imprint is published by Springer Nature
The registered company is Springer International Publishing AG
The registered company address is: Gewerbestrasse 11, 6330 Cham, Switzerland

Contents

1 Introduction 1
Kazuo Nishimura, Alain Venditti and Nicholas C. Yannelis

Part I Self-Fulfilling Expectations and Sunspots

**2 Assessing the Local Stability Properties of Discrete
Three-Dimensional Dynamical Systems: A Geometrical
Approach with Triangles and Planes and an Application
with Some Cones** 15
Jean-Paul Barinci and Jean-Pierre Drugeon

**3 From Sunspots to Black Holes: Singular Dynamics
in Macroeconomic Models** 41
Paulo B. Brito, Luís F. Costa and Huw D. Dixon

**4 Sunspot Fluctuations in Two-Sector Models
with Variable Income Effects** 71
Frédéric Dufourt, Kazuo Nishimura, Carine Nourry
and Alain Venditti

**5 From Self-Fulfilling Mistakes to Behavioral
Learning Equilibria** 97
Cars Hommes

**6 Regime-Switching Sunspot Equilibria in a One-Sector
Growth Model with Aggregate Decreasing Returns
and Small Externalities** 125
Takashi Kamihigashi

**7 Equilibrium Dynamics in a Two-Sector OLG Model
with Liquidity Constraint** 147
Antoine Le Riche and Francesco Magris

8   Homoclinic Orbit and Stationary Sunspot Equilibrium
     in a Three-Dimensional Continuous-Time Model
     with a Predetermined Variable............................ 175
     Hiromi Murakami, Kazuo Nishimura and Tadashi Shigoka

Part II   Bubbles and Stabilizing Policy

9   Rational Land and Housing Bubbles in Infinite-Horizon
     Economies 203
     Stefano Bosi, Cuong Le Van and Ngoc-Sang Pham

10   The Stabilizing Virtues of Monetary Policy on Endogenous
     Bubble Fluctuations...................................... 231
     Lise Clain-Chamosset-Yvrard and Thomas Seegmuller

11   Can Consumption Taxes Stabilize the Economy
     in the Presence of Consumption Externalities?.................. 259
     Teresa Lloyd-Braga and Leonor Modesto

Part III   Growth

12   Uncertainty and Sentiment-Driven Equilibria................... 281
     Jess Benhabib, Pengfei Wang and Yi Wen

13   Technological Progress, Employment and the Lifetime
     of Capital... 305
     Raouf Boucekkine, Natali Hritonenko and Yuri Yatsenko

14   Nonbalanced Growth in a Neoclassical Two-Sector
     Optimal Growth Model.................................... 339
     Harutaka Takahashi

Part IV   General Equilibrium

15   An Argument for Positive Nominal Interest 363
     Gaetano Bloise and Herakles Polemarchakis

16   Winners and Losers from Price-Level Volatility:
     Money Taxation and Information Frictions 387
     Guido Cozzi, Aditya Goenka, Minwook Kang and Karl Shell

17   A Note on Information, Trade and Common Knowledge......... 403
     Leonidas C. Koutsougeras and Nicholas C. Yannelis

Chapter 1
Introduction

Kazuo Nishimura, Alain Venditti and Nicholas C. Yannelis

We are delighted to present this book honoring Professor Jean-Michel Grandmont. Over his illustrious career, Jean-Michel has made a number of highly important contributions to economic theory, in particular on non-linear dynamics and sunspots. These topics have been the core of the International Conference on *Instability and public policies in a globalized world: Conference in honor of Jean-Michel Grandmont*, organized at Aix-Marseille School of Economics and GREQAM on June 6–8, 2013. This book presents the state-of-art in non-linear dynamics and sunspots by his colleagues, students and friends that have been influenced by and are admirers of his work.

Jean-Michel already stood out as a brilliant figure when he graduated at Polytechnique in 1962 among the very best, 10th out of more than 300, and he entered one of the most prestigious Corps, the Corps des Ponts et Chaussées. Fortunately, Jean-Michel went to research in economics although getting a business position is the standard career of french engineers from prestigious schools.

K. Nishimura (✉)
Research Institute for Economics and Business Administration,
Kobe University, Kobe, Japan
e-mail: nishimura@rieb.kobe-u-ac.jp

A. Venditti
Aix-Marseille University (Aix-Marseille School of Economics)-
CNRS-EHESS, Marseille, France
e-mail: alain.venditti@univ-amu.fr

A. Venditti
EDHEC Business School, Nice, France

N.C. Yannelis
University of Iowa, Iowa, IA, USA
e-mail: nicholasyannelis@gmail.com

© Springer International Publishing AG 2017
K. Nishimura et al. (eds.), *Sunspots and Non-Linear Dynamics*,
Studies in Economic Theory 31, DOI 10.1007/978-3-319-44076-7_1

1

Jean-Michel obtained his Ph.D. in Berkeley in 1971 where he worked "On the Temporary Competitive Equilibrium" under the supervision of Gérard Debreu, Nobel Prize in Economics in 1983. Over the last 40 years, Jean-Michel contributed to diverse areas of economics ranging from the general equilibrium theory to monetary theory, learning, aggregation, non-linear dynamics and sunspots. On every count, his work has been characterized by insight, originality and technical clarity as well as rigor.

Jean-Michel came back to France in 1970 and joined the CEPREMAP in Paris where he stayed until 1996. There, he published pioneering papers on general equilibrium with money, temporary equilibria, learning, aggregation, and of course bifurcations and chaos in economic dynamics which strongly improved our understanding of the interrelationships between business cycles and economic growth. In 1996, he moved to the Centre de Recherche en Economie et Statistique (CREST), where he has remained ever since. He has been Associate Professor from 1977 to 1992 and Professor of Economics from 1992 to 2004 at Ecole Polytechnique. He is also affiliated to the Department of Economics of Ca' Foscari University of Venice since 2004 and has been nominated a fellow of the Research Institute for Economics and Business Administration at Kobe University in 2008.

From 1996 to 2000, Jean-Michel served as the chairman of the Department of Economics at Ecole Polytechnique. During this period he was responsible for recruiting new faculty members and building an excellent undergraduate program, tasks in which he showed initiative and executive ability. As a result, the quality of the teaching at Ecole Polytechnique has been significantly improved and the number of students that the program attracted has been strongly increased.

Jean-Michel also got involved in promoting research in quantitative economics, mainly through the Econometric Society. He has been nominated Chairman in 1977, Vice President in 1988 and then President in 1990 of the Econometric Society.

He has received the Honorary Degree from Keio University in 2007 for his contributions to economic theory. He was also selected a foreign member of the American Academy of Arts and Sciences (AAAS) in 1992. He received the Palmes Académiques from the French Ministry of Education in 1995.

In addition, he has served on the editorial boards of many top journals in economics. In particular, he became an associate editor of Econometrica from 1976 to 1983 and of the Journal of Economic Theory from 1973 to 2015. Since 1995, he is a co-Editor of the International Journal of Economic Theory.

His major contributions can be classified into five groups of topics that remain however strongly linked through a general and comprehensive view of Economics.

1. **General equilibrium with money.** The 1983 book *Money and Value* provides a thorough and rigorous treatment of topics as various as the quantity theory of money, the classical dichotomy, short run and long run effects of monetary policy and expectations. See also [4, 5, 6, 7, 8, 12].
2. **Temporary equilibrium.** The papers [1, 2, 11] provide all the basic necessary tools to discuss the role of expectations in the determination of equilibrium. A number of applications followed: temporary competitive equilibria, temporary

Keynesian equilibria, monetary policy in the short and in the long run [9, 10, 13, 14, 15, 16, 18, 26].

3. **Learning.** The rational expectations assumption naturally finds its justification into its relationships to the learning processes that lead the agents to revise their beliefs until they hold these expectations. Sufficient conditions on these processes, either for convergence or divergence, in non linear deterministic models, appear in [23, 28, 29, 30, 36].

4. **Non linear cycles and sunspots.** This important field was initiated in [19], and developed in [20, 22, 24, 25, 31, 34, 37]. It is still quite active today: non linear dynamics and chaos make most of the program of this book.

5. **Aggregation.** Aggregation is a basic question raised by the microeconomic foundations of macroeconomics. It is a difficult subject that Jean Michel has studied in different ways. The first paper [17] on this topic shows the boundaries of the Arrow theorem and provides extremely useful conditions for the applicability of the median voter theorem. Later studies deal with aggregate demand [21, 32], learning [33], and more recently finance [38].

Jean-Michel Grandmont also chose to devote a large fraction of his time to teaching and thesis management. Many students, in several institutions (CORE, University of Aix-Marseille, Bonn, Lisbon, Strasbourg, Paris, Venice, Yale, École Polytechnique), have benefited of Jean Michel's advices and encouragements. A number of them figure prominently among the contributors to this volume.

One of the co-editors of this book has clear recollections on his first meeting with Jean-Michel. He was at the very end of his year of military service and, he believed, at the very end of his Ph.D. Thesis. He had been advised by one of his co-advisor, the late Louis-André Gérard-Varet, to meet Jean-Michel in order to get advices on his work. So he obtained very easily an appointment in Paris at CEPREMAP. When arriving in his office, it was quite obvious that Jean-Michel did not have time to look at the material of the thesis before the appointment. But then, after an extremely quick and highly efficient reading of the main parts of the chapters, Jean-Michel clearly explained that much more additional work was necessary in order to get an acceptable document. He provided a large number of remarks to improve many results and asked the disappointed young co-editor to come back after having revised his document. While strongly depressed, he decided to follow Jean-Michel's advices and a few months later the thesis had indeed significantly improved! This is just one anecdote but it could almost literally apply to many former students from CORE, Lisbon, Paris, Marseille, Strasbourg or Yale who are contributors to this book.

1.1 Overview of the Papers

The 16 contributions presented in this book are grouped into four different topics on which Jean-Michel has made a large number of contributions, as presented previously.

1.1.1 Self-fulfilling Expectations and Sunspots

The opening paper by Jean-Paul Barinci and Jean-Pierre Drugeon, "*Assessing the Local Stability Properties of Discrete Three-Dimensional Dynamical Systems: a Geometrical Approach with Triangles and Planes and an Application with some Cones*", provides a general analysis of the determinacy properties of three-dimensional discrete-time dynamical systems. The authors introduce a new geometrical argument which brings about a complete typology of the eigenvalues moduli and then provide a new apparatus for assessing from a geometrical standpoint the emergence of local bifurcations for parameterized economies. This general methodology is illustrated through the extensive characterization of the stability properties of a standard overlapping-generations (OLG) model with endogenous labor.

In the second paper, "*From Sunspots to Black Holes: Singular Dynamics in Macroeconomic Models*", Paulo Brito, Luis Costa and Huw Dixon present conditions for the emergence of singularities in dynamic general equilibrium (DGE) models. The concept of *impasse singularity* is introduced to exhibit new types of DGE dynamics, in particular temporary determinacy/indeterminacy. These results are illustrated through two simple models: the Benhabib and Farmer (1994) one-sector model with aggregate externalities and one with a cyclical fiscal policy rule.

Brito, Costa and Dixon's paper is followed by "*Sunspot Fluctuations in Two-Sector Models with Variable Income Effects*". Here, Frédéric Dufourt, Kazuo Nishimura, Carine Nourry and Alain Venditti analyse a version of the Benhabib and Farmer (1996) two-sector model with sector-specific externalities in which they consider a class of utility functions inspired from the one considered in Jaimovich and Rebelo (2009) which is flexible enough to encompass varying degrees of income effect. They first show that local indeterminacy and sunspot fluctuations occur under plausible configurations regarding all structural parameters—in particular regarding the intensity of income effects. Second, they prove that, for any given size of income effect, there is a non-empty range of values for the Frisch elasticity of labor and the elasticity of intertemporal substitution in consumption such that indeterminacy occurs.

In the fourth paper of this group, "*A Survey on Self-Fulfilling Mistakes*", Cars Hommes links some of his own work on expectations, learning and bounded rationality to the inspiring ideas of Jean-Michel Grandmont. In particular, his work on consistent expectations and behavioral learning equilibria may be seen as formalizations of Jean-Michel's ideas of self-fulfilling mistakes contained in [36]. Some of his learning-to-forecast laboratory experiments with human subjects have also been strongly influenced by Jean-Michel's ideas. Key features of self-fulfilling mistakes are multiple equilibria, excess volatility and persistence amplification.

This survey paper is followed by the contribution of Takashi Kamihigashi, "*Regime-Switching Sunspot Equilibria in a One-Sector Growth Model with Aggregate Decreasing Returns and Small Externalities*". It is shown that regime-switching sunspot equilibria, in which labor supply is positive in one state while it is zero in the other, easily arise in a standard one-sector growth model with aggregate decreasing returns and arbitrarily small externalities. Regime-switching sunspot equilibria are explicitly constructed in the case where the utility function of consumption is linear. A stochastic optimal growth model whose optimal process appears to be a regime-switching sunspot equilibrium of the original economy without capital externality is also constructed.

The sixth paper of this group is provided by Antoine Le Riche and Francesco Magris, "*Equilibrium Dynamics in a Two-Sector OLG Model with Liquidity Constraint*". They study a two-sector OLG economy in which a share of old age consumption expenditures must be paid out of money balances. It is first shown that the competitive equilibrium is dynamically efficient if and only if the share of capital on total income is large enough while a steady state capital per capita above its Golden Rule level is not consistent with a binding liquidity constraint. Assuming gross substitutability in consumption, they show that the dynamic efficiency property ensures the local determinacy of equilibrium and, as a consequence, rule out sunspot fluctuations. In addition, they provide a detailed analysis of the dynamic properties of the equilibrium focusing on flip and Hopf bifurcations under different sectoral capital intensity configurations.

The last contribution of this group is "*Sunspots and Homoclinic Bifurcations in Continuous-Time Endogenous Growth Models*". Here, Hiromi Murakami, Kazuo Ni-shimura and Tadashi Shigoka consider a three-dimensional continuous-time general stationary model that includes one predetermined variable and two non-predetermined variables. Assuming that the model has a two-dimensional invariant manifold and that the manifold includes a one-dimensional closed curve that could be either a homoclinic orbit or a closed orbit, sunspot equilibria are constructed. These results are then applied to two-sector endogenous growth models that are variants of Lucas (1988) model.

1.1.2 Bubbles and Stabilizing Policy

In "*Rational Land and Housing Bubbles in Infinite-Horizon Economies*", Stefano Bosi, Cuong Le Van and Ngoc-Sang Pham consider rational land and housing bubbles in an infinite-horizon general equilibrium model. Land is an input to produce while the house is a (durable) good to consume. It is shown that dividends on both these long-lived assets are endogenous and their sequences are computed. Then different concepts of bubbles, including individual and strong bubbles, are introduced and studied.

This paper is followed by "*The Stabilizing Virtues of Monetary Policy on Endogenous Bubble Fluctuations*" where Lise Clain-Chamosset-Yvrard and Thomas Seegmuller explore the stabilizing role of monetary policy on the existence of endogenous fluctuations when the economy experiences a rational bubble. Considering an OLG model, expectation-driven fluctuations are based on the co-existence of portfolio choices between three assets (capital, bonds and money), credit market imperfections and a collateral effect, and are shown to occur under a positive bubble on bonds. Then, the stabilizing role of a monetary policy managed by a (standard) Taylor rule is studied.

The last paper of this set is by Teresa Lloyd-Braga and Leonor Modesto, "*Can Consumption Taxes Stabilize the Economy in the Presence of Consumption Externalities?*". The stabilization role of consumption taxes under a balanced-budget rule and in the presence of consumption externalities of the "keeping up with the Joneses" type is discussed in a finance constrained economy. Departing from a situation where sufficiently strong externalities make the steady state indeterminate, it is shown that sufficiently procyclical consumption tax rates are able to ensure local saddle path stability. However, government intervention with stabilization purposes may not be successful as this procyclicality leads to the appearance of another possibly indeterminate steady state with lower levels of output.

1.1.3 Growth

In "*Uncertainty and Sentiment-Driven Equilibria*", Jess Benhabib, Pengfei Wang and Yi Wen construct a simple neoclassical model to capture the Keynesian idea that equilibrium aggregate supply is determined by aggregate demand and thus influenced by consumer sentiments about aggregate income. They show that when firms' production and employment decisions must be based on expectations of aggregate demand and that realized demand follows from firms' production and employment decisions through market-clearing mechanisms, rational expectations about aggregate demand can lead to stochastic sentiment-driven equilibrium despite the absence of production externalities, incomplete financial markets, strategic complementarity or any non-convexities in the model.

This contribution is followed by "*Technological Progress, Employment and the Lifetime of Capital*", in which Raouf Boucekkine, Natali Hritonenko and Yuri Yatsenko study the impact of technological progress on the level of employment in a vintage capital model where capital and labor are gross complement, labor supply is endogenous and indivisible, there is full employment, and the rate of labor-saving technological progress is endogenous. The stationary distributions of vintage capital goods and the corresponding equilibrium values for employment and capital lifetime are characterized. It is shown that both variables are non-monotonic functions of technological progress indicators.

The last paper of this group is "*Nonbalanced Growth in a Neoclassical Two-sector Optimal Growth Model*". Here, Harutaka Takahashi considers a neoclassical two-sector optimal growth model with Cobb-Douglas technologies and sector specific technological progress. Non-balanced growth, i.e. the fact that the two sectors are characterized by different long-run balanced growth rates, is characterized. This result refers to the recent literature showing that Kaldor and Kuznets facts are compatible in standard growth theory.

1.1.4 General Equilibrium

In "*An Argument for Positive Nominal Interest*", Gaetano Bloise and Herakles Polemarchakis consider a dynamic OLG economy in which money provides liquidity as a medium of exchange. A central bank, that sets the nominal rate of interest and distributes its profit to shareholders as dividends is characterized by shares that are traded in the asset market. It is shown that nominal rates of interest that tend to zero, but do not vanish, eliminate equilibrium allocations that do not converge to a Pareto optimal allocation.

The second paper of this set is "*Winners and Losers from Price-Level Volatility: Money Taxation and Information Frictions*" by Guido Cozzi, Aditya Goenka, Minwook Kang and Karl Shell which analyzes an economy with taxes and transfers denominated in dollars and an information friction that allows for volatility in equilibrium prices and allocations. When the price level is expected to be stable, the competitive equilibrium allocation is Pareto optimal. When the price level is volatile, it is not Pareto optimal, but the stable equilibrium allocations do not necessarily dominate the volatile ones, and there can be winners and losers from volatility.

The final paper of this book, "*A Note on Information, Trade and Common Knowledge*", by Leonidas Koutsougeras and Nicholas C. Yannelis, is dealing with the well known no trade result in Milgrom-Stokey (1982) using the appropriate definition of efficiency among several available in the asymmetric information framework. It is shown that if Pareto efficiency is understood in the private information sense, i.e., allocations and possible reallocations are adapted to the private information of each individual, then the no trade result is valid.

All of us who have contributed to this book consider it a privilege to be able to honor Jean-Michel in this way. As colleagues, former students and friends, we have benefited from Jean-Michel's insights, encouragement and perseverance. His inspiration, his friendship, his generosity, and his contributions to economics are distinguished and invaluable. We are delighted to contribute our papers to this book in his honor and we look forward to our continuing relationships with Jean-Michel in the years to come.

Writing for the friends and colleagues of Jean-Michel
Kazuo Nishimura, Alain Venditti and Nicholas Yannelis

1.2 Main Scientific Writings of Jean-Michel Grandmont

Book

Money and Value, Econometric Society Monographs, Cambridge University Press, 1983. French version, Economica, 1986.

Articles

1. "Continuity Properties of a von Neumann-Morgenstern Utility", *Journal of Economic Theory*, 1972.
2. "On the Short Run Equilibrium in a Monetary Economy", in J. Drze (Ed.), *Allocation under Uncertainty, Equilibrium and Optimality*, McMillan, 1974.
3. "A Technical Note on Classical Gains from Trade", *Journal of International Economics*, 1972, with D. McFadden.
4. "On the Role of Money and the Existence of a Monetary Equilibrium", *Review of Economic Studies*, 1972, with Y. Younès.
5. "On the Efficiency of a Monetary Equilibrium", *Review of Economic Studies*, 1973, with Y. Younès.
6. "Sur la demande de monnaie de court terme et de long terme", *Annales de l'INSEE*, 1972. English version "On the Short Run and Long Run Demand for Money", *European Economic Review*, 1973.
7. "Sur les taux d'intérêt en France", *Revue Économique*, 1973, with G. Neel.
8. "Money in the Pure Consumption Loan Model", *Journal of Economic Theory*, 1973, with G. Laroque.
9. "Foreign Exchange Markets: A Temporary General Equilibrium Approach", CORE DP, 1973, Catholic University of Louvain, with A.P. Kirman.
10. "Monnaie et Banque Centrale", *Annales de l'INSEE*, 1973, with G. Laroque. English version: "On Money and Banking", *Review of Economic Studies*, 1975.
11. "Stochastic Processes of Temporary Equilibria", *Journal of Mathematical Economics*, 1974, with W. Hildenbrand.
12. "On the Liquidity Trap", *Econometrica*, 1976, with G. Laroque.
13. "On Temporary Keynesian Equilibria", *Review of Economic Studies*, 1976, with G. Laroque.
14. "Temporary General Equilibrium Theory", *Econometrica*, 1977. Invited Lecture at the World Congress of the Econometric Society, Toronto, 1975. Shorter version in M. Intriligator (Ed.), *Frontiers of Quantitative Analysis*, North Holland, 1977. French version: "Théorie de l'équilibre temporaire", *Revue Économique*, 1976.
15. "Equilibrium with Quantity Rationing and Recontracting", *Journal of Economic Theory*, 1978, with G. Laroque and Y. Younès.
16. "The Logic of the Fix Price Method", *Scandinavian Journal of Economics*, 1977. Reprinted in S. Ström and L. Werin (Eds), *Topics in Disequilibrium Economics*, McMillan, 1978.
17. "Intermediate Preferences and the Majority Rule", *Econometrica*, 1978. Reprinted in *Aggregation and Revelation of Preferences*, J.J. Laffont (Ed.), North Holland, 1979.

18. "Classical and Keynesian Unemployment in the IS-LM Model", in *Monetary Theory and Institutions*, M. de Cecco and J.P. Fitoussi (Eds), McMillan, 1986. Spanish version in *Analysis Economico*, 1984, Mexico.

19. "On Endogenous Competitive Business Cycles", Walras-Bowley Lecture at the North American Summer meetings of the Econometric Society, Stanford, 1984. Published in *Econometrica*, 1985. Summary in H. Sonnenschein (Ed.), *Models of Economic Dynamics*, Springer-Verlag, 1986. French Version: "Cycles concurrentiels endognes", *Cahiers du Séminaire d'Économétrie*, 1985.

20. "Periodic and Aperiodic Behaviour in Discrete Onedimensional Dynamic Systems", in *Contributions to Mathematical Economics*, W. Hildenbrand and A. Mas-Colell (Eds), North Holland, 1986.

21. "Distributions of Preferences and the Law of Demand", *Econometrica*, 1987.

22. "Stabilizing Competitive Business Cycles", *Journal of Economic Theory*, 1986. Reprinted in *Nonlinear Economic Dynamics*, J.M. Grandmont (Ed.), Academic Press, 1987.

23. "Stability of Cycles and Expectations", *Journal of Economic Theory*, 1986, with G. Laroque, Reprinted in *Nonlinear Economic Dynamics*, J.M. Grandmont (Ed.), Academic Press, 1987.

24. "Nonlinear Difference Equations, Bifurcations and Chaos: An Introduction", CEPREMAP, June 1988, and Stanford Technical Reports, 1988.

25. "Local Bifurcations and Stationary Sunspots", in W.A. Barnett, J. Geweke and K. Shell (Eds), *Economic Complexity: Chaos, Sunspots, Bubbles and Nonlinearity*, Cambridge University Press, 1989.

26. "Keynesian Issues and Economic Theory", *Scandinavian Journal of Economics*, June 1989.

27. "Report on M. Allais' Scientific Work", *Scandinavian Journal of Economics*, January 1989.

28. "Stability, Expectations and Predetermined Variables", with G. Laroque, in P. Champsaur et alii (Eds), *Essays in Honour of E. Malinvaud*, Vol. I, MIT Press, 1990.

29. "Temporary Equilibrium: Money, Expectations and Dynamics", in L. McKenzie and S. Zamagni (Eds.), *Value and Capital. Fifty Years later*, McMillan, 1991.

30. "Economic Dynamics with Learning: Some Instability Examples", with G. Laroque, in Barnett, W., Cornet, B. et alii (Eds.), *Equilibrium Theory and Applications*, Cambridge University Press, 1991.

31. "Expectations Driven Business Cycles", *European Economic Review*, April 1991.

32. "Transformations of the Commodity Space, Behavioral Heterogeneity and the Aggregation Problem", *Journal of Economic Theory*, June 1992.

33. "Aggregation, Learning and Rationality", *Proceedings of the 1992 World Congress of the International Economic Association* (Moscow), McMillan, 1996.

34. "Expectations Driven Nonlinear Business Cycles", *Proceedings of the German Academy of Sciences*, Wesdeutscher Verlag 1993, and in *FIEF Studies on Busi-*

ness Cycles, Oxford University Press, 1994. Shorter version in *MITA Journal of Economics* (Tokyo), 1994.

35. "Behavioral Heterogeneity and Cournot Oligopoly Equilibrium", *Ricerche Economiche*, 1993.
36. "Expectations Formation and Stability of Large Socioeconomic Systems", *Econometrica*, 1998.
37. "Capital Labor Substitution and Nonlinear Endogenous Business Cycles", *Journal of Economic Theory*, 1998, with P. Pintus and R. De Vilder.
38. "Heterogenous Probabilities in Complete Asset Markets", *Advances in Mathematical Economics*, 1999 (Springer Verlag, S. Kusuoka and T. Maruyama, Eds), with L. Calvet and I. Lemaire. Japanese translation in *Mita Journal of Economics*, Tokyo, 1999.

1.3   List of Papers

Self-fulfilling Expectations and Sunspots

1. Jean-Paul Barinci and Jean-Pierre Drugeon: "Assessing the Local Stability Properties of Discrete Three-Dimensional Dynamical Systems: a Geometrical Approach with Triangles and Planes & an Application with some Cones".

2. Paulo Brito, Luis Costa, and Huw Dixon: "From Sunspots to Black Holes: Singular Dynamics in Macroeconomic Models".

3. Frédéric Dufourt, Kazuo Nishimura, Carine Nourry and Alain Venditti: "Sunspot Fluctuations in Two-Sector Models with Variable Income Effects".

4. Cars Hommes: "From Self-Fullling Mistakes to Behavioral Learning Equilibria".

5. Takashi Kamihigashi: "Regime-Switching Sunspot Equilibria in a One-Sector Growth Model with Aggregate Decreasing Returns and Small Externalities".

6. Antoine Le Riche and Francesco Magris: "Equilibrium Dynamics in a Two-Sector OLG Model with Liquidity Constraint".

7. Hiromi Murakami, Kazuo Nishimura and Tadashi Shigoka: "Homoclinic orbit and stationary sunspot equilibrium in a three-dimensional continuous-time model with a predetermined variable".

Bubbles and Stabilizing Policy

8. Stefano Bosi Cuong Le Van and Ngoc-Sang Pham: "Rational Land and Housing Bubbles in Infinite-Horizon Economies".

9. Lise Clain-Chamosset-Yvrard and Thomas Seegmuller: "The Stabilizing Virtues of Monetary Policy on Endogenous Bubble Fluctuations".

10. Teresa Lloyd-Braga and Leonor Modesto: "Can Consumption Taxes Stabilize the Economy in the Presence of Consumption Externalities?".

Growth

11. Jess Benhabib, Pengfei Wang and Yi Wen: "Uncertainty and Sentiment-Driven Equilibria".

12. Raouf Boucekkine, Natali Hritonenko and Yuri Yatsenko: "Technological Progress, Employment and the Lifetime of Capital".

13. Harutaka Takahashi: "Nonbalanced Growth in a Neoclassical Two-sector Optimal Growth Model".

General Equilibrium

14. Gaetano Bloise and Herakles Polemarchakis: "An Argument for Positive Nominal Interest".

15. Guido Cozzi, Aditya Goenka, Minwook Kang and Karl Shell: "Winners and Losers from Price-Level Volatility: Money Taxation and Information Frictions".

16. Leonidas Koutsougeras and Nicholas C. Yannelis: "A Note on Information, Trade and Common Knowledge".

Part I
Self-Fulfilling Expectations and Sunspots

Chapter 2
Assessing the Local Stability Properties of Discrete Three-Dimensional Dynamical Systems: A Geometrical Approach with Triangles and Planes and an Application with Some Cones

Jean-Paul Barinci and Jean-Pierre Drugeon

Abstract The difficulties associated with the appraisal of the determinacy properties of a three-dimensional system are circumvented by the introduction of a new
geometrical argument. It brings about a complete typology of the eigenvalues moduli in discrete time three-dimensional dynamical systems and then provides a new
apparatus for assessing from a geometrical standpoint the emergence of local bifurcations for parameterised economies. The argument is considered through the extensive
characterisation of the stability properties of a benchmark model of inter-temporal
economic analysis.

Keywords Discrete time dynamical systems · Geometrical approach · Parameterised economies

JEL Classification: E32 · C62

This research was completed thanks to the supports of the Novo Tempus research grant, ANR-
12-BSH1-0007, Program BSH1-2012, and of the Labex MME-DII. The authors would like
to thank Jean-Michel Grandmont for the numerous conversations they benefitted from after a
seminar they held at Crest. They are also indebted to the anonymous referee for his suggestions.
The usual disclaimer nonetheless applies.

J.-P. Barinci
E.P.E.E., Université d'Evry Val d'Essonne, Évry, France

J.-P. Drugeon (✉)
Paris School of Economics and Centre National de la Recherche Scientifique,
Paris, France
e-mail: jpdrug@parisschoolofeconomics.eu

2.1 Introduction

The difficulties associated with the appraisal of the local determinacy properties
of a three-dimensional discrete time dynamical system have long deterred a more
widespread use of the associated setups in economic theory. This contribution is
intended to introduce graphical methods for assessing the stability and the scope
for local bifurcations within such systems. It also provides some illustrations in
parameterised economies.

As they reconsider the role of factors substitutability in competitive economies,
Grandmont (1998) and Grandmont et al. (1998) have come to introduce a tractable
graphical way of assessing local uniqueness or local indeterminacy for dynamical
systems of order two. Their approach is based upon a graphical partition of the
$(\mathscr{T}\mathscr{D})$-plane defined from the two coefficients $\mathscr{T}$ and $\mathscr{D}$ of the second-order characteristic polynomial $P(z) = z^2 - \mathscr{T}z + \mathscr{D}$ that is associated with a two-dimensional
dynamical system in the neighbourhood of some steady state, these coefficients $\mathscr{T}$
and $\mathscr{D}$ being assumed to depend upon a range of n parameters $\{\lambda_1, \lambda_2, \ldots, \lambda_{n-1}, \lambda_n\}$.
Such a partition is then completed by drawing the linear critical loci associated to
the occurrence of real and complex eigenvalues with unitary modulus, respectively
two straight-lines (AB) and (AC) of slopes $+1$ and -1 and a horizontal segment
[BC] over the plane defined from the coefficients $\mathscr{T}$ and $\mathscr{D}$. These critical loci
feature boundaries between stability and instability zones, a *full stability*—all the
moduli of the eigenvalues are less than one—zona being noticeably depicted by
the interior of the triangle (ABC). A given economy—a set of fundamental prefer-
ences and technological parameterisations $\{\lambda_1, \lambda_2, \ldots, \lambda_{n-1}, \lambda_n\}$—was then to be
understood as a point over that plane whilst the appraisal of its local dynamics sum-
marised to the localisation of this point. Letting one of its building parameters, say
some $\lambda_i \in \{\lambda_1, \lambda_2, \ldots, \lambda_{n-1}, \lambda_n\}$, vary gives rise to a family of economies, namely a
curve $_{\lambda_i}\Delta$, over that plane the localisation of which provided insights about the associated qualitative changes undergone by the dynamical properties of the economy.
The crux interest of this construction for economic theory stems from its explicit consideration of meaningful and generic concepts without having to resort to specific
parametric formulations. That graphical method was remarkable from its tractability
and its potential for significantly easing the appraisal of otherwise complex formal
structures.

Two key difficulties however quickly emerge as being associated with the extension of the above approach to three-dimensional dynamical systems and the elab-
oration of a graphical partition of a three-dimensional $(\mathscr{T}\mathscr{M}\mathscr{D})$-space defined
from the three coefficients $\mathscr{T}$, $\mathscr{M}$ and $\mathscr{D}$ of the third-order characteristic polynomial $Q(z) = -z^3 + \mathscr{T}z^2 - \mathscr{M}z + \mathscr{D}$ in the neighbourhood of some steady state.
Firstly, the intricacies of three-dimensional graphs and the geometry of a three-
dimensional $(\mathscr{T}\mathscr{M}\mathscr{D})$-space are far more difficult to grasp than the aforementioned
two-dimensional pictures in a two-dimensional $(\mathscr{T}\mathscr{D})$-space. Secondly, the elabora-
tion of a graphical partition is anchored on the introduction of critical loci associated
with the occurrence of eigenvalues with unitary modulus: whilst linearity keeps on

being an attribute of the critical loci associated with real eigenvalues—this results in planes in the $(\mathcal{T}\mathcal{M}\mathcal{D})$-space, one is now faced with the *uprise of a nonlinear critical locus* in order to picture the occurrence of complex eigenvalues with unitary modulus. The first of these issues shall be circumvented by apprehending the original three-dimensional $(\mathcal{T}\mathcal{M}\mathcal{D})$-space through a collection of sections along the $\mathcal{D}$ coordinate and thus of $(\mathcal{T}\mathcal{M})_{\mathcal{D}}$ planes parameterized by $\mathcal{D}$. Fortunately enough, such an approach also entails linear definitions for the three parameterised critical loci: one indeed recovers two straight-lines $(A_{\mathcal{D}}B_{\mathcal{D}})$ and $(A_{\mathcal{D}}C_{\mathcal{D}})$ of slopes $+1$ and -1 and a segment $[B_{\mathcal{D}}C_{\mathcal{D}}]$—its slope is now to vary according to $\mathcal{D}$—over a finite collection of planes $(\mathcal{T}\mathcal{M})_{\mathcal{D}}$ that are also *parameterised* by the coefficient $\mathcal{D}$ and defined for $|\mathcal{D}| < 1$, $\mathcal{D} < -1$ and $\mathcal{D} > 1$, the interior of the parameterised triangle $(A_{\mathcal{D}}B_{\mathcal{D}}C_{\mathcal{D}})$ being accordingly changed from a *full stability* area—all the moduli of the eigenvalues are less than one—to a *full instability* one—all the moduli of the eigenvalues are greater than one.

Assuming further that the coefficients $\mathcal{T}$ and $\mathcal{M}$ depend upon a range of n parameters $\{\lambda_1, \lambda_2, \ldots, \lambda_{n-1}, \lambda_n\}$ while the coefficient $\mathcal{D}$ depends upon *at most* $n - 1$ such parameters, say the $n - 1$ first ones, $\{\lambda_1, \lambda_2, \ldots, \lambda_{n-1}\}$, the appraisal of the range of configurations admissible for a given economy can anew be completed over a parameterised plane. Letting indeed the parameter λ_n vary gives rise to a family of economies, namely a curve $_{\lambda_n}\Delta$, over a parameterised plane—it is *uniquely* defined for a given $\mathcal{D}$—the localisation of which provides insights about the associated qualitative changes undergone by the dynamical properties of the economy when λ_n spans its interval of admissible values.

As an illustration of the appropriateness of this approach, an overlapping generations model is subsequently analysed, the two-period overlapping generations model having become a workhorse for the theory of descriptive fluctuations. The literature on the subject has focused on both the Samuelson's (1958) pure exchange framework, e.g., Grandmont (1985), and the Diamond's (1965) setting with productive capital, e.g., Reichlin (1986). The present contribution more precisely considers a slightly modified Diamond's setting in order to illustrate the easiness of use of the graphical method and the tools it introduces. The first departure from the original framework lies in the consideration of a labor-leisure arbitrage in the first-period of agent's life. The requirement of capital-wealth equality is further relaxed, i.e., it is not any longer assumed that the consumer's wealth is equal to the value of the capital stock. Here private wealth—the sum of private assets–and capital are assumed to be separate entities. This is an extension of the pure-exchange Gale's (1972) model in which the equilibrium value of the private wealth could be nonzero. Taking over the Gale's terminology, economies for which private wealth is smaller, respectively greater, than capital are labelled as *Classical*, respectively *Samuelson*. The local dynamics nearby the steady state of both type of economies are characterized thanks to the aforementioned tools and the parallel use of some infinite cones whose generatrices are defined by a boundary featuring strict concavity and a boundary denoting the border between gross substitutability and complementarity for leisure and first-period consumption.

Beyond the specifics of that setup, the current class of techniques can be applied to quite a large range of parameterised environments that would result in third-order dynamical systems. In models of economics, the applicability of the whole approach revolves around the identification of some fundamental parameter that would not appear into the coefficient $\mathcal{D}$, i.e., the one that corresponds to the product of the eigenvalues. Even though such a qualification may sound as being restrictive, the computation of that coefficient being commonly the most difficult for a given Jacobian Matrix, it typically uncovers a rather simplified analytical form with respect to $\mathcal{T}$ and $\mathcal{M}$ and thus a dependence with respect to a fewer range of coefficients, a property that should prove useful in potential future applications.

The geometrical techniques are introduced in Sect. 2.2. Section 2.3 builds upon an extension of the pure-exchange model of overlapping generations. Some formal details are provided in a final appendix.

2.2   A Geometrical Argument for the Appraisal of the Local Stability Properties of Three-Dimensional Dynamical Systems

2.2.1   A Simple Typology for the Eigenvalues of a Discrete Three-Dimensional Dynamical System

Letting the equilibrium dynamics of an economy be described by a system: $y_{t+1} = G(y_t)$, $y_t \in \mathbb{R}^3_+$, steady states equilibria are the roots of $\bar{y} - G(\bar{y}) = 0$. The characterisation of the local dynamics nearby a given steady equilibrium proceeds from the appraisal of an associated linear map $\zeta_{t+1} = \mathcal{J}\zeta_t$, for $\mathcal{J} := DG(\bar{y})$ the Jacobian matrix of $G(\cdot)$ evaluated at $\bar{y}$ and $\zeta_t := y_t - \bar{y}$ the deviation from the steady state. The eigenvalues of the matrix $\mathcal{J}$ are the zeroes of the following third order polynomial:

$$
\begin{aligned}
Q(z) &= (z_1 - z)(z_2 - z)(z_3 - z) \qquad\qquad\qquad\qquad (2.1)\\
&= -z^3 + (z_1 + z_2 + z_3)z^2 - (z_1z_2 + z_1z_3 + z_2z_3)z + z_1z_2z_3 \\
&= -z^3 + \mathcal{T}z^2 - \mathcal{M}z + \mathcal{D},
\end{aligned}
$$

for $\mathcal{T}$, $\mathcal{M}$ and $\mathcal{D}$ that respectively denote the trace, the sum of the principal minors of order two and the determinant of the Jacobian matrix $\mathcal{J} := DG(\bar{y})$.

The locus such that the coefficients $\mathcal{T}$, $\mathcal{M}$, $\mathcal{D}$ satisfy $Q(+1) = 0$ is a plane—henceforward referred to as the *saddle-node critical plane*—of the $(\mathcal{T}\mathcal{M}\mathcal{D})$-space whose characteristic equation is given by:

$$
-1 + \mathcal{T} - \mathcal{M} + \mathcal{D} = 0. \qquad\qquad\qquad (2.2)
$$

Generically, a saddle-node bifurcation[1] will occur when the triple $(\mathscr{T}, \mathscr{M}, \mathscr{D})$ crosses this plane and the uniqueness properties of the steady state will be lost. Similarly, the locus such that the coefficients $\mathscr{T}, \mathscr{M}, \mathscr{D}$ satisfy $Q(-1) = 0$ is a plane—henceforth mentioned as the *flip critical plane*—of the $(\mathscr{T}\mathscr{M}\mathscr{D})$-space whose characteristic equation is given by:

$$1 + \mathscr{T} + \mathscr{M} + \mathscr{D} = 0. \tag{2.3}$$

A flip bifurcation is bound to occur in its neighbourhood when the triple $(\mathscr{T}, \mathscr{M}, \mathscr{D})$ crosses this plane and two-period cycles will emerge.

Lastly, when a pair of nonreal characteristic roots exhibiting an unitary norm occurs, the remaining eigenvalue, e.g., z_3, summarises to the product of the eigenvalues $\mathscr{D}$.[2] The latter coefficient thus becomes a characteristic root, i.e., $Q(\mathscr{D}) = 0$. Solving, the characteristic polynomial hence restates as $Q(z) = (\mathscr{D} - z)P(z)$, for $P(z) = z^2 - (\mathscr{T} - \mathscr{D})z + \mathscr{M} - (\mathscr{T} - \mathscr{D})\mathscr{D}$. A standard analysis of $P(\cdot)$ then indicates that the locus of coefficients $\mathscr{T}, \mathscr{M}$ and $\mathscr{D}$ such that two roots are complex conjugate with unitary modulus is given by:

$$\mathscr{M} - 1 - (\mathscr{T} - \mathscr{D})\mathscr{D} = 0, \tag{2.4a}$$

$$|\mathscr{T} - \mathscr{D}| < 2, \tag{2.4b}$$

Eq. (2.4a) being associated with $P(\cdot)$ that assumes a pair of roots with a product equal to 1 whereas Eq. (2.4b) follows from the restriction for a negative sign for the discriminant associated to $P(\cdot)$. This locus defines a hyperbolic paraboloid in the $(\mathscr{T}\mathscr{M}\mathscr{D})$-space. A Poincaré-Hopf bifurcation will occur when the triple $(\mathscr{T}, \mathscr{M}, \mathscr{D})$ crosses the complex interior component of the critical surface (2.4a)–(2.4b) and quasi-periodic equilibria will emerge in its neighbourhood.

For a given $\mathscr{D}$, the depiction of these three critical surfaces is going to be facilitated[3] by the ensued consideration of a collection of sections along the $\mathscr{D}$ coordinate, henceforth denoted as $(\mathscr{T}\mathscr{M})_{\mathscr{D}}$, any of the aforementioned critical loci being then represented through *a straight-line or a segment*.

More explicitly and first introducing the benchmark case $\mathscr{D} = 0$ on Fig. 2.1, the set of coefficients $(\mathscr{T}, \mathscr{M})$ such that $Q(+1) = 0$ and $Q(-1) = 0$ respectively correspond to the saddle-node and flip critical lines (A_0C_0) and (A_0B_0)—the index 0 refers to the value of the parameter $\mathscr{D}$ under which the whole picture is drawn—whilst the corresponding set for two nonreal eigenvalues with unitary norm is depicted by the horizontal Poincaré-Hopf critical segment $[B_0C_0]$. This gives rise to a construction familiar from the two-dimensional analysis, namely the triangle $(A_0B_0C_0)$ defined by $|\mathscr{T}| < |1 + \mathscr{M}|$ and $|\mathscr{M}| < 1$.

---

[1] *Vide* Devaney (1986) or Grandmont (2008) for an extensive typology of local bifurcations.

[2] Letting z_1 and z_2 be two eigenvalues with a unitary norm, it is indeed obtained that $z_1 z_2 z_3 = |z| z_3 = \mathscr{D}$.

[3] Equation (2.4a) depicts a *ruled surface*, i.e., a surface generated by straight-lines in $\mathscr{T}$ and $\mathscr{M}$ for a given value of $\mathscr{D}$.

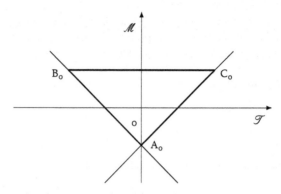

Fig. 2.1 Benchmark case $\mathscr{D} = 0$

The two panels of Fig. 2.2 then assess the status of this construction for various fixed values of $\mathscr{D}$ in the neighbourhood of the benchmark case $\mathscr{D} = 0$, respectively for $\mathscr{D} < 0$ and $\mathscr{D} > 0$. As $\mathscr{D}$ is decreased over $\mathbb{R}_-$ or increased over $\mathbb{R}_+$, the slopes of $(A_{\mathscr{D}} C_{\mathscr{D}})$ and $(A_{\mathscr{D}} B_{\mathscr{D}})$ are let unmodified. In opposition to this, the segment $[B_{\mathscr{D}} C_{\mathscr{D}}]$, of slope $\mathscr{D}$, respectively follows a translated clockwise rotation for $\mathscr{D} < 0$ and a translated counter-clockwise rotation for $\mathscr{D} > 0$.

The expressions of the *parameterised* coordinates of $A_{\mathscr{D}}$, $B_{\mathscr{D}}$ and $C_{\mathscr{D}}$ that underlie the definition of the triangle $(A_{\mathscr{D}} B_{\mathscr{D}} C_{\mathscr{D}})$ can readily be computed from the solving of (2) and (3), (3) and (4), (2) and (4) and list as:

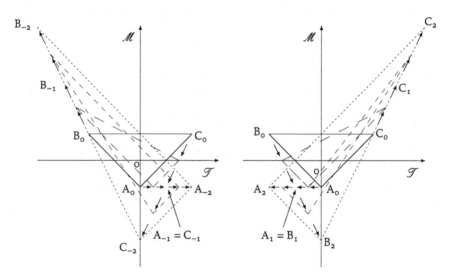

Fig. 2.2 Translated rotations of the benchmark triangle $(A_{\mathscr{D}} B_{\mathscr{D}} C_{\mathscr{D}})$

$$\left(\mathscr{T}_{A_{\mathscr{D}}}, \mathscr{M}_{A_{\mathscr{D}}}\right) = (-\mathscr{D}, -1), \tag{2.5a}$$

$$\left(\mathscr{T}_{B_{\mathscr{D}}}, \mathscr{M}_{B_{\mathscr{D}}}\right) = (-2 + \mathscr{D}, 1 - 2\mathscr{D}), \tag{2.5b}$$

$$\left(\mathscr{T}_{C_{\mathscr{D}}}, \mathscr{M}_{C_{\mathscr{D}}}\right) = (2 + \mathscr{D}, 1 + 2\mathscr{D}). \tag{2.5c}$$

Worthwhile noticing is also the non-generic occurrence, for $\mathscr{D} = -1$ and $\mathscr{D} = 1$, of $A_{-1} = B_{-1}$ and $A_1 = B_1$: the Poincaré critical segment becomes respectively part of the flip and the saddle node critical loci. Such occurrences imply that the formal definition of the triangle $\left(A_{\mathscr{D}}B_{\mathscr{D}}C_{\mathscr{D}}\right)$ is modified as $|\mathscr{D}|$ goes through one, namely:

$$\begin{cases} \mathscr{M} < 1 + (\mathscr{T} - \mathscr{D})\mathscr{D} \\ |\mathscr{T} + \mathscr{D}| < 1 + \mathscr{M} \end{cases} \quad \text{for} \quad |\mathscr{D}| < 1, \tag{2.6a}$$

$$\begin{cases} \mathscr{M} > 1 + (\mathscr{T} - \mathscr{D})\mathscr{D} \\ |1 + \mathscr{M}| > \mathscr{T} + \mathscr{D} \end{cases} \quad \text{for} \quad \mathscr{D} < -1, \tag{2.6b}$$

$$\begin{cases} \mathscr{M} > 1 + (\mathscr{T} - \mathscr{D})\mathscr{D} \\ |1 + \mathscr{M}| < \mathscr{T} + \mathscr{D} \end{cases} \quad \text{for} \quad \mathscr{D} > 1. \tag{2.6c}$$

The case $|\mathscr{D}| = 1$ is however non-generic and merely two generic configurations, namely $|\mathscr{D}| < 1$ and $|\mathscr{D}| > 1$, are to be considered, making use, as illustrated by Figs. 2.3 and 2.4, of a finite collection of $(\mathscr{T}\mathscr{M})_{\mathscr{D}}$ planes. Putting this into perspective and as made clear by Figs. 2.3 and 2.4, there will be *no loss of generality* in considering, for a given sign of $\mathscr{D}$, a finite collection of sections $(\mathscr{T}\mathscr{M})$ of the space $(\mathscr{T}\mathscr{M}\mathscr{D})$ will fully describe the set of admissible geometric configurations.

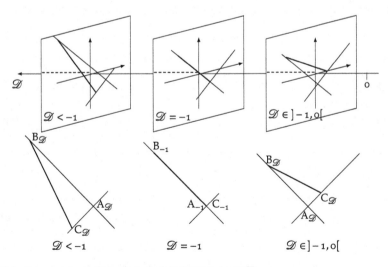

Fig. 2.3 Two generic configurations as $\mathscr{D}$ is decreased over $\mathbb{R}_-$

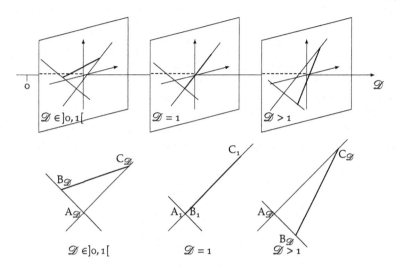

Fig. 2.4 Two generic configurations as $\mathscr{D}$ is increased over $\mathbb{R}_+$

It then remains to characterise any of the generic configurations of Figs. 2.3 and 2.4 in terms of the cardinality of the set of stable roots. It is first noticed that the origin $(\mathscr{T}, \mathscr{M}) = (0, 0)$ belongs to $(A_{\mathscr{D}}B_{\mathscr{D}}C_{\mathscr{D}})$ for any $\mathscr{D} \in \mathbb{R} \setminus \{-1, +1\}$: this appears from the translated rotations of Fig. 2.2 but this is also rapidly checked from the analytical definitions (2.6a)–(2.6c) of the triangles for $|\mathscr{D}| < 1$, $\mathscr{D} < -1$ or $\mathscr{D} > 1$. This geometric property translates as the satisfaction of $Q(z) = -z^3 + \mathscr{D} = 0$ by the characteristic polynomial, hence $z^3 = \mathscr{D}$ and the occurrence of a *triple* real eigenvalue at the origin. This will assume an absolute value greater than one for $|\mathscr{D}| > 1$ and an absolute value less than one for $|\mathscr{D}| < 1$.

As long as the system is maintained in the interior of the triangle $(A_{\mathscr{D}}B_{\mathscr{D}}C_{\mathscr{D}})$, its stability properties are left unaltered with respect to the ones the origin $(0, 0)$, that eventually establishes the corresponding number of stable eigenvalues between parenthesis for both configurations on Fig. 2.5—equivalently, the dimension of the local stable manifold.

Considering then a perturbation that occasions on Fig. 2.5 the leave from the *unstable* triangle $(A_{\mathscr{D}}B_{\mathscr{D}}C_{\mathscr{D}})$ for $|\mathscr{D}| > 1$. A crossing of the Poincaré-Hopf critical segment $[B_{\mathscr{D}}C_{\mathscr{D}}]$ would then imply that the modulus of the complex eigenvalues enters into the unit circle and an area characterized by two eigenvalues with a norm that is less than one. When such a leave from the *unstable* triangle $(A_{\mathscr{D}}B_{\mathscr{D}}C_{\mathscr{D}})$ with zero stable roots rather proceeds through the crossing of the saddle-node critical line $(A_{\mathscr{D}}C_{\mathscr{D}})$ or the flip critical line $(A_{\mathscr{D}}B_{\mathscr{D}})$, a unique eigenvalue with respect to the unit circle will be modified and the system falls in an area with one stable eigenvalue. Finally, the crossing of the flip critical line after having crossed the saddle node critical line or the reversed sequence will lead the system within an area that exhibits a pair of moduli within the unit circle. A related line of reasoning can straightforwardly

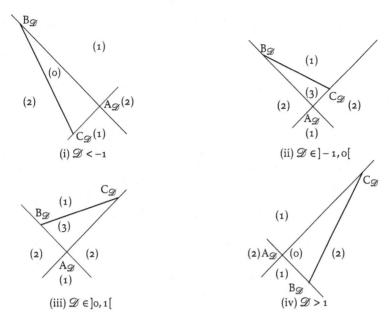

Fig. 2.5 Critical loci and typologies of stable eigenvalues for $|\mathscr{D}| < 1$ and $|\mathscr{D}| > 1$

be completed for the typology of stable eigenvalues associated with $|\mathscr{D}| < 1$ and the *stable* definition of the triangle $\left(A_{\mathscr{D}}B_{\mathscr{D}}C_{\mathscr{D}}\right)$.

2.2.2   Assessing the Stability Properties of Parameterised Economies

This section shall argue that Fig. 2.5 equips the analysis with a range of tools that are going to facilitate the undertaking of a sensitivity analysis in actual parameterised economies. Consider indeed some characteristic polynomial $Q(z) = -z^3 + \mathscr{T}z^2 - \mathscr{M}z + \mathscr{D}$ in the neighbourdhood of some steady state and assume that the coefficients $\mathscr{T}$ and $\mathscr{M}$ depend upon a list of four parameters whilst $\mathscr{D}$ merely depends upon one parameter, say the fourth of the list, whence some formulations $\mathscr{T}(\lambda_1, \lambda_2, \lambda_3, \lambda_4)$, $\mathscr{M}(\lambda_1, \lambda_2, \lambda_3, \lambda_4)$ and $\mathscr{D}(\lambda_4)$. From Sect. 2.2.1, an appraisal of the local stability properties of the economy $\{\lambda_1, \lambda_2, \lambda_3, \lambda_4\}$ will be available from the features of parameterised $(\mathscr{T}\mathscr{M})_{\mathscr{D}}$-planes. For illustration purposes, consider the range of values of the coefficient λ_4 for which the coefficient $\mathscr{D}$ is such that $\mathscr{D}(\lambda_4) > 1$, the typology of the eigenvalues being available from Fig. 2.5(iv). Further let the coefficients $\mathscr{T}$ and $\mathscr{D}$ both assume the same positive linear dependency with respect to parameters λ_1 and λ_2 whose domains are restricted to the positive real line. Letting, e.g., the parameter λ_1 vary, this will result in a parameterised half-line $_{\lambda_1}\Delta$—arrowed on

Fig. 2.6—with a slope of $+1$ that is parallel to $(A_{\mathscr{D}}C_{\mathscr{D}})$. It is positioned over the plane $(\mathscr{TM})_{\mathscr{D}}$ by considering how its *origin*, defined for $\lambda_1 = 0$, would vary with the remaining parameters λ_2 and λ_3, the value of λ_4 being, by definition, given over a plane $(\mathscr{TM})_{\mathscr{D}}$. Figure 2.6 describes a configuration where the dashed locus $_{\lambda_3}\Lambda_{\lambda_1=0}$ follows a counterclockwise rotation whilst the parameter λ_2 is increased, that in turn allows for introducing a dotted locus $_{\lambda_2}\Lambda_{\lambda_1=0,\lambda_3=0}$ that is, by assumption, also parallel to $(A_{\mathscr{D}}C_{\mathscr{D}})$.

Otherwise stated and for the range of values of λ_4 such that $\mathscr{D}(\lambda_4) > 1$, the dotted locus $_{\lambda_2}\Lambda_{\lambda_1=0,\lambda_3=0}$ is located above both of the loci $\left(A_{\mathscr{D}}B_{\mathscr{D}}\right)$ and $\left(A_{\mathscr{D}}C_{\mathscr{D}}\right)$. As this is clear from Fig. 2.6, for arbitrary small values of λ_3 and whatever the value of λ_2, the parameterised straight-line $_{\lambda_1}\Delta$ under consideration will locate in the same area with a unique modulus inside the unit circle, that would, e.g., correspond to a *determinacy property* for a system with a unique predetermined variable. In opposition to this and for larger values of λ_3, the origin of $_{\lambda_1}\Delta$ will locate below the locus $\left(A_{\mathscr{D}}C_{\mathscr{D}}\right)$. If it is also considered for arbitrary small values of λ_2, that origin will be found below $\left(A_{\mathscr{D}}B_{\mathscr{D}}\right)$ and within an area with, again, one modulus inside the unit circle. The straight-line $_{\lambda_1}\Delta$ will assume an intersection with $\left(A_{\mathscr{D}}B_{\mathscr{D}}\right)$ for larger values of λ_1: if this takes place by the left of $B_{\mathscr{D}}$, there will exist a range of values of λ_1 for which $_{\lambda_1}\Delta$ is located in the interior of the triangle $\left(A_{\mathscr{D}}B_{\mathscr{D}}C_{\mathscr{D}}\right)$ with no modulus inside the unit circle and an *instability* configuration for a system with a unique predetermined

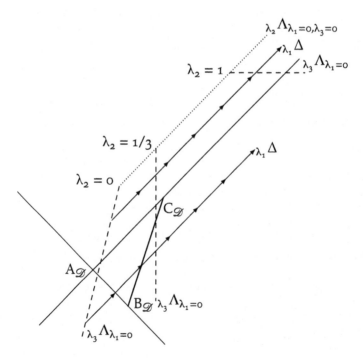

Fig. 2.6 The loci $_{\lambda_1}\Delta$, $_{\lambda_3}\Lambda_{\lambda_1=0}$ and $_{\lambda_2}\Lambda_{\lambda_1=0,\lambda_3=0}$ for $\mathscr{D}(\lambda_4) > 1$

variable. Finally, for still larger values of λ_1 and beyond the segment $\left[B_{\mathscr{D}}C_{\mathscr{D}}\right]$, the parameterised line $_{\lambda_1}\Delta$ will end up in an area with two moduli inside the unit circle that would correspond to an *indeterminacy* configuration for a system with a unique predetermined variable.

2.3 A Simple Parameterised Economy: The Golden Rule in the Model of Overlapping Generations

This section will consider an economy populated by generations of agents living for two periods. The representative agent works ℓ hours and consumes c when young but then solely consumes c' when old. His preferences are described by a separably additive utility function, i.e., $\gamma_c U_1(c) + U_2(c') - \gamma_\ell V(\ell)$, where $\gamma_c > 0$ and $\gamma_\ell > 0$ are scaling parameters. In the sequel it will be assumed that both $U_1(\cdot)$ and $U_2(\cdot)$ are increasing and concave whilst $V(\cdot)$ is increasing and convex. At date $t \geq 1$, the young agent of generation t chooses a consumption vector (c_t^t, c_{t+1}^t), a supply of labor ℓ_t^t and savings x_{t+1}^t so as to maximize his utility subject to:

$$c_t^t + x_{t+1}^t = w_t \ell_t^t,$$
$$c_{t+1}^t = \mathscr{R}_{t+1}x_{t+1}^t,$$

and $c_t^t \geq 0, c_{t+1}^t \geq 0, \ell_t^t \geq 0$, for w_t the wage rate and $\mathscr{R}_{t+1}$ the gross return on savings. The single good is produced by a constant returns neoclassical production function $AF(K, L)$, where K and L are respectively the productive capital and the labor employed, $A > 0$ being a scaling parameter. In every period, competitive firms maximize profits, given the wage rate and the rental rate—for the sake of simplicity it will be assumed that capital fully depreciates on use within the period. The FOC of the optimisation problem for the young agent, that are necessary and sufficient under the assumed properties of the utility function, list as:

$$\gamma_\ell \frac{\partial V}{\partial \ell}\left(\ell_t^t\right) = w_t \mathscr{R}_{t+1} \frac{\partial U_2}{\partial c'}\left(c_{t+1}^t\right),$$
$$\gamma_\ell \frac{\partial V}{\partial \ell}\left(\ell_t^t\right) = w_t \gamma_c \frac{\partial U_1}{\partial c}\left(c_t^t\right),$$
$$c_t^t + x_{t+1}^t = w_t \ell_t^t,$$
$$c_{t+1}^t = \mathscr{R}_{t+1}x_{t+1}.$$

In the lines of Gale (1973), the agent's sum of the assets will be allowed to differ from the total amount of productive capital. In other words, it will be assumed that capital is not the only channel of inter-temporal exchange. In that perspective, let B_t denote the difference between the savings willingness of the young and the stock of capital, i.e., $B_t := x_{t+1}^t - K_{t+1}$. Taking into account the market-clearing conditions for the factor markets, it can be established that, in reduced form, a competitive equilibrium is a sequence (K_t, L_t, B_t) satisfying:

$$\gamma_\ell \frac{\partial V}{\partial L}(L_t) = A\frac{\partial F}{\partial L}(K_t, L_t)A\frac{\partial F}{\partial K}(K_{t+1}, L_{t+1})\frac{\partial U_2}{\partial c'}\left(A\frac{\partial F}{\partial K}(K_{t+1}, L_{t+1})\right.$$

$$\left. \times \left[A\frac{\partial F}{\partial L}(K_t, L_t)L_t - c_t^t\right]\right), \quad (2.7a)$$

$$\gamma_\ell \frac{\partial V}{\partial \ell}(L_t) = A\frac{\partial F}{\partial L}(K_t, L_t)\gamma_c\frac{\partial U_1}{\partial c}(c_t^t), \quad (2.7b)$$

$$K_{t+1} + B_t + c_t^t = A\frac{\partial F}{\partial L}(K_t, L_t)L_t, \quad (2.7c)$$

$$B_{t+1} = A\frac{\partial F}{\partial K}(K_{t+1}, L_{t+1})B_t, \quad (2.7d)$$

for (2.7b) that defines, for $\partial^2 U_1/\partial c^2 \neq 0$, c_t^t as a function of (K_t, L_t). Two distinct inter-temporal transfer institutions with a distinct interpretations for the parameter B, can then be considered on top of productive capital: either fiat money in the line of Samuelson (1958), or public debt following the work of Diamond (1965), both giving rise to a reduced form (2.7a)–(2.7d). From the former perspective, assume that *two* assets are available: the productive capital, that is remunerated at rate $\mathscr{R}$, plus outside money with unitary price Q. The maximization problem of the young would then have a solution if and only if the no-arbitrage condition between the two assets holds, i.e., $Q_{t+1}/Q_t = \mathscr{R}_{t+1}$. The capital-money portfolio choice being then indeterminate, $x_{t+1}^t \equiv K_{t+1} + M_{t+1}Q_t$. Assuming that the stock of money is in constant supply M, the money market clearing condition writes $M_t = M$ for all $t \geq 0$. In this monetary interpretation, $B_t \equiv MQ_t$ and the no-arbitage equation (2.7d) is thus determining the equilibrium price of money. From the second perspective, assume that the government has issued at date t a debt G_t to the younger generation. This debt has a one-period maturity and will be repaid with interest at the same rate on return on capital. At date $t + 1$, the debt burden is $\mathscr{R}_{t+1}G_t$. Provided that the policy followed is to maintain a constant zero deficit, the government budget constraint then implies $G_{t+1} = \mathscr{R}_{t+1}G_t$. The equilibrium savings willingness of the young must adequate the demand of capital from firms and the debt issued by the government, hence $x_{t+1}^t = K_{t+1} + G_t$. For this public debt interpretation, $B_t \equiv G_t$ and Eq. (2.7d) hence gives the equilibrium value of the public debt.

2.3.1   The Golden Rule Steady State: Existence and Normalisation

Under the previous assumptions, the system (2.7a)–(2.7d) defines an implicit three-dimensional dynamical system $(K_{t+1}, L_{t+1}, B_{t+1})' = \Upsilon(K_t, L_t, B_t)$. A steady state is a fixed point of the map $\Upsilon(\cdot)$, i.e., a triple $\left(K^\star, L^\star, B^\star\right)$ such that $\left(K^\star, L^\star, B^\star\right)' = \Upsilon\left(K^\star, L^\star, B^\star\right)$. The economy under study possesses two steady states, namely the wealth-capital or balanced steady state in which $B^\star = 0$ and the *Golden Rule*—its understanding is detailed below—in which $B^\star \neq 0$. In the sequel, the focus will be exclusively upon the latter. Formally, the *Golden Rule* is a triple $(K^\star, L^\star, B^\star) \in \mathbb{R}^\star_+ \times \mathbb{R}^\star_+ \times \mathbb{R} \setminus \{0\}$ such that:

$$\gamma_\ell \frac{\partial V}{\partial \ell}(L^\star) = A \frac{\partial F}{\partial L}(K^\star, L^\star) \frac{\partial U_2}{\partial c'} \left[A \frac{\partial F}{\partial L}(K^\star, L^\star) L^\star - c^\star \right], \qquad (2.8a)$$

$$\gamma_\ell \frac{\partial V}{\partial \ell}(L^\star) = A \frac{\partial F}{\partial L}(K^\star, L^\star) \gamma_c \frac{\partial U_1}{\partial c}(c^\star), \qquad (2.8b)$$

$$K^\star + B^\star + c^\star = A \frac{\partial F}{\partial L}(K^\star, L^\star) L^\star, \qquad (2.8c)$$

$$1 = A \frac{\partial F}{\partial K}(K^\star, L^\star). \qquad (2.8d)$$

Noticing that (2.8a), (2.8b) and (2.8d) are the FOC of a constrained stationary second-best program:

$$\max_{\{c,c',L,K\}} \gamma_c U_1(c) + U_2(c') - \gamma_\ell V(L) \quad \text{subject to} \quad c + c' \leq AF(K, L) - K,$$

the above defined competitive equilibrium steady state is then efficient, i.e., it coincides with the Golden Rule. The quantity of money or the government debt required to sustain the Golden Rule is given by (2.8c). Following the terminology coined by Gale (1973), an economy in which this quantity is negative, respectively positive, is termed *Classical*, respectively *Samuelson*.[4] Currently, the fact that $B < 0$ means that the sum of the young agents' assets, namely their savings, is smaller that the total amount of capital; sustaining the Golden Rule stock of capital hence requires transfers towards the young agents. Oppositely, for $B > 0$, the sustainment of the Golden Rule stock of capital requires transfers from the young.

In order to simplify the analysis, and making use of the scaling parameters, conditions for the existence of a *normalised* Golden Rule will be explicitly detailed. Let then

$$\alpha_c\left(K^\star, L^\star, c^\star\right) := c^\star \Big/ A\frac{\partial F}{\partial L}(K^\star, L^\star)L^\star,$$

$$s\left(K^\star, L^\star\right) := A\frac{\partial F}{\partial K}(K^\star, L^\star)K^\star \Big/ F\left(K^\star, L^\star\right),$$

respectively denote the share of first-period consumption in wage income and the share of capital in output, both being evaluated at the steady state and henceforward compactly referred to as α_c and s.

---

[4]See, e.g., the enlightening discussion in Weil (2008).

Now, fix arbitrarily a vector $(K^\star, L^\star, c^\star) \in \mathbb{R}_+^* \times \mathbb{R}_+^* \times \mathbb{R}_+^*$. Solving the Eqs. (2.8a), (2.8b) and (2.8d) in $(\gamma_c, \gamma_\ell, A)$, it is obtained that:

$$\gamma_c^\star = \frac{\partial U_2}{\partial c'} \left[\frac{(1-s)K^\star}{s} - c^\star \right] \bigg/ \frac{\partial U_1}{\partial c}(c^\star), \tag{2.9a}$$

$$\gamma_\ell^\star = (1-s)K^\star \frac{\partial U_2}{\partial c'} \left[\frac{(1-s)K^\star}{s} - c^\star \right] \bigg/ L^\star \frac{\partial V}{\partial L}(L^\star)s, \tag{2.9b}$$

$$A^\star = 1 \bigg/ \frac{\partial F}{\partial K}(K^\star, L^\star). \tag{2.9c}$$

The steady state value $B^\star$ then follows from (2.8c):

$$B^\star = \left[\frac{1-s}{s}(1-\alpha_c) - 1 \right] K^\star.$$

Aside from $B^\star$, whose sign is unrestricted, and under the earlier assumptions on preferences, $\gamma_c^\star$, $\gamma_\ell^\star$ and $A^\star$ are unambiguously positive. It follows that a unique restriction is to be imposed on the arbitrary choice of $(K^\star, L^\star, c^\star) \in \mathbb{R}_+^* \times \mathbb{R}_+^* \times \mathbb{R}_+^*$, in order to ensure the existence of the normalized steady state, namely: $c'^\star = (1-s)K^\star/s - c^\star > 0$. Note however that the latter is equivalent to the holding of the restriction $\alpha_c < 1$ that will be hereafter assumed to prevail. To sum up, choose arbitrarily $(K^\star, L^\star, c^\star) \in \mathbb{R}_+^* \times \mathbb{R}_+^* \times \mathbb{R}_+^*$ such that $\alpha_c \in [0, 1[$. Let $(\gamma_c, \gamma_\ell, A) = (\gamma_c^\star, \gamma_\ell^\star, A^\star)$. By construction, $(K^\star, L^\star, B^\star) \in \mathbb{R}_+^* \times \mathbb{R}_+^* \times \mathbb{R}^*$ is a Golden Rule steady state. In the sequel, the local dynamics will be characterised in the neighbourhood of the normalised steady state.

2.3.2 Some Parameterised Curves

The coefficients of the characteristic polynomial list, letting $V(\ell) = \ell$, along:

$$\mathcal{T} = 1 + \frac{1-s}{s} + \frac{s}{1-s} + (1-\mathcal{D})\frac{s}{1-s}\frac{1}{1-\alpha_c}\left[\frac{(1+\eta_c)\alpha_c}{\eta_c} - \frac{\varsigma}{s} \right], \tag{2.10a}$$

$$\mathcal{M} = \mathcal{T} - (1-\mathcal{D})\left(\frac{1}{s} - \frac{1}{1-\alpha_c} \right), \tag{2.10b}$$

$$\mathcal{D} = \frac{1}{1+\eta_{c'}}. \tag{2.10c}$$

for

$$\varsigma := \frac{\partial F}{\partial K} \cdot \frac{\partial F}{\partial L} \Big/ F \cdot \frac{\partial^2 F}{\partial K \partial L}, \quad 1 - s := \frac{\partial F}{\partial L} \cdot L \Big/ F, \quad s := \frac{\partial F}{\partial K} \cdot K \Big/ F,$$

$$\eta_c := c_t^t \frac{\partial^2 U_1}{\partial \left(c_t^t\right)^2} \Big/ \frac{\partial U_1}{\partial c_t^t}, \quad \eta_{c'} := c_{t+1}^t \frac{\partial^2 U_2}{\partial \left(c_{t+1}^t\right)^2} \Big/ \frac{\partial U_2}{\partial c_{t+1}^t},$$

that are evaluated at $\left(K^\star, L^\star, c^\star\right)$ and where ς denotes the elasticity of substitution between the productive factors. While concavity assumptions ensure that $\eta_c < 0$ and $\eta_{c'} < 0$, gross substitutability properties would correspond to $1 + \eta_c > 0$ and $1 + \eta_{c'} > 0$. It is also worth emphasising the gross substitutability on second-period consumption $1 + \eta_{c'} > 0$ translates as $\mathscr{D} > 1$ whereas its violation would result in $\mathscr{D} < 0$. Let further, and for convenience, $1/\eta_1 := \eta_c \ 1/\eta_2 := \eta_{c'}$. In a more concise form, the coefficients of the characteristic polynomial may be understood as a triple of functions of structural parameters, namely

$$\left\{ \mathscr{T}\left(\eta_1, \eta_2, \varsigma, \alpha_c, s\right), \mathscr{M}\left(\eta_1, \eta_2, \varsigma, \alpha_c, s\right), \mathscr{D}\left(\eta_2\right) \right\},$$

These functions can be seen as *parametric equations* for curves in the $(\mathscr{T}\mathscr{M}\mathscr{D})$-space. It is then noticed that $\mathscr{D}$ does neither depend upon the parameters describing the technology, namely s and ς, nor on the ones that relate to first-period consumption, namely α_c and η_1. The current approach being based upon diagrams over the planes $(\mathscr{T}\mathscr{M})_\mathscr{D}$, the following parameterised curve is *generated* by the variations of η_1 for fixed $\left(\cdot, \eta_2, \varsigma, \alpha_c, s\right)$ and hence leaves unaffected the coefficient $\mathscr{D}$:

$$_{\eta_1}\Delta := \left\{ \left(\mathscr{T}\left(\eta_1, \eta_2, \varsigma, \alpha_c, s\right), \mathscr{M}\left(\eta_1, \eta_2, \varsigma, \alpha_c, s,\right), \mathscr{D}\left(\eta_2\right) \right) : \eta_1 \in \left] -\infty, 0\right[\right\}. \tag{2.11}$$

From (2.10b), it is observed that $_{\eta_1}\Delta$ is a half-line starting from $_0\Delta$ that is parallel to $\left(A_\mathscr{D}C_\mathscr{D}\right)$ and whose direction vector is available from:

$$\mathscr{T}' = \mathscr{M}' = (1 - \mathscr{D})\frac{s}{1-s}\frac{\alpha_c}{1-\alpha_c} \gtreqless 0 \quad \text{for} \quad 1 - \mathscr{D} \gtreqless 0. \tag{2.12}$$

It is further noticed that the position of this straight-line with respect to the locus $\left(A_\mathscr{D}C_\mathscr{D}\right)$ is ruled by the sign of:

$$(1 - \mathscr{D})\frac{1}{1-\alpha_c}\left[\frac{1-s}{s}(1-\alpha_c) - 1\right]. \tag{2.13}$$

One is then to undertake a sensitivity analysis on $_0\Delta$ and as the share of first period consumption α_c spans its interval $[0, 1[$, hence the locus:

$$_{\alpha_c}\Lambda_{\eta_1=0} := \left\{ \left(\mathcal{T}(0, \eta_2, \varsigma, \alpha_c, s), \mathcal{M}(0, \eta_2, \varsigma, \alpha_c, s), \mathcal{D}(\eta_2) \right) : \alpha_c \in [0, 1[\right\},$$

for $\quad \mathcal{T}(0, \eta_2, \varsigma, \alpha_c, s) = 1 + \dfrac{1-s}{s} + \dfrac{s}{1-s} + (1-\mathcal{D}) \dfrac{s}{1-s} \dfrac{1}{1-\alpha_c} \left(\alpha_c - \dfrac{\varsigma}{s} \right),$

$$\mathcal{M}(0, \eta_2, \varsigma, \alpha_c, s) = \mathcal{T}(0, \varsigma, \alpha_c, s, \eta_2) - (1-\mathcal{D}) \left(\dfrac{1}{s} - \dfrac{1}{1-\alpha_c} \right).$$

It is further established in Appendix A that this locus does in turn correspond to a half-line that starts from $\left(\mathcal{T}(0, \eta_2, \varsigma, 0, s), \mathcal{M}(0, \eta_2, \varsigma, 0, s) \right)$ and assumes a slope of $(1-\varsigma)/(s-\varsigma)$.

It is finally also appropriate to first clarify how the half-line points $_{\alpha_c}\Lambda_{\eta_1=0}$ moves while the parameter ς spans its interval $[0, +\infty]$. This $_\varsigma\Lambda_{\eta_1=0, \alpha_c=0}$ locus is formally defined as follows:

$$_\varsigma\Lambda_{\eta_1=0, \alpha_c=0} := \left\{ \left(\mathcal{T}(0, \eta_2, \varsigma, 0, s), \mathcal{M}(0, \eta_2, \varsigma, 0, s), \mathcal{D}(\eta_2) \right) : \varsigma \in]0, +\infty[\right\},$$

for $\quad \mathcal{T}(0, \eta_2, \varsigma, 0, s) = 1 + \dfrac{1-s}{s} + \dfrac{s}{1-s} + (1-\mathcal{D}) \dfrac{s}{1-s} \left(-\dfrac{\varsigma}{s} \right),$

$$\mathcal{M}(0, \eta_2, \varsigma, 0, s) = \mathcal{T}(0, \eta_2, \varsigma, 0, s) - (1-\mathcal{D}) \dfrac{1}{s}.$$

This, once again, results in the obtention of a half-line starting from $\left(\mathcal{T}(0, \eta_2, 0, 0, s), \mathcal{M}(0, \eta_2, 0, 0, s) \right)$ whose properties are detailed in Appendix B.

The diagrams on Figs. 2.7 and 2.8 picture the three half-lines $_{\eta_1}\Delta$, $_{\alpha_c}\Lambda_{\eta_1=0}$ and $_\varsigma\Lambda_{\eta_1=0, \alpha_c=0}$. They clarify their dependency with respect to the admissible values of $\mathcal{D}$. They allow at a glance to picture the dependency of $_{\eta_1}\Delta$ with respect to ς and α_c. Letting ς be fixed sums up to select a point of a point of $_\varsigma\Lambda_{\eta_1=0, \alpha_c=0}$, that will in its turn gives rise to a specific half-line $_{\alpha_c}\Lambda_{\eta_1=0}$. Subsequently selecting a value of α_c, i.e., a point of $_{\alpha_c}\Lambda_{\eta_1=0}$, eventually defines the starting point of a *particular* set of economies, i.e., the $_{\eta_1}\Delta$ associated to these given values of α_c and ς. Diagrams such as Figs. 2.7 and 2.8 hence allow for contemplating the whole set of such economies for the range of admissible values of α_c and ς: as a simple illustration and for $\mathcal{D} > 1$, a larger substitutability between the factors would, e.g., uniformly translate into a north-east move for the economies.

It is finally of interest to introduce a last, economically relevant, locus that separates, in the course of the line $_{\eta_1}\Delta$, the areas where gross substitutability prevails between first-period consumption and leisure from the ones where it is gross complementarity that prevails. As for the origin of the parameterized line $_{\alpha_c}\Lambda_{\eta_1=0}$ and from Appendix C, the border between these two areas emerges as being described by the curve $\left(\mathcal{T}(-1, \eta_2, \varsigma, \alpha_c, s), \mathcal{M}(-1, \eta_2, \varsigma, \alpha_c, s) \right)$ as α_c spans its interval $[0, 1[$, namely:

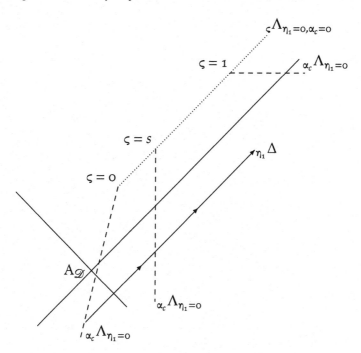

Fig. 2.7 The loci $_{\eta_1}\Delta$, $_{\alpha_c}\Lambda_{\eta_1=0}$ and $_\varsigma\Lambda_{\eta_1=0,\alpha_c=0}$ for $\mathscr{D}>1$

$$_{\alpha_c}\Lambda_{\eta_1=-1}:=\left\{\left(\mathscr{T}\left(-1,\eta_2,\varsigma,\alpha_c,s\right),\mathscr{M}\left(-1,\eta_2,\varsigma,\alpha_c,s\right),\mathscr{D}(\eta_2):\alpha_c\in[0,1[\right\},\right.$$

$$\text{for}\quad \mathscr{T}\left(-1,\eta_2,\varsigma,\alpha_c,s\right)=1+\frac{1-s}{s}+\frac{s}{1-s}-(1-\mathscr{D})\frac{1-s}{s}\frac{1}{1-\alpha_c}\frac{\varsigma}{s},$$

$$\mathscr{M}\left(-1,\eta_2,\varsigma,\alpha_c,s\right)=\mathscr{T}\left(-1,\eta_2,\varsigma,\alpha_c,s\right)-(1-\mathscr{D})\left(\frac{1}{s}-\frac{1}{1-\alpha_c}\right).$$

As this is formally established in Appendix C and pictured on Figs. 2.9 and 2.10, the loci $_{\alpha_c}\Lambda_{\eta_1=0}$ and $_{\alpha_c}\Lambda_{\eta_1=-1}$ share a common origin for $\alpha_c=0$, that further defines an infinite cone with an apex at that point and two generatrices that correspond to the two loci. The interior of a cone corresponds to an area where gross complementarity prevails between first-period consumption and leisure. In opposition to this, the gross substitutability property prevails beyond the locus $_{\alpha_c}\Lambda_{\eta_1=-1}$. Consider then an economy starting from given point on $_{\alpha_c}\Lambda_{\eta_1=0}$: while, by construction, it is first characterised by gross complementarity, as soon as it crosses $_{\alpha_c}\Lambda_{\eta_1=-1}$, it falls into an area where gross substitutability is recovered.

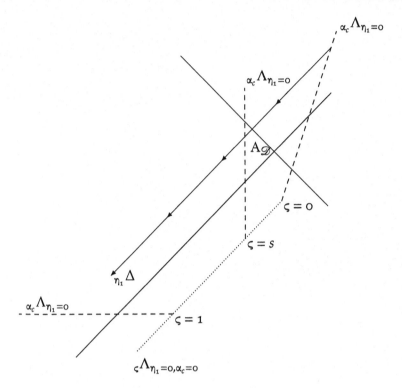

Fig. 2.8 The loci $_{\eta_1}\Delta$, $_{\alpha_c}\Lambda_{\eta_1=0}$ and $_\varsigma\Lambda_{\eta_1=0,\alpha_c=0}$ for $\mathscr{D}<0$

2.3.3 A Local Stability Analysis

Taking advantage of the previous constructions, a global qualitative picture becomes available from the mere localisation of the apex of the cone when it is defined for $\varsigma = 0$. Its position with respect to the critical lines $(A_\mathscr{D}C_\mathscr{D})$ and $(A_\mathscr{D}B_\mathscr{D})$ in turn derives from:

$$-1 + \mathscr{T}(0,\eta_2,0,0,s) - \mathscr{M}(0,\eta_2,0,0,s) + \mathscr{D} = (1-\mathscr{D})\left(\frac{1-s}{s}-1\right),$$
(2.14a)

$$1 + \mathscr{T}(0,\eta_2,0,0,s) + \mathscr{M}(0,\eta_2,0,0,s) + \mathscr{D} = (1+\mathscr{D})\frac{1}{s} + 2\frac{1}{1-s}.$$
(2.14b)

Fig. 2.9 The cones for
$\mathscr{D} > 1$

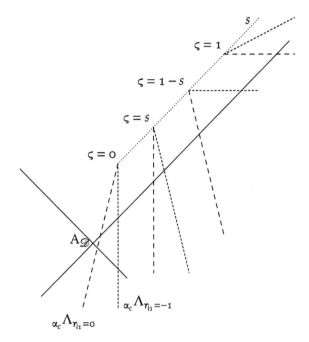

Fig. 2.10 The cones for
$\mathscr{D} < 0$

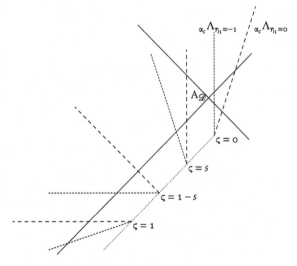

2.3.3.1 Gross Substitutability Between Second-Period Consumption and Leisure ($\mathscr{D} > 1$)

Under a gross substitutability assumption, $\mathscr{D} > 1$, whence, from (2.14a), $Q(+1) < 0$, and, from (2.14b) and for $s < 1/2$, $Q(-1) > 0$. Otherwise stated, the apex of the cone for $\varsigma = 0$ is located above both $\left(A_{\mathscr{D}}C_{\mathscr{D}}\right)$ and $\left(A_{\mathscr{D}}B_{\mathscr{D}}\right)$ on Fig. 2.11. From

Fig. 2.11 The $_{\eta_1}\Delta$ line for
$\mathscr{D} > 1$

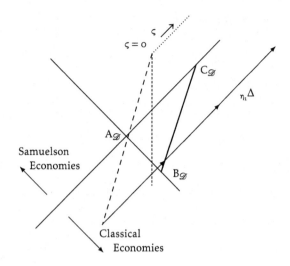

Appendices A and C, the direction vectors of the generatrices of the cone are of negative sign: the generatrices $_{\alpha_c}\Lambda_{\eta_1=0}$ and $_{\alpha_c}\Lambda_{\eta_1=-1}$ are then to cross $\left(A_{\mathscr{D}}C_{\mathscr{D}}\right)$. It however remains to check on which side of $A_{\mathscr{D}}$ this is to occur.

For that purpose, observe from (2.13) that, if $\alpha_c \in [0, 1[$ is such that $1/(1 - \alpha_c) = (1 - s)/s$ and the pair $\left(\mathscr{T}(0, \eta_2, 0, \alpha_c, s), \mathscr{M}(0, \eta_2, 0, \alpha_c, s)\right)$ is on $\left(A_{\mathscr{D}}C_{\mathscr{D}}\right)$ that currently corresponds to the borderline between classical and Samuelson economies, then

$$\mathscr{T}(0, \eta_2, 0, \alpha_c, s) = 1 + \frac{1 - s}{s} + \frac{s}{1 - s} + (1 - \mathscr{D})\left[1 - \frac{s}{1 - s}\right].$$

Recalling that, from (2.5a), $\mathscr{T}_{A_{\mathscr{D}}} = -\mathscr{D}$, it is derived that:

$$\mathscr{T}(0, \eta_2, 0, \alpha_c, s) - \mathscr{T}_{A_{\mathscr{D}}} = 1 + \frac{1 - s}{s} + 1 + \mathscr{D}\frac{s}{1 - s} > 0,$$

that eventually implies that the generatrices $_{\alpha_c}\Lambda_{\eta_1=0}$ and $_{\alpha_c}\Lambda_{\eta_1=-1}$ intersect $\left(A_{\mathscr{D}}C_{\mathscr{D}}\right)$ on the right hand side of $A_{\mathscr{D}}$.

In the same vein, while it is known that the generatrices are to intersect $\left(A_{\mathscr{D}}B_{\mathscr{D}}\right)$, it remains to check on which side of $B_{\mathscr{D}}$ this is to occur. To get some insight about it, consider now

$$\mathscr{T}(-1, \eta_2, 0, 0, s) = 1 + \frac{1 - s}{s} + \frac{s}{1 - s}$$

and compare this with $\mathscr{T}_{B_{\mathscr{D}}} = -2 + \mathscr{D}$ that derives from Eq. (2.5b). It is obtained that:

$$\mathscr{T}(-1, \eta_2, 0, 0, s) - \mathscr{T}_{B_{\mathscr{D}}} = 1 + \frac{1 - s}{s} + \frac{s}{1 - s} + 2 - \mathscr{D}.$$

Figure 2.11 depicts a configuration with sufficiently large values of $\mathscr{D}$ for which $\mathscr{T}\left(-1, \eta_2, 0, 0, s\right) < \mathscr{T}_{B_{\mathscr{D}}}$. In such a case, the curve $_{\eta_1}\Delta$ will be associated with a succession of flip and Poincaré-Hopf bifurcations that will both take place below the line $\left(A_{\mathscr{D}}C_{\mathscr{D}}\right)$, i.e., from (2.13), for classical economies. Interestingly, both bifurcation phenomena take place in an area with a gross substitutability property between first-period consumption and leisure. It is however to be recalled that these conclusions hold for $\varsigma = 0$, that is for fixed coefficients Leontief-type technologies. Letting ς undergo positive values and from Fig. 2.9, the apexes of the cones are to move north-east along the dotted line while the generatrices are to follow a counter clockwise translation. The scope for bifurcations is then first to shrink and then to disappear with an increased substitutability between the productive factors.

Focusing then on the uniqueness issue and taking advantage of the cardinality of the stable eigenvalues available from Fig. 2.4, Samuelson economies will unambiguously be associated with a unique modulus inside the unit circle and a local uniqueness property. Scenarios for Classical economies are in their turn conditional to both α_c and ς, their key-feature being that locally indeterminate steady states are admissible under a gross substitutability property between leisure and first-period consumption.

To put these results into the perspective of the endogenous fluctuations literature, an inter-temporal consumption arbitrage on top of an inter-temporal consumption-leisure arbitrage has been proved to generate a new *degree* of instability for Classical economies. In spite of a gross substitutability assumption on preferences, flip cycles are indeed allowed for sufficiently large values of the share of first-period consumption. With that regard, it is worth recalling that for a formulation without first-period consumption, savings must be a decreasing function of the interest rate in order for flip cycles to exist—*vide* the extensive discussion in Benhabib and Laroque (1988, Proposition III.1, p.154). The retainment of a gross substitutability assumption on preferences would have then uniformly ruled out any area for flip cycles. As for the Poincaré-Hopf bifurcation, though it is uniformly precluded for Samuelson economies, it reveals as a robust phenomenon in Classical economies.

2.3.3.2 Gross Complementarity Between Second-Period Consumption and Leisure ($\mathscr{D} < 0$)

Under a gross complementarity assumption, $\mathscr{D} < 0$: from (2.14a), $Q(+1) > 0$. From (2.14b), the sign of $Q(-1)$ remains ambiguous. Henceforward focusing on the most interesting case with $\mathscr{D} > -1$—the interior of the triangle $\left(A_{\mathscr{D}}B_{\mathscr{D}}C_{\mathscr{D}}\right)$ is now associated with a strong indeterminacy configuration three moduli inside the unit circle, $Q(-1) > 0$ and the apex of the cone is located below $\left(A_{\mathscr{D}}C_{\mathscr{D}}\right)$ but above $\left(A_{\mathscr{D}}B_{\mathscr{D}}\right)$ on Fig. 2.12. From Appendices A and C, the direction vectors of the generatrices are of positive sign and they will cross $\left(A_{\mathscr{D}}C_{\mathscr{D}}\right)$ from below. In addition to this, as $\mathscr{T}\left(-1, \eta_2, 0, 0, s\right) > \mathscr{T}_{C_{\mathscr{D}}}$, these crossings will occur on the right hand side of $C_{\mathscr{D}}$. This configuration with strong indeterminacies is represented on Fig. 2.12.

Fig. 2.12 The $_{\eta_1}\Delta$ approach
for $\mathscr{D} \in\ ]-1, 0[$

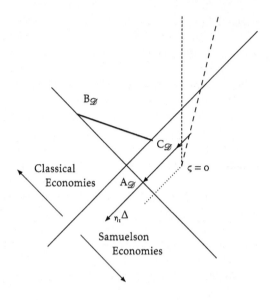

To sum up, the allowance for gross complementaries between leisure and second-period consumption had dramatic implications since Samuelson economies, even though they kept on being associated with an gross substitutability between leisure and first-period consumption, were no longer immune to local indeterminacies and the occurrence of a flip bifurcation. Classical economies were then still more prone to multiplicities with configurations with *two* degrees of indeterminacy, the latters being further potentially available under a gross complementarity between leisure and second-period consumption. Extrapolating the south west move of the cone along Fig. 2.10 and as ς undergoes positive values, the generatrices are to follow a counter-clockwise translation. This strong indeterminacy phenomenon and the scope for bifurcations are to disappear.

Appendix A: The Origin of $_{\eta_1}\Delta$

The starting point of $_{\eta_1}\Delta$ is defined from the following set parameterised by α_c:

$$\left\{\left(\mathscr{T}(0, \eta_2, \varsigma, \alpha_c, s), \mathscr{M}(0, \eta_2, \varsigma, \alpha_c, s), \mathscr{D}(\eta_2)\right) : \alpha_c \in [0, 1[\right\},$$

for
$$\mathscr{T}(0, \eta_2, \varsigma, \alpha_c, s) = 1 + \frac{1-s}{s} + \frac{s}{1-s} + (1-\mathscr{D})\frac{s}{1-s}\frac{1}{1-\alpha_c}\left(\alpha_c - \frac{\varsigma}{s}\right),$$

$$\mathscr{M}(0, \eta_2, \varsigma, \alpha_c, s) = \mathscr{T}(0, \eta_2, \varsigma, \alpha_c, s) - (1-\mathscr{D})\left(\frac{1}{s} - \frac{1}{1-\alpha_c}\right).$$

Letting $\mathscr{T}|_{\eta_1=0} := \mathscr{T}(0, \eta_2, \varsigma, \alpha_c, s)$ and $\mathscr{M}|_{\eta_1=0} := \mathscr{M}(0, \eta_2, \varsigma, \alpha_c, s)$ and from their above expressions, their dependency with respect to α_c are available as:

$$\left(\mathscr{T}|_{\eta_1=0}\right)'_{\alpha_c} = (1 - \mathscr{D})\frac{s}{1-s}\frac{1}{(1-\alpha_c)^2}\left(1 - \frac{\varsigma}{s}\right) \gtrless 0,$$

$$\left(\mathscr{M}|_{\eta_1=0}\right)'_{\alpha_c} = (1 - \mathscr{D})\frac{1}{(1-\alpha_c)^2}\frac{1}{1-s}(1 - \varsigma) \gtrless 0,$$

whence a slope available as

$$_{\alpha_c}\Lambda'_{\eta_1=0} = \frac{\left(\mathscr{M}|_{\eta_1=0}\right)'_{\alpha_c}}{\left(\mathscr{T}|_{\eta_1=0}\right)'_{\alpha_c}} = \frac{1-\varsigma}{s-\varsigma}.$$

This again indicates that $_{\alpha_c}\Lambda_{\eta_1=0}$ depicts a straight-line, its origin being derived by letting $\alpha_c = 0$:

$$\mathscr{T}|_{\eta_1=0,\alpha_c=0} = 1 + \frac{1-s}{s} + \frac{s}{(1-s)} - (1 - \mathscr{D})\frac{\varsigma}{1-s},$$

$$\mathscr{M}|_{\eta_1=0,\alpha_c=0} = \mathscr{T}|_{\eta_1=0,\alpha_c=0} - (1 - \mathscr{D})\frac{1-s}{s}.$$

Appendix B: A ς-Sensitivity Analysis

The origin is in its turn fully described by letting ς span its interval $[0, +\infty]$ through $_\varsigma\Lambda_{\eta_0=0,\alpha_c=0}$. As it is readily checked that both $\mathscr{T}|_{\eta_1=0,\alpha_c=0}$ and $\mathscr{M}|_{\eta_1=0,\alpha_c=0}$ increase as functions of ς for $\mathscr{D} > 1$ but decrease as functions of ς for $\mathscr{D} < 0$:

$$\left(\mathscr{T}|_{\eta_1=0,\alpha_c=0}\right)'_\varsigma = -(1 - \mathscr{D})\frac{1}{1-s},$$

$$\left(\mathscr{M}|_{\eta_1=0,\alpha_c=0}\right)'_\varsigma = -(1 - \mathscr{D})\frac{1}{1-s},$$

whence a slope available as

$$_\varsigma\Lambda'_{\eta_1=0,\alpha_c=0} = \frac{\left(\mathscr{M}|_{\eta_1=0,\alpha_c=0}\right)'_\varsigma}{\left(\mathscr{T}|_{\eta_1=0,\alpha_c=0}\right)'_\varsigma}$$

$$= 1,$$

this latter curve being parallel to $(A_{\mathscr{D}}C_{\mathscr{D}})$, its origin $_0\Lambda_{\eta_0=0,\alpha_c=0}$ being given by:

$$\mathscr{T}|_{\eta_1=0,\alpha_c=0}^{\varsigma=0} = 1 + \frac{1-s}{s} + \frac{s}{1-s},$$

$$\mathscr{M}|_{\eta_1=0,\alpha_c=0}^{\varsigma=0} = \mathscr{T}|_{\eta_1=0,\alpha_c=0}^{\varsigma=0} - (1-\mathscr{D})\frac{1-s}{s}.$$

Appendix C: The Border Between Gross Substitutability and Gross Complementarity

The gross substitutability and gross complementarity zones are separated by the following set:

$$\left\{\left(\mathscr{T}(-1,\eta_2,\varsigma,\alpha_c,s),\mathscr{M}(-1,\eta_2,\varsigma,\alpha_c,s),\mathscr{D}(\eta_2)\right) : \alpha_c \in [0,1[\right\},$$

$$\text{for} \quad \mathscr{T}(-1,\eta_2,\varsigma,\alpha_c,s) = 1 + \frac{1-s}{s} + \frac{s}{1-s} - (1-\mathscr{D})\frac{1-s}{s}\frac{1}{1-\alpha_c}\frac{\varsigma}{s},$$

$$\mathscr{M}(-1,\eta_2,\varsigma,\alpha_c,s) = \mathscr{T}(-1,\eta_2,\varsigma,\alpha_c,s) - (1-\mathscr{D})\left(\frac{1}{s} - \frac{1}{1-\alpha_c}\right).$$

Letting $\mathscr{T}|_{\eta_1=-1} := \mathscr{T}(-1,\eta_2,\varsigma,\alpha_c,s)$ and $\mathscr{M}|_{\eta_1=-1} := \mathscr{M}(-1,\eta_2,\varsigma,\alpha_c,s)$ and from their above expressions, their dependency with respect to α_c is in turn available from:

$$\left(\mathscr{T}|_{\eta_1=-1}\right)'_{\alpha_c} = -(1-\mathscr{D})\frac{\varsigma}{1-s}\frac{1}{(1-\alpha_c)^2} \lessgtr 0 \quad \text{for} \quad \mathscr{D} \lessgtr 1,$$

$$\left(\mathscr{M}|_{\eta_1=-1}\right)'_{\alpha_c} = (1-\mathscr{D})\frac{1}{(1-\alpha_c)^2}\left(1 - \frac{\varsigma}{1-s}\right) \gtrless 0 \quad \text{for} \quad \mathscr{D} \lessgtr 1,$$

that in turn indicates a slope of

$$\frac{\left(\mathscr{M}|_{\eta_1=-1}\right)'_{\alpha_c}}{\left(\mathscr{T}|_{\eta_1=-1}\right)'_{\alpha_c}} = 1 - \frac{1-s}{\varsigma}$$

for the locus $_{\alpha_c}\Lambda_{\eta_1=-1}$, the latter being a straight-line with a nil slope for $\varsigma = 1-s$, that provides a first critical threshold value for ς. The coordinates of the origin of the latter derive by letting $\alpha_c = 0$:

$$\mathscr{T}|_{\eta_1=-1,\alpha_c=0} = 1 + \frac{1-s}{s} + \frac{s}{(1-s)} - (1-\mathscr{D})\frac{\varsigma}{1-s},$$

$$\mathscr{M}|_{\eta_1=-1,\alpha_c=0} = \mathscr{T}|_{\eta_1=-1,\alpha_c=0} - (1-\mathscr{D})\frac{1-s}{s}.$$

Interestingly, the origins of $_{\alpha_c}\Lambda_{\eta_1=-1}$ and of $_{\alpha_c}\Lambda_{\eta_1=0}$ coincide along $_0\Lambda_{\eta_1=-1} \equiv {}_0\Lambda_{\eta_1=0}$, this latter locus corresponding to the apex of the cone defined by the generatrices $_{\alpha_c}\Lambda_{\eta_1=-1}$ and $_{\alpha_c}\Lambda_{\eta_1=0}$ for a given ς.

References

Benhabib, J., & Laroque, G. (1988). On competitive cycles in productive economies. *Journal of Economic Theory, 45*, 145–170.

Devaney, R. (1986). *Introduction to chaotic dynamical systems*. Addison-Wesley.

Diamond, P. (1965). National debt in a neoclassical growth model. *American Economic Review, 35*, 1126–1150.

Gale, D. (1973). Pure exchange equilibrium of dynamic economic models. *Journal of Economic Theory, 6*, 12–36.

Grandmont, J. -M. (1985). On endogenous business cycles. *Econometrica, 6*, 12–36.

Grandmont, J. -M. (1998) Market psychology and business cycles. *Journal of Economic Theory, 83*, 122–177.

Grandmont, J. -M. (2008). Nonlinear difference equations, bifurcations and chaos: An introduction. *Research in Economics, 62*, 122–177.

Grandmont, J. -M., Pintus, P., & de Vilder, R. (1998). Capital-labour substitution and competitive endogenous business cycles. *Journal of Economic Theory, 80*, 14–59.

Reichlin, P. (1986). Equilibrium cycles in an overlapping generations economy with production. *Journal of Economic Theory, 40*, 89–102.

Samuelson, P. A. (1958). An exact consumption-loan model of interest with or without the social contrivance of money. *Journal of Political Economy, 66*, 467–482.

Weil, P. Overlapping generations: The first jubilee. *Journal of Economic Perspectives, 22*, 115–134.

Chapter 3
From Sunspots to Black Holes: Singular Dynamics in Macroeconomic Models

Paulo B. Brito, Luís F. Costa and Huw D. Dixon

Abstract We present conditions for the emergence of singularities in DGE models. We distinguish between slow-fast and impasse singularity types, review geometrical methods to deal with both types of singularity and apply them to DGE dynamics. We find that impasse singularities can generate new types of DGE dynamics, in particular temporary determinacy/indeterminacy. We illustrate the different nature of the two types of singularities and apply our results to two simple models: the Benhabib and Farmer (1994) model and one with a cyclical fiscal policy rule.

Keywords Slow-fast singularities · Impasse singularities · Macroeconomic dynamics · Temporary indeterminacy

JEL codes C62 · D43 · E32

3.1 Introduction

Jean-Michel Grandmont has been interested in the rigorous mathematical analysis of nonlinear dynamic systems in economic applications throughout his long career. In particular, he has often focussed on the less standard aspects of dynamics revealed

---

We acknowledge, without implicating, the useful comments by Alain Venditti, Nuno Barradas, and an anonymous referee. This work is part of UECE's strategic project PEst-OE/EGE/UI0436/2014. UECE has financial support from national funds by FCT (Fundação para a Ciência e a Tecnologia).

P.B. Brito (✉) · L.F. Costa
ISEG - Lisbon School of Economics and Management, Universidade de Lisboa,
Rua Do Quelhas 6, 1200-781 Lisboa, Portugal
e-mail: pbrito@iseg.ulisboa.pt

P.B. Brito · L.F. Costa
UECE - Research Unit on Complexity and Economics, ISEG, Rua Miguel Lupi 20,
1249-078 Lisboa, Portugal

H.D. Dixon
Cardiff Business School, Cardiff University, Aberconway Building, Colum Drive,
Cardiff CF10 3EU, UK

© Springer International Publishing AG 2017
K. Nishimura et al. (eds.), *Sunspots and Non-Linear Dynamics*,
Studies in Economic Theory 31, DOI 10.1007/978-3-319-44076-7_3

by the mathematics, such as chaos and sunspots—Grandmont (1985, 1998, 2008), and Grandmont et al. (1998) provide us some fine examples of his contribution to our understanding on how endogenous fluctuations may emerge in competitive economies. This line of research has been associated with the study of models with multiple equilibria, local indeterminacy, and sunspot equilibria, which have became part of mainstream macroeconomics in the last three decades.

In this chapter we explore something that Grandmont did not get around to studying: the possibility of singular dynamics in economic models due to infinite eigenvalues. We believe that infinite eigenvalues are more than a simple curiosity. With finite eigenvalues the resulting non-singular dynamics display the property that the dimension of the stable manifold is always the same, so that there is either permanent determinacy or permanent indeterminacy of equilibrium paths. However, we believe that this is unrealistic: casual empiricism suggests that economies can pass through periods when they are more volatile (corresponding to indeterminacy) and periods when they are less volatile (corresponding to determinate dynamics). Of course, there might be many explanations of this, but the presence of infinite eigenvalues opens up the possibility that the dimension of the stable manifold can change at a point in time resulting in a "natural" and endogenous change in the determinacy of the equilibrium path in real (finite) time. The determinacy (indeterminacy) of an economy can thus be a temporary phenomenon and the economy can switch between determinate and indeterminate dynamics along its equilibrium path without the intervention of changes in underlying parameters.

We provide a general analysis of a generic dynamic general equilibrium (DGE) model and show that two different types of singularity can arise. One is the case where a parameter varies and at a particular value renders the eigenvalues infinite everywhere. This is a *singular perturbation* and calls for a method of solution dealing with *slow-fast* systems. The second is a case of *impasses* for which there is a one-dimensional impasse set where eigenvalues become locally infinite. This creates a barrier which can only be crossed at a specific point or isolated points at which particular properties are satisfied. This is the case where eigenvalues change sign from plus infinity to minus infinity (or the other way around). In this case, if an equilibrium path crosses this barrier, it does so at a particular point in time. The determinacy properties will switch at that point in time to reflect the change in sign of the eigenvalue. For example, if we have a path that approaches the barrier with two negative eigenvalues, one of which goes to infinity, this will exert a strong pull towards a point on the barrier, which will be reached in finite time. If that point on the barrier satisfies certain conditions (see Proposition 4 in Sect. 3.2.4), the path will emerge at the other side, but with one positive eigenvalue and one negative at which point it can continue to the steady-state. Hence, there is a period of temporary indeterminacy: prior to reaching the barrier, the point acts as a sink and the dynamics are indeterminate. Once the barrier is passed, the dynamics become determinate with a saddle path. In effect, impasses can give rise to a change in the dimension of the stable manifold along the equilibrium path. In this chapter we characterize the conditions required for such a crossing to be possible.

In Sect. 3.3 we are able to provide two generic economic examples of DGE models that can display singular perturbations and impasses. One is the model of Benhabib and Farmer (1994) in which they noted the possibility that "one root passes through minus infinity and re-emerges as a positive real root." As we demonstrate, this case corresponds to a perturbation singularity and has to be analyzed as a slow-fast dynamic system. The other case is a Ramsey DGE model with endogenous labor in which there is a cyclical fiscal policy rule and distortionary taxation, where we show that not only are impasses possible, but also that they can display the crossing behavior which gives rise to temporary indeterminacy.

Whilst physics and engineering have explored the implications of singularities for some time (e.g. in the study of Black holes), economists have tended to look the other way. There are a few exceptions: Barnett and He (2010) and He and Barnett (2006) are, to our knowledge, the only papers dealing with the importance of singularities in economic models (in their case associated to the introduction of feedback policy rules). Singularities play a large role in other scientific fields and we believe that we also need to understand their potential role in macroeconomic dynamics.

The aim of this chapter is to present some results on the geometrical properties of local dynamics in a neighborhood of a singularity and to apply them to simple DGE structures. In Sect. 3.2 we provide an intuition for their existence in a general DGE macro model with endogenous employment. Section 3.3 provides one example for each type of singularity by applying the methods presented in Sect. 3.2.

3.2   A General DGE Model with Singularities

3.2.1   The General Structure of DGE Models

Several DGE models, extending the Ramsey model, feature a semi-explicit differential-algebraic equations (DAE) system in three variables—capital (K), consumption (C), and labor (L):

$$\dot{K} = y(K, L, \varphi) - C, \tag{3.1}$$

$$\dot{C} = C(r(K, L, \varphi) - \rho)\theta(C, L), \tag{3.2}$$

$$0 = u_L(C, L) + u_C(C, L)w(K, L) \equiv v(K, C, L), \tag{3.3}$$

together with the conditions

$$K(0) = K_0, \tag{3.4}$$

$$0 = \lim_{t \to \infty} u_C(C(t), L(t))K(t)e^{-\rho t}, \tag{3.5}$$

where $y(\cdot)$ is an aggregate (net) production function, $r(\cdot)$ is the aggregate (net) return-on-capital function, $\theta(C, L) \equiv -u_C/(Cu_{CC}) > 0$ is the elasticity of intertemporal substitution in consumption that corresponds to the felicity (or utility-flow) function $u(C, L)$,[1] $\rho > 0$ is the discount rate, and $w(\cdot)$ is the aggregate (net) wage-rate function. The parameter vector, including ρ, is generically denoted by φ.

Equation (3.1) is the instantaneous budget constraint, Eq. (3.2) is the Euler equation, and Eq. (3.3) is the arbitrage condition between consumption and labor supply (the leisure-consumption trade off). Equations (3.4) is the initial condition for the stock of capital, where K_0 is a known positive number and (3.5) is the transversality condition.

Definition 1 (*DGE path*) A DGE path is a function $(K(t), C(t), L(t))$ mapping **all** $t \in [0, \infty)$ into a subset of $\mathbb{R}^3_{++}$ which is a solution to the DAE system (3.1)–(3.3) such that the initial and the transversality conditions, (3.4) and (3.5), hold.

A necessary condition for a solution to (3.1)–(3.3) to be a DGE path is that it exists and it is positive *for all* $t \in [0, \infty)$. Solutions of system (3.1)–(3.3) that only exist for a finite interval $t \in [0, t_s)$, with t_s finite, cannot be DGE paths.

A steady-state of system (3.1)–(3.3) is a point $(\bar{K}, \bar{C}, \bar{L}) \in \mathbb{R}^3_{++}$ such that $\dot{K} = \dot{C} = 0$. Steady-states are fixed points of the non-linear equation system

$$\begin{cases} y(K, L, \varphi) = C, \\ r(K, L, \varphi) = \rho, \\ v(K, C, L) = 0. \end{cases} \tag{3.6}$$

A *stationary DGE* is a DGE path such that the variables are permanently at their steady-state levels: $(K(t), C(t), L(t)) = (\bar{K}, \bar{C}, \bar{L})$, for all $t \in [0, \infty)$. It can only exist if $K_0 = \bar{K}$.

An *asymptotic-stationary DGE path* is a DGE path that converges asymptotically to a steady-state, i.e. $\lim_{t \to \infty} (K(t), C(t), L(t)) = (\bar{K}, \bar{C}, \bar{L})$.

From now on, we restrict the analysis to stationary and asymptotic stationary DGE paths, by introducing the following assumption[2]:

Assumption 1 There is at least one steady-state for system (3.1)–(3.3).

The stable manifold associated to steady-state $(\bar{K}, \bar{C}, \bar{L})$, $\mathcal{W}^s(\bar{K}, \bar{C}, \bar{L})$, is defined as the set of initial points such that the DGE path is asymptotically stationary:

$$\mathcal{W}^s(\bar{K}, \bar{C}, \bar{L}) \equiv \{(K, C, L) \in \mathbb{R}^3_{++} : \lim_{t \to \infty} (K(t), C(t), L(t)) = (\bar{K}, \bar{C}, \bar{L})\}.$$

Considering that both types of DGE paths (asymptotic-stationary and stationary) satisfy the transversality condition, their existence and uniqueness (or multiplicity) can be assessed by the characteristics of the stable manifolds associated to a steady

---

[1] We use the notation for derivatives $f_x \equiv \frac{\partial y}{\partial x}$ and $f_{xy} \equiv \frac{\partial^2 y}{\partial x \partial y}$.

[2] For simplicity we exclude the existence of periodic solutions, but the analysis can easily be extended to that case.

state $(\bar{K}, \bar{C}, \bar{L})$: (a) If the stable manifold is empty and $K_0 = \bar{K}$, then a DGE path exists and it is stationary; (b) If the stable manifold is non-empty and K_0 belongs to it, then a DGE exists and it is asymptotic-stationary.

Asymptotic-stationary DGE paths can be classified further according to their degree of determinacy (or multiplicity). For this purpose we need to define the local stable manifold $(\mathcal{W}^s_{loc})$ associated to a steady-state $(\bar{K}, \bar{C}, \bar{L})$ as the set of points, belonging to a vicinity of $(\bar{K}, \bar{C}, \bar{L})$, that asymptotically converge to that steady-state:

$$\mathcal{W}^s_{loc}(\bar{K}, \bar{C}, \bar{L}) \equiv \{(K, C, L) \in N : \lim_{t \to \infty} (K(t), C(t), L(t)) = (\bar{K}, \bar{C}, \bar{L})\}.$$

where N is a neighborhood containing the steady-state such that the Euclidean distance $\|(K, C, L) - (\bar{K}, \bar{C}, \bar{L})\| < \delta$ for a small number $\delta > 0$.

Definition 2 (*Asymptotic- determinate and -indeterminate DGE paths*) A DGE path is asymptotic-determinate if the local stable manifold $\mathcal{W}^s_{loc}(\bar{K}, \bar{C}, \bar{L})$ is one-dimensional. A DGE path is asymptotic-indeterminate if the local stable manifold $\mathcal{W}^s_{loc}(\bar{K}, \bar{C}, \bar{L})$ is two-dimensional.

This means that, given an initial point for K (K_0) sufficiently close to a steady-state, we say there is determinacy if a single DGE path converges to that steady-state and we say there is indeterminacy if an infinite number of paths converge to it.

In Sect. 3.3 we will show that one possible consequence of the existence of singularities in system (3.1)–(3.3) is that the determinacy properties of DGE paths can change along the transition path. Therefore we distinguish further:

Definition 3 (*Temporary-determinate and temporary-indeterminate DGE paths*) A DGE path is temporary-(in)determinate if, for a finite t, it belongs to a subset of the stable manifold $\mathcal{W}^s(\bar{K}, \bar{C}, \bar{L})$ that is locally one-dimensional (two-dimensional).

Definition 4 (*Permanent-determinate and permanent-indeterminate DGE paths*) A DGE path displays permanent determinacy (indeterminacy) if the stable manifold is one-dimensional (two-dimensional) for all $t \in [0, \infty)$.

Two cases are possible in the presence of singularities. First, if the stable manifold has the same degree of determinacy *globally*, then we have permanent determinacy or indeterminacy, depending on the dimension of the local stable manifold. This is the case with regular models, i.e. models without singularities. Second, if the dimension of the stable manifold for points sufficiently far away from the steady-state is *different* from that of the local stable manifold, then we have *temporary* determinacy or indeterminacy or both. In models with particular types of singularities, knowing the dimension of the local stable manifold at the steady-state does not allow us to characterize the determinacy properties of DGE paths sufficiently far away from the steady-state.

Independently from generating a new type of indeterminacy, the existence of singularities can also confine the existence of DGE paths to a subset of the domain of (K, C, L).

3.2.2 DGE Paths in the Presence of Singularities

Assume that consumption and leisure are substitutes, so that $v(\cdot)$ is monotonic in C. Due to the existence of externalities or any other type of distortion, let $v(\cdot)$ be non-monotonic in L. This property prevents us from eliminating L in Eq. (3.3). Instead, by using the implicit-function theorem, we solve Eq. (3.3) for C as a function of the other variables: $C = c(K, L)$.

Differentiating $c(K(t), L(t))$ with respect to time and using Eqs. (3.1) and (3.2) allows us to obtain a reduced constrained ordinary differential equation (ODE) system in (K, L):

$$\dot{K} = s(K, L, \varphi),$$
$$c_L(K, L, \varphi)\, \dot{L} = c(K, L, \varphi)\, (R(K, L, \varphi) - \rho)\, \theta(K, L, \varphi), \qquad (3.7)$$

where $s(K, L, \varphi) \equiv y(K, L, \varphi) - c(K, L, \varphi)$ is a savings function and

$$R(K, L, \varphi) \equiv r(K, L, \varphi) - \frac{c_K(K, L, \varphi)}{c(K, L, \varphi)\theta(K, L, \varphi)}\, s(K, L, \varphi),$$

is a modified return-on-capital function. Henceforth, we deal with the solutions of system (3.7), seen as mappings $t \mapsto (K(t), L(t))$, for $[0, \infty) \to \Omega$, where

$$\Omega \equiv \{\, (K, L) \in \mathbb{R}^2_{++} : c(K, L) > 0 \}.$$

Again, notice that system (3.7) depends on the parameter vector $\varphi \in \Phi$ (representing endowments, preferences, and technologies), where Φ is model-specific.

In this chapter, we explore models in which the consumption function takes the specific form

$$C = c(K, L, \varphi) = z\, (K, \mathcal{L}, \varphi), \text{ with } \mathcal{L} \equiv L^{\epsilon(\varphi)}.$$

Therefore, $c_K(K, L, \varphi) = z_K(K, L^{\epsilon(\varphi)}, \varphi)$ and $c_L(K, L, \varphi) = \epsilon(\varphi) z_{\mathcal{L}}(K, L^{\epsilon(\varphi)}, \varphi) L^{\epsilon(\varphi)-1}$. Using this specification it is convenient to rewrite the system (3.7) as

$$\dot{K} = f_1(K, L, \varphi),$$
$$\epsilon(\varphi)\, \delta(K, L, \varphi)\, \dot{L} = f_2(K, L, \varphi), \qquad (3.8)$$

where

$$f_1(K, L, \varphi) \equiv s(K, L, \varphi) \equiv y(K, L, \varphi) - z(K, L^{\epsilon(\varphi)}, \varphi), \qquad (3.9)$$
$$f_2(K, L, \varphi) \equiv z(K, L^{\epsilon(\varphi)}, \varphi)\, (R(K, L, \varphi) - \rho)\, \theta\, (K, L, \varphi)\, L^{\epsilon(\varphi)-1}, \qquad (3.10)$$
$$\delta(K, L, \varphi) \equiv z_{\mathcal{L}}(K, L^{\epsilon(\varphi)}, \varphi). \qquad (3.11)$$

Steady-states of system (3.7) are points $(\bar{K}, \bar{L})$ belonging to set:

$$\Gamma_E \equiv \{ (K, L) \in \Omega : f_1(K, L, \varphi) = f_2(K, L, \varphi) = 0 \},$$

which can have one element or more. The stationarity conditions are similar to those in the benchmark competitive DGE model

$$s(\bar{K}, \bar{L}, \varphi) = 0, \quad r(\bar{K}, \bar{L}, \varphi) = \rho,$$

i.e. steady-state savings are zero and the return on capital equals the rate of time preference, since there is no depreciation.

Let us denote a solution to system (3.8) for a given time t starting from a point $(K^*, L^*) \in \Omega$ by $\varphi_t(K^*, L^*)$. Define the stable manifold associated to a steady-state $(\bar{K}, \bar{L}) \in \Gamma_E$ as

$$\mathcal{W}^s(\bar{K}, \bar{L}) \equiv \{ (K^*, L^*) \in \Omega : \lim_{t \to \infty} \varphi_t(K^*, L^*) = (\bar{K}, \bar{L}) \}.$$

Therefore, as a consequence of Assumption 1, a DGE path is a trajectory $(K(t), L(t))_{t \in [0, \infty)}$ such that $(K(t), L(t)) \in \mathcal{W}^s(\bar{K}, \bar{L})$ for every $t \in [0, \infty)$. In the case in which $\mathcal{W}^s(\bar{K}, \bar{L})$ is empty, the DGE path is stationary and it exists only if $K(0) = \bar{K}$.[3]

We introduce the three following assumptions. First, function $z(\cdot)$ is monotonic in K, which implies that $c_K(K, L, \epsilon) \neq 0$ for all $(K, L, \varphi) \in \Omega \times \Phi$. Second, ϵ is a primitive parameter or a function of the primitive parameters, $\epsilon = \epsilon(\varphi)$, such that it is a small number centered around zero, e.g. $-1 < \epsilon < 1$ or $-1 < \epsilon(\varphi) < 1$. Third, function $z(\cdot)$ is non-monotonic in L and there is one value $L_s = L_s(K, \varphi)$ such that $z_L(K, L_s^\epsilon, \varphi) = 0$.

The two last assumptions imply that the consumption function can have a zero derivative with respect to labor from two different sources: (i) when the parameter vector is such that $\epsilon(\varphi_0) = 0$ or (ii) when $L = L_s(K, \varphi)$. Those are the two main origins of singularities in our model.

Let us introduce the following definitions to distinguish two different types of singularities:

Definition 5 A point $(K, L, \varphi) \in \Omega \times \Phi$ is a regular point if $\epsilon(\varphi) \delta(K, L, \varphi) \neq 0$. A singularity exists if point $(K, L, \varphi) \in \Omega \times \Phi$ is such that $\epsilon(\varphi) \delta(K, L, \varphi) = 0$. We call singular perturbation to a point $\varphi^p \in \Phi$ such that $\epsilon(\varphi^p) = 0$. We call impasse set to a subset of points $(K^s, L^s) \in \Omega$ such that $\delta(K^s, L^s, \varphi) = 0$.

Henceforth, we assume for simplicity that there is either a singular perturbation or impasses, not both:

Assumption 2 There are no points (K^*, L^*, φ^*) such that $\epsilon(\varphi^*) = 0$ and $\delta(K^*, L^*, \varphi^*) = 0$ simultaneously.

---

[3] Observe that we are now referring to a two-dimensional projection in (K, L) of a three-dimensional system in (K, L, C).

Before we characterize DGE paths in the presence of singularities, we can show that they have a fundamentally different nature depending on whether they arise from a singular perturbation or from an impasse.

Proposition 1 *Singular-perturbed steady-states are generic points of set Ω while impasse steady-states are non-generic points of set Ω.*

This is easy to see if we take into account that singular-perturbed steady states belong to set:

$$\Gamma_E^p \equiv \{ (K, L) \in \Omega : f_1(K, L, \varphi^p) = f_2(K, L, \varphi^p) = 0\},$$

where $\epsilon(\varphi^p) = 0$ and impasse steady-states belong to set

$$P_E \equiv \{ (K, L, \varphi) \in \Omega \times \Phi : f_1(K, L, \varphi) = f_2(K, L, \varphi) = \delta(K, L, \varphi) = 0\}, \tag{3.12}$$

in which one parameter should take a particular value.

Proposition 2 *Singular perturbations have a permanent character (i.e. they characterize the entire DGE paths), because they occur when a parameter exhibits a particular critical value (or the parameters satisfy a specific relationship). An impasse singularity generically constrains the existence of DGE paths, or changes their dynamic properties, after a finite time.*

In order to prove this we denote the vector field in system (3.7) by

$$F(K, L, \varphi) = \begin{pmatrix} f_1(K, L, \varphi) \\ f_2(K, L, \varphi) \\ \epsilon(\varphi)\delta(K, L, \varphi) \end{pmatrix}. \tag{3.13}$$

The Jacobian of $F(\cdot)$, evaluated at any point $(K, L) \in \Omega$, is

$$DF(K, L, \varphi) =$$
$$\begin{pmatrix} f_{1,K}(K, L, \varphi) & f_{1,L}(K, L, \varphi) \\ \frac{f_{2,K}(K, L, \varphi)}{\epsilon(\varphi)\delta(K, L, \varphi)} - \frac{f_2(K, L, \varphi)\delta_K(K, L, \varphi)}{\epsilon(\varphi)\delta^2(K, L, \varphi)} & \frac{f_{2,L}(K, L, \varphi)}{\epsilon(\varphi)\delta(K, L, \varphi)} - \frac{f_2(K, L, \varphi)\delta_L(K, L, \varphi)}{\epsilon(\varphi)\delta^2(K, L, \varphi)} \end{pmatrix}, \tag{3.14}$$

has trace and determinant given by

$$\text{tr}DF(K, L, \varphi) = f_{1,K}(K, L, \varphi) + \frac{f_{2,L}(K, L, \varphi)\delta(K, L, \varphi) - f_2(K, L, \varphi)\delta_L(K, L, \varphi)}{\epsilon(\varphi)\delta^2(K, L, \varphi)},$$

and

$$\det DF(K, L, \varphi) = \frac{f_{1,K}(K, L, \varphi) f_{2,L}(K, L, \varphi) - f_{1,L}(K, L, \varphi) f_{2,K}(K, L, \varphi)}{\epsilon(\varphi) \delta(K, L, \varphi)} +$$

$$+ f_2(K, L, \varphi) \left(\frac{f_{1,K}(K, L, \varphi)\delta_L(K, L, \varphi) - f_{1,L}(K, L, \varphi)\delta_K(K, L, \varphi)}{\epsilon(\varphi) \delta^2(K, L, \varphi)} \right).$$

$$(3.15)$$

There is a point in common between the two types of singularities: generically after the singularity is crossed the dimension of the stable manifold changes because one eigenvalue of $DF(K, L, \varphi)$ changes sign by passing through plus or minus infinity. However, while for singular perturbations the singularity crossing is associated to a change of a parameter away from a critical value, for impasse singularities it is associated to a change in the value of variables (K, L) away from a specific relationship, as we see next.

For singular perturbations we have:

$$\lim_{\varphi \to \varphi^p} \mathrm{tr} DF(K, L, \varphi) = \lim_{\varphi \to \varphi^p} \det DF(K, L, \varphi) = \pm\infty,$$

and if φ_1 and φ_2 belong to the vicinity of φ^p and $\epsilon(\varphi_1) < 0 < \epsilon(\varphi_2)$, then

$$\mathrm{sign}\,(\det DF(K, L, \varphi_1)) \neq \mathrm{sign}\,(\det DF(K, L, \varphi_2)).$$

For impasse points we have:

$$\lim_{(K,L) \to (K^s, L^s)} \mathrm{tr} DF(K, L, \varphi) = \lim_{(K,L) \to (K^s, L^s)} \det DF(K, L, \varphi) = \pm\infty,$$

and if (K_1, L_1) and (K_2, L_2) belong to a small neighborhood of (K^s, L^s) and $\delta(K_1, L_1, .) < 0 < \delta(K_2, L_2, .)$, then $\mathrm{sign}(\det DF(K_1, L_1, \varphi)) \neq \mathrm{sign}(\det DF (K_2, L_2, \varphi))$.

The Jacobian of $F(\cdot)$ evaluated at a steady-state $(\bar{K}, \bar{L}) \in \Gamma_E$, for a given parameter value φ, is

$$DF(\bar{K}, \bar{L}, \varphi) = \begin{pmatrix} f_{1,K}(\bar{K}, \bar{L}, \varphi) & f_{1,L}(\bar{K}, \bar{L}, \varphi) \\ \dfrac{f_{2,K}(\bar{K}, \bar{L}, \varphi)}{\epsilon(\varphi)\,\delta(\bar{K}, \bar{L}, \varphi)} & \dfrac{f_{2,L}(K, L, \varphi)}{\epsilon(\varphi)\,\delta(\bar{K}, \bar{L}, \varphi)} \end{pmatrix}, \qquad (3.16)$$

having the trace and the determinant given by

$$\mathrm{tr} DF(\bar{K}, \bar{L}, \varphi) = f_{1,K}(\bar{K}, \bar{L}, \varphi) + \frac{f_{2,L}(K, L, \varphi)}{\epsilon(\varphi)\,\delta(K, L, \varphi)},$$

$$\det DF(\bar{K}, \bar{L}, \varphi) = \frac{f_{1,K}(\bar{K}, \bar{L}, \varphi) f_{2,L}(\bar{K}, \bar{L}, \varphi) - f_{1,L}(\bar{K}, \bar{L}, \varphi) f_{2,K}(\bar{K}, \bar{L}, \varphi)}{\epsilon(\varphi)\,\delta(\bar{K}, \bar{L}, \varphi)}.$$

$$(3.17)$$

For singular perturbations we still have:

$$\lim_{\varphi \to \varphi^p} \text{tr} DF(\bar{K}, \bar{L}, \varphi) = \lim_{\varphi \to \varphi^p} \det DF(\bar{K}, \bar{L}, \varphi) = \pm\infty.$$

When impasse singularities exist and if the steady-state is not an impasse-singular steady-state, that is if $(\bar{K}, \bar{L}) \notin P_E$ and $\varphi \neq \varphi^p$, then both $\text{tr} DF(\bar{K}, \bar{L}, \varphi)$ and $\det DF(\bar{K}, \bar{L}, \varphi)$ are finite, which means that the eigenvalues are finite and the steady-state is regular.

Therefore, there is a major difference between the two types of singularity. The singular perturbation changes the signs of the eigenvalues of the Jacobian $DF(K, L, \varphi)$ at each and every point $(K, L) \in \Omega$, including at steady-states, when the parameter φ passes through the critical point φ^p. The impasse singularity, in contrast, generates infinitely-valued eigenvalues only when the variables (K, L) satisfy the condition $\delta(K, L) = 0$, which generically occurs at points which are not steady-states, and steady-states are regular points. To put it simply, while DGE paths are everywhere pointwise-singular for singular perturbations, DGE paths are almost everywhere pointwise-regular and are pointwise-singular for a small number of impasse points for impasse singularities.

Next, we enumerate and characterize the types of DGE paths that can occur in the presence of a singularity.

3.2.3 DGE Paths in the Presence of Singular Perturbations

In this section, we assume there is a critical value for φ, $\varphi^p \in \Phi$, such that $\epsilon(\varphi^p) = 0$. From Assumption 2 we also have $\delta(K, L, \varphi^p) \neq 0$ for all $(K, L) \in \Omega$. We study the DGE dynamics associated to varying φ in the vicinity of φ^p such that $\epsilon(\varphi) \in (-1, 1)$ and $\delta(K, L, \epsilon(\varphi)) \neq 0$.

For convenience, we write system (3.8) in the following equivalent form:

$$\begin{aligned} \dot{K} &= f_1(K, L, \epsilon), \\ \epsilon \dot{L} &= f_2^s(K, L, \epsilon), \end{aligned} \tag{3.18}$$

where $f_2^s(K, L, \epsilon) \equiv f_2(K, L, \epsilon)/\delta(K, L, \epsilon)$.

Systems of type (3.18), with $\epsilon \in (-1, 1)$, are called *singular-perturbed* or *slow-fast* systems.[4] That designation is justified because the two variables have two different time scales: K has a slow time scale, and L has a fast time scale. Time t is called slow time and time $\tau = t/\epsilon$ is called fast time. If ϵ is close to zero we see that the adjustment of L is very fast along a curve $f_2^s(K, L, \epsilon) \approx 0$.

As time derivatives in system (3.18) refer to the dynamics along the slow time scale we call it the slow system. The dynamics along the fast time scale is

---

[4]See Kuehn (2015) for a recent textbook presentation.

$$K' = \frac{dK}{d\tau} = \epsilon f_1(K, L, \epsilon),$$

$$L' = \frac{dL}{d\tau} = f_2^s(K, L, \epsilon).$$

(3.19)

This pair of systems (3.18), (3.19) is called a *slow-fast system*.

The slow-fast vector fields associated to the slow-fact system (3.18), (3.19) are, respectively,

$$F(K, L, \epsilon) = \begin{pmatrix} f_1(K, L, \epsilon) \\ \dfrac{f_2^s(K, L, \epsilon)}{\epsilon} \end{pmatrix} \text{ and } F^f(K, L, \epsilon) = \begin{pmatrix} \epsilon f_1(K, L, \epsilon) \\ f_2^s(K, L, \epsilon) \end{pmatrix}.$$

If $\epsilon \neq 0$, the dynamics and bifurcations of system (3.18) are both regular and well-known. If functions $f_1(K, L)$ and $f_2(K, L)$ are differentiable then, given any initial value $K_0 \in \Omega$ a solution to system (3.18), (3.19) exists, it is unique, and it is continuous in t, for all $t \in [0, \infty)$.

Furthermore, at a regular equilibrium point $(\bar{K}, \bar{L}) \in \Gamma_E$ (for $\Gamma_E \neq \varnothing$), the local dynamics is characterized by the eigenvalues of the Jacobian for the slow-system evaluated at that steady-state, $DF(\bar{K}, \bar{L})$. Clearly, there may exist values for ϵ such that one or more eigenvalues can have zero real parts (and therefore regular bifurcation can exist) or can change from real to complex (or back).[5]

The Jacobian of the slow-vector field, F, evaluated at a steady-state

$$DF(\bar{K}, \bar{L}, \epsilon) = \begin{pmatrix} f_{1,K}(\bar{K}, \bar{L}, \epsilon) & f_{1,L}(\bar{K}, \bar{L}, \epsilon) \\ \dfrac{f_{2,K}^s(\bar{K}, \bar{L}, \epsilon)}{\epsilon} & \dfrac{f_{2,L}^s(\bar{K}, \bar{L}, \epsilon)}{\epsilon} \end{pmatrix},$$

has trace and determinant given by

$$\text{tr} DF(\bar{K}, \bar{L}, \epsilon) = f_{1,K}(\bar{K}, \bar{L}, \epsilon) + \frac{f_{2,L}^s(\bar{K}, \bar{L}, \epsilon)}{\epsilon}, \tag{3.20}$$

$$\det DF(\bar{K}, \bar{L}, \epsilon) = \frac{f_{1,K}(\bar{K}, \bar{L}, \epsilon) f_{2,L}^s(\bar{K}, \bar{L}, \epsilon) - f_{1,L}(\bar{K}, \bar{L}, \epsilon) f_{2,K}^s(\bar{K}, \bar{L}, \epsilon)}{\epsilon}. \tag{3.21}$$

The local stable manifold is one-dimensional for $\det DF(\bar{K}, \bar{L}, \epsilon) < 0$ and it is two-dimensional (or zero-dimensional) for $\det DF(\bar{K}, \bar{L}, \epsilon) > 0$ and $\text{tr} DF(\bar{K}, \bar{L}, \epsilon) < 0$ (or >0).

If $\epsilon = 0$ we already know that the eigenvalues of the Jacobian for the slow system evaluated at any point of Ω become infinite. In particular, the eigenvalues of the Jacobian of the slow system evaluated at the steady-state also become infinite because

---

[5]Major references for continuous-time regular dynamics and bifurcations are Guckenheimer and Holmes (1990) or Kuznetsov (2005).

$$\lim_{\epsilon \to 0} \mathrm{tr} DF(\bar{K}, \bar{L}, \epsilon) = \lim_{\epsilon \to 0} \det DF(\bar{K}, \bar{L}, \epsilon) = \pm\infty.$$

However, the local dynamics in the neighborhood of a singular-perturbation point can be revealed by characterising the dynamics of the fast system in the neighbourhood of a steady-state. For any value of ϵ, the Jacobian of the fast vector field F^s evaluated at a steady-state $(\bar{K}, \bar{L}, \epsilon)$ is

$$DF^s(\bar{K}, \bar{L}, \epsilon) = \begin{pmatrix} \epsilon\, f_{1,K}(\bar{K}, \bar{L}, \epsilon) & \epsilon\, f_{1,L}(\bar{K}, \bar{L}, \epsilon) \\ f^s_{2,K}(\bar{K}, \bar{L}, \epsilon) & f^s_{2,L}(\bar{K}, \bar{L}, \epsilon) \end{pmatrix}.$$

As the trace and determinant are

$$\mathrm{tr} DF^s(\bar{K}, \bar{L}, \epsilon) = \epsilon\, f_{1,K}(\bar{K}, \bar{L}, \epsilon) + f^s_{2,L}(\bar{K}, \bar{L}, \epsilon), \tag{3.22}$$

$$\det DF^s(\bar{K}, \bar{L}, \epsilon) = \epsilon \left[f_{1,K}(\bar{K}, \bar{L}, \epsilon) f^s_{2,L}(\bar{K}, \bar{L}, \epsilon) - f_{1,L}(\bar{K}, \bar{L}, \epsilon) f^s_{2,K}(\bar{K}, \bar{L}, \epsilon) \right], \tag{3.23}$$

then

$$\mathrm{tr} DF^s(\bar{K}, \bar{L}, 0) = f^s_2(\bar{K}, \bar{L}, 0), \ \det DF^s(\bar{K}, \bar{L}, 0) = 0.$$

we see that the fast vector field evaluated in the vicinity of a singular perturbation, $\epsilon = 0$, has the characteristics of a regular-bifurcation point, i.e. the determinant changes sign by passing through zero, instead of passing through infinite as is the case for the slow vector field.

In order to understand the local dynamics when $\epsilon = 0$ it is convenient to write the slow-fast system as

$$\begin{aligned} \dot{K} &= f_1(K, L, 0), & K' &= 0. \\ 0 &= f^s_2(K, L, 0), & L' &= f^s_2(K, L, 0). \end{aligned} \tag{3.24}$$

We define a *singular-perturbed critical subset* by[6]

$$\mathcal{S}^p = \{(K, L) \in \Omega : f_2(K, L, 0) = 0\}.$$

We say that point $(K^p, L^p) \in \mathcal{S}^p$ is a *slow-fast singular* for $f^s_{2,L}(K^p, L^p, 0) = 0$. Point $(K^*, L^*) \neq (K^p, L^p) \in \mathcal{S}^p$ is *slow-fast regular* for $f^s_{2,L}(K^*, L^*, 0) \neq 0$.

At a slow-fast regular point we can solve equation $f^s_2(K, L, 0) = 0$ for L as a function of K, $L = h(K)$, by applying the implicit-function theorem. Locally the slope of function $h(K)$ is

$$h_K(K) = -\frac{f^s_{2,K}(K, L, 0)}{f^s_{2,L}(K, L, 0)} = -\frac{f_{2,K}(K, L, 0)}{f_{2,L}(K, L, 0)} \ \text{for } (K, L) \in \mathcal{S}^p.$$

---

[6]Recall we are assuming that $\delta(K, L, \epsilon) \neq 0$.

This means that the dynamics evolves along the surface $\mathcal{S}^p$, which is geometrically represented by the isocline associated to L.

We call slow-fast regular point to a point (K, L) such that $f_{2,L}(K, L, 0) \neq 0$ and slow-fast singular point to a point (K', L') such that $f_{2,L}(K', L', 0) = 0$. If slow-fast singular points do not exist, then that surface contains only one of two types of points: (i) *slow-fast regular attracting* points if $f_{2,L}^s(K, L, 0) = f_{2,L}(K, L, 0)/\delta(K, L, 0) < 0$ or (ii) *slow-fast regular repelling* points if $f_{2,L}^s(K, L, 0) = f_{2,K}(K, L, 0)/\delta(K, L, 0) > 0$, for $(K, L) \in \mathcal{S}^p$.

One of the most important results on the mathematics of slow-fast systems, the Fenichel (1979) theorem, states that in the neighborhood of surface $\mathcal{S}^p$ the dimension of the stable manifold is not changed by a small variation of the parameter ϵ in one of the neighborhoods of $\epsilon = 0$, that is either when $\epsilon \to 0^+$ or when $\epsilon \to 0^-$.

We denote the steady-state of system (3.24) by $(\bar{K}^p, \bar{L}^p)$. A taxonomy for the generic local dynamics in a neighborhood of a steady-state for system (3.18), (3.19) can be built:

1. Let $f_{1,K}(\bar{K}, \bar{L}, \epsilon) f_{2,L}^s(\bar{K}, \bar{L}, \epsilon) - f_{1,L}(\bar{K}, \bar{L}, \epsilon) f_{2,K}^s(\bar{K}, \bar{L}, \epsilon) > 0$ in a small neighborhood of ϵ centred around zero, i.e., for $\epsilon \in (0^-, 0^+)$:

   (a) If $f_{2,L}^s(\bar{K}, \bar{L}, \epsilon) < 0$, then (i) the steady state is a stable node for $\epsilon \to 0^+$, as we have $\mathrm{tr} DF^s(\bar{K}, \bar{L}, 0^+) < 0$ and $\det DF^s(\bar{K}, \bar{L}, 0^+) > 0$; (ii) set $\mathcal{S}^p$ only contains slow-fast regular attractor points converging to steady-state $(\bar{K}^p, \bar{L}^p)$ for $\epsilon = 0$, since $f_{2,L}^s(\bar{K}^p, \bar{L}^p, 0) < 0$; and (iii) the regular steady-state is a saddle point for $\epsilon \to 0^-$, as $\det DF^s(\bar{K}, \bar{L}, 0^-) < 0$.

   (b) If $f_{2,L}^s(\bar{K}, \bar{L}, \epsilon) > 0$, then (i) the steady state is an unstable node for $\epsilon \to 0^+$, as we have $\mathrm{tr} DF^s(\bar{K}, \bar{L}, 0^+) > 0$ and $\det DF^s(\bar{K}, \bar{L}, 0^+) > 0$; (ii) set $\mathcal{S}^p$ only contains slow-fast regular repeller points diverging from steady-state $(\bar{K}^p, \bar{L}^p)$ for $\epsilon = 0$, since $f_{2,L}^s(\bar{K}^p, \bar{L}^p, 0) > 0$; and (iii) the regular steady-state is a saddle point for $\epsilon \to 0^-$, as $\det DF^s(\bar{K}, \bar{L}, 0^-) < 0$.

2. For $f_{1,K}(\bar{K}, \bar{L}, \epsilon) f_{2,L}^s(\bar{K}, \bar{L}, \epsilon) - f_{1,L}(\bar{K}, \bar{L}, \epsilon) f_{2,K}^s(\bar{K}, \bar{L}, \epsilon) < 0$ in a small neighborhood of ϵ centred around zero, i.e. for $\epsilon \in (0^-, 0^+)$:

   (a) If $f_{2,L}^s(\bar{K}, \bar{L}, \epsilon) < 0$, then (i) the steady state is a saddle point for $\epsilon \to 0^+$, as we have $\det DF^s(\bar{K}, \bar{L}, 0^+) < 0$; (ii) set $\mathcal{S}^p$ only contains slow-fast regular attractor points converging to steady-state $(\bar{K}^p, \bar{L}^p)$ for $\epsilon = 0$, since $f_{2,L}^s(\bar{K}^p, \bar{L}^p, 0) < 0$; and (iii) the regular steady-state is an unstable node for $\epsilon \to 0^-$, as $\mathrm{tr} DF^s(\bar{K}, \bar{L}, 0^-) > 0$ and $\det DF^s(\bar{K}, \bar{L}, 0^-) > 0$.

   (b) If $f_{2,L}^s(\bar{K}, \bar{L}, \epsilon) > 0$, then (i) the steady state is a saddle point for $\epsilon \to 0^+$, as $\det DF^s(\bar{K}, \bar{L}, 0^+) < 0$; (ii) set $\mathcal{S}^p$ only contains slow-fast regular repeller points diverging from steady-state $(\bar{K}^p, \bar{L}^p)$ for $\epsilon = 0$, since $f_{2,L}^s(\bar{K}^p, \bar{L}^p, 0) > 0$; and (iii) the regular steady-state is a stable node for $\epsilon \to 0^-$, as $\mathrm{tr} DF^s(\bar{K}, \bar{L}, 0^-) < 0$ and $\det DF^s(\bar{K}, \bar{L}, 0^-) > 0$.

We can summarize the previous discussion in the following Proposition 3 which describes the types of DGE paths that can exist in the presence of a singular perturbation:

Proposition 3 (DGE paths in the presence of a singular perturbation) *Assume that there is a singular-perturbation and that $K_0 \in \mathcal{W}^s(\bar{K}, \bar{L})$, if $\mathcal{W}^s(\bar{K}, \bar{L})$ is non-empty, or that $K_0 = \bar{K}$, if $\mathcal{W}^s(\bar{K}, \bar{L})$ is empty. Then, only two generic cases exist:*

1. *If the DGE path exhibits permanent indeterminacy for $\epsilon \to 0^+$ ($\epsilon \to 0^-$), it is determinate for both $\epsilon = 0$ and $\epsilon \to 0^-$ ($\epsilon = 0^+$).*
2. *If the DGE is stationary for $\epsilon \to 0^+$ ($\epsilon \to 0^-$), it is still stationary for $\epsilon = 0$ and it is determinate for $\epsilon = 0^-$ ($\epsilon \to 0^+$).*

The cases presented above always produce stable or unstable nodes, but not stable or unstable foci. This is due to the fact that in generic cases, for $\epsilon \approx 0$, the trace of the Jacobian $DF(\bar{K}, \bar{L}, 0^{\pm})$ becomes very large in absolute value, which implies that the eigenvalues have to be real. However, for values of ϵ in a wider range around zero, the trace of the Jacobian tends to decrease which implies that the discriminant

$$\Delta DF(\bar{K}, \bar{L}, \epsilon) = \frac{1}{4}\left[\left(f_{1,K}(\bar{K}, \bar{L}, \epsilon) - \frac{f_{2,L}^s(\bar{K}, \bar{L}, \epsilon)}{\epsilon}\right)^2 - \frac{4 f_{1,L}(\bar{K}, \bar{L}, \epsilon)\, f_{2,K}^s(\bar{K}, \bar{L}, \epsilon)}{\epsilon}\right]$$

can become negative, and eigenvalues may become complex, leading to oscillatory dynamics.

A geometric argument for the existence of real eigenvalues close to the singular critical set containing only slow-fast regular points is that the solution path will tend to evolve along the isocline for L, where $f_2(K, L, 0) = 0$, which is a monotonic surface.

Whilst we have only considered the cases in which slow-fast singular points exists for non-zero $f_{2,L}(K, L, 0)$, there are also further results which hold for the zero case (for bifurcation results in slow-fast systems for the zero case, see Kuehn (2015, Chaps. 3 and 8).

3.2.4 DGE Paths in the Presence of Impasse Singularities

In this section we assume that $\epsilon(\varphi) \neq 0$ for all $\varphi \in \Phi$ and define the *impasse set* as

$$S = \{(K, L) \in \Omega : \delta(K, L, \varphi) = 0\}. \tag{3.25}$$

Assumption 2 implies that set S is non-empty, which means that it introduces a partition over state Ω, such that $\Omega = \Omega_- \cup S \cup \Omega_+$, where we define:

$$\Omega_- \equiv \{(K, L) : \delta(K, L, \varphi) < 0\} \text{ and } \Omega_+ \equiv \{(K, L) : \delta(K, L, \varphi) > 0\},$$

which are both open subsets containing exclusively regular points.

From now on we assume there are only regular impasse points, that is points $(K, L) \in S$ such that $\nabla \delta(K, L) \neq \mathbf{0}$. Singular impasse points are points satisfying $\nabla \delta(K, L) = \mathbf{0}$.

Zhitomirskii (1993) and Llibre et al. (2002) prove that regular impasse points $(K^s, L^s) \in S$, such that $\delta(K^s, L^s) = 0$, $\nabla \delta(K^s, L^s) \neq \mathbf{0}$, and $f_1(K^s, L^s) \neq 0$, can be of the following types[7]:

1. $(K^s, L^s) \in S$ is an *impasse-repeller point* if $\delta_L(K^s, L^s, \varphi) \neq 0$, $f_2(K^s, L^s, \varphi)$ $\neq 0$ and $\delta_L(K^s, L^s, \varphi) f_2(K^s, L^s, \varphi) > 0$. At an impasse-repeller point there are two trajectories that are repelled away from S, one to the interior of Ω_- and another to the interior of Ω_+. This means that, if those paths are solutions of system (3.8) they can, in most generic cases, be continued until $t \to \infty$, thus satisfying a necessary condition for being DGE paths. We denote the sets of impasse-repeller points by $S_+ \equiv \{(K, L) \in S : \delta_L(K, L, \varphi) f_2(K, L, \varphi) > 0\}$.

2. $(K^s, L^s) \in S$ is an *impasse-attractor point* if $\delta_L(K^s, L^s, \varphi) \neq 0$, $f_2(K^s, L^s, \varphi) \neq 0$ and $\delta_L(K^s, L^s, \varphi) f_2(K^s, L^s, \varphi) < 0$. An impasse-attractor point attracts two trajectories, one coming from the interior of Ω_- and another from the interior of Ω_+. This means that, if those paths are solutions of system (3.8), they are only defined for $t \in [0, t^s)$, where t^s is the time of collision with S. Therefore, they cannot be DGE paths, as these trajectories cannot be continued until $t \to \infty$. We denote the sets of impasse-attractor points by $S_- \equiv \{(K, L) \in S : \delta_L(K, L, \varphi) f_2(K, L, \varphi) < 0\}$.

3. $(K^s, L^s) \in S$ is a *impasse-tangent point* if $\delta_L(K^s, L^s, \varphi) = 0$ and $f_2(K^s, L^s, \varphi) \neq 0$. At such a point, there is one trajectory coming from the interior of Ω_+ (or Ω_-) that is tangent to S at $t = t^s < \infty$ and has a continuation in the interior of the same subset Ω_+ (or Ω_-). This means that, if those paths are solutions of system (3.8) they can, in most generic cases, be continued until $t \to \infty$ within the same subset Ω_+ (or Ω_-). Thus, a necessary condition for being considered DGE paths holds. We denote the set of impasse-tangent points by $\Gamma_K \equiv \{(K, L) \in S : \delta_L(K, L, \varphi) = 0, \ f_2(K, L, \varphi) \neq 0\}$.

4. $(K^s, L^s) \in S$ is a *impasse-transversal point* if $\delta_L(K^s, L^s, \varphi) \neq 0$ and $f_2(K^s, L^s, \varphi) = 0$. We denote the set of these points by $\Gamma_I \equiv \{(K, L) \in S : f_2(K, L, \varphi) = 0, \ \delta_L(K, L, \varphi) \neq 0\}$. Trajectories that cross through impasse-transversal may or may not exist. In other words, we may obtain trajectories that originate in the interior of one or both subsets, Ω_+ or Ω_-, are transversal to S, and have a continuation for $t \in (t^s, \infty)$ in the interior of the other subset. We may have two additional cases: (i) paths that remain within the same subset Ω_+ or Ω_- for all $t \in [0, \infty)$ and satisfy a necessary condition to be DGE paths and (ii) paths colliding with S, but which have no continuation cannot be DGE paths, as solutions do not exist for $t \in (t_s, \infty)$. Zhitomirskii (1993) demonstrates that three types of impasse-transversal points exist: (i) *impasse-transversal nodes*, at which an infinite number of trajectories crossing S exist, both traveling from the interior of one subset (Ω_+ or Ω_-) to the interior of the other subset, and no trajecto-

---

[7]A singular impasse point is a point in S such that $\nabla \delta = \mathbf{0}$.

ries going in the opposite direction exist; (ii) *impasse-transversal saddles*, at which two trajectories crossing $\mathcal{S}$ exist, one traveling from the interior of Ω_+ to the interior of Ω_- and another traveling from the interior of Ω_- to the interior of Ω_+; and (iii) *impasse-transversal foci*, at which there can be no crossing of $\mathcal{S}$.

In order to determine the type of a impasse-transversal point, it is convenient to consider the dynamics for the desingularized system bellow:

$$
\begin{aligned}
\dot{K} &= \epsilon(\varphi)\,\delta(K, L, \varphi)\, f_1(K, L, \varphi),\\
\dot{L} &= f_2(K, L, \varphi),
\end{aligned}
\tag{3.26}
$$

which has the same integral curves as system (3.8), but in which the direction field within subspace Ω_- is inverted and the singularity at $\mathcal{S}$ is removed. We define the desingularized vector field for system (3.26) as

$$
F^r(K, L, \varphi) = \begin{pmatrix} \epsilon(\varphi)\,\delta(K, L, \varphi)\, f_1(K, L, \varphi)\\ f_2(K, L, \varphi) \end{pmatrix}.
\tag{3.27}
$$

By denoting an impasse-transversal point as $(K^i, L^i) \in \Gamma_I$, we can readily see that they are fixed points of the desingularized vector field, due to the fact that $\delta(K^i, L^i, \varphi) = f_2(K^i, L^i, \varphi) = 0$, i.e. $F^r(K^i, L^i, \varphi) = \mathbf{0}$, where $\delta_L(K^i, L^i, \varphi) \neq 0$. Therefore, the Jacobian of the desingularized vector field $F^r(\cdot)$, evaluated at (K^i, L^i), is given by

$$
DF^r(K^i, L^i, \varphi) = \begin{pmatrix} \epsilon(\varphi)\,\delta_K(K^i, L^i, \varphi)\, f_1(K^i, L^i, \varphi) & \epsilon(\varphi)\,\delta_L(K^i, L^i, \varphi)\, f_1(K^i, L^i, \varphi)\\ f_{2,K}(K^i, L^i, \varphi) & f_{2,L}(K^i, L^i, \varphi) \end{pmatrix}.
\tag{3.28}
$$

Its trace and determinant are, respectively,

$$
\operatorname{tr} DF^r(K^i, L^i, \varphi) = \epsilon(\varphi)\,\delta_K(K^i, L^i, \varphi)\, f_1(K^i, L^i, \varphi) + f_{2,L}(K^i, L^i, \varphi),
$$

and

$$
\begin{aligned}
&\det DF^r(K^i, L^i, \varphi) =\\
&= \epsilon(\varphi)\, f_1(K^i, L^i, \varphi)\left(\delta_K(K^i, L^i, \varphi)\, f_{2,L}(K^i, L^i, \varphi) - \delta_L(K^i, L^i, \varphi)\, f_{2,K}(K^i, L^i, \varphi)\right),
\end{aligned}
\tag{3.29}
$$

which implies that the discriminant is

$$
\Delta DF^r(K^i, L^i, \varphi) = (\operatorname{tr} DF^r)^2(K^i, L^i, \varphi) - \det DF^r(K^i, L^i, \varphi).
$$

The eigenvalues of Jacobian (3.28) allows us to determine the type of the impasse-transversal point:

1. (K^i, L^i) is an *impasse-transversal node* if the eigenvalues of (3.28) are both real and share the same sign, i.e. if (K^i, L^i) belongs to $\Gamma_{IN} \equiv \{(K, L) \in \Gamma_I :$ $\det DF^r(K, L, \varphi) > 0, \Delta DF^r(K, L, \varphi) > 0\}$.
2. (K^i, L^i) is an *impasse-transversal saddle* if the eigenvalues of (3.28) are both real and exhibit opposite signs, i.e. if (K^i, L^i) belongs to $\Gamma_{IS} \equiv \{(K, L) \in \Gamma_I :$ $\det DF^r(K, L, \varphi) < 0\}$.
3. (K^i, L^i) is an *impasse-transversal focus* if the eigenvalues of (3.28) are complex-conjugate, i.e. is if (K^i, L^i) belongs to $\Gamma_{IF} \equiv \{(K, L) \in \Gamma_I : \det DF^r(K, L, \varphi)$ $> 0, \Delta DF^r(K, L, \varphi) < 0\}$.

Let us consider that (K^i, L^i) is an impasse-transversal point and $\mathcal{W}^s(K^i, L^i)$ is the stable manifold associated to it. We denote by $\mathcal{W}^s_+(K^i, L^i)$ the stable sub-manifold which is contained in Ω_+ and by $\mathcal{W}^s_-(K^i, L^i)$ the stable sub-manifold which is contained in Ω_-.

If point (K^i, L^i) is an impasse-transversal node and $\mathrm{tr} DF^r(K^i, L^i, .) < 0 (> 0)$, then it is attracting (repelling) from side Ω_+ (Ω_-) and it is repelling from side Ω_- (Ω_+). Thus, there is an infinite number of trajectories coming from the interior of Ω_+ (Ω_-) that are attracted to point (K^i, L^i) and are all repelled from (K^i, L^i) to the interior of Ω_- (Ω_+). Furthermore, the basin of attraction of (K^i, L^i) is a two-dimensional stable manifold contained in subset Ω_+ (Ω_-), i.e. $\mathcal{W}^s(K^i, L^i) = \mathcal{W}^s_+(K^i, L^i)$ $(\mathcal{W}^s(K^i, L^i) = \mathcal{W}^s_-(K^i, L^i))$.

If point (K^i, L^i) is an impasse-transversal saddle, there are only two trajectories converging to (K^i, L^i) in finite time, one converging from the interior or Ω_+ and another converging from the interior of Ω_-. These trajectories belong to different integral curves passing through (K^i, L^i), which means that the eigenspaces tangent to the stable manifolds, from both sides Ω_+ and Ω_-, are not collinear. All the remaining integral curves passing trough (K^i, L^i) contain repelling trajectories to the interior of Ω_+ or Ω_-. Therefore, the stable manifold associated with (K^i, L^i) has two one-dimensional branches $\mathcal{W}^s_+(K^i, L^i) \subset \Omega_+$ and $\mathcal{W}^s_-(K^i, L^i) \subset \Omega_-$, and $\mathcal{W}^s(K^i, L^i) = \mathcal{W}^s_+(K^i, L^i) \cup \mathcal{W}^s_-(K^i, L^i)$.

While impasse-repeller and impasse-attractor points belong to a surface in (K, L),[8] the sets of impasse-tangent and -transversal points are isolated points in (K, L).[9] If no impasse-tangent or -transversal points exist, then set S only contains repeller or attractor points. To put it simply, if both Γ_I and Γ_K are empty, then only one of two cases is possible: either $S = S_+$ and $S_- = \varnothing$, or $S = S_-$ and $S_+ = \varnothing$. However, if an impasse-tangent or -transversal point exists, then the impasse set contains two open subsets of impasse-repeller and impasse-attractor points. There are two cases: (i) if Γ_I is empty and Γ_K is not, then $S = S_- \cup \Gamma_K \cup S_+$ or (ii) if Γ_K is empty and Γ_I is not, then $S = S_- \cup \Gamma_I \cup S_+$.

The possibility of crossing through the impasse set S is another important consequence from the existence of impasse-transversal points. This crossing behavior cannot exist in models without impasse singularities.

---

[8] Indeed, a one-dimensional manifold.

[9] A zero-dimensional manifold.

Lemma 1 (Existence of crossing trajectories) *Trajectories crossing the impasse surface S, from the interior of Ω_+ or Ω_-, can only exist if there is an impasse-transversal point that is either an impasse-transversal node or an impasse-transversal saddle, i.e. if Γ_{IN} or Γ_{IS} are non-empty.*[10]

Non-generic regular impasse points are points $(K, L, \varphi) \in (\Gamma_I \cup \Gamma_K) \times \Phi$ such that a parameter (or relation between parameters) satisfies a critical condition. These points correspond to one-parameter impasse bifurcations—for a complete characterization see Llibre et al. (2002). One example are points satisfying $\delta(K, L, \varphi) = f_1(K, L, \varphi) = f_2(K, L, \varphi) = 0$, which are members of the set P_E defined in Eq. (3.12). Clearly $P_E = \Gamma_E \cap \Gamma_I$. Non-generic points belonging to P_E, satisfying further critical conditions on the parameters, are sometimes called *singularity-induced bifurcation points*.

We call *singular steady-state* to a fixed point located on the impasse surface, S, i.e. to a member of set P_E. Therefore, a *regular steady-state* is a member of the set $\Gamma_E \setminus P_E$. No singular steady-states exist if P_E is empty, although impasse-transversal points (i.e. is non-empty Γ_I) may exist.

Next, we assume that there is a regular steady-state $(\bar{K}, \bar{L}) \in \Gamma_E$ in the neighbourhood of an impasse surface S. We will see next that the types of impasse points not only determine which types of steady-states can exist, depending on their location as regards S, but also the confinement, or not, of the stable manifold $W^s(\bar{K}, \bar{L})$ to one of the subspaces (Ω_+ or Ω_-).

We start with cases in which S only contains *generic regular impasse points*, such that $\delta_L(K^s, L^s) f_2(K^s, L^s) \neq 0$: that is S contains only impasse-repeller or impasse-attractor points.

Lemma 2 (Local dynamics in the neighborhood of S containing only generic points) *Assume there is a regular steady-state in a neighborhood of set S such that $\delta_L(K^s, L^s) f_2(K^s, L^s) \neq 0$ for all $(K^s, L^s) \in S$.*

1. *If the impasse set contains only impasse-attractor points, i.e., if $S = S_-$, then the stable manifold $W^s(\bar{K}, \bar{L})$ is empty or it is one-dimensional;*
2. *If the impasse set contains only impasse-repeller points, i.e., if $S = S_+$, then the stable manifold $W^s(\bar{K}, \bar{L})$ is one- or it is two-dimensional;*
3. *If the stable manifold is not empty it is allways contained in same subset as the steady-state: if $(\bar{K}, \bar{L}) \in \Omega_+$, then $W^s(\bar{K}, \bar{L}) = W^s_+(\bar{K}, \bar{L}) \subset \Omega_+$ and if $(\bar{K}, \bar{L}) \in \Omega_-$, then $W^s(\bar{K}, \bar{L}) = W^s_-(\bar{K}, \bar{L}) \subset \Omega_-$;*

Proof If a steady-state exists in a neighbourhood of an impasse-attractor point then it is either a saddle point or an unstable node or an unstable focus. If a steady-state exists in a neighbourhood of an impasse-repeller point then it is either a saddle point or an stable node or a stable focus. Because there are no crossing trajectories through S, if it only contains attractor or repeller points, and if the steady-state is a saddle point or is a stable node or focus, then the stable manifold $W^s(\bar{K}, \bar{L})$ is a subset of the subspace containing the steady-state $(\bar{K}, \bar{L})$.   □

---

[10]For a proof, see Cardin et al. (2012) *inter alia*.

Next we assume that the impasse set S contains only one *non-generic regular impasse point* that partitions it into two open sets of impasse-repeller and impasse-attractor points.

Lemma 3 (Local dynamics in the neighborhood of S containing one non-generic point) *Assume there is one regular steady-state $(\bar{K}, \bar{L})$ and that set S contains one non-generic impasse point such that $\delta_L(K^s, L^s) f_2(K^s, L^s) = 0$ for $(K^s, L^s) \in S$. Then:*

1. *If there is one impasse-tangent point $(K^k, L^k) \in \Gamma_K$ then the stable manifold $\mathcal{W}^s(\bar{K}, \bar{L})$ can be empty, one-dimensional or two-dimensional. In the last two cases it is contained in the interior (or in the closure containing the impasse-tangent point) of the subspace containing the steady-state: if $(\bar{K}, \bar{L}) \in \Omega_+$ then $\mathcal{W}^s(\bar{K}, \bar{L}) \supseteq \mathcal{W}^s_+(\bar{K}, \bar{L})$, or if $(\bar{K}, \bar{L}) \in \Omega_-$ then $\mathcal{W}^s(\bar{K}, \bar{L}) \supseteq \mathcal{W}^s_-(\bar{K}, \bar{L})$.*
2. *If there is one impasse-transversal focus point $(K^i, L^i) \in \Gamma_{IF}$ then the stable manifold $\mathcal{W}^s(\bar{K}, \bar{L})$ is one-dimensional and it belongs to the subspace containing the steady-state: if $(\bar{K}, \bar{L}) \in \Omega_+$ then $\mathcal{W}^s(\bar{K}, \bar{L}) = \mathcal{W}^s_+(\bar{K}, \bar{L})$, or if $(\bar{K}, \bar{L}) \in \Omega_-$ then $\mathcal{W}^s(\bar{K}, \bar{L}) = \mathcal{W}^s_-(\bar{K}, \bar{L})$.*
3. *If there is one impasse-transversal saddle point $(K^i, L^i) \in \Gamma_{IS}$ either the stable manifold $\mathcal{W}^s(\bar{K}, \bar{L})$ is empty or it is non-empty. If it is non-empty two cases are possible: (i) it can be contained in the interior of one subspace, that is if $(\bar{K}, \bar{L}) \in \Omega_+$, then $\mathcal{W}^s(\bar{K}, \bar{L}) = \mathcal{W}^s_+(\bar{K}, \bar{L})$, or if $(\bar{K}, \bar{L}) \in \Omega_-$, then $\mathcal{W}^s(\bar{K}, \bar{L}) = \mathcal{W}^s_-(\bar{K}, \bar{L})$ and it is two-dimensional; (ii) it can have elements in the two subspaces, that is $\mathcal{W}^s(\bar{K}, \bar{L}) = \mathcal{W}^s_+(\bar{K}, \bar{L}) \cup \{(K^i, L^i)\} \cup \mathcal{W}^s_-(\bar{K}, \bar{L})$ such that the sub-manifold coinciding with the stable manifold associated to (K^i, L^i) is one-dimensional and the sub-manifold belonging to the same space of $(\bar{K}, \bar{L})$ is two-dimensional.*
4. *If there is one impasse-transversal node $(K^i, L^i) \in \Gamma_{IN}$ the stable manifold $\mathcal{W}^s(\bar{K}, \bar{L})$ is allways non-empty. Two cases are possible: (i) it can be contained in the interior of one subspace, that is if $(\bar{K}, \bar{L}) \in \Omega_+$, then $\mathcal{W}^s(\bar{K}, \bar{L}) = \mathcal{W}^s_+(\bar{K}, \bar{L})$, or if $(\bar{K}, \bar{L}) \in \Omega_-$, then $\mathcal{W}^s(\bar{K}, \bar{L}) = \mathcal{W}^s_-(\bar{K}, \bar{L})$ and it is one-dimensional; (ii) it can have elements in the two subspaces, that is $\mathcal{W}^s(\bar{K}, \bar{L}) = \mathcal{W}^s_+(\bar{K}, \bar{L}) \cup \{(K^i, L^i)\} \cup \mathcal{W}^s_-(\bar{K}, \bar{L})$ such that the sub-manifold coinciding with the stable manifold associated to (K^i, L^i) is two-dimensional and the sub-manifold belonging to the same space of $(\bar{K}, \bar{L})$ is one-dimensional.*

Proof Assume S contains one impasse-tangent point, i.e. $(K^k, L^k) \in \Gamma_K$. We proved that $S = S_+ \cup \Gamma_K \cup S_-$ and that there is one trajectory which is tangent to S in finite time. Since the steady-state is regular, we have $(\bar{K}, \bar{L}) \neq (K^k, L^k)$. However, an integral curve passing through $(\bar{K}, \bar{L})$ and (K^k, L^k) may exist, along which trajectories that can converge to or diverge from $(\bar{K}, \bar{L})$. In addition, from Lemma 2, we can distinguish two cases: (i) if the steady-state $(\bar{K}, \bar{L})$ is close to S_-, it is an unstable node, an unstable focus or a saddle point or (ii) if it is close to S_+, it is a saddle point, a stable node or a stable focus. In the first case, the tangent trajectory diverges from the steady-state and in the second case, it can be converging or diverging. In any case, no crossing of surface S is possible. Therefore, assuming that the stable manifold is non-empty, if the tangent trajectory is diverging, then either $\mathcal{W}^s(\bar{K}, \bar{L}) = \mathcal{W}^s_+(\bar{K}, \bar{L})$ or $\mathcal{W}^s(\bar{K}, \bar{L}) = \mathcal{W}^s_-(\bar{K}, \bar{L})$, and if the trajectory is converging, then either $\mathcal{W}^s(\bar{K}, \bar{L}) = \mathcal{W}^s_+(\bar{K}, \bar{L}) \cup \{(K^k, L^k)\}$ or $\mathcal{W}^s(\bar{K}, \bar{L}) = \mathcal{W}^s_-(\bar{K}, \bar{L}) \cup \{(K^k, L^k)\}$.

Assume S contains one impasse-transversal focus point, i.e. $(K^i, L^i) \in \Gamma_{IF}$. We proved that $S = S_+ \cup \Gamma_{IF} \cup S_-$ and that there are no trajectories crossing S in finite time. Since the steady-state is regular, we have $(\bar{K}, \bar{L}) \neq (K^i, L^i)$. Applying Lemma 2, we potentially have the following cases: (i) if the steady-state $(\bar{K}, \bar{L})$ is close to S_-, it can be an unstable node or focus or a saddle point or (ii) if it is close to S_+, it can be a saddle point or a stable node or focus. However, from Lemma 4 the steady-state can only be a saddle point. Considering that there are no trajectories crossing the surface S then, if the stable manifold is non-empty, either $\mathcal{W}^s(\bar{K}, \bar{L}) = \mathcal{W}^s_+(\bar{K}, \bar{L})$ or $\mathcal{W}^s(\bar{K}, \bar{L}) = \mathcal{W}^s_-(\bar{K}, \bar{L})$ and $\mathcal{W}^s(\bar{K}, \bar{L})$ is one-dimensional.

Assume S contains one impasse-transversal saddle point, i.e. $(K^i, L^i) \in \Gamma_{IS}$. We proved that $S = S_+ \cup \Gamma_{IS} \cup S_-$ and that (see Lemma 1) there are two non-collinear converging trajectories, originated in both subsets Ω_+ and Ω_-, and passing through (K^i, L^i). Associated to the trajectories passing through (K^i, L^i) there is a one-dimensional stable manifold, spanning the two subspaces Ω_+ and Ω_-, and therefore $\mathcal{W}^s(K^i, L^i) = \mathcal{W}^s_+(K^i, L^i) \cup \mathcal{W}^s_-(K^i, L^i)$. Since the steady-state is regular, we have $(\bar{K}, \bar{L}) \neq (K^i, L^i)$. However, an integral curve passing through $(\bar{K}, \bar{L})$ and (K^i, L^i) can exist, along which trajectories may converge to or diverge from $(\bar{K}, \bar{L})$. Again, using Lemmas 2 and 4, we have: (i) if the steady-state $(\bar{K}, \bar{L})$ is close to S_-, it is an unstable node or an unstable focus, or (ii) if it is close to S_+ it is a stable node or a stable focus. An important distinction is related to the direction of trajectories flowing along integral curves passing through the two points $(\bar{K}, \bar{L})$ and (K^i, L^i): trajectories may diverge from $(\bar{K}, \bar{L})$ or converge to it. In the first case, the stable manifolds $\mathcal{W}^s_+(K^i, L^i)$ and $\mathcal{W}^s(\bar{K}, \bar{L})$ are disjoint and in the second case, they exhibit a non-empty intersection. This allows us enumerate the characteristics of the stable manifold $\mathcal{W}^s(\bar{K}, \bar{L})$, when it is non-empty. If integral curves passing through (K^i, L^i) and $(\bar{K}, \bar{L})$ diverge from $(\bar{K}, \bar{L})$ and the stable manifold is non-empty, then either $\mathcal{W}^s(\bar{K}, \bar{L}) = \mathcal{W}^s_+(\bar{K}, \bar{L})$ or $\mathcal{W}^s(\bar{K}, \bar{L}) = \mathcal{W}^s_-(\bar{K}, \bar{L})$ and $\mathcal{W}^s(\bar{K}, \bar{L})$ is two-dimensional. If integral curves passing through (K^i, L^i) and $(\bar{K}, \bar{L})$ converge to $(\bar{K}, \bar{L})$ and the stable manifold is non-empty, then the stable manifold $\mathcal{W}^s(\bar{K}, \bar{L})$ contains points in both subsets Ω_+ and Ω_-, i.e. $\mathcal{W}^s(\bar{K}, \bar{L}) = \mathcal{W}^s_-(\bar{K}, \bar{L}) \cup \{(K^i, L^i)\} \cup \mathcal{W}^s_+(\bar{K}, \bar{L})$. Thus, two cases are possible: (i) if $(\bar{K}, \bar{L}) \in \Omega_+$, then $\mathcal{W}^s_-(\bar{K}, \bar{L}) = \mathcal{W}^s_-(K^i, L^i)$ is one-dimensional and $\mathcal{W}^s_+(\bar{K}, \bar{L}) \not\subset \mathcal{W}^s_+(K^i, L^i)$ is two-dimensional and (ii) if $(\bar{K}, \bar{L}) \in \Omega_-$, then $\mathcal{W}^s_+(\bar{K}, \bar{L}) = \mathcal{W}^s_+(K^i, L^i)$ is one-dimensional and $\mathcal{W}^s_-(\bar{K}, \bar{L}) \not\subset \mathcal{W}^s_-(K^i, L^i)$ is two-dimensional.

Assume S contains one impasse-transversal node point, i.e. $(K^i, L^i) \in \Gamma_{IN}$. we proved that $S = S_+ \cup \Gamma_{IN} \cup S_-$ and that (see Lemma 1) there is an infinite number of trajectories coming from only one subspaces, Ω_+ or Ω_-, and passing through (K^i, L^i). There is a two-dimensional stable manifold associated to

(K^i, L^i), $\mathcal{W}^s(K^i, L^i) = \mathcal{W}^s_+(K^i, L^i) \subseteq \Omega_+$ or $\mathcal{W}^s(K^i, L^i) = \mathcal{W}^s_-(K^i, L^i) \subseteq \Omega_-$. Since the steady-state is regular, we have $(\bar{K}, \bar{L}) \neq (K^i, L^i)$, but there is an integral curve passing through $(\bar{K}, \bar{L})$ and (K^i, L^i), over which trajectories may converge to or diverge from $(\bar{K}, \bar{L})$. Lemmas 2 and 4 imply that the steady-state will always be a saddle point independently from being close to $\mathcal{S}_-$ or to $\mathcal{S}_+$. Two main cases can be distinguished: First, if both $(\bar{K}, \bar{L})$ and $\mathcal{W}^s(K^i, L^i)$ are in the same subset, Ω_+ or Ω_-, or if they are in two different subsets, but along any integral curve joining (K^i, L^i) and $(\bar{K}, \bar{L})$, trajectories do not converge to $(\bar{K}, \bar{L})$, then either $\mathcal{W}^s(\bar{K}, \bar{L}) = \mathcal{W}^s_+(\bar{K}, \bar{L})$ or $\mathcal{W}^s(\bar{K}, \bar{L}) = \mathcal{W}^s_-(\bar{K}, \bar{L})$ and $\mathcal{W}^s(\bar{K}, \bar{L})$ is one-dimensional. Second, if $(\bar{K}, \bar{L})$ and $\mathcal{W}^s(K^i, L^i)$ are in different subsets and there are integral curves joining (K^i, L^i) and $(\bar{K}, \bar{L})$ where trajectories converge to $(\bar{K}, \bar{L})$, then the stable manifold $\mathcal{W}^s(\bar{K}, \bar{L})$ contains points in both subsets, Ω_+ and Ω_-, i.e. $\mathcal{W}^s(\bar{K}, \bar{L}) = \mathcal{W}^s_-(\bar{K}, \bar{L}) \cup \{(K^i, L^i)\} \cup \mathcal{W}^s_+(\bar{K}, \bar{L})$. Two cases are possible: (i) if $(\bar{K}, \bar{L}) \in \Omega_+$, then $\mathcal{W}^s_-(\bar{K}, \bar{L}) = \mathcal{W}^s_-(K^i, L^i)$ is two-dimensional and $\mathcal{W}^s_+(\bar{K}, \bar{L})$ is one-dimensional, or (ii) if $(\bar{K}, \bar{L}) \in \Omega_-$, then $\mathcal{W}^s_+(\bar{K}, \bar{L}) = \mathcal{W}^s_+(K^i, L^i)$ is two-dimensional and $\mathcal{W}^s_-(\bar{K}, \bar{L})$ is one–dimensional. $\qquad \square$

The results above allow us to classify DGE paths according to two dimensions: (i) according to their crossing through $\mathcal{S}$ and (ii) according to the evolution of their determinacy properties over time. We call *regular DGE path* to a DGE path that does not cross $\mathcal{S}$ and *singular DGE path* to a DGE path does it. We say *determinacy is permanent* if the dimension of the stable manifold is the same throughout $\mathcal{W}^s(\bar{K}, \bar{L})$, with the exception of an impasse point, and *determinacy is temporary* if the dimension of the stable manifold is not the same throughout $\mathcal{W}^s(\bar{K}, \bar{L})$.

Proposition 4 *Let set $\mathcal{S}$ be non-empty and assume there is one regular steady-state. Then, the following types of DGE paths are possible in the presence of impasse singularities:*

1. *Stationary DGE paths if the stable manifold associated to $(\bar{K}, \bar{L})$ is empty and $K_0 = \bar{K}$.*
2. *Regular asymptotic stationary DGE paths, which are permanently determinate or indeterminate, and are confined to subset Ω_+ (Ω_-), if $(\bar{K}, \bar{L}) \in \Omega_+$ (Ω_-) and $K_0 \in \mathcal{W}^s_+(\bar{K}, \bar{L})$ $(\mathcal{W}^s_-(\bar{K}, \bar{L}))$.*
3. *Singular asymptotic stationary DGE paths, which are not permanently determinate or indeterminate, if the stable manifold has different dimensions on subsets Ω_- and Ω_+, and if the initial point K_0 belongs to one branch of the stable manifold and the steady-state $(\bar{K}, \bar{L})$ belongs to its complement. Two cases are possible: (1) the DGE path can be initially temporarily determinate and asymptotically indeterminate only if there is an impasse-transversal saddle point; or (2) the DGE path can be initially temporarily indeterminate and asymptotically determinate only if there is an impasse-transversal node.*

3.3   Singular Macrodynamics

In this section we present two models with the structure of the benchmark model presented in Sect. 3.2 to illustrate the two types of singularity. The first is the well known Benhabib and Farmer (1994) model and the second is a DGE model with a government following a cyclical fiscal policy rule.

3.3.1   The Benhabib and Farmer (1994) Model

Benhabib and Farmer (1994) provide us a fine example of the slow-fast singularity. In this case, the production side is represented by $y(K, L) \equiv K^\alpha L^\beta$, $\alpha, \beta > 0$ with $\alpha + \beta > 1$, $r(K, L) \equiv aK^{\alpha-1}L^\beta$, and $w(K, L) \equiv bK^\alpha L^{\beta-1}$, where $a, b \in (0, 1)$. Since the utility function is $u(C, L) = \log C - (1 + \chi)^{-1} L^{1+\chi}$ with $\chi \geq 0$, we have $\theta(K, L) = 1$ and $v(C, K, L) \equiv -L^{\beta-1-\epsilon} + bC^{-1}K^\alpha L^{\beta-1}$, where $\epsilon(\varphi) \equiv \beta - (1 + \chi)$ and $\varphi \equiv \left(a\ b\ \alpha\ \beta\ \rho\right)^{\mathsf{T}}$.

To apply the results from Sect. 3.2.3 we write the slow-fast system as

$$\dot{K} = f_1(K, L, \epsilon) \equiv K^\alpha L^\beta (1 - bL^{\epsilon-\beta}),$$
$$\epsilon \dot{L} = f_2^s(K, L, \epsilon) \equiv L\left[K^{\alpha-1}L^\beta \left(a - \alpha(1 - bL^{\epsilon-\beta})\right) - \rho\right],$$

and

$$K' = \epsilon f_1(K, L, \epsilon) \equiv \epsilon K^\alpha L^\beta \left(1 - bL^{\epsilon-\beta}\right),$$
$$L' = f_2^s(K, L, \epsilon) \equiv L\left[K^{\alpha-1}L^\beta \left(a - \alpha\left(1 - bL^{\epsilon-\beta}\right)\right) - \rho\right],$$

where it is clear that a slow-fast singularity exists for $\epsilon = 0$ and there are no impasse singularities. There is a unique positive steady-state $(\bar{K}, \bar{L})$ where $\bar{K} = (a/\rho)^{\frac{1}{1-\alpha}} \bar{L}^{\frac{\beta}{1-\alpha}}$ and $\bar{L} = b^{\frac{1}{\beta-\epsilon}}$. The Jacobian for the slow system has trace and determinant given by

$$\mathrm{tr}DF(\bar{K}, \bar{L}) = \frac{\rho^2(1 - \alpha)(\beta - \epsilon)}{a\epsilon}, \quad \det DF(\bar{K}, \bar{L}) = \frac{\rho\left(\beta(a - \alpha) + \epsilon\alpha\right)}{a\epsilon}.$$

We can easily see that both these quantities can take infinite values for $\epsilon = 0$.

Henceforth, we use the authors' assumptions on the parameter values, i.e. we suppose that $\alpha > a$ and $\beta > b$. Thus, we find that the dimension of the stable manifold depends on ϵ: for $\epsilon < 0$ the steady-state is either a stable focus or node and for $\epsilon > 0$ it is a saddle.[11]

---

[11] Benhabib and Farmer (1994, p. 34) already noted this behavior for particular values of the parameters: "As χ moves below -0.015 the roots both become real but remain negative until at (approximately) $\chi = -0.05$ [i.e. $\epsilon = 0$] one root passes through minus infinity and reemerges as a positive real root". To compare with our results please note that we introduced a slight change in notation: while the authors set χ as non-positive we set χ as non-negative.

Using the theory presented in Sect. 3.2.3 we can draw further conclusions. First, the slow-fast subset (defined by $f_2(K, L, 0) = 0$) is $\mathcal{S}^p = \{(K, L) : L = h(K)\}$, where $h(K) \equiv \left(\frac{\rho K^{1-\alpha} - \alpha b}{a - \alpha}\right)^{\frac{1}{\beta}}$. Second, all points belonging to set $\mathcal{S}^p$ are slow-fast regular and attracting due to the fact that $f_{2,L}(K, L_p) = \beta(a - \alpha)K^{\alpha-1}L_p^{\beta-1} < 0$ and there are no singular slow-fast points.

Figure 3.1 illustrates the dynamic behavior of the Benhabib and Farmer (1994) model in the (K, L) space. On the left-hand-side (LHS) panel, we represent the phase diagram for $\epsilon > 0$, but small. We can observe that there is a unique steady-state equilibrium represented by point Γ_E. Since there are two negative eigenvalues associated with this stationary point, all DGE paths converge asymptotically to Γ_E. However, the steady-state is locally and globally indeterminate, as there is an infinite number of initial points for a given K_0, leading to the long-run equilibrium. We can also see that L adjusts very fast so that the trajectory quickly approaches the isocline $\dot{L} = 0$ and then K starts adjusting more slowly until the steady-state is reached.

On the right-hand-side (RHS) panel, we represent the phase diagram for $\epsilon < 0$, also small. Now, the unique steady-state is locally and globally determinate, as there is one positive and one negative eigenvalue associated with it. For each initial level for the capital stock, K_0, there is only one value of L, such that convergence to the steady-state is asymptotically attained and the transversality condition holds. Notice that the stable manifold associated to the steady-state, $\mathcal{W}^s(\Gamma_E)$, stays very close to the $\dot{L} = 0$ isocline, meaning that labor adjusts faster than capital, as in the case $\epsilon > 0$.

For $\epsilon = 0$, we obtain a degenerate case where the adjustment of L is automatic, so that the stable manifold coincides with the $\dot{L} = 0$ isocline, and this curve is the geometrical analog of set $\mathcal{S}^p$.

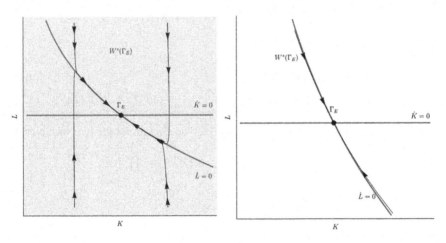

Fig. 3.1 Phase diagrams for the Benhabib and Farmer (1994) model for $\epsilon > 0$ (LHS panel) and $\epsilon < 0$ (RHS panel)

3.3.2 Singularities Generated by Cyclical Fiscal Policy Rules

In this section we present a DGE model for an economy with a government that follows a cyclical fiscal policy rule.[12] The production side is represented by a production function with constant returns to scale, i.e. $y(K, L) = K^\alpha L^{1-\alpha}$, with $\alpha \in (0, 1)$. Thus, we obtain $r(K, L) \equiv \alpha(L/K)^{1-\alpha}$, and $w(K, L) \equiv (\lambda - \alpha)(K/L)^\alpha$. The utility function is the same as in Benhabib and Farmer (1994).

The Government imposes a distortionary income tax on households with a flat rate $T \in (0, 1)$. Thus, the relevant input prices for households are $(1 - T) r(K, L)$ and $(1 - T) w(K, L)$. The tax revenue, $Ty(K, L)$, is used *within each period* to finance government final spending (G).[13]

We assume *that the distortionary tax rule takes* the form $T(K, L) = \phi\, y(K, L)^\mu$, where $\mu < 0$, i.e. the rule is countercyclical, such that the constraint $0 < T(K, L) < 1$ holds.[14]

From the assumptions above and considering Eq. (3.3), we obtain the consumption function $C = c(L, K) \equiv (1 - T(K, L))w(K, L)L^{-\chi}$, which is a non-monotonic function of L, as $c_L \lesseqgtr 0$ if and only if $T(K, L) \lesseqgtr T_s \in (0, 1)$ where

$$T_s \equiv \frac{\alpha + \chi}{\epsilon}, \quad \epsilon(\varphi) \equiv \alpha + \chi - (1 - \alpha)\mu, \tag{3.30}$$

where $\varphi \equiv (\alpha\ \chi\ \phi\ \mu)^\top \in \Phi = (0, 1) \times \mathbb{R}^3_{++}$ and $\epsilon > 0$ for all parameter values, thus preventing the existence of slow-fast singularities. The system (3.8) takes the form:

$$\dot{K} = f_1(K, L) \equiv (1 - T(K, L))\, y(K, L)\, (1 - \ell(L)),$$
$$\delta(K, L)\dot{L} = f_2(K, L) \equiv L(1 - T(K, L))\, (R(K, L) - \rho), \tag{3.31}$$

where

$$\delta(K, L) \equiv \epsilon(T(K, L) - T_s), \tag{3.32}$$
$$R(K, L) \equiv r(K, L)\, (\mu T(K, L) + (1 - (1 + \mu)T(K, L))\ell(L)), \tag{3.33}$$

and $\ell(L) = (L^*/L)^{1+\chi}$, for $L^* = (1 - \alpha)^{1+\chi}$. The domain for the two variables is

$$\Omega = \left\{ (K, L) \in \mathbb{R}^2_{++} : T(K, L) < 1 \right\}.$$

If we write the dynamic system as

---

[12]For another example in a macro model with imperfect competition see Brito et al. (2016).

[13]Considering that we have an infinitely-lived representative household, it would act as if budget was balanced at all moments in time, i.e. Ricardian equivalence holds.

[14]One special case is given by $\phi = G$ and $\mu = -1$, corresponding to setting the expenditure level.

$$\dot{K} = (1 - T(K, L))y(K, L)(1 - \ell(L)),$$
$$\dot{L} = \frac{L(1 - T(K, L))(R(K, L) - \rho)}{\epsilon\delta(K, L)}, \quad (3.34)$$

it is clear that impasse-type singularities exist for values of (K, L) such that $\delta(K, L) = 0$, i.e. for $T(K, L) = T_s$.

The manifolds $T(K, L) = 1$ and $T(K, L) = T_s$ both limit the set of admissible values for (K, L), which is partitioned by the impasse set $S = \{(K, L) : T(K, L) = T_s\}$ into a set of low tax rates $\Omega_- \equiv \{(K, L) : 0 < T(K, L) < T_s\}$ and high tax rates $\Omega_+ \equiv \{(K, L) : T_s < T(K, L) < 1\}$. Notice that, for a given L the high (low) tax rate set corresponds to lower (higher) values for K.

The steady-states $(\bar{K}, \bar{L})$ of system (3.34) cannot be determined explicitly. However, we always have $\bar{L} = L^*$, then $\bar{\ell} = \ell(\bar{L}) = 1$ and $\bar{K} \in \{K : (1 - T(K, \bar{L}))r(K, \bar{L}) = \rho\}$. Since $\rho > 0$, then the steady-state constraint $\bar{T} = T(\bar{K}, \bar{L}) < 1$ always holds. Let us define a critical value for ϕ:

$$\phi^* \equiv T^* \left[\left(\frac{\alpha(1 - T^*)}{\rho} \right)^{\frac{\alpha}{1-\alpha}} \bar{L} \right]^{-\mu}, \text{ for } T^* \equiv \frac{1 - \alpha + \alpha\mu}{1 - \alpha}.$$

Thus, we have two possible cases:

1. For $\phi = \phi^*$, the steady-state is unique (there could be a local regular bifurcation), i.e. $\Gamma_E = \{(\bar{K}^*, \bar{L})\}$ with $\bar{K}^* \equiv \left(\frac{\alpha(1-T^*)}{\rho} \right)^{\frac{1}{1-\alpha}} \bar{L}$.
2. For $\phi < \phi^*$, two steady-states exist, i.e. $\Gamma_E = \{(\bar{K}_L, \bar{L}), (\bar{K}_H, \bar{L})\}$ such that $\bar{K}_L < \bar{K}^* < \bar{K}_H$. In this case we obtain the following relations between the steady-state tax rates: $\bar{T}_H < T^* < \bar{T}_L$, for $\bar{T}_i = T(\bar{K}_i, \bar{L})$ with $i = H, L$.

The Jacobian of the reduced system, (3.8), evaluated at any steady-state $(\bar{K}, \bar{L})$, has the trace and determinant given by

$$\text{tr} DF(\bar{K}, \bar{L}) = -\frac{\rho(\alpha + \chi)(1 - (1+\mu)\bar{T})}{\epsilon(\bar{T} - T_s)}, \quad \det DF(\bar{K}, \bar{L}) = \frac{\rho^2(1+\chi)(\mu^* - \mu)(T^* - \bar{T})}{\epsilon(\bar{T} - T_s)},$$

where $\mu^* \equiv (1 - \alpha)/\alpha$ and $T^* \equiv \mu^*/(\mu^* - \mu)$. Let us define another critical value for ϕ:

$$\phi_i \equiv T_s \left[(\ell_i)^{-\frac{1}{1+\chi}} \left(\frac{\alpha(\mu T_i + (1 - (1+\mu)T_i)\ell_i)}{\rho} \right)^{\frac{1}{1+\chi^*}} \bar{L} \right]^{-\mu}$$

where $\ell_i \equiv \frac{(1 + \chi^*)(\alpha + \chi)}{(1 + \chi)(\chi^* - \chi)}$ and $1 + \chi^* \equiv \frac{1 - \alpha}{\alpha}$.

Now, consider the generic case $\phi < \phi^*$, for which two steady-states exist. Three generic cases and one non-generic case are possible, concerning the local dynamic properties:

1. For $\phi < \phi_s$, two saddle steady-state equilibria exist such that the associated tax rates are $\bar{T}_L < \min\{T^*, T_s\} < \max\{T^*, T_s\} < \bar{T}_H$. The low tax rate steady-state is located in set Ω_- and the high tax rate steady-state is in set Ω_+.
2. For $\chi > \chi^*$ and $\phi^i < \phi < \phi^*$, one saddle steady-state equilibrium exists at $\bar{T}_L$ and one unstable node or focus exists at $\bar{T}_H$. Both steady-states are located in the low tax rate subset Ω_-.
3. For $\chi < \chi^*$ and $\phi^i < \phi < \phi^*$, one saddle steady-state equilibrium exists at $\bar{T}_H$ and one stable node or focus exists at $\bar{T}_L$. Both steady-states are located in the high tax rate subset Ω_+.
4. For $\phi = \phi_i$, one impasse-singular steady-state equilibrium exists at $\bar{T}_L$ ($\bar{T}_H$) if $\chi < \chi^*$ (($\chi)\chi^*$).

As far as impasse points (i.e. elements of set $\mathcal{S}$) are concerned, we have the following results. First, no impasse-tangent points exist, i.e. Γ_K is empty. Second, impasse-transversal points exist, i.e. $\Gamma_I = \{(K, L) \in \Omega : T(K, L) = T_s, R(K, L) = \rho\} \neq \varnothing$. Set Γ_I partitions set $\mathcal{S}$ into two (not necessarily compact) subsets: (i) the set of impasse-repeller points $\mathcal{S}_+ = \{(K, L) \in \mathcal{S} : R(K, L) < \rho\}$ and (ii) the set of impasse-atractor points $\mathcal{S}_- = \{(K, L) \in \mathcal{S} : R(K, L) < \rho\}$. In addition, Γ_I has only one element for $\chi > \chi^*$ and it has two elements for $\chi < \chi^*$ and $\phi < \phi_i$.

Considering that we can obtain two steady-states and two impasse-transversal points, this model goes beyond the conditions stated in Sect. 3.2.4. In order to illustrate a type of DGE paths that can occur in models with singularities, but do not occur in regular models, Fig. 3.2 presents a phase diagram illustrating a case for which $\chi < \chi^*$. In this case, two saddle steady-states (labelled H and L) exist and two impasse-transversal points (one shown in Fig. 3.2, as Γ_I) exist. Steady-state $(\bar{K}_L, \bar{L})$ is located in subset Ω_+, steady-state $(\bar{K}_H, \bar{L})$ is located in subset Ω_- and the stable manifolds associated with both are represented by $\mathcal{W}_L^s$ and $\mathcal{W}_H^s$. While

Fig. 3.2 Phase diagram for the endogenous fax rule model

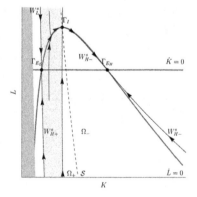

$\mathcal{W}_L^s$ is located entirely on Ω_+, the stable manifold $\mathcal{W}_H^s$ is located in the union of two subsets together with the impasse-transversal node point Γ_I: $\mathcal{W}_H^s \supset \mathcal{W}_{H+}^s \cup \mathcal{W}_{H-}^s$ where $\mathcal{W}_{H+}^s$ is two-dimensional and $\mathcal{W}_{H-}^s$ is one-dimensional. This means that the DGE path $(K_H(t), L(t))_{t \in [0,\infty)}$ converging to the high tax rate steady-state displays *temporary indeterminacy*. That is, if $K_H(0) = K_0 \in \mathcal{W}_{H+}^s$ there is an infinite number of values for $L_H(0)$ consistent with rational expectations equilibrium dynamics. All trajectories in this area converge to a point Γ_I, at a finite moment in time t_s, depending upon $L_H(0)$, and after that moment converge through a common path to the steady-state $(\bar{K}_H, \bar{L})$.

Finally, notice that the stable manifold $\mathcal{W}_L^s$ defines the boundary to $\mathcal{W}_{H+}^s$. This clearly indicates that a local analysis using the usual approach, whilst appropriate for standard DGE models, is clearly misleading here.

3.4 Conclusion

3.4.1 A Tale with Illustrative Metaphors

Before we conclude this chapter, the reader may find the usage of some metaphors helpful to understand the role of singularities in DGE models, especially that of impasse singularities.[15] For that purpose, we will make use of expressions used in astrophysics that became popular for the general audience, namely *black holes, white holes and wormholes*.

However, a word of caution is due. We do not intend to emulate gravitational fields here and we do not claim that there is an isomorphism from physical concepts to the ones used here. Despite the fact that mathematical structures behind physics models and those presented here are different, there is a strong analogy between the consequences of singularities in both types of model.

This is not new in economics. When Cass and Shell (1983) coined the phrase *sunspots* to represent random shocks that do not affect fundamentals, but end up affecting economic activity, they were not trying to emulate magnetic fields on the surface of stars. These are metaphors that help us understand the mathematics.

Attractor-impasse points, i.e. elements of set $\mathcal{S}_-$, can be seen as singularities within *black holes*, as trajectories within their basin of attraction cannot escape their strong "gravitational" pull. We saw that infinitely-living rational agents do not choose entering this area of space as they would eventually arrive at the singularity and remain there forever. Unless the singularity is a steady-state equilibrium a rational agent would never choose this path. Repeller-impasse points, i.e. elements of set $\mathcal{S}_+$, can be seen as singularities within *white holes*, as trajectories are drawn away from it. Impasse-transversal nodes work as a conduit between a *black-hole* and a *white-hole* singularity, thus linking different areas of space. Let us call them *wormholes*. Point

---

[15]We thank an anonymous referee and Nuno Barradas for suggesting this clarification.

Γ_I in Fig. 3.2 provides such an example. It works like a *black hole* from the point of view of the (light) shaded area on the LHS of it. However, it also works like a *white hole* from the point of view of the area on its RHS of the impasse-transversal point, by expelling the trajectories onto a single determinate path towards Γ_{E_H}. Thus, it functions as a "portal" linking this otherwise two "parallel universes": one governed by high tax rates and another one by low taxes rates.

3.4.2   Final Remarks

The test of a good macro-model is not whether it predicts a little better in "normal" times, but whether it anticipates abnormal times and describes what happens then. Black holes "normally" don't occur. Standard economic methodology would therefore discard physics models in which they play a central role. In Stiglitz (2011, p. 17).

Singularities play a large role in other scientific fields, including physics and electrical engineering (where they are related to systemic shut-downs). In economics, the phenomenon has remained largely unexplored or, in the spirit of Stiglitz's critique, just put aside as an "abnormal time." In this chapter we have hoped to rectify this "black hole" in economics.

Despite the simplicity of the DGE models we have looked at in Sect. 3.3 of this chapter, we have found singular dynamics were possible. There are also a large number of dynamic models with externalities, rules, and distortions which may well also give rise to singularities. We find it surprising that singular dynamics have been absent from the macroeconomic dynamics literature—a veritable "black hole" in macroeconomic theory. We presented conditions for the emergence of singularities, described two types of singularities, slow-fast (perturbation) and impasse singularities, presented geometrical methods to deal with both of them, and applied our analysis to two simple cases, the Benhabib and Farmer (1994) model and a DGE model with a government following a cyclical fiscal policy rule. Because researchers have not known how to deal with singularities, we believe that they have either been ignored or avoided. We now have the tools for analyzing singularities and hope that this will mean that their implications within existing models can now be explored properly.

We hope that this chapter may provide a contribution to bring these *black holes* and the techniques associated with them to the mainstream discourse in economics. Strange as they may sound now, so did *sunspots* in the 1980s. Only the test of time can tell if this hope will materialise. Nonetheless, there is no shadow of a doubt over how Jean-Michel Grandmont's contributions to economics has stood that test of time.

Appendix

To simplify notation let $x = (x_1, x_2) \equiv (K, L)$ and consider functions $f_1(x)$, $f_2(x)$, and $\delta(x)$. In addition, consider the two vector fields $F(x)$, as in Eq. (3.13), and $F^r(x)$, as in Eq. (3.27). Then, at a regular steady-state $(\bar{x})$, we have $f_1(\bar{x}) = f_2(\bar{x}) = 0$ and $\delta(\bar{x}) \neq 0$. Furthermore, at a generic impasse-transversal point (x^i), we have $\delta(x^i) = f_2(x^i) = 0$ and $f_1(x^i) \neq 0$.

Lemma 4 *Assume there is one generic impasse-transversal point x^i and one regular steady-state $\bar{x}$, both belonging to a set X, such that $f_{2,x_2}(x)$ has the same sign for any point $x \in X$. Therefore,* $\text{sign}\left(\det DF^r(x^i)\right) = -\text{sign}\left(\det DF(\bar{x})\right)$.

Proof For sake of simplicity, let us set $\epsilon(\varphi) = 1$. The determinants of the Jacobian for $F(x)$ evaluated at the steady-state and $F^r(x)$ at an impasse-transversal point are respectively given by

$$\det DF(\bar{x}) = \frac{f_{1,x_1}(\bar{x})\, f_{2,x_2}(\bar{x}) - f_{1,x_2}(\bar{x})\, f_{2,x_1}(\bar{x})}{\delta(\bar{x})},$$
$$\det DF^r(x^i) = f_1(x^i)\left(\delta_{x_1}(x^i)\, f_{2,x_2}(x^i) - \delta_{x_2}(x^i)\, f_{2,x_1}(x^i)\right).$$

First, note that both points share a common condition $f_2(x_1, x_2) = 0$. If this function is differentiable, we can write $\nabla f_2(x) \cdot dx = 0$. By computing Taylor approximations to $f_1(x^i)$ in a neighbourhood of $\bar{x}$ and to $\delta(\bar{x})$ in a neighbourhood of x^i, and considering the differentiability of $f_2(\cdot)$, we obtain:

$$f_1(x^i) = \frac{\det DF(\bar{x})\delta(\bar{x})}{f_{2,x_2}(\bar{x})}(x_1^i - \bar{x}_1),$$
$$\delta(\bar{x}) = \frac{\det DF^r(x^i)}{f_1(x^i) f_{2,x_2}(x^i)}(\bar{x}_1 - x_1^i).$$

Thus,

$$\frac{\det DF^r(x^i)}{\det DF(\bar{x})} = -(\delta(\bar{x}))^2 \frac{f_{2,x_2}(x^i)}{f_{2,x_2}(\bar{x})}.$$

$\square$

References

Barnett, W. A., & He S. (2010). Existence of singularity bifurcation in an Euler-equations model of the United States economy: Grandmont was right. *Economic Modelling, 27*(6), 1345–1354.

Benhabib, J., & Farmer, R. (1994). Indeterminacy and increasing returns. *Journal of Economic Theory, 63*, 19–41.

Brito, P., Costa, L., & Dixon, H. (2016). *Singular macroeconomic dynamics and temporary indeterminacy*. Mimeo.

Cardin, P. T., da Silva, P. R., & Teixeira, M. A. (2012). Implicit differential equations with impasse singularities and singular perturbation problems. *Israel Journal of Mathematics, 189*, 307–322.

Cass, D., & Shell, K. (1983). Do sunspots matter? *Journal of Political Economy, 91*(2), 193–227.

Fenichel, N. (1979). Geometric singular perturbation theory for ordinary differential equations. *Journal of Difference Equations and Applications, 31*, 53–98.

Grandmont, J.-M. (1985). On endogenous competitive business cycles. *Econometrica, 53*, 995–1045.

Grandmont, J. -M. (1998). Expectations formation and stability of large socioeconomic systems. *Econometrica, 66*(4), 741–782.

Grandmont, J. -M. (2008). Nonlinear difference equations, bifurcations and chaos: An introduction. *Research in Economics, 62*(3), 122–177.

Grandmont, J.-M., Pintus, P., & de Vilder, R. (1998). Capital-labor substitution and competitive nonlinear endogenous business cycles. *Journal of Economic Theory, 80*(1), 14–59.

Guckenheimer, J., & Holmes, P. (1990). *Nonlinear oscillations and bifurcations of vector fields* (2nd ed.). Springer.

He, Y., & Barnett, W. A. (2006). Singularity bifurcations. *Journal of Macroeconomics, 28*, 5–22.

Kuehn, C. (2015). *Multiple time scale dynamics* (1st ed.). Applied Mathematical Sciences 191. Springer.

Kuznetsov, Y. A. (2005). *Elements of applied bifurcation theory* (3rd ed.). Springer.

Llibre, J., Sotomayor, J., & Zhitomirskii, M. (2002). Impasse bifurcations of constrained systems. *Fields Institute Communications, 31*, 235–256.

Stiglitz, J. (2011). Rethinking macroeconomics: What failed, and how to repair it. *Journal of the European Economic Association, 9*(4), 591–645.

Zhitomirskii, M. (1993). Local normal forms for constrained systems on 2-manifolds. *Boletim da Sociedade Brasileira de Matemática, 24*, 211–232.

Chapter 4
Sunspot Fluctuations in Two-Sector Models with Variable Income Effects

Frédéric Dufourt, Kazuo Nishimura, Carine Nourry and Alain Venditti

Abstract We analyze a version of the Benhabib and Farmer (1996) two-sector model with sector-specific externalities in which we consider a class of utility functions inspired from the one considered in Jaimovich and Rebelo (2009) which is flexible enough to encompass varying degrees of income effect. First, we show that local indeterminacy and sunspot fluctuations occur in 2-sector models under plausible configurations regarding all structural parameters—in particular regarding the intensity of income effects. Second, we prove that there even exist some configurations for which local indeterminacy arises *under any degree of income effect.* More precisely, for any given size of income effect, we show that there is a non-empty range of values for the Frisch elasticity of labor and the elasticity of intertemporal substitution in consumption such that indeterminacy occurs. This contrasts with the results obtained in one-sector models in both Nishimura et al. (2009), in which it is shown that indeterminacy cannot occur under either GHH and KPR preferences, and in Jaimovich (2008) in which local indeterminacy only arises for intermediary income effects.

This work has been carried out thanks to the support of the A*MIDEX project (no. ANR-11-IDEX-0001-02) funded by the "Investissements d'Avenir" French Government program, managed by the French National Research Agency (ANR). We would like to thank an anonymous referee together with X. Raurich and T. Seegmuller for useful comments and suggestions.

F. Dufourt · C. Nourry · A. Venditti (✉)
Aix-Marseille University (Aix-Marseille School of Economics)-CNRS-EHESS,
Marseille, France
e-mail: alain.venditti@univ-amu.fr

C. Nourry
e-mail: carine.nourry@univ-amu.fr

F. Dufourt
e-mail: frederic.dufourt@univ-amu.fr

K. Nishimura
Research Institute for economics and business administration, Kobe University,
Kobe, Japan
e-mail: nishimura@rieb.kobe-u.ac.jp

A. Venditti
EDHEC Business School, Nice, France

© Springer International Publishing AG 2017
K. Nishimura et al. (eds.), *Sunspots and Non-Linear Dynamics*,
Studies in Economic Theory 31, DOI 10.1007/978-3-319-44076-7_4

Keywords Indeterminacy · Sunspots · Income and substitution effects · Sector-specific externalities · Infinite-horizon two-sector model

Journal of Economic Literature Classification Numbers C62 · E32 · O41

4.1   Introduction

It is well-known since Benhabib and Farmer (1994, 1996) that local indeterminacy and sunspot fluctuations arise in two-sector models under much more empirically plausible configurations regarding structural parameters than in their one-sector equivalents. In particular, indeterminacy occurs for calibrations consistent with a low degree of increasing returns to scale and a standard (negatively sloped) equilibrium labor demand function. Besides, it has been shown that two-sector models submitted to correlated sunspot and technological shocks are able to account for many empirical regularities regarding the comovements of consumption and investment over the business cycle, and regarding the allocation of labor across these two sectors (Dufourt et al. 2015). Yet, these results were obtained under a specification of individual preferences derived from Greenwood et al. (1988) (thereafter GHH), which implies that there is no income effect on labor supply.

From a theoretical point of view, one may thus wonder whether results obtained under GHH preferences can be extended to a framework in which the magnitude of the income effect on labor supply differs from zero. While this issue has been the subject of particular attention in one-sector models (see in particular Jaimovich 2008), no systematic study of the role of income effects in two-sector models has been provided so far.[1] The aim of this chapter is to undertake such an analysis.

We analyze a version of the Benhabib and Farmer (1996) two-sector model with sector-specific externalities in which we consider a class of utility functions which is flexible enough to encompass varying degrees of income effect. Our specification of individual preferences is inspired from—but slightly differs from—the one considered in Jaimovich and Rebelo (2009) (JR). This specification admits as particular (and polar) cases the GHH formulation without income effect and the canonical specification of King et al. (1988) (KPR) used in many DSGE models. We analyze how the local stability properties of the model change when we vary the parameter governing the intensity of the income effect, and we determine the conditions under which local indeterminacy arises. We perform this analysis for different configurations regarding the other structural parameters influencing the wage elasticity of labor supply, the elasticity of intertemporal substitution (EIS) in consumption, and the degree of increasing returns to scale (IRS).

---

[1] Nishimura and Venditti (2010) show that local indeterminacy can occur under both GHH and KPR preferences—the latter displaying positive income effect—but there is no clear picture of the impact of the income effect on the occurrence of sunspot fluctuations.

Our main results can be described as follows. First, we show that local indeterminacy and sunspot fluctuations occur in 2-sector models under plausible configurations regarding all structural parameters—in particular regarding the intensity of income effects. Second, we show that there even exist some configurations for which local indeterminacy arises *under any degree of income effect*. More precisely, for any given size of income effect, we show that there is a non-empty range of values for the Frisch elasticity of labor and the elasticity of intertemporal substitution in consumption such that indeterminacy occurs. This contrasts with the results obtained in one-sector models in both Nishimura et al. (2009), in which it is shown that indeterminacy cannot occur under either GHH and KPR preferences as long as realistic parameter values are considered, in particular when the slope of the labor demand function is negative, and in Jaimovich (2008) in which local indeterminacy only arises for intermediary income effects.

The rest of this Chapter is organized as follows. We present the model and we characterize the intertemporal equilibrium in the next Section. In Sect. 4.3, we prove the existence of a unique steady state and we provide the expression of the characteristic polynomial. The complete set of conditions for local indeterminacy are derived in Sect. 4.4. Section 4.5 provides some numerical illustrations, while economic intuitions underlying our main theoretical results are given in Sect. 4.6. Some concluding remarks are stated in Sect. 4.7, whereas all the technical details are given in a final Appendix.

4.2 The Model

We consider a standard two-sector infinite-horizon model with productive externalities and JR-type preferences (see Jaimovich 2008 and Jaimovich and Rebelo 2009). Households are infinitely-lived, accumulate capital, and derive utility from consumption and leisure. Firms produce differentiated consumption and investment goods using capital and labor, and sell them to consumers. All markets are perfectly competitive.

4.2.1 The Production Structure

Firms in the consumption sector produce output $Y_c(t)$ according to a Cobb-Douglas production function:

$$Y_c(t) = K_c(t)^\alpha L_c(t)^{1-\alpha} \tag{4.1}$$

where $K_c(t)$ and $L_c(t)$ are capital and labor allocated to the consumption sector.

In the investment sector, output $Y_I(t)$ is also produced according to a Cobb-Douglas production function but which is affected by a productive externality

$$Y_I(t) = A(t)K_I(t)^\alpha L_I(t)^{1-\alpha} \tag{4.2}$$

where $K_I(t)$ and $L_I(t)$ are the numbers of capital and labor units used in the production of the investment good, and $A(t)$ is the externality parameter. Following Benhabib and Farmer (1996), we assume that the externality is sector-specific and depends on the average levels $\bar{K}_I(t)$ and $\bar{L}_I(t)$ of capital and labor used in the investment sector, such that:

$$A(t) = \bar{K}_I(t)^{\alpha\Theta}\bar{L}_I(t)^{(1-\alpha)\Theta} \tag{4.3}$$

with $\Theta \geq 0$.[2] These economy-wide averages are taken as given by individual firms. Assuming that factor markets are perfectly competitive and that capital and labor inputs are perfectly mobile across the two sectors, the first order conditions for profit maximization of the representative firm in each sector are:

$$r(t) = \frac{\alpha Y_c(t)}{K_c(t)} = p(t)\frac{\alpha Y_I(t)}{K_I(t)}, \tag{4.4}$$

$$w(t) = \frac{(1-\alpha)Y_c(t)}{L_c(t)} = p(t)\frac{(1-\alpha)Y_I(t)}{L_I(t)} \tag{4.5}$$

where r, p and w are respectively the rental rate of capital, the price of the investment good and the real wage rate at time t, all in terms of the price of the consumption good.

4.2.2 Households' Behavior

We consider an economy populated by a continuum of unit mass of identical infinitely-lived agents. The representative agent enters each period t with a capital stock $k(t)$ inherited from the past. He then supplies elastically an amount $l(t) \in [0, \bar{l})$ of labor (with $\bar{l} > 0$ his exogenous time endowment), rents its capital stock $k(t)$ to the representative firms in the consumption and investment sectors, consumes $c(t)$, and invests $i(t)$ in order to accumulate capital.

Denoting by $y(t)$ the GDP, the budget constraint faced by the representative household is

$$c(t) + p(t)i(t) = r(t)k(t) + w(t)l(t) \equiv y(t) \tag{4.6}$$

---

[2]We do not consider externalities in the consumption good sector as they do not play any crucial role in the existence of multiple equilibria.

Assuming that capital depreciates at rate $\delta \in (0, 1)$ in each period, the law of motion of the capital stock is:

$$\dot{k}(t) = i(t) - \delta k(t) \tag{4.7}$$

The intertemporal optimization problem of the representative household is then given by:

$$\max_{\{c(t),i(t),l(t)\}} \int_0^{+\infty} U(c(t), (\bar{l} - l(t)))e^{-\rho t} dt$$
$$s.t. \quad c(t) + p(t)i(t) \equiv y(t) = r(t)k(t) + w(t)l(t) \tag{4.8}$$
$$\dot{k}(t) = i(t) - \delta k(t)$$
$$k(0) \, given$$

where $\rho \geq 0$ is the discount rate.

The Hamiltonian in current value is given by:

$$\mathcal{H} = U(c(t), (\bar{l} - l(t))) + \lambda(t) \left[r(t)k(t) + w(t)l(t) - c(t) - p(t)i(t)\right] + q(t) \left[i(t) - \delta k(t)\right]$$

with $q(t)$ the co-state variable which corresponds to the utility price of the capital good in current value and $\lambda(t)$ the Lagrange multiplier associated with the budget constraint. The first order conditions of problem (4.8) are given by the following equations:

$$U_1(c(t), (\bar{l} - l(t))) = \lambda(t) \tag{4.9}$$
$$U_2(c(t), (\bar{l} - l(t))) = w(t)\lambda(t) \tag{4.10}$$
$$q(t) = p(t)\lambda(t) \tag{4.11}$$
$$\dot{q}(t) = (\delta + \rho)q(t) - r(t)\lambda(t) \tag{4.12}$$

An equilibrium path also satisfies the transversality condition

$$\lim_{t \to +\infty} e^{-\rho t} U_1(c(t), (\bar{l} - l(t)))p(t)k(t) = 0. \tag{4.13}$$

Following Jaimovich (2008) and Jaimovich and Rebelo (2009), we assume a JR-type utility function which is flexible enough to encompass varying degrees of income effect. Denoting leisure as $\mathcal{L} = \bar{l} - l$, let

$$U(c, \mathcal{L}) = \frac{\left[c - \frac{(\bar{l} - \mathcal{L})^{1+\chi}}{1+\chi} c^\gamma\right]^{1-\sigma} - 1}{1 - \sigma} \tag{4.14}$$

with $\sigma \geq 0$, $\chi \geq 0$ and $\gamma \in [0, 1]$. This utility function satisfies the standard normality condition between consumption and leisure. In the following, we will also introduce some parameter restrictions ensuring that concavity holds at the steady-state.[3] This specification nests as particular cases the Greenwood-Hercovitz-Huffman (1988) (GHH) formulation (obtained when $\gamma = 0$), characterized by the lack of any income effect on labor supply, and the King-Plosser-Rebelo (1988) (KPR) formulation (obtained when $\gamma = 1$), characterized by a large income effect compatible with endogenous growth. We are then able to control the magnitude of the income effect by varying the calibration for γ between these two extremes.

Remark 1 Jaimovich (2008) and Jaimovich and Rebelo (2009) actually consider discrete-time models with a slightly different specification such that

$$U(c_t, \mathcal{L}_t, X_t) = \frac{\left[c_t - \frac{(\bar{l}-\mathcal{L}_t)^{1+\chi}}{1+\chi} X_t \right]^{1-\sigma} - 1}{1-\sigma} \tag{4.15}$$

with $X_t = c_t^\gamma X_{t-1}^{1-\gamma}$. When $\gamma \in (0, 1)$, the income effect depends on the dynamics of this additional state variable X_t. Such a formulation allows to get more persistence of income effects during the transition, but focusing on such a property is out of the scope of this paper.

Remark 2 Using this specification for the utility function, from Eqs. (4.9)–(4.10) we can write the first order condition that drives the trade-off between consumption and leisure as follows

$$\frac{(1+\chi)l^\chi c^\gamma}{1+\chi-\gamma l^{1+\chi} c^{\gamma-1}} = w \tag{4.16}$$

Denoting $\mathcal{I}$ the total income of the representative agent and normalizing the price of consumption to 1, we consider the static budget constraint

$$c + w\mathcal{L} = \mathcal{I} \tag{4.17}$$

Considering that $\mathcal{L} = \bar{l} - l$, solving Eqs. (4.16)–(4.17) gives demand functions for consumption and leisure, namely $c = c(w, \mathcal{I})$ and $\mathcal{L} = \mathcal{L}(w, \mathcal{I})$. Assuming a constant wage, considering that $d\mathcal{L} = -dl$ and deriving the ratio wl/c from (4.16), we then get the following derivatives that describe the income effect for any $\gamma \in [0, 1]$:

$$\varepsilon_{c\mathcal{I}} \equiv \frac{dc}{d\mathcal{I}} = \left[1 + \gamma \frac{(1+\chi)l^{1+\chi}c^{\gamma-1}}{1+\chi-\gamma l^{1+\chi}c^{\gamma-1}} \frac{1+\chi-l^{1+\chi}c^{\gamma-1}}{(1+\chi)\chi + \gamma l^{1+\chi}c^{\gamma-1}} \right]^{-1}$$

$$\varepsilon_{l\mathcal{I}} \equiv \frac{dl}{d\mathcal{I}} = -\frac{d\mathcal{L}}{d\mathcal{I}} = -\gamma \frac{1+\chi-l^{1+\chi}c^{\gamma-1}}{(1+\chi)\chi + \gamma l^{1+\chi}c^{\gamma-1}} \left[1 + \gamma \frac{(1+\chi)l^{1+\chi}c^{\gamma-1}}{1+\chi-\gamma l^{1+\chi}c^{\gamma-1}} \frac{1+\chi-l^{1+\chi}c^{\gamma-1}}{(1+\chi)\chi + \gamma l^{1+\chi}c^{\gamma-1}} \right]^{-1} \tag{4.18}$$

---

[3]It is important to note that when $\gamma \neq 0$, this utility function may not be concave. This characteristic is well-known for the KPR specification with $\gamma = 1$ for which additional restrictions on σ and χ are required to guarantee concavity (see for instance Hintermaier 2003). However, in order to avoid technical and cumbersome assumptions, we will only focus with Lemma 1 below on the conditions for local concavity properties around the steady state. Precise general conditions for global concavity can be provided upon request.

These expressions clearly show that in the GHH case with $\gamma = 0$ there is no income effect as $\varepsilon_{l\mathcal{I}} = 0$ and $\varepsilon_{c\mathcal{I}} = 1$ while in the KPR case with $\gamma = 1$ we get some income effect with $\varepsilon_{l\mathcal{I}} \in (-1, 0)$ and $\varepsilon_{c\mathcal{I}} \in (0, 1)$. In the intermediary case with $\gamma \in (0, 1)$, the income effect lies in between these two extremes.

4.2.3 Intertemporal Equilibrium

We consider symmetric equilibria which consist of prices $\{r(t), p(t), w(t)\}_{t\geq 0}$ and quantities $\{c(t), l(t), i(t), k(t), Y_c(t), Y_I(t), K_c(t), K_I(t), L_c(t), L_I(t)\}_{t\geq 0}$ that satisfy the household's and the firms' first-order conditions as given by (4.4)–(4.5) and (4.9)–(4.12), the technological and budget constraints (4.1)–(4.3) and (4.6)–(4.7), the good market equilibrium conditions

$$c = Y_c, \ i = Y_I,$$

the market clearing conditions for capital and labor

$$K_c + K_I = k, \ L_c + L_I = l$$

and the transversality condition (4.13).[4]

All firms in the investment sector being identical, we have $\bar{K}_I = K_I$ and $\bar{L}_I = L_I$. At the equilibrium, the production function in the investment good sector is then given by

$$Y_I = K_I^{\alpha(1+\Theta)} L_I^{(1-\alpha)(1+\Theta)} \tag{4.19}$$

We thus have increasing social returns which size is measured by Θ.

4.3 Steady State and Characteristic Polynomial

After a few manipulations, the intertemporal equilibrium described above can be reduced to a dynamic system of two equations in two variables, k and p. From the firms' first-order conditions (4.4)–(4.5), we derive that the equilibrium capital-labor ratios in the consumption and investment sectors are identical and equal to $a \equiv k/l = K_c/L_c = K_I/L_I = \alpha w/((1-\alpha)r)$, with $w = (1-\alpha)a^\alpha$ and $r = \alpha a^{\alpha-1}$. Combining these results with (4.1)–(4.2), we get $pA = 1$ with $A = K_I(k/l)^{-(1-\alpha)\Theta}$ and thus

$$K_I = (k/l)^{1-\alpha} p^{-1/\Theta} \tag{4.20}$$

---

[4]When there is no possible confusion, the time index (t) is not mentioned.

Moreover, substituting these expressions into the production functions (4.1)–(4.2), we also derive:

$$i = Y_I = p^{-\frac{1+\Theta}{\Theta}} \equiv Y_I(p) \tag{4.21}$$

$$c = Y_c = \left(\frac{k}{l}\right)^{\alpha-1}\left[k - \left(\frac{k}{l}\right)^{1-\alpha} p^{-1/\Theta}\right] \tag{4.22}$$

Combining Eqs. (4.9)–(4.10), describing the labor-leisure trade-off at the equilibrium, with (4.22) allows to write consumption and labor as functions of the capital stock k and the price of the investment good p, namely $c = c(k, p)$, and $l = l(k, p)$. It follows therefore that

$$\begin{aligned} a &= k/l(k, p) \equiv a(k, p) \\ w &= (1 - \alpha)(a(k, p))^{\alpha} \equiv w(k, p) \\ r &= \alpha(a(k, p))^{\alpha-1} \equiv r(k, p) \end{aligned} \tag{4.23}$$

Let us introduce the following elasticities:

$$\epsilon_{cc} = -\frac{U_1(c,\mathcal{L})}{U_{11}(c,\mathcal{L})c}, \quad \epsilon_{lc} = -\frac{U_2(c,\mathcal{L})}{U_{21}(c,\mathcal{L})c}, \quad \epsilon_{cl} = -\frac{U_1(c,\mathcal{L})}{U_{12}(c,\mathcal{L})l}, \quad \epsilon_{ll} = -\frac{U_2(c,\mathcal{L})}{U_{22}(c,\mathcal{L})l} \tag{4.24}$$

Note that ϵ_{cc} corresponds to the elasticity of intertemporal substitution in consumption while the Frisch elasticity of the labor supply is given by

$$\epsilon_{lw} = \left[\frac{1}{\epsilon_{ll}} - \frac{1}{\epsilon_{cl}}\right]^{-1} \tag{4.25}$$

Combining (4.9)–(4.12) with (4.20)–(4.23), the equations of motion are finally derived as

$$\dot{k} = Y_I(p) - \delta k$$

$$\dot{p} = \frac{(\delta+\rho)p - r(k,p) + \left[\frac{1}{\epsilon_{cc}}\frac{\partial c}{\partial k}\frac{p}{c(k,p)} - \frac{1}{\epsilon_{cl}}\frac{\partial l}{\partial k}\frac{p}{l(k,p)}\right](Y_I(p)-\delta k)}{E(k,p)} \tag{4.26}$$

with

$$E(k, p) = 1 - \left[\frac{1}{\epsilon_{cc}}\frac{\partial c}{\partial p}\frac{p}{c(k,p)} - \frac{1}{\epsilon_{cl}}\frac{\partial l}{\partial p}\frac{p}{l(k,p)}\right] \tag{4.27}$$

Any solution $\{k(t), p(t)\}_{t\geq 0}$, with $k(0)$ given, that also satisfies the transversality condition (4.13) is called an equilibrium path.

A steady state of the dynamical system (4.26) is defined by a pair (k^*, p^*) solution of

$$Y_I(p) = \delta k, \quad r(k, p) = (\delta + \rho)p \tag{4.28}$$

We then derive:

Proposition 1 *Assume that* $\chi[1 - \alpha(1 + \Theta)] + \gamma(1 - \alpha) - \alpha\Theta \neq 0$. *Then there exists a unique steady state* (k^*, p^*) *such that* $Y_I(p^*) = \delta k^*$ *and* $r(k^*, p^*) = (\delta + \rho)p^*$.

Proof See Appendix "Proof of Proposition 1".   □

Remark 3 Using a continuity argument we derive from Proposition 1 that there exists an intertemporal equilibrium for any initial capital stock $k(0)$ in the neighborhood of k^*. Moreover, any solution of (4.26) that converges to the steady state satisfies the transversality condition (4.13) and is an equilibrium. Therefore, given $k(0)$, if there is more than one initial price $p(0)$ in the stable manifold of the steady state, the equilibrium path from $k(0)$ is not unique and we have local indeterminacy.

Remind also from footnote 3 that the JR-type utility function as given by (4.14) may not be concave. Since we focus on the local stability properties of equilibria around the steady state, we provide a local condition for concavity.

Lemma 1 *The JR-type utility function as given by (4.14) is concave in a neighborhood of the steady state if and only if*

$$\sigma \geq \sigma_c(\gamma) \equiv \frac{\gamma C(\gamma + \chi)[1 + \chi - (1-\gamma)C]}{(1+\chi)^2 \left[\chi + \gamma C \left(2 - \frac{C(1-\gamma)}{1+\chi}\right)\right]} \tag{4.29}$$

with $C = [(1-\alpha)(\delta + \rho)]/[\rho + \delta(1-\alpha)](< 1)$.

Remark 4 When evaluated at the steady state, the income effect (4.18), the elasticity of intertemporal substitution in consumption as defined in (4.24) and the Frisch elasticity of labor (4.25) become:[5]

$$\varepsilon_{cI} = \frac{\chi + \gamma C}{\chi + \gamma C[2 - C(1-\gamma)]}, \quad \varepsilon_{lI} = -\frac{\gamma[1 - C(1-\gamma)]}{\chi + \gamma C[2 - C(1-\gamma)]} \tag{4.30}$$

$$\epsilon_{cc} = \left[\sigma \frac{1+\chi}{1 + \chi - C(1-\gamma)} - \gamma(1-\gamma)\frac{C}{1+\chi}\right]^{-1}, \quad \epsilon_{lw} = \frac{1}{\chi + \gamma C} \tag{4.31}$$

Considering (4.21) and linearizing the dynamical system (4.26) around the steady state leads to the characteristic polynomial

$$\mathcal{P}(\lambda) = \lambda^2 - \mathcal{T}\lambda + \mathcal{D} \tag{4.32}$$

with

$$\mathcal{D} = \frac{-\delta\left(\delta + \rho - \frac{\partial r}{\partial p}\right) - \frac{\delta(1+\Theta)}{\Theta}\frac{\partial r}{\partial k}\frac{k^*}{p^*}}{E(k^*, p^*)}$$

$$\mathcal{T} = \frac{\rho + \delta\left(\frac{1}{\epsilon_{cc}}\frac{\partial c}{\partial p}\frac{p^*}{c^*} - \frac{1}{\epsilon_{cl}}\frac{\partial l}{\partial p}\frac{p^*}{l^*}\right) - \frac{\partial r}{\partial p} - \frac{\delta(1+\Theta)}{\Theta}\left(\frac{1}{\epsilon_{cc}}\frac{\partial c}{\partial k}\frac{k^*}{c^*} - \frac{1}{\epsilon_{cl}}\frac{\partial l}{\partial k}\frac{k^*}{l^*}\right)}{E(k^*, p^*)} \tag{4.33}$$

Most of these partial derivatives are functions of ϵ_{cc}, ϵ_{cl}, ϵ_{lc} and ϵ_{ll}. The role of ϵ_{lc} and ϵ_{ll} occurs through the presence of endogenous labor but remains implicit at this stage.

System (4.26) has one state variable and one control variable. As is well known, if (4.32) has two roots with negative real parts, there is a continuum of converging paths and thus a continuum of equilibria: the steady state is locally indeterminate

---

[5]See Appendix "Proof of Lemma 1".

and there exist expectation-driven endogenous fluctuations. Local indeterminacy therefore requires that $\mathcal{D} > 0$ and $\mathcal{T} < 0$. Obviously saddle-point stability is obtained when $\mathcal{D} < 0$, while total instability holds (with both eigenvalues having positive real parts) if $\mathcal{D} > 0$ and $\mathcal{T} > 0$.[6] In the following, we will focus on locally indeterminate equilibria and we will also look for the existence of a Hopf bifurcation, occurring when $\mathcal{T} = 0$ while $\mathcal{D} > 0$, which leads to periodic cycles.

4.4 Local Indeterminacy with Variable Income Effects

Deriving the local stability properties of system (4.26) in the most general case (without additional parameter restrictions) is very cumbersome, as a lot of different configurations may arise. In order to reduce the number of possible configurations, we now introduce the following parameter restrictions:

Assumption 1 $\alpha < 1/2$, $\delta = 0.025$, $\rho > 0.005$, $\chi \leq 3$ and $\Theta \in (0, \bar{\Theta})$ with $\bar{\Theta} = (1 - \alpha)/\alpha$.

The calibration for δ is common to many studies in the DSGE literature and corresponds to an annual capital depreciation rate of 10%. The restriction on α is innocuous as capital shares are typically less than 50% of GDP in industrialized economies. Likewise, the assumption on the rate of time preference ρ is not very restrictive as the standard calibration for this parameter is $\rho = 0.01$. The restriction on χ allows to consider realistic values for the Frisch elasticity of labor ϵ_{lw} as given in (4.31) (see Sect. 4.5). Finally, using a benchmark calibration for the US economy at quarterly frequency, namely $(\alpha, \rho, \delta) = (0.3, 0.01, 0.025)$, Assumption 1 implies $\bar{\Theta} \approx 2.33$. This bound defines an interval for Θ which largely covers the range of available estimates for the degree of IRS in the investment sector, since empirical studies typically conclude for values around 0.3.[7] We obtain:

Proposition 2 *Under Assumption 1, consider the following critical values of* σ, Θ *and* χ:

$$
\sigma^{sup}(\gamma) \equiv \frac{[1+\chi-\mathcal{C}(1-\gamma)]\left\{\left[\alpha+\chi+\gamma\mathcal{C}\left(2-\frac{\mathcal{C}(1-\gamma)}{1+\chi}\right)\right]\Theta[\rho+\delta(1-\alpha)]+\frac{\gamma\mathcal{C}\alpha\delta}{1+\chi}[\alpha+\chi+\gamma(1-\alpha)]\right\}}{(1+\chi)\alpha\delta\left[\alpha+\chi+\gamma\mathcal{C}\left(2-\frac{\mathcal{C}(1-\gamma)}{1+\chi}\right)\right]}
$$

$$
\sigma^{H}(\gamma) \equiv \frac{[1+\chi-\mathcal{C}(1-\gamma)]\left\{\left[\alpha+\chi+\gamma\mathcal{C}\left(2-\frac{\mathcal{C}(1-\gamma)}{1+\chi}\right)\right]\rho\Theta[\rho+\delta(1-\alpha)]+\gamma\mathcal{C}\alpha\delta\left[\frac{\rho[\alpha+\chi+\gamma(1-\alpha)]}{1+\chi}+\alpha\delta\Theta\right]\right\}}{(1+\chi)\alpha\delta[\rho+\Theta(\delta+\rho)]\left[\alpha+\chi+\gamma\mathcal{C}\left(2-\frac{\mathcal{C}(1-\gamma)}{1+\chi}\right)\right]}
$$

$$
\tilde{\Theta}(\gamma) \equiv \frac{\gamma^{2}\mathcal{C}\alpha^{2}\delta\left[1-(1-\gamma)\mathcal{C}\left(2-\frac{\mathcal{C}(1-\gamma)}{1+\chi}\right)\right]}{\left[\chi+\gamma\mathcal{C}\left(2-\frac{\mathcal{C}(1-\gamma)}{1+\chi}\right)\right]\left[\alpha+\chi+\gamma\mathcal{C}\left(2-\frac{\mathcal{C}(1-\gamma)}{1+\chi}\right)\right][\rho+\delta(1-\alpha)]} \in (0, \bar{\Theta})
$$

$$
\underline{\chi}(\gamma) \equiv \frac{\alpha\Theta-\gamma(1-\alpha)}{1-\alpha-\alpha\Theta} \in (0, 3)
$$

(4.34)

---

[6] We will show in this case that there exists a Hopf bifurcation leading to the existence of periodic cycles.

[7] For example, Basu and Fernald (1997) obtain a point estimate for the degree of IRS in the durable manufacturing sector in the US economy of 0.33, with standard deviation 0.11.

with $\sigma^{sup}(\gamma) > \sigma^H(\gamma)$. Let $\sigma^{inf}(\gamma) = \max\{\sigma^H(\gamma), \sigma_c(\gamma)\}$. Then the steady state (k^*, p^*) is locally indeterminate if and only if $\chi > \underline{\chi}(\gamma)$, $\Theta \in (\tilde{\Theta}(\gamma), \bar{\Theta})$ and $\sigma \in (\sigma^{inf}(\gamma), \sigma^{sup}(\gamma))$, while saddle-point stability holds if $\Theta \in (\tilde{\Theta}(\gamma), \bar{\Theta})$ and $\sigma > \sigma^{sup}(\gamma)$ or $\Theta < \tilde{\Theta}(\gamma)$.

Proof See Appendix "Proof of Proposition 2". □

Remark 5 Some comments on the occurrence of saddle-point stability are in order here. As shown in Appendix "Proof of Proposition 2", under Assumption 1, $\mathcal{D}$ is positive and local indeterminacy may arise if and only if $\sigma < \sigma^{sup}(\gamma)$. But Lemma 1 shows that the JR utility function is locally concave if and only if $\sigma \geq \sigma_c(\gamma)$. The compatibility of these two conditions is ensured if and only if $\Theta > \tilde{\Theta}(\gamma)$. Therefore, $\mathcal{D}$ is positive and saddle-point stability holds in two cases: (i) when $\Theta \in (\tilde{\Theta}(\gamma), \bar{\Theta})$ and $\sigma > \sigma^{sup}(\gamma)$, or (ii) when $\Theta < \tilde{\Theta}(\gamma)$ which implies $\sigma > \sigma^{sup}(\gamma)$ under the concavity condition.

Proposition 2 shows that for any intensity $\gamma \in [0, 1]$ of income effects, there is a non-empty range of values for the parameter σ such that indeterminacy occurs. This conclusion is in sharp contrast with the results obtained in one-sector models. For example, Nishimura et al. (2009) show that indeterminacy is ruled out in such models under both GHH ($\gamma = 0$) and KPR ($\gamma = 1$) preferences, as long as realistic parameter values are considered. Likewise, Jaimovich (2008) shows in a calibrated version of the aggregate infinite-horizon model with increasing returns that local indeterminacy arises for *intermediary values* of γ, while it is ruled out when the income effect is too low (γ close to 0) or too large (γ close to 1).

Proposition 2 also implies that a Hopf bifurcation exists in the parameter space, provided that $\sigma^H(\gamma) > \sigma_c(\gamma)$. One can complete the proposition by deriving conditions under which this inequality is satisfied.

Corollary 1 *Under Assumption 1, let $\chi > \underline{\chi}(\gamma)$ and consider the critical values as given by (4.34) together with the following one:*

$$\hat{\Theta}(\gamma) = \frac{\gamma^2 C \alpha^2 \delta \rho \left[1 - \frac{(1-\gamma)C}{1+\chi}\left(2 - \frac{C(1-\gamma)}{1+\chi}\right)\right]}{\left[\alpha + \chi + \gamma C\left(2 - \frac{C(1-\gamma)}{1+\chi}\right)\right]\left[\rho[\rho + \delta(1-\alpha)]\left[\chi + \gamma C\left(2 - \frac{C(1-\gamma)}{1+\chi}\right)\right] - \frac{\gamma C \alpha \delta(\gamma+\chi)(\delta+\rho)}{1+\chi}\right] + \gamma C(\alpha\delta)^2\left[\chi + \gamma C\left(2 - \frac{C(1-\gamma)}{1+\chi}\right)\right]} \in (0, \bar{\Theta})$$

Denote $\underline{\Theta}(\gamma) = \max\{\hat{\Theta}(\gamma), \tilde{\Theta}(\gamma)\}$. If $\Theta \in (\underline{\Theta}(\gamma), \bar{\Theta})$, the steady state (k^*, p^*) is saddle-point stable when $\sigma > \sigma^{sup}(\gamma)$, locally indeterminate when $\sigma \in (\sigma^H(\gamma), \sigma^{sup}(\gamma))$ and totally unstable when $\sigma \in (\sigma_c(\gamma), \sigma^H(\gamma))$. When σ crosses $\sigma^H(\gamma)$ from above a Hopf bifurcation generically occurs and gives rise to the existence of locally indeterminate (totally unstable) periodic cycles in a left (right) neighborhood of $\sigma^H(\gamma)$.

Proof See Appendix "Proof of Corollary 1". □

Remark 6 Corollary 1 shows that local indeterminacy arises when $\sigma \in (\sigma^H(\gamma), \sigma^{sup}(\gamma))$ with the occurrence of a pair of purely imaginary complex eigenvalues when

$\sigma = \sigma^H(\gamma)$. The Hopf bifurcation Theorem (see Grandmont 2008) then implies that there exist periodic cycles for σ in a left or right neighborhood of $\sigma^H(\gamma)$ depending on whether the bifurcation is super or sub-critical. In the super-critical case, the periodic cycles occur when the steady-state is totally unstable which implies that the periodic cycles are stable, i.e. locally indeterminate. On the contrary, in the sub-critical case, the periodic cycles occur when the steady-state is locally indeterminate. This means that the periodic cycles are totally unstable and define a corridor of stability for the steady-state and thus for the existence of an equilibrium. Indeed, any path starting from the outside of the area defined by a periodic cycle is a divergent one that will violate the transversality condition and cannot be an equilibrium.

Remind that σ and χ are the crucial parameters influencing the degree of intertemporal substitution in consumption and the Frisch elasticity of labor supply. Proposition 2 and Corollary 1 then provide clear-cut conclusions about the conditions required for local indeterminacy and the existence of sunspot-driven fluctuations in canonical two-sector models. Local indeterminacy occurs, for any given degree $\gamma \in [0, 1]$ of income effects, provided that the degree of IRS is not too small, the wage elasticity of labor supply is not too large, and the EIS in consumption is in an intermediary range. Note that the interval of values for the amount of externalities Θ compatible with local indeterminacy is quite large under the benchmark calibration $(\alpha, \rho, \delta) = (0.3, 0.01, 0.025)$ as $\underline{\Theta}(\gamma) \in [0, 0.0323)$ for $\gamma \in [0, 1]$ and $\chi \geq 0$ while $\bar{\Theta} \approx 2.33$.

As an illustration to Proposition 2, Fig. 4.1 plots the relevant bifurcation loci and the local indeterminacy areas in the three-dimensional space with axes given by (χ, γ, σ) in panel (a), and by $(\chi, \gamma, \epsilon_{cc})$ in panel (b). The critical values obtained for the EIS in consumption ϵ_{cc} in panel (b) are derived from the analytical expression relating σ to ϵ_{cc} at the steady-state as given by (4.31). Moreover, panels (c) and (d) in Fig. 4.2 display, for each pair (χ, γ), the corresponding values for the Frisch elasticity of labor supply (panel (c)), and the income effect on labor supply (panel (d)), both evaluated at the steady state (see (4.30) and (4.31)). All these graphs are computed using the benchmark calibration $(\alpha, \rho, \delta) = (0.3, 0.01, 0.025)$ and a degree of IRS in the investment sector of $\Theta = 0.33$, the point estimate obtained by Basu and Fernald (1997).

As can be seen, local indeterminacy occurs for a wide range of values for the EIS in consumption ϵ_{cc}, typically ranging between 0.61 and 1.95 using our benchmark

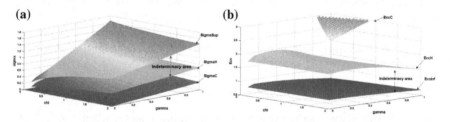

Fig. 4.1 **a** Bifurcations in the (χ, γ, σ) plane; **b** bifurcations in the $(\chi, \gamma, \epsilon_{cc})$ plane.

(c)   **(d)**

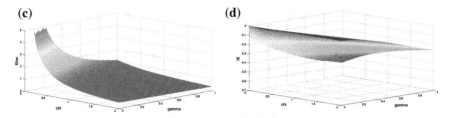

Fig. 4.2   c Wage elasticity of labor supply ϵ_{lw}; **d** income effect on labor supply $\varepsilon_{l\mathcal{I}}$.

calibration. This is in line with the empirical literature, which provides estimates typically ranging between 0 and 2. Moreover, combining the results displayed in panel (b) with panels (c) and (d), it can be observed that when γ is small (close to 0), indeterminacy emerges for a wide range of values for the Frisch labor supply elasticity (ranging between 0.5 and more than 5), a wide range of values for the EIS in consumption (ranging between 0.7 and 1.7), but a moderate intensity of income effects on labor supply (ranging between 0 and -0.15). Conversely, when γ tends to 1, indeterminacy can emerge under much more significant income effects (up to a value of -0.55, obtained when χ is close to its minimum value $\underline{\chi}(1)$ consistent with indeterminacy). Yet, the maximal value for the wage-elasticity of labor is now relatively small (with a maximum value given by $\epsilon_{lw} = 1.2$).

However, it is worth pointing out that the interval of values for the structural parameter σ given in Proposition 2 and Corollary 1 varies with the size of the income effect γ. But usually we consider a constant value for σ, e.g. $\sigma = \boldsymbol{\sigma}$ for any $\gamma \leq 1$. In such a case, one may wonder whether local indeterminacy may arise for any size of the income effect $\gamma \in [0, 1]$. The answer to this question depends on the values of $\sigma^{sup}(0)$ and $\sigma^{H}(1)$. Clearly, a positive answer requires $\sigma^{sup}(0) > \sigma^{H}(1)$. We need also to satisfy the necessary condition for local indeterminacy exhibited in Proposition 2 and Corollary 1, namely $\chi > \underline{\chi}(\gamma)$ and $\Theta > \underline{\Theta}(\gamma)$, for any $\gamma \in [0, 1]$. Noting that the maximal values of $\underline{\chi}(\gamma)$ and $\underline{\Theta}(\gamma)$ are respectively $\underline{\chi}(0) = \alpha\Theta/(1 - \alpha - \alpha\Theta)$ and $\underline{\Theta}(1)$,[8] this property is satisfied if $\chi > \underline{\chi}(0)$ and $\Theta > \underline{\Theta}(1)$. We then get the following Lemma.

Lemma 2 *Under Assumption 1, let $\chi > \underline{\chi}(0)$. Then there exist $\bar{\bar{\Theta}} \in (0, \bar{\Theta})$ and $\bar{\chi} > \underline{\chi}(0)$ such that when $\Theta \in (0, \bar{\bar{\Theta}})$, $\sigma^{sup}(0) - \sigma^{H}(1) \lessgtr 0$ if and only if $\chi \lessgtr \bar{\chi}$.*

Proof See Appendix "Proof of Lemma 2". ☐

Let us introduce an additional technical assumption:

Assumption 2 $\bar{\bar{\Theta}} > \underline{\Theta}(1)$ and $\bar{\chi} < 3$.

Considering again $(\alpha, \rho, \delta) = (0.3, 0.01, 0.025)$, this Assumption easily holds as $\bar{\bar{\Theta}} \approx 0.67$, $\underline{\Theta}(1) < 0.0323$ and $\bar{\chi} \approx 1.39$. We can then finally derive the following Corollary:

---

[8]It can be shown indeed that $\underline{\Theta}(\gamma)$ is an increasing function of γ while $\underline{\chi}(\gamma)$ is a decreasing function.

Corollary 2 *Under Assumptions 1–2, let $\Theta \in (\underline{\Theta}(1), \bar{\bar{\Theta}})$. Then the following cases occur:*

(a) *If $\chi \in (\underline{\chi}(0), \bar{\chi})$, there exist $0 < \underline{\gamma} < \bar{\gamma}$ such that the steady state is locally indeterminate in the following cases:*

   (i) when $\sigma \in (\sigma^H(0), \sigma^{sup}(0))$ and $\gamma \in [0, \underline{\gamma})$;
   (ii) when $\sigma \in (\sigma^{sup}(0), \sigma^H(1))$ and $\gamma \in (\underline{\gamma}, \bar{\gamma})$;
   (iii) when $\sigma \in (\sigma^H(1), \sigma^{sup}(1))$ and $\gamma \in (\bar{\gamma}, 1]$.

(b) *If $\chi > \bar{\chi}$, the steady state is locally indeterminate for any $\gamma \in [0, 1]$ when $\sigma \in (\sigma^H(1), \sigma^{sup}(0))$. Moreover, there exist $0 < \underline{\gamma} < \bar{\gamma}$ such that local indeterminacy also holds in the following cases:*

   (i) when $\sigma \in (\sigma^H(0), \sigma^H(1))$ and $\gamma \in [0, \underline{\gamma})$;
   (ii) when $\sigma \in (\sigma^{sup}(0), \sigma^{sup}(1))$ and $\gamma \in (\bar{\gamma}, 1]$.

Corollary 2 shows that there is a trade-off between the values of χ, σ and γ for the existence of local indeterminacy. When χ is low, i.e. the Frisch elasticity of labor is large, the lower (higher) the values of σ, the lower (higher) the values of γ must be for local indeterminacy to arise. The same type of results partially arises when χ is large enough, i.e. the Frisch elasticity of labor is low enough, as low (high) values of σ still require low (high) values of γ. However, local indeterminacy may also arise for any $\gamma \in [0, 1]$ as long as σ admits intermediary values. As σ is inversely related to the EIS in consumption, we conclude that the size of the income effect necessary for the existence of self-fulfilling expectations strongly depends on the way the representative agent adjusts his intertemporal consumption profile. Such a conclusion is important as there is no clear evidence of the empirically realistic values of the size γ. Khan and Tsoukalas (2011) provide some estimates in favor of a large income effect with $\gamma > 0.5$, while Schmitt-Grohé and Uribe (2012) conclude for evidences in favor of a low income effect with values of γ close to zero. It is therefore necessary to explore our main results on a numerical basis in order to evaluate the magnitude of each structural parameter that affects the occurrence of expectations-driven fluctuations.

4.5 Numerical Illustrations

We have shown in Proposition 2 and Corollary 1 that local indeterminacy arises under different scenarios for the values of the Frisch elasticity of labor ϵ_{lw}, the EIS in consumption ϵ_{cc} and the size of income effect. There is no consensus in the literature about ϵ_{lw} and ϵ_{cc}. Concerning ϵ_{lw}, Rogerson and Wallenius (2009) and Prescott and Wallenius (2011) recommend values around 3 to calibrate business cycle

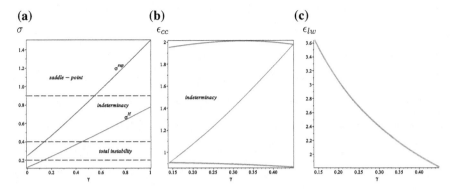

Fig. 4.3 **a** Indeterminacy areas for $\chi = 0.15$; **b** and **c** variations of ϵ_{cc} and ϵ_{lw} when $\sigma = 0.4$

models, based on both theoretical considerations and cross-country tax analysis.[9] More recently, Chetty et al. (2012) recommend on the contrary an aggregate Frisch elasticity of 0.5 on the intensive margin for labor supply. Concerning ϵ_{cc}, while early studies suggest quite low values, e.g. Campbell (1999) and Kocherlakota (1996), more recent estimates provide a much more contrasted view. Indeed, Mulligan (2002) and Vissing-Jorgensen and Attanasio (2003) repeatedly obtained estimates above unity, typically in the range 1.1–2.1.

Let us now provide some numerical illustrations in order to check whether macroeconomic fluctuations based on self-fulfilling expectations may arise under realistic calibrations for these parameters. When $(\alpha, \rho, \delta) = (0.3, 0.01, 0.025)$, we get $(\underline{\Theta}(1), \underline{\chi}(0), \bar{\chi}) \approx (0.0263, 0.1475, 1.39)$. Let us then assume $\chi = 0.15$ and $\Theta = 0.3$ so that for any given $\gamma \in [0, 1]$, $\Theta \in (\underline{\Theta}(\gamma), \bar{\Theta})$ and $\chi \in (\underline{\chi}(\gamma), \bar{\chi})$. In this configuration, we are in case (a) of Corollary 2 with $\sigma^{sup}(0) < \sigma^{H}(1)$. We then get the following Fig. 4.3 covering different possible values of σ.
It follows that local indeterminacy occurs:

(i) when $\sigma = 0.2$ if $\gamma \in [0, 0.136)$,
(ii) when $\sigma = 0.4$ if $\gamma \in (0.14, 0.449)$,
(iii) when $\sigma = 0.9$ if $\gamma \in (0.549, 1]$.

As an illustration of configuration (ii), and according to Fig. 4.1, we find values for the EIS in consumption in line with the more recent estimates provided by Mulligan (2002) and Vissing-Jorgensen and Attanasio (2003). Moreover, the values for the Frisch elasticity of labor match the recommendations of Rogerson and Wallenius (2009) and Prescott and Wallenius (2011).

Considering now $\chi = 1.7$, we are in case (b) of Corollary 2 and we get $\sigma^{H}(1) \approx 0.6774$ and $\sigma^{sup}(0) \approx 0.737$. It follows therefore that if $\sigma = 0.7$, local indeterminacy arises for any $\gamma \in [0, 1]$. We have indeed the following Fig. 4.4.

---

[9]See Prescott and Wallenius (2011) for a discussion of the factors that make the wage elasticity of aggregate labor supply significantly differ from the corresponding elasticity at the micro level.

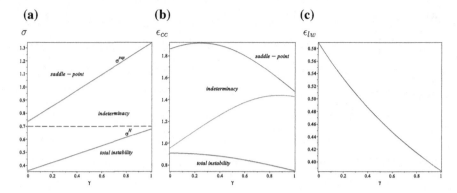

Fig. 4.4 **a** Indeterminacy for any $\gamma \in [0, 1]$; **b** and **c** variations of ϵ_{cc} and ϵ_{lw} when $\sigma = 0.7$

Moreover, the EIS again belongs to an empirically realistic interval compatible with the estimates of Mulligan (2002) and Vissing-Jorgensen and Attanasio (2003) and the Frisch elasticity of labor is now in line with the recommendation of Chetty et al. (2012).

Our results then prove that the existence of sunspot fluctuations can be obtained for any size of income effect as long as the values of the Frisch elasticity of labor and the EIS in consumption are adequately chosen. Moreover, in any cases, these values can be in line with the estimates provided by the recent literature.

4.6 Economic Interpretations

The general intuition for the existence of indeterminacy in a one-sector model is quite simple.[10] Starting from the steady state, let us assume that agents expect a faster rate of accumulation. To be an equilibrium this new path would require a higher return on investment. If higher anticipated stocks of future capital raise the marginal product of capital by drawing labor out of leisure, the expected higher rate of return may be self-fulfilling. When there is a sufficient amount of increasing returns based on externalities and the Frisch elasticity of the labor supply is large enough, the movement of labor into production may be strong enough to boost the rate of return leading to self-fulfilling expectations and multiple equilibria. However, depending on the utility function, if we consider as suggested by the empirical evidence that the labor demand function is decreasing with respect to wage, there is an upper bound for the size of externalities and such a mechanism may not be admissible. As shown by Hintermaier (2001, 2003), with a KPR utility function ($\gamma = 1$), the concavity restrictions prevent the occurrence of this mechanism. Similarly, Nishimura et al. (2009) prove the same impossibility result with a GHH utility function ($\gamma = 0$).

---

[10]See Benhabib and Farmer (1994).

In this case, the argument is not based on concavity but on the absence of income effect. As externalities are not strong enough and labor is not affected by the increased income ($\varepsilon_{lT} = 0$), the expected increase of the marginal product of capital does not generate a sufficient adjustment of labor and the expectations cannot be self-fulfilling. This explains why Jaimovich (2008) obtains the existence of local indeterminacy for intermediary values of γ.[11]

In two-sector models, the story is different. As shown by Benhabib and Farmer (1996), when external effects in each sector depend on the aggregate output of their own sector, factor reallocations across sectors can have strong effects on marginal products. It follows that local indeterminacy can occur with much smaller externalities than those required in the one sector case, a standard positive slope for the labor demand function and under a lower variability of labor. Our main conclusions are of course compatible with a decreasing labor demand function which is obtained as soon as $\Theta < \alpha/(1 - \alpha)(< \bar{\Theta})$.

We prove that the existence of sunspot fluctuations depends on a trade-off between the values of the Frisch elasticity of labor, the EIS in consumption and the size of income effect. To understand such a trade-off, let us start with Corollary 2-(a). It is shown that for a given low value of χ, i.e. a large value of ϵ_{lw}, the larger the income effect, the lower the EIS in consumption for indeterminacy to arise. In order to get an intuition for such conclusions, consider the expressions of ε_{cT} and ε_{lT} evaluated at the steady state as given in (4.30). It is easy to check that ε_{cT} and ε_{lT} are decreasing in γ while increasing in χ.

Starting from the steady state, let us assume as previously that agents expect an increase in the future marginal return on capital leading then to a decrease in current consumption in order to invest more today and at the same time an increase in future income. As ε_{cT} is decreasing in γ, the lower the income effect, the larger the increase in future consumption following the larger expected income. This effect therefore generates a large fluctuation of consumption and the expectation can be self-fulfilling provided the EIS is large enough. Since the two-sector structure requires lower external effects, the adjustment of labor is sufficient to get multiple equilibria even under a low income effect. Similarly, when the income effect is larger, the increase of future consumption following a larger expected income is weaker, and the expectations can now be self-fulfilling under a lower EIS in consumption.

Let us now consider Corollary 2-(b). We show here that if the value of χ is large enough, i.e. the value of ϵ_{lw} is low enough, local indeterminacy may arise for any size of income effect when the EIS in consumption has intermediary values. Following the same intuition, as ε_{cT} is increasing in χ, the larger expected future income implies a significant increase in future consumption that can be compatible with the decrease of present consumption if the EIS is sufficiently high. But now, as the income effect is increased by the large value of χ, this impact can be large enough no matter what is the value of $\gamma \in [0, 1]$.

---

[11]Recall however that his utility formulation contains an additional state variable X_t which may play a significant role for these result.

4.7 Concluding Comments

Although one-sector infinite horizon models are known to require very specific positive amount of income effect for the existence of local indeterminacy,[12] two-sector models have been shown to generate sunspot-driven business cycles under no-income effect preferences. Dufourt et al. (2015) indeed show that when properly calibrated, the model solves several empirical puzzles traditionally associated with two-sector RBC models.[13] However, there is not yet a complete analysis of the impact of various income effects on the occurrence of local indeterminacy.

This paper provides such an analysis. We have shown that for any given size of income effect, there is a non-empty range of values for the Frisch elasticity of labor and the EIS in consumption such that indeterminacy occurs. This is in contrast to the results obtained for aggregate models both in Hintermaier (2001, 2003) and Nishimura et al. (2009), in which it is shown that indeterminacy cannot occur under GHH and KPR preferences, and in Jaimovich (2008), in which local indeterminacy only arises for intermediary income effects.

More precisely, we have proved that for a large enough Frisch elasticity of labor, the larger the income effect, the lower the EIS in consumption for indeterminacy to arise. On the contrary, when the Frisch elasticity of labor is low enough, local indeterminacy may arise for any size of income effect when the EIS in consumption has intermediary values. We then exhibit a clear trade-off between all these structural parameters that characterize preferences and that affect the occurrence of expectations-driven fluctuations. Such a conclusion appears as important as there is yet no clear empirical estimates of the size of income effect.[14]

4.8 Appendix

4.8.1 Proof of Proposition 1

Consider the steady state with $Y_I = \delta k$ and $r = (\delta + \rho)p$. Since $r = p\alpha Y_I/K_I$, we get

$$K_I = \frac{\alpha\delta}{\delta + \rho}k \tag{4.35}$$

Using the production function (4.19) for the investment good we derive

$$Y_I = \left(\tfrac{k}{l}\right)^{(\alpha-1)(1+\Theta)}\left(\tfrac{\alpha\delta}{\delta+\rho}k\right)^{1+\Theta} = \delta k$$

Solving this equation yields

---

[12] See Jaimovich (2008), Nishimura et al. (2009).
[13] See also Guo and Harrison (2010), Nishimura and Venditti (2010).
[14] See Khan and Tsoukalas (2011), Schmitt-Grohé and Uribe (2012).

$$k^* = l^{\frac{(1-\alpha)(1+\Theta)}{1-\alpha(1+\Theta)}} \left(\frac{\alpha}{\delta+\rho}\right)^{\frac{1+\Theta}{1-\alpha(1+\Theta)}} \delta^{\frac{\Theta}{1-\alpha(1+\Theta)}} \equiv l^{\frac{(1-\alpha)(1+\Theta)}{1-\alpha(1+\Theta)}} \kappa^* \tag{4.36}$$

Substituting this expression into (4.22) we get

$$c^* = l^{\frac{1-\alpha}{1-\alpha(1+\Theta)}} \frac{\delta(1-\alpha)+\rho}{\delta+\rho} \kappa^{*\alpha} \equiv l^{\frac{1-\alpha}{1-\alpha(1+\Theta)}} \psi^* \tag{4.37}$$

Recall that the trade-off between consumption and leisure is described by

$$\frac{(1+\chi)l^\chi c^\gamma}{1+\chi - \gamma l^{1+\chi} c^{\gamma-1}} = w \tag{4.38}$$

Using (4.23) with (4.36)–(4.37) we get

$$(1+\chi)l^{\chi + \frac{\gamma(1-\alpha)}{1-\alpha(1+\Theta)}} \psi^{*\gamma} = (1-\alpha)l^{\frac{\alpha\Theta}{1-\alpha(1+\Theta)}} \kappa^{*\alpha} \left[1 + \chi - \gamma l^{1+\chi - \frac{(1-\gamma)(1-\alpha)}{1-\alpha(1+\Theta)}} \psi^{*\gamma-1}\right]$$

If $\chi[1 - \alpha(1+\Theta)] + \gamma(1-\alpha) - \alpha\Theta \neq 0$, solving this equation yields

$$l^* = \left\{\frac{(1-\alpha)\kappa^*}{\psi^{*\gamma}}\left[1 + \frac{(1-\alpha)\kappa^*\gamma}{(1+\chi)\psi^*}\right]^{-1}\right\}^{\frac{1-\alpha(1+\Theta)}{\chi[1-\alpha(1+\Theta)]+\gamma(1-\alpha)-\alpha\Theta}}$$

We finally derive from (4.23)

$$p^* = \alpha(k^*/l^*)^{\alpha-1}$$

$\square$

4.8.2 Proof of Lemma 1

Using (4.24) and the first order conditions (4.9)–(4.10), we get $\epsilon_{cl} = \epsilon_{lc}(c/wl)$. Using the expression of w given in (4.23) together with the values of k^* and l^* provided in Sect. "Proof of Proposition 1" we find $wl/c = (1-\alpha)(\delta+\rho)/[\delta(1-\alpha)+\rho]$. Then at the steady state we get

$$\epsilon_{cl} = \frac{\delta(1-\alpha)+\rho}{(1-\alpha)(\delta+\rho)}\epsilon_{lc} \tag{4.39}$$

Using (4.24), we compute for the utility function as given by (4.14) the following elasticities:

$$\frac{1}{\epsilon_{cc}} = \sigma\frac{c-\gamma\frac{l^{1+\chi}}{1+\chi}c^\gamma}{c-\frac{l^{1+\chi}}{1+\chi}c^\gamma} - \gamma(1-\gamma)\frac{\frac{l^{1+\chi}}{1+\chi}c^\gamma}{c-\gamma\frac{l^{1+\chi}}{1+\chi}c^\gamma}, \qquad \frac{1}{\epsilon_{lc}} = \sigma\frac{c-\gamma\frac{l^{1+\chi}}{1+\chi}c^\gamma}{c-\frac{l^{1+\chi}}{1+\chi}c^\gamma} - \gamma$$

$$\frac{1}{\epsilon_{cl}} = \frac{l^{1+\chi}c^\gamma}{c-\gamma\frac{l^{1+\chi}}{1+\chi}c^\gamma}\left[\sigma\frac{c-\gamma\frac{l^{1+\chi}}{1+\chi}c^\gamma}{c-\frac{l^{1+\chi}}{1+\chi}c^\gamma} - \gamma\right], \qquad \frac{1}{\epsilon_{ll}} = \sigma\frac{l^{1+\chi}c^\gamma}{c-\frac{l^{1+\chi}}{1+\chi}c^\gamma} + \chi \tag{4.40}$$

Obviously, normality holds as we derive from these expressions that

$$\frac{1}{\epsilon_{cc}} - \frac{1}{\epsilon_{lc}} \geq 0 \text{ and } \frac{1}{\epsilon_{cl}} - \frac{1}{\epsilon_{ll}} \geq 0 \tag{4.41}$$

Consider now Eq. (4.39) together with the expressions given by (4.40). We then derive that

$$\frac{l^{1+\chi}c^{\gamma-1}}{1 - \gamma\frac{l^{1+\chi}}{1+\chi}c^{\gamma-1}} = \frac{(1-\alpha)(\delta+\rho)}{\delta(1-\alpha)+\rho} \tag{4.42}$$

Denoting $C = [(1 - \alpha)(\delta + \rho)]/[\delta(1 - \alpha) + \rho] < 1$, solving this equation yields

$$l^{1+\chi}c^{\gamma-1} = \frac{C(1+\chi)}{1+\chi+\gamma C} \tag{4.43}$$

and thus

$$\frac{c - \gamma\frac{l^{1+\chi}}{1+\chi}c^{\gamma}}{c - \frac{l^{1+\chi}}{1+\chi}c^{\gamma}} = \frac{1+\chi}{1+\chi-C(1-\gamma)}, \quad \frac{l^{1+\chi}c^{\gamma}}{c - \frac{l^{1+\chi}}{1+\chi}c^{\gamma}} = \frac{C(1+\chi)}{1+\chi-C(1-\gamma)} \tag{4.44}$$

Using these expressions we then derive from (4.40):

$$\frac{1}{\epsilon_{cc}} = \sigma\frac{1+\chi}{1+\chi-C(1-\gamma)} - \gamma(1-\gamma)\frac{C}{1+\chi}, \quad \frac{1}{\epsilon_{lc}} = \sigma\frac{1+\chi}{1+\chi-C(1-\gamma)} - \gamma$$

$$\frac{1}{\epsilon_{cl}} = \frac{C}{\epsilon_{lc}}, \qquad\qquad\qquad\qquad\qquad \frac{1}{\epsilon_{ll}} = \sigma\frac{(1+\chi)C}{1+\chi-C(1-\gamma)} + \chi \tag{4.45}$$

Concavity of the utility function requires

$$\frac{1}{\epsilon_{cc}\epsilon_{ll}} - \frac{1}{\epsilon_{lc}\epsilon_{cl}} \geq 0 \quad \text{and} \quad \frac{1}{\epsilon_{cc}} \geq 0$$

Straightforward computations show that these two inequalities are satisfied if and only if

$$\sigma \geq \sigma_c(\gamma) \equiv \frac{\gamma C(\gamma+\chi)[1+\chi-(1-\gamma)C]}{(1+\chi)^2\left[\chi+\gamma C\left(2-\frac{C(1-\gamma)}{1+\chi}\right)\right]}$$

□

4.8.3   Proof of Proposition 2

We start by the computation of $\mathcal{D}$ and $\mathcal{T}$ using a general formulation for $U(c, \mathcal{L})$. Consider the consumption-labor trade-off as described by (4.9)–(4.10) together with the expressions of wage and consumption as given by (4.22) and (4.23). We get the following two equations

$$U_2(c, \ell - l)l^{\alpha} = (1-\alpha)k^{\alpha}U_1(c, \ell - l) \tag{4.46}$$

$$cl^{\alpha-1} = k^{\alpha-1}\left[k - \left(\frac{k}{l}\right)^{1-\alpha} p^{-1/\Theta}\right] \qquad (4.47)$$

Total differentiation of (4.46) gives

$$\frac{dc}{c}\left(\frac{1}{\epsilon_{cc}} - \frac{1}{\epsilon_{lc}}\right) + \frac{dl}{l}\left(\frac{1}{\epsilon_{ll}} - \frac{1}{\epsilon_{cl}} + \alpha\right) = \alpha \frac{dk}{k} \qquad (4.48)$$

Total differentiation of (4.47) gives

$$\frac{dc}{c} - (1-\alpha)\frac{dl}{l} = -(1-\alpha)\frac{dk}{k} + \frac{k^*}{k^*-K_I^*}\frac{dk}{k} - \frac{K_I^*}{k^*-K_I^*}\left[(1-\alpha)\left(\frac{dk}{k} - \frac{dl}{l}\right) - \frac{1}{\Theta}\frac{dp}{p}\right] \qquad (4.49)$$

At the steady state we know that $(\delta + \rho)p = r$ with $r = p\alpha Y_I/K_I = p\alpha\delta k/K_I$. We then derive $K_I^* = \alpha\delta k^*/(\delta + \rho)$ and thus

$$\frac{k^*}{k^*-K_I^*} = \frac{\delta+\rho}{\rho+\delta(1-\alpha)}, \quad \frac{K_I^*}{k^*-K_I^*} = \frac{\alpha\delta}{\rho+\delta(1-\alpha)}$$

Equation (4.48) then becomes:

$$[\rho + \delta(1-\alpha)]\frac{dc}{c} - (1-\alpha)(\delta+\rho)\frac{dl}{l} = \alpha(\delta+\rho)\frac{dk}{k} + \frac{\alpha\delta}{\Theta}\frac{dp}{p} \qquad (4.50)$$

From (4.48) we derive

$$\frac{dl}{l} = -\frac{dc}{c}\frac{\frac{1}{\epsilon_{cc}} - \frac{1}{\epsilon_{lc}}}{\frac{1}{\epsilon_{ll}} - \frac{1}{\epsilon_{cl}}+\alpha} + \frac{dk}{k}\frac{\alpha}{\frac{1}{\epsilon_{ll}} - \frac{1}{\epsilon_{cl}}+\alpha} \qquad (4.51)$$

Substituting this expression into (4.50) gives

$$\frac{dc}{c} = \frac{\alpha(\delta+\rho)\left(\frac{1}{\epsilon_{ll}} - \frac{1}{\epsilon_{cl}}+1\right)}{[\rho+\delta(1-\alpha)]\left(\frac{1}{\epsilon_{ll}} - \frac{1}{\epsilon_{cl}}+\alpha\right) + (1-\alpha)(\delta+\rho)\left(\frac{1}{\epsilon_{cc}} - \frac{1}{\epsilon_{lc}}\right)}\frac{dk}{k}$$

$$+ \frac{\alpha\delta\left(\frac{1}{\epsilon_{ll}} - \frac{1}{\epsilon_{cl}}+\alpha\right)}{\Theta\left[[\rho+\delta(1-\alpha)]\left(\frac{1}{\epsilon_{ll}} - \frac{1}{\epsilon_{cl}}+\alpha\right) + (1-\alpha)(\delta+\rho)\left(\frac{1}{\epsilon_{cc}} - \frac{1}{\epsilon_{lc}}\right)\right]}\frac{dp}{p} \qquad (4.52)$$

Substituting (4.52) into (4.51) finally gives

$$\frac{dl}{l} = -\frac{\alpha\left[(\delta+\rho)\left(\frac{1}{\epsilon_{cc}} - \frac{1}{\epsilon_{lc}}\right) - [\rho+\delta(1-\alpha)]\right]}{[\rho+\delta(1-\alpha)]\left(\frac{1}{\epsilon_{ll}} - \frac{1}{\epsilon_{cl}}+\alpha\right) + (1-\alpha)(\delta+\rho)\left(\frac{1}{\epsilon_{cc}} - \frac{1}{\epsilon_{lc}}\right)}\frac{dk}{k}$$

$$- \frac{\alpha\delta\left(\frac{1}{\epsilon_{cc}} - \frac{1}{\epsilon_{lc}}\right)}{\Theta\left[[\rho+\delta(1-\alpha)]\left(\frac{1}{\epsilon_{ll}} - \frac{1}{\epsilon_{cl}}+\alpha\right) + (1-\alpha)(\delta+\rho)\left(\frac{1}{\epsilon_{cc}} - \frac{1}{\epsilon_{lc}}\right)\right]}\frac{dp}{p} \qquad (4.53)$$

We then conclude from this

$$\frac{\partial c}{\partial k}\frac{k^*}{c^*} = \frac{\alpha(\delta+\rho)\left(\frac{1}{\epsilon_{ll}} - \frac{1}{\epsilon_{cl}}+1\right)}{[\rho+\delta(1-\alpha)]\left(\frac{1}{\epsilon_{ll}} - \frac{1}{\epsilon_{cl}}+\alpha\right)+(1-\alpha)(\delta+\rho)\left(\frac{1}{\epsilon_{cc}} - \frac{1}{\epsilon_{lc}}\right)}$$

$$\frac{\partial c}{\partial p}\frac{p^*}{c^*} = \frac{\alpha\delta\left(\frac{1}{\epsilon_{ll}} - \frac{1}{\epsilon_{cl}}+\alpha\right)}{\Theta\left[[\rho+\delta(1-\alpha)]\left(\frac{1}{\epsilon_{ll}} - \frac{1}{\epsilon_{cl}}+\alpha\right)+(1-\alpha)(\delta+\rho)\left(\frac{1}{\epsilon_{cc}} - \frac{1}{\epsilon_{lc}}\right)\right]}$$

$$\frac{\partial l}{\partial k}\frac{k^*}{l^*} = -\frac{\alpha\left[(\delta+\rho)\left(\frac{1}{\epsilon_{cc}} - \frac{1}{\epsilon_{lc}}\right) - [\rho+\delta(1-\alpha)]\right]}{[\rho+\delta(1-\alpha)]\left(\frac{1}{\epsilon_{ll}} - \frac{1}{\epsilon_{cl}}+\alpha\right)+(1-\alpha)(\delta+\rho)\left(\frac{1}{\epsilon_{cc}} - \frac{1}{\epsilon_{lc}}\right)}$$

$$\frac{\partial l}{\partial p}\frac{p^*}{l^*} = -\frac{\alpha\delta\left(\frac{1}{\epsilon_{cc}} - \frac{1}{\epsilon_{lc}}\right)}{\Theta\left[[\rho+\delta(1-\alpha)]\left(\frac{1}{\epsilon_{ll}} - \frac{1}{\epsilon_{cl}}+\alpha\right)+(1-\alpha)(\delta+\rho)\left(\frac{1}{\epsilon_{cc}} - \frac{1}{\epsilon_{lc}}\right)\right]}$$

(4.54)

Recall now that $r = \alpha(k/l)^{\alpha-1}$ and $Y_I = p^{-(1+\Theta)/\Theta}$. Using again the steady state relationships $Y_I = \delta k$ and $(\delta+\rho)p = r$, we derive

$$\frac{dY_I}{dp}\frac{p^*}{Y_I^*} = -\frac{1+\Theta}{\Theta}, \quad \frac{dr}{dk}\frac{Y_I^*}{p^*} = -\delta(1-\alpha)(\delta+\rho)\left(1 - \frac{dl}{dk}\frac{k^*}{l^*}\right), \quad \frac{dr}{dp} = (1-\alpha)(\delta+\rho)\frac{dl}{dp}\frac{p^*}{l^*}$$

(4.55)

Linearizing the dynamical system (4.26) around the steady state leads to the following Jacobian matrix

$$J = \begin{pmatrix} -\delta & -\frac{1+\Theta}{\Theta}\frac{Y_I^*}{p^*} \\ -\frac{\frac{\partial r}{\partial k}+\delta\left[\frac{1}{\epsilon_{cc}}\frac{\partial c}{\partial k}\frac{p^*}{c^*} - \frac{1}{\epsilon_{lc}}\frac{\partial l}{\partial k}\frac{p^*}{l^*}\right]}{E(k^*,p^*)} & \frac{\delta+\rho - \frac{\partial r}{\partial p} - \frac{1+\Theta}{\Theta}\frac{Y_I^*}{p^*}\left[\frac{1}{\epsilon_{cc}}\frac{\partial c}{\partial c}\frac{p^*}{c^*} - \frac{1}{\epsilon_{lc}}\frac{\partial l}{\partial k}\frac{p^*}{l^*}\right]}{E(k^*,p^*)} \end{pmatrix}$$

with $E(k, p)$ as given by (4.27). The associated characteristic polynomial is then given by (4.32) with the Determinant and Trace of the Jacobian matrix as defined by (4.33). Using (4.31), (4.54) and (4.55) we finally derive after straightforward simplifications

$$\mathcal{D}(\gamma) = \frac{\delta(\delta+\rho)(1+\chi+\gamma C)[\rho+\delta(1-\alpha)]\left[\frac{(1-\alpha)(\gamma+\chi)}{1+\chi} - \alpha\Theta\right]}{\left[\alpha+\chi+\gamma C\left(2 - \frac{C(1-\gamma)}{1+\chi}\right)\right]\left[\Theta[\rho+\delta(1-\alpha)] - \frac{\sigma(1+\chi)\alpha\delta}{1+\chi-C(1-\gamma)}\right] + \frac{\gamma C\alpha\delta}{1+\chi}[\alpha+\chi+\gamma(1-\alpha)]}$$

$$\mathcal{T}(\gamma) = \frac{\left[\alpha+\chi+\gamma C\left(2 - \frac{C(1-\gamma)}{1+\chi}\right)\right]\left[\rho\Theta[\rho+\delta(1-\alpha)] - \frac{\sigma(1+\chi)\alpha\delta}{1+\chi-C(1-\gamma)}[\rho+\Theta(\delta+\rho)]\right] + \gamma C\alpha\delta\left[\frac{\rho[\alpha+\chi+\gamma(1-\alpha)]}{1+\chi} + \alpha\delta\Theta\right]}{\left[\alpha+\chi+\gamma C\left(2 - \frac{C(1-\gamma)}{1+\chi}\right)\right]\left[\Theta[\rho+\delta(1-\alpha)] - \frac{\sigma(1+\chi)\alpha\delta}{1+\chi-C(1-\gamma)}\right] + \frac{\gamma C\alpha\delta}{1+\chi}[\alpha+\chi+\gamma(1-\alpha)]}$$

(4.56)

Note that if $\Theta = 0$ we conclude under the concavity condition $\sigma \geq \sigma_c(\gamma)$ that $\mathcal{D} < 0$, and the steady state is always saddle-point stable, i.e. locally determinate.

Assume first that

$$\frac{(1-\alpha)(\gamma+\chi)}{1+\chi} - \alpha\Theta > 0 \quad \text{or equivalently} \quad \chi > \frac{\alpha\Theta - \gamma(1-\alpha)}{1-\alpha-\alpha\Theta} \equiv \underline{\chi}(\gamma) \quad (4.57)$$

To keep reasonable values for the external effect we assume from here that $\Theta < \bar{\Theta} \equiv (1-\alpha)/\alpha$ and thus $\underline{\chi}(\gamma) > 0$. Then $\mathcal{D} > 0$ if and only if its denominator is positive, namely if and only if

$$\sigma < \sigma^{sup}(\gamma) \equiv \frac{[1+\chi - C(1-\gamma)]\left\{\left[\alpha+\chi+\gamma C\left(2-\frac{C(1-\gamma)}{1+\chi}\right)\right]\Theta[\rho+\delta(1-\alpha)] + \frac{\gamma C\alpha\delta}{1+\chi}[\alpha+\chi+\gamma(1-\alpha)]\right\}}{(1+\chi)\alpha\delta\left[\alpha+\chi+\gamma C\left(2-\frac{C(1-\gamma)}{1+\chi}\right)\right]}$$

(4.58)

But then local indeterminacy arises if and only if $\mathcal{T}(\gamma) < 0$, namely if and only if its numerator is negative, i.e.

$$\sigma > \sigma^{H}(\gamma) \equiv \frac{[1+\chi - C(1-\gamma)]\left\{\left[\alpha+\chi+\gamma C\left(2-\frac{C(1-\gamma)}{1+\chi}\right)\right]\rho\Theta[\rho+\delta(1-\alpha)] + \gamma C\alpha\delta\left[\frac{\rho[\alpha+\chi+\gamma(1-\alpha)]}{1+\chi}+\alpha\delta\Theta\right]\right\}}{(1+\chi)\alpha\delta[\rho+\Theta(\delta+\rho)]\left[\alpha+\chi+\gamma C\left(2-\frac{C(1-\gamma)}{1+\chi}\right)\right]}$$

(4.59)

Obvious computations show that $\sigma^{sup}(\gamma) > \sigma^{H}(\gamma)$ for any $\gamma \in [0, 1]$. We need however to check that $\sigma^{sup}(\gamma) > \sigma_c(\gamma)$ in order to be able to have a compatibility between the concavity property of the utility function at the steady state $\sigma \geq \sigma_c(\gamma)$ and the condition for local indeterminacy $\sigma < \sigma^{sup}(\gamma)$. Tedious but straightforward computations yield $\sigma^{sup}(\gamma) > \sigma_c(\gamma)$ if and only if

$$\Theta > \tilde{\Theta}(\gamma) \equiv \frac{\gamma^2 C\alpha^2\delta\left[1 - \frac{(1-\gamma)C}{1+\chi}\left(2-\frac{C(1-\gamma)}{1+\chi}\right)\right]}{\left[\chi+\gamma C\left(2-\frac{C(1-\gamma)}{1+\chi}\right)\right]\left[\alpha+\chi+\gamma C\left(2-\frac{C(1-\gamma)}{1+\chi}\right)\right][\rho+\delta(1-\alpha)]}$$

(4.60)

Under Assumption 1 we have $\tilde{\Theta}'(\gamma) > 0$ and $\tilde{\Theta}(\gamma) < \bar{\Theta}$ for any $\gamma \in [0, 1]$.

Denoting $\sigma^{inf}(\gamma) = \max\{\sigma^{H}(\gamma), \sigma_c(\gamma)\}$, we have proved that under condition (4.57), for any given $\gamma \in [0, 1]$, local indeterminacy occurs if and only if $\Theta \in (\tilde{\Theta}(\gamma), \bar{\Theta})$ and $\sigma \in (\sigma^{inf}(\gamma), \sigma^{sup}(\gamma))$. Obviously, recalling that Lemma 1 shows that the JR utility function is locally concave if and only if $\sigma \geq \sigma_c(\gamma)$, we derive from (4.58) and (4.60) that $\mathcal{D}$ is negative and the steady state (k^*, p^*) is saddle-point stable in two cases: (i) when $\Theta \in (\tilde{\Theta}(\gamma), \bar{\Theta})$ and $\sigma > \sigma^{sup}(\gamma)$, or (ii) when $\Theta < \tilde{\Theta}(\gamma)$ which implies $\sigma (\geq \sigma_c(\gamma)) > \sigma^{sup}(\gamma)$.

Let us consider now the case in which

$$\frac{(1-\alpha)(\gamma+\chi)}{1+\chi} - \alpha\Theta < 0 \text{ or equivalently } \chi < \frac{\alpha\Theta - \gamma(1-\alpha)}{1-\alpha-\alpha\Theta} \equiv \underline{\chi}(\gamma)$$

(4.61)

We need to assume here that $\Theta > \gamma(1 - \alpha)/\alpha$ and thus that $\gamma < 1$ to get a compatibility with the assumption $\Theta < \bar{\Theta}$. Following the same argument as previously, we conclude now that local indeterminacy arises if $\sigma > \sigma^{sup}(\gamma)$ and $\sigma < \sigma^{H}(\gamma)$. But such a configuration is not possible as $\sigma^{sup}(\gamma) > \sigma^{H}(\gamma)$ for any $\gamma \in [0, 1]$. It follows that under condition (4.61), the steady state (k^*, p^*) is saddle-point stable when $\sigma < \sigma^{sup}(\gamma)$, totally unstable when $\sigma > \sigma^{sup}(\gamma)$ and is ruled out.

We conclude therefore that for any given $\gamma \in [0, 1]$, local indeterminacy arises if and only if $\chi > \underline{\chi}(\gamma)$, $\Theta \in (\tilde{\Theta}(\gamma), \bar{\Theta})$ and $\sigma \in (\sigma^{inf}(\gamma), \sigma^{sup}(\gamma))$. $\qquad\square$

4.8.4   Proof of Corollary 1

Taking into account the concavity condition as given in Lemma 1, the existence of a Hopf bifurcation requires the bound $\sigma^{H}(\gamma)$ as given in (4.34) to be larger than $\sigma_c(\gamma)$. We then get $\sigma^{H}(\gamma) > \sigma_c(\gamma)$ if and only if

$$\Theta g(\rho, \gamma, \chi) > \gamma^2 C \alpha^2 \delta \rho \left[1 - \frac{(1-\gamma)C}{1+\chi} \left(2 - \frac{C(1-\gamma)}{1+\chi} \right) \right]$$

with

$$g(\rho, \gamma, \chi) = \left[\alpha + \chi + \gamma C \left(2 - \frac{C(1-\gamma)}{1+\chi} \right) \right] \left[\rho[\rho + \delta(1-\alpha)] \left[\chi + \gamma C \left(2 - \frac{C(1-\gamma)}{1+\chi} \right) \right] \right.$$

$$\left. - \frac{\gamma C \alpha \delta (\gamma + \chi)(\delta + \rho)}{1+\chi} \right] + \gamma C (\alpha \delta)^2 \left[\chi + \gamma C \left(2 - \frac{C(1-\gamma)}{1+\chi} \right) \right]$$

Under Assumption 1 we have $g(\rho, \gamma, \chi) > 0$ for any $\gamma \in [0, 1]$. It follows that $\sigma^H(\gamma) > \sigma_c(\gamma)$ if and only if

$$\Theta > \hat{\Theta}(\gamma) \equiv \frac{\gamma^2 C \alpha^2 \delta \rho \left[1 - \frac{(1-\gamma)C}{1+\chi} \left(2 - \frac{C(1-\gamma)}{1+\chi} \right) \right]}{g(\rho, \gamma, \chi)}$$

Assumption 1 also implies $\hat{\Theta}'(\gamma) > 0$ and $\hat{\Theta}(\gamma) < \bar{\Theta}$ for any $\gamma \in [0, 1]$. The result follows from Proposition 2 considering $\underline{\Theta}(\gamma) = \max\{\hat{\Theta}(\gamma), \tilde{\Theta}(\gamma)\}$.   □

4.8.5   Proof of Lemma 2

The maximal value of $\underline{\chi}(\gamma)$ is $\underline{\chi}(0) = \alpha \Theta / (1 - \alpha - \alpha \Theta)$. We then assume $\chi > \underline{\chi}(0)$ in order to ensure $\chi > \underline{\chi}(\gamma)$ for any $\gamma \in [0, 1]$. Let us consider the following two critical values

$$\sigma^{sup}(0) \equiv \frac{\Theta\{\alpha \rho + \chi[\rho + \delta(1-\alpha)]\}}{(1+\chi)\alpha \delta}$$

$$\sigma^H(1) \equiv \frac{(\alpha + \chi + 2C)\rho\Theta[\rho + \delta(1-\alpha)] + C\alpha\delta(\rho + \alpha\delta\Theta)}{\alpha\delta[\rho + \Theta(\delta+\rho)](\alpha + \chi + 2C)} \tag{4.62}$$

We easily get

$$\lim_{\chi \to +\infty} \sigma^{sup}(0) = \frac{\Theta[\rho + \delta(1-\alpha)]}{\alpha\delta} > \lim_{\chi \to +\infty} \sigma^H(1) = \frac{\rho\Theta[\rho + \delta(1-\alpha)]}{\alpha\delta[\rho + \Theta(\delta+\rho)]} \tag{4.63}$$

Similarly, we have

$$\sigma^{sup}(0)|_{\chi = \underline{\chi}(0)} = \frac{\Theta[\rho + \Theta(\delta + \rho)]}{\delta}$$

$$\sigma^H(1)|_{\chi = \underline{\chi}(0)} \equiv \frac{\rho\Theta[\rho + \delta(1-\alpha)]}{\alpha\delta[\rho + \Theta(\delta+\rho)]} + \frac{C(1 - \alpha - \alpha\Theta)(\rho + \alpha\delta\Theta)}{[\rho + \Theta(\delta+\rho)][\alpha(1-\alpha)(1+\Theta) + 2C(1 - \alpha - \alpha\Theta)]}$$

It follows obviously that

$$\lim_{\Theta \to 0} \sigma^{sup}(0)|_{\chi = \underline{\chi}(0)} = 0 < \lim_{\Theta \to 0} \sigma^H(1)|_{\chi = \underline{\chi}(0)} \equiv \frac{C}{\alpha + 2C}$$

while

$$\lim_{\Theta \to \bar{\Theta}} \sigma^{sup}(0)|_{\chi=\underline{\chi}(0)} = \frac{(1-\alpha)[\rho+\delta(1-\alpha)]}{\alpha^2\delta} > \lim_{\Theta \to \bar{\Theta}} \sigma^{H}(1)|_{\chi=\underline{\chi}(0)} \equiv \frac{(1-\alpha)\rho}{\alpha\delta}$$

Therefore, there exists $\bar{\bar{\Theta}} \in (0, \bar{\Theta})$ such that if $\Theta \in (0, \bar{\bar{\Theta}})$, then $\sigma^{sup}(0)|_{\chi=\underline{\chi}(0)} < \sigma^{H}(1)|_{\chi=\underline{\chi}(0)}$. Based on this result and using (4.63), we conclude that there also exists $\bar{\chi} \in (\underline{\chi}(0), +\infty)$ such that when $\Theta \in (0, \bar{\bar{\Theta}})$, $\sigma^{sup}(0) - \sigma^{H}(1) \lessgtr 0$ if and only if $\chi \lessgtr \bar{\chi}$.   □

References

Basu, S., & Fernald, J. (1997). Returns to scale in US production: Estimates and implications. *Journal of Political Economy, 105*, 249–283.
Benhabib, J., & Farmer, R. (1994). Indeterminacy and increasing returns. *Journal of Economic Theory, 63*, 19–41.
Benhabib, J., & Farmer, R. (1996). Indeterminacy and sector specific externalities. *Journal of Monetary Economics, 37*, 397–419.
Campbell, J. (1999). Asset prices, consumption and the business cycle. In J. B. Taylor & M. Wood-ford (Eds.), *Handbook of macroeconomics* (pp. 1231–1303). Amsterdam: North-Holland.
Chetty, R., Guren, A., Manoli, D., & Weber, A. (2012). Does indivisible labor explain the difference between micro and macro elasticities? A meta-analysis of extensive margin elasticities. *NBER Working Paper No. 16729.*
Dufourt, F., Nishimura, K., & Venditti, A. (2015). Indeterminacy and sunspots in two-sector RBC models with generalized no-income-effect preferences. *Journal of Economic Theory, 157*, 1056–1080.
Grandmont, J. -M. (2008). Nonlinear difference equations, bifurcations and chaos: An introduction. *Research in Economics, 62*, 122–177.
Greenwood, J., Hercovitz, Z., & Huffman, G. (1988). Investment, capacity utilization and the real business cycle. *American Economic Review, 78*, 402–417.
Guo, J. T., & Harrison, S. (2010). Indeterminacy with no-income effect preferences and sector-specific externalities. *Journal of Economic Theory, 145*, 287–300.
Hintermaier, T. (2001). Lower bounds on externalities in sunspot models. *Working Paper EUI.*
Hintermaier, T. (2003). On the minimum degree of returns to scale in sunspot models of business cycles. *Journal of Economic Theory, 110*, 400–409.
Jaimovich, N. (2008). Income effects and indeterminacy in a calibrated one-sector growth model. *Journal of Economic Theory, 143*, 610–623.
Jaimovich, N., & Rebelo, S. (2009). Can news about the future drive the business cycles? *American Economic Review, 99*, 1097–1118.
Khan, H., & Tsoukalas, J. (2011). Investment shocks and the comovement problem. *Journal of Economic Dynamics and Control, 35*, 115–130.
King, R., Plosser, C., & Rebelo, S. (1988). Production, growth and business cycles. *Journal of Monetary Economics, 21*, 191–232.
Kocherlakota, N. (1996). The equity premium: It's still a puzzle. *Journal of Economic Literature, 36*, 42–71.
Mulligan, C. (2002). Capital interest and aggregate intertemporal substitution. *NBER Working Paper 9373.*
Nishimura, K., & Venditti, A. (2010). Indeterminacy and expectation-driven fluctuations with non-separable preferences. *Mathematical Social Sciences, 60*, 46–56.

Nishimura, K., Nourry, C., & Venditti, A. (2009). Indeterminacy in aggregate models with small externalities: An interplay between preferences and technology. *Journal of Nonlinear and Convex Analysis, 10*(2), 279–298.

Prescott, E., & Wallenius, J. (2011). *Aggregate labor supply* (p. 457). Research Department Staff Report: Federal Reserve Bank of Minneapolis.

Rogerson, R., & Wallenius, J. (2009). Micro and macro elasticities in a life cycle model with taxes. *Journal of Economic Theory, 144*, 2277–2292.

Schmitt-Grohé, S., & Uribe, M. (2012). What's news in business cycles. *Econometrica, 80*, 2733–2764.

Vissing-Jorgensen, A., & Attanasio, O. (2003). Stock-market participation, intertemporal substitution and risk aversion. *American Economic Review Papers and Proceedings, 93*, 383–391.

Chapter 5
From Self-Fulfilling Mistakes to Behavioral Learning Equilibria

Cars Hommes

Abstract This essay links some of my own work on expectations, learning and bounded rationality to the inspiring ideas of Jean-Michel Grandmont. In particular, my work on consistent expectations and behavioral learning equilibria may be seen as formalizations of JMG's ideas of self-fulfilling mistakes. Some of our learning-to-forecast laboratory experiments with human subjects have also been strongly influenced by JMG's ideas. Key features of self-fulfilling mistakes are multiple equilibria, excess volatility and persistence amplification.

Keywords Expectations · Learning · Chaos · Almost self-fulfilling equilibria · Laboratory experiments

JEL Classification D84 · D83 · E32 · C92

5.1 Introduction

The ideas of Jean-Michel Grandmont have inspired the work of many young scholars in economics. During my own Ph.D. thesis work on chaos in economic models (Hommes 1991), I have for example been studying his seminal contribution on chaos in overlapping generations models (Grandmont 1985). For many years thereafter, another seminal contribution Grandmont (1998)[1] on expectations formation and stability in large socio-economic systems has provided inspiration for my work on expectations, learning and bounded rationality in the last two decades (see e.g., Hommes 2013a, b).

---

[1] An essential part of this work was already presented at JMG's Presidential address at the World Meetings of the Econometric Society, Barcelona, 1990.

C. Hommes (✉)
CeNDEF, University of Amsterdam and Tinbergen Institute, Amsterdam, Netherlands
e-mail: C.H.Hommes@uva.nl

© Springer International Publishing AG 2017
K. Nishimura et al. (eds.), *Sunspots and Non-Linear Dynamics*,
Studies in Economic Theory 31, DOI 10.1007/978-3-319-44076-7_5

Let me start off by quoting JMG at length (Grandmont 1998, pp. 776–777):

> Complex "learning equilibria" may be at first sight good candidates to explain why agents keep making significant and recurrent mistakes when trying to predict the fate of socioeconomic systems in which they participate. To be acceptable, however, the observed patterns along such "learning equilibria" should display some reasonable degree of consistency with the agents' beliefs. One might envision situations in which agents do believe (wrongly) that the world is relatively simple (e.g. linear) but subject to random shocks, and in which the corresponding (deterministic) "learning equilibria" are complex enough to make the agents' forecasting mistakes "self-fulfilling" in a well defined sense. For instance, the agents might be assumed to have at their disposal a reasonably wide, but nevertheless limited, battery of statistical tests ("bounded rationality") which would not allow them to reject the hypothesis that their recurrent forecasting mistakes are attributable to random disturbances ... It is not quite clear to me at this stage whether such a program can actually generate operational results or is even feasible (for a first step, see Sorger (1998), Hommes and Sorger (1998). Yet progress on this front, if possible, might provide an interesting alternative to our current paradigms, which rely very heavily on extreme, and often criticized, rationality axioms.

This essay summarizes some of my work emphasizing how it has been following these ideas. Section 5.2 starts off from the concept of a consistent expectations equilibrium (CEE), as introduced in Hommes and Sorger (1998), which may be seen as a formalization of Grandmont's idea of a *self-fulfilling mistake*. Along a self-fulfilling mistake agents incorrectly believe that the economy follows a stochastic process, whereas the actual dynamics is generated by a deterministic chaotic process which is indistinguishable from the former (stochastic) process by linear statistical tests. The concept of CEE was motivated by the fact that piecewise linear asymmetric tent maps generate deterministic chaotic time series with exactly the same autocorrelations structure as a stochastic AR(1) process. Along a (chaotic) CEE agents use a simple linear, AR(1) forecasting rule and, given this belief, the economy follows a nonlinear chaotic asymmetric tent map dynamics with the same autocorrelation structure. Hommes and Sorger (1998) showed the existence of chaotic CEE in the cobweb "hog cycle" model with a backward bending supply curve. They also studied the stability of CEE under learning, introducing *sample autocorrelation* (SAC-)*learning*, where agents learn the two parameters of the AR(1) forecasting rule by the observed sample average and (first-order) sample autocorrelation coefficient.

Section 5.3 discusses an application of CEE in Hommes and Rosser (2001), in a fishery model with backward bending supply. They simulated stochastic nonlinear models where agents *learn to believe in chaos*, that is, the system converges to a noisy chaotic system, with SAC-learning parameters converging to sample average and sample autocorrelations. This situation qualifies as a *self-fulfilling mistake*: agents can not reject the hypothesis that the economy follows and AR(1) process, while the true law of motion of the economy follows a noisy chaotic process. Section 5.4 discusses more recent work of Hommes et al. (2013) on *stochastic consistent expectations equilibria* (SCEE), generalizing the notion of CEE to a nonlinear stochastic framework. A SCEE is a self-fulfilling mistake where agents learn the correct AR(1) rule, in terms of sample average and sample autocorrelations, in a nonlinear stochastic environment.

A CEE may be viewed as an early example of a *Restricted Perceptions Equilibrium* (RPE), as in Evans and Honkapohja (2001), based on the idea that agents have misspecified beliefs, but within the context of their forecasting model they are unable to detect their misspecification.[2]

In Sect. 5.5 we discuss recent work of Hommes and Zhu (2014), who apply the idea of SCEE in a stochastic *linear* modeling framework. The idea here is that agents use a simple (misspecified) univariate AR(1) forecasting rule in a higher dimensional linear framework. A *behavioral learning equilibrium* (BLE) or, more precisely, a first-order stochastic consistent expectations equilibrium (SCEE), arises when the sample average and the first-order autocorrelations of the AR(1) rule coincide with observed realizations. Hence, along a BLE the parameters of the AR(1) rule are not free, but pinned down by two simple observable statistics, the sample average and the first-order sample autocorrelation. Such a simple, parsimonious learning equilibrium may be a more plausible outcome of the coordination process of individual expectations in large complex socio-economic systems. An interesting feature of BLE is that *multiple equilibria* may arise in very simple settings. Section 5.6 discusses laboratory experiments on expectations, stressing the empirical relevance of coordination on almost self-fulfilling equilibria in positive feedback systems (Heemeijer et al. 2009) and recent experiments of Arifovic et al. (2016) in a complex overlapping generations framework a la Grandmont (1985). The final section concludes.

5.2 Consistent Expectations Equilibrium

Consider an expectations feedback system of the form

$$p_t = F(p_t^e), \tag{5.1}$$

where p_t is the state (or price) of the economy, p_t^e the forecast of the price in period t and F the *actual law of motion* of the economy. In general, the map F may be complex and nonlinear. A well known example of (5.1) is the classical cobweb "hog cycle" model, where $F = D^{-1}S$ is the composition of inverse demand and supply curves.

Throughout this paper, we assume that agents are boundedly rational and do *not* know the law of motion F of the economy. Rather agents form a belief about the price generating process. Assume that all agents believe that prices are generated by a stochastic AR(1) process, that is, their perceived law of motion (PLM) is given by

$$p_t = \alpha + \beta(p_{t-1} - \alpha) + \delta_t, \tag{5.2}$$

---

[2]In his survey Branch (2006) argues that the RPE is a natural alternative to rational expectation equilibrium (REE) because it is to some extent consistent with Muth's original hypothesis of REE, while allowing for bounded rationality by restricting the class of the perceived law of motion.

where α and $\beta \in [-1, 1]$ represent the long run mean and the first-order autocorrelations coefficient of the PLM, and δ_t is an IID noise term. Given the PLM (5.2) and prices known up to p_{t-1}, the optimal forecast, that is, the prediction for p_t minimizing the mean squared prediction error, is[3]

$$p_t^e = \alpha + \beta(p_{t-1} - \alpha). \tag{5.3}$$

Given that agents use the linear forecast (5.3), the *implied actual law of motion* becomes

$$p_t = F_{\alpha,\beta}(p_{t-1}) := D^{-1}S(\alpha + \beta(p_{t-1} - \alpha)). \tag{5.4}$$

The (observable) *sample average* of a time series $(p_t)_{t=0}^{\infty}$ is

$$\bar{p} = \lim_{T \to \infty} \frac{1}{T+1} \sum_{t=0}^{T} p_t \tag{5.5}$$

and the (observable) *sample autocorrelation* coefficients are given by

$$\rho_j = \lim_{T \to \infty} \frac{c_{j,T}}{c_{0,T}}, \qquad j \geq 1, \tag{5.6}$$

where

$$c_{j,T} = \frac{1}{T+1} \sum_{t=0}^{T-j} (p_t - \bar{p})(p_{t+j} - \bar{p}), \qquad j \geq 0. \tag{5.7}$$

A consistent expectations equilibrium (CEE) is defined as (Hommes and Sorger 1998).[4]

Definition A triple $\{(p_t)_{t=0}^{\infty}; \alpha, \beta\}$, where $(p_t)_{t=0}^{\infty}$ is a sequence of prices and α and β are real numbers, $\beta \in [-1, 1]$, is called a *consistent expectations equilibrium* (CEE) if

1. the sequence $(p_t)_{t=0}^{\infty}$ satisfies the implied actual law of motion (5.4) and is bounded,
2. the sample average $\bar{p}$ in (5.5) exists and is equal to α, and
3. the sample autocorrelation coefficients ρ_j, $j \geq 1$, in (5.6) exist and one of the following is true:

---

[3]More generally, for a (nonlinear) stochastic process Y the optimal forecast conditional on X minimizing the mean squared error is the conditional expectation $E(Y|X)$; see e.g. Hamilton (1994) for a discussion and a proof.

[4]Extensions and applications of CEE include Sögner and Mitlöhner (2002), Tuinstra (2003), Branch and McGough (2005), Lansing (2009, 2010), Bullard et al. (2008, 2010).

    a. if $(p_t)_{t=0}^\infty$ is a convergent sequence, then $\text{sgn}(\rho_j) = \text{sgn}(\beta^j)$, $j \geq 1$;
    b. if $(p_t)_{t=0}^\infty$ is not convergent, then $\rho_j = \beta^j$, $j \geq 1$.

A CEE is a price sequence together with AR(1) belief parameters α and β, such that expectations are *self-fulfilling* in terms of the *observable* sample average and sample autocorrelations. The two parameters α and β of the AR(1) forecasting rule are not free, but pinned down by simple observable statistics. Along a CEE expectations are thus correct in a linear statistical sense. Hommes and Sorger (1998) showed that, given an AR(1) belief, there are (at least) three different types of CEE:

- a *steady state CEE* in which the price sequence $(p_t)_{t=0}^\infty$ converges to a steady state p^*, with $\alpha = p^*$ and $\beta = 0$;
- a *2-cycle CEE* in which the price sequence $(p_t)_{t=0}^\infty$ converges to a period two cycle $\{p_1^*, p_2^*\}$, $p_1^* \neq p_2^*$, with $\alpha = (p_1^* + p_2^*)/2$ and $\beta = -1$;
- a *chaotic CEE* in which the price sequence $(p_t)_{t=0}^\infty$ is chaotic, with sample average α and autocorrelations β^j.

A steady state CEE is a REE (at least in the long run) corresponding to some fixed point where demand D and supply S intersect. A 2-cycle CEE also is a REE, where the price jumps back and forth between two different intersection points of the demand and supply curves. A chaotic CEE is a non-rational equilibrium, where agents believe in a linear stochastic law of motion, while the true law of motion is nonlinear (e.g., a piecewise linear tent map) and chaotic. Which of these cases occurs in the cobweb model depends on the implied actual law of motion, i.e. upon the composite mapping $D^{-1}S$ in (5.4), determined by demand and supply curves. In general, different types of CEE may co-exist as will be discussed below.

5.2.1 Sample Autocorrelation Learning

The notion of CEE involves an AR(1) belief with fixed parameters α and β, which have been pinned down by two simple statistics, the sample average and the (first order) sample autocorrelation. But how would agents learn these parameters? Assume that agents use *adaptive learning* to update their belief parameters α_t and β_t, as additional observations become available. There is a large literature on adaptive learning in macroeconomics, see e.g. Sargent (1993) and Evans and Honkapohja (2001) for extensive discussion and overviews. Many adaptive learning algorithms use standard econometrics/statistical tools such as (recursive) ordinary least squares.

    Hommes and Sorger (1998) proposed another natural and simple learning scheme called *sample autocorrelation learning* (SAC-learning), with parameters based upon sample average and first order sample autocorrelation coefficient[5]:

---

[5] Although not identical, SAC-learning is closely related to the ordinary least squares (OLS-)learning scheme; see the discussion in Hommes and Sorger (1998). A convenient feature of the SAC estimate β_t in (5.9) is that it always lies in the interval $[-1, 1]$, reflecting the fact that the firs order autocorrelation coefficient is not explosive, while the OLS-estimate may be outside this interval

$$\alpha_t = \frac{1}{t+1} \sum_{i=0}^{t} p_i , \qquad t \geq 1, \tag{5.8}$$

$$\beta_t = \frac{\sum_{i=0}^{t-1}(p_i - \alpha_t)(p_{i+1} - \alpha_t)}{\sum_{i=0}^{t}(p_i - \alpha_t)^2}, \qquad t \geq 1. \tag{5.9}$$

When, in each period, the belief parameters are updated according to (5.8) and (5.9) the (temporary) law of motion (5.4) becomes

$$p_{t+1} = F_{\alpha_t,\beta_t}(p_t) = D^{-1}S(\alpha_t + \beta_t(p_t - \alpha_t)), \qquad t \geq 0. \tag{5.10}$$

One can also rewrite SAC-learning in recursive form. Define

$$R_t = \frac{1}{t+1} \sum_{i=0}^{t} (p_i - \alpha_t)^2,$$

then the SAC-learning is equivalent to the following recursive dynamical system (Hommes and Sorger 1998).

$$\begin{cases} \alpha_t = \alpha_{t-1} + \dfrac{1}{t+1}(p_t - \alpha_{t-1}), \\ \beta_t = \beta_{t-1} + \dfrac{1}{t+1}R_t^{-1}\Big[(p_t - \alpha_{t-1})\Big(p_{t-1} + \dfrac{p_0}{t+1} - \dfrac{t^2+3t+1}{(t+1)^2}\alpha_{t-1} - \dfrac{1}{(t+1)^2}p_t\Big) \\ \quad - \dfrac{t}{t+1}\beta_{t-1}(p_t - \alpha_{t-1})^2\Big], \\ R_t = R_{t-1} + \dfrac{1}{t+1}\Big[\dfrac{t}{t+1}(p_t - \alpha_{t-1})^2 - R_{t-1}\Big]. \end{cases}$$
$$\tag{5.11}$$

An important feature of CEE and SAC-learning is that both have a simple, intuitive *behavioral interpretation*. In a CEE agents use a linear forecasting rule with two parameters, the mean α and the first-order autocorrelation β. Both can be observed from past observations by inferring the average price level and the (first-order) persistence of the time series. For example, $\beta = 0.5$ means that, on average, prices mean revert toward their long-run mean by 50 %. The linear univariate AR(1) rule and the SAC-learning process are examples of simple forecasting heuristics that can be used without any knowledge of statistical techniques, simply by observing a time series and roughly "guestimating" its sample average and its first-order persistence.[6]

Which type of CEE exist in the nonlinear cobweb model and to which of them will the SAC-learning dynamics converge? Hommes and Sorger (1998) show that in the simplest case, when demand is decreasing and supply is increasing, the *only* CEE

---

[6]In learning-to-forecast laboratory experiments for many subjects forecasting behavior is well described by simple rules, such as a simple AR(1) rule; see Sect. 5.6.

is the REE steady state price p^*. This means that, even when the underlying market equilibrium equations are *not* known, agents will be able to learn and coordinate on the REE price if they learn the correct sample average and sample autocorrelations. Hence, in a nonlinear cobweb economy with monotonic demand and supply, boundedly rational agents should, at least in theory, be able to learn the unique REE from time series observations.[7]

Hommes and Sorger (1998) study CEE in a cobweb model with linear demand and a non-monotonic, piecewise linear backward bending supply curve. They present examples of 2-cycle and chaotic CEE, where, given an AR(1) perceived law of motion, the implied actual law of motion is a chaotic piecewise linear tentmap. Different types of CEE, steady state, 2-cycle and chaotic, may co-exist and the SAC-learning dynamics exhibits *path-dependence*, with the long run CEE depending upon initial states. In the long run however, the SAC-learning always settles down to one of the CEE, where agents have learned the correct sample average and sample autocorrelation.

5.3 Learning to Believe in (Noisy) Chaos

Hommes and Rosser (2001) consider another example of a cobweb model with a backward bending supply curve having its origin in a fishery model. This example serves to illustrate how agents may *"learn to believe in chaos"* in a stylized nonlinear, stochastic environment. That is, in an unknown nonlinear environment agents learn the parameters of a simple, linear AR(1) forecasting rule, while the law of motion of the economy is nonlinear and chaotic. In the long run, the sample mean and first order autocorrelation coefficient of the AR(1) rule converge to the observed sample means and first order autocorrelation of the unknown nonlinear chaotic process. Under SAC-learning, the two parameters of the AR(1) rule thus converge, while the implied actual law of motion of the economy converges to a chaotic map. Moreover, agents can not reject the null hypothesis of a stochastic AR(1) process through statistical hypothesis testing and therefore in a linear statistical sense beliefs and realizations coincide.[8]

---

[7]For the cobweb model, Bray and Savin (1986) show that OLS learning also converges to the REE steady state. In the laboratory experiments of Hommes et al. (2007) however, prices do not always converge to the REE steady state but exhibit excess volatility when the cobweb model is strongly unstable.

[8]The notion *learning to believe in chaos* has been introduced in Hommes (1998, p. 360), and the first examples have been given by Sorger (1998) and Hommes and Sorger (1998). For related work on the instability of OLS learning, see e.g. Bullard (1994) and Grandmont (1998). Schönhofer (1999) has used the notion of learning to believe in chaos in a somewhat different context, namely when the entire OLS-learning process fluctuates chaotically. In Schönhofers' examples belief parameters of the OLS-learning scheme do *not* converge, but keep fluctuating chaotically and at the same time, due to inflation, prices diverge to infinity, so that agents are in fact running an OLS-regression on a non-stationary time series. Tuinstra and Wagener (2007) consider the same model with heterogeneous expectations, with agents switching between different OLS-estimation methods.

SAC-learning of a chaotic CEE may be seen as an example of an *approximate rational expectations equilibrium* (Sargent 1999) or a Restricted Perceptions Equilibrium (RPE) (Evans and Honkapohja 2001; Branch 2006). Agents have misspecified beliefs, but within the context of their forecasting model they are unable to detect their misspecification, and they learn the optimal misspecified forecasts. We also would like to stress the *behavioural rationality* interpretation of CEE and SAC-learning, because the simple AR(1) rule is intuitively plausible and SAC-learning may be seen as a learning heuristic through guestimating the sample average and first order sample autocorrelation.

We generalize the law of motion to a nonlinear stochastic system. SAC-learning is given by (5.8), (5.9), as before, but we add a noise term to the implied actual law of motion, i.e.

$$p_{t+1} = F_{\alpha_t, \beta_t}(p_t) + \epsilon_t = D^{-1}S(\alpha_t + \beta_t(p_t - \alpha_t)) + \epsilon_t, \qquad t \geq 0, \qquad (5.12)$$

where ϵ_t is an independently identically distributed (IID) random process.

Figure 5.1 illustrates an example of *learning to believe in noisy chaos*. Under SAC-learning the belief parameters $\alpha_t \to \alpha^* \approx 5400$ and $\beta_t \to \beta^* \approx -0.87$ converge to constants, while the underlying law of motion F_{α^*, β^*} converges to a chaotic map. Prices then keep fluctuating chaotically with noise. Recall that our boundedly rational agents have no knowledge about underlying market equilibrium equations, and therefore do not know the implied actual law of motion. They only observe time series and update their forecasting parameters based upon simple statistics, the sample average and the first order sample autocorrelation coefficient. Would agents in the long run be satisfied with their linear forecasting rules and stick to their AR(1) belief?

Figure 5.1e shows that the forecasting errors under SAC-learning are uncorrelated. Agents therefore do not make systematic mistakes, or at least there is no linear structure in their forecasting errors. As a next step, one could do statistical hypothesis testing of the linear forecasting rule. Would boundedly rational agents be able to reject their stochastic AR(1) belief or perceived law of motion by linear statistical hypothesis testing? Hommes and Rosser (2001) show that in this example the null hypothesis that prices follow a stochastic AR(1) process can not be rejected at the 10 % level. Agents thus *learn to believe in noisy chaos*.

These equilibria are persistent with respect to dynamic noise. In fact, the presence of noise may increase the probability of convergence to such learning equilibria. Agents are using a simple, but misspecified model to forecast an unknown, possibly complicated actual law of motion. Without noise, boundedly rational agents using time series analysis might be able to detect the misspecification and improve their forecast model. In the presence of dynamic noise however, misspecification becomes harder to detect and boundedly rational agents using linear statistical techniques can do no better than stick to their optimal, simple linear model of the world. This clearly satisfies Grandmont's earlier quote of a self-fulfilling mistake (Grandmont 1998, pp. 776–777):

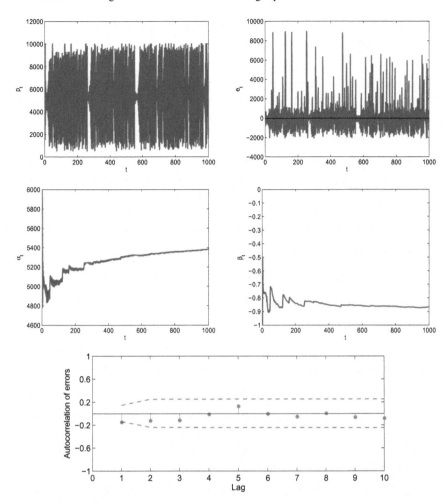

Fig. 5.1 Learning to believe in noisy chaos. In the presence of noise, SAC learning converges to a (noisy) chaotic CEE, with chaotic prices fluctuations (*top left*) and at the same time convergence of the belief parameters $\alpha_t \to \alpha^* \approx 5400$ (*mid left*) and $\beta_t \to \beta^* \approx -0.87$ (*mid right*). Forecasting errors (*top right*) are (noisy) chaotic and seemingly unpredictable. The ACF of forecasting errors (*bottom plot*) shows that errors are uncorrelated

One might envision situations in which agents do believe (wrongly) that the world is relatively simple (e.g. linear) but subject to random shocks, and in which the corresponding (deterministic) "learning equilibria" are complex enough to make the agents' forecasting mistakes "self-fulfilling" in a well defined sense. For instance, the agents might be assumed to have at their disposal a reasonably wide, but nevertheless limited, battery of statistical tests ("bounded rationality") which would not allow them to reject the hypothesis that their recurrent forecasting mistakes are attributable to random disturbances …

5.4 Stochastic Consistent Expectations Equilibrium (SCEE)

Hommes et al. (2013) have generalized the notion of consistent expectations equilibrium to a stochastic setting. Let the law of motion of an economic system be given by the stochastic system

$$x_t = f(x_{t+1}^e, \ u_t), \tag{5.13}$$

where x_t is the state of the system at date t, x_{t+1}^e is the expected value of x at date $t + 1$, $\{u_t\}$ is an IID noise process with mean zero and f is a continuous (nonlinear) function. Note that the timing is different and (5.13) has the form of a *temporary equilibrium* map, with the state x_t depending on the expected future state x_{t+1}^e. As before, agents are boundedly rational and do not know the exact form of the (nonlinear) law of motion (5.13), but rather agents' perceived law of motion is a stochastic AR(1) process. Given this perceived law of motion, the 2-period ahead forecast x_{t+1}^e that minimizes the mean-squared forecasting error is

$$x_{t+1}^e = \alpha + \beta^2(x_{t-1} - \alpha). \tag{5.14}$$

Here we use the convention of the learning literature that x_t in (5.13) is not yet observable when the forecast x_{t+1}^e is made. Combining the forecast (5.14) and the law of motion of the economy (5.13), we obtain the implied actual law of motion (ALM)

$$x_t = f(\alpha + \beta^2(x_{t-1} - \alpha), \ u_t). \tag{5.15}$$

Hommes et al. (2013) define a *first-order stochastic consistent expectations equilibrium* (SCEE) as follows.

Definition 5.4.1 A triple (μ, α, β), where μ is a probability measure and α and β are real numbers with $\beta \in (-1, 1)$, is called a first-order stochastic consistent expectations equilibrium (SCEE) if the following three conditions are satisfied:

S1 The probability measure μ is a nondegenerate invariant measure for the stochastic difference equation (5.15);
S2 The stationary stochastic process defined by (5.15) with the invariant measure μ has unconditional mean α, that is, $E_\mu(x) = \int x \, d\mu(x) = \alpha$;
S3 The stationary stochastic process defined by (5.15) with the invariant measure μ has unconditional first-order autocorrelation coefficient β.

A first-order SCEE is thus characterized by two consistency requirements: the unconditional mean and the unconditional first-order autocorrelation coefficient generated by the actual (unknown) stochastic process (5.15) coincide with the corresponding statistics of the perceived linear AR(1) process. Along a SCEE the two parameters α and β of the AR(1) forecasting rule are thus not free, but pinned down by two simple observable statistics. This means that along a first-order SCEE agents correctly perceive the mean and the first-order autocorrelation (i.e., the persistence) of the stochastic state of the economy, without fully understanding its (nonlinear) structure.

Under SAC-learning the actual law of motion becomes

$$x_t = f(\alpha_{t-1} + \beta_{t-1}^2(x_{t-1} - \alpha_{t-1}), \ u_t), \tag{5.16}$$

with time-varying parameters α_t, β_t as before in (5.8, 5.9).

Hommes et al. (2013) study SAC-learning of SCEE in the highly nonlinear, chaotic overlapping generations model of Grandmont (1985) of the form

$$p_t = g(p_{t+1}^e) + \epsilon_t, \tag{5.17}$$

where g is a non-monotonic map with infinitely many periodic and chaotic equilibria. An interesting finding is that SAC-learning always converges to a simple equilibrium, either a steady state or a 2-cycle, as illustrated in Fig. 5.4. In such a complex OLG-economy, SAC-learning of an AR(1) rule thus leads to learning-to-believe in a steady state or learning-to-believe in a two-cycle.

The nonlinear framework for SCEE is very general. A drawback of the nonlinear framework however is that computation of first-order autocorrelations is typically not analytical tractable. The next section presents a simpler *linear* framework for SCEE, where agents use a simple, but *misspecified* univariate AR(1) rule in a higher dimensional linear framework.

5.5 Behavioral Learning Equilibria

Hommes and Zhu (2014) apply the first order SCEE to a *linear* framework, in which the univariate AR(1) forecasting rule is *misspecified*. The simplest class of models arises when the actual law of motion of the economy is a one-dimensional linear stochastic process x_t, driven by an exogenous AR(1) process y_t. More precisely, the actual law of motion of the economy is given by

$$x_t = f(x_{t+1}^e, \ y_t, \ u_t) = b_0 + b_1 x_{t+1}^e + b_2 y_t + u_t, \tag{5.18}$$
$$y_t = a + \rho y_{t-1} + \varepsilon_t, \tag{5.19}$$

with parameters $b_0 > 0$, b_1 in the interval $(-1, 1)$,[9] $b_2 > 0$, $a > 0$ and $0 < \rho < 1$; u_t are IID shocks (Fig. 5.2).

---

[9]This assumption is made to ensure stationarity; for $|b_1| > 1$ the dynamics under learning easily becomes explosive.

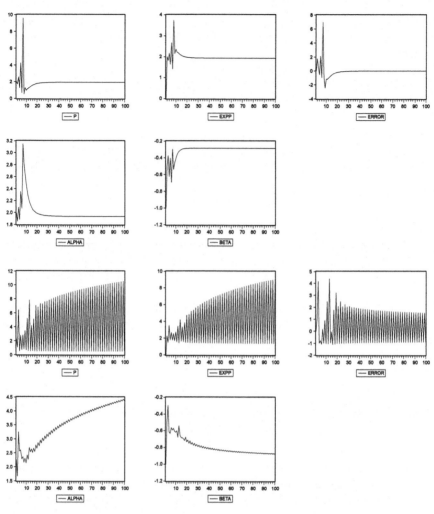

Fig. 5.2 Convergence of SAC-learning in the OLG-model of Grandmont (1985). The model has infinitely many periodic and chaotic equilibria, but SAC-learning always selects a simple equilibrium outcome, either a steady state (*top panels*) or a 2-cycle (*bottom panels*). The plots show the price (P), expected price (EXPP), forecast error and the time-varying parameters α_t and β_t

The rational expectations equilibrium x_t^* of (5.18, 5.19) is a linear function of the driving variable y_t, and is given by

$$ x_t^* = \frac{b_0}{1 - b_1} + \frac{ab_1b_2}{(1 - b_1\rho)(1 - b_1)} + \frac{b_2}{1 - b_1\rho}y_t + u_t. \qquad (5.20) $$

Its unconditional mean and first-order autocorrelation are given by Hommes and Zhu (2014):

$$\overline{x^*} := E(x_t^*) = \frac{b_0(1-\rho)+ab_2}{(1-b_1)(1-\rho)}, \tag{5.21}$$

$$Corr(x_t^*, x_{t-1}^*) = \frac{\rho b_2^2}{b_2^2 + (1-b_1\rho)^2(1-\rho^2)\frac{\sigma_u^2}{\sigma_\varepsilon^2}}. \tag{5.22}$$

Note that in the special case $\sigma_u = 0$, the above expression reduces to *Corr* $(x_t^*, x_{t-1}^*) = \rho$, that is, when there is no exogenous noise u_t in (5.18), the persistence of the REE coincides exactly with the persistence of the exogenous driving force y_t.

Hommes and Zhu (2014) show that for the linear system (5.18, 5.19) at least one nonzero first-order SCEE (α^*, β^*) exists, with $\alpha^* = x^*$ and $0 < \beta < 1$. They call this equilibrium a *behavioral learning equilibrium (BLE)*, since it provides a simple, parsimonious forecasting rule, with the parameters pinned down by the simple statistics sample average and sample autocorrelation, on which a population of agents in large socio-economic systems may coordinate. Two important applications of this general framework are an asset pricing model driven by AR(1) dividends and a New Keynesian Phillips Curve (NKPC) where inflation is driven by an AR(1) process for marginal costs.

5.5.1   Asset Pricing Model

Consider an asset pricing model with a risky asset that pays stochastic dividends y_t following an AR(1) process. The equilibrium price of the risky asset p_t is given by

$$p_t = \frac{1}{R}\left[p_{t+1}^e + a + \rho y_t\right], \tag{5.23}$$

where $R > 1$ is the gross risk free rate of return. Compared to the general framework (5.18), we have $b_0 = \frac{a}{R}$, $b_1 = \frac{1}{R}$, $b_2 = \frac{\rho}{R}$ and $\sigma_u = 0$.

Using (5.20), the rational expectations equilibrium p_t^* becomes

$$p_t^* = \frac{aR}{(R-1)(R-\rho)} + \frac{\rho}{R-\rho}y_t. \tag{5.24}$$

In particular, if $\{y_t\}$ is IID, i.e., $a = \bar{y}$ and $\rho = 0$, then $p_t^* \equiv \frac{a}{R-1} = \frac{\bar{y}}{R-1}$ is constant. The corresponding mean, variance and first-order autocorrelation coefficient of the rational expectation price p_t^* are given by, respectively,

$$\overline{p^*} := E(p_t^*) = \frac{a}{(R-1)(1-\rho)} = \frac{\bar{y}}{R-1}, \tag{5.25}$$

$$Var(p_t^*) = E(p_t^* - \overline{p^*})^2 = \frac{\rho^2 \sigma_\varepsilon^2}{(R-\rho)^2(1-\rho^2)} \tag{5.26}$$

$$Corr(p_t^*, p_{t-1}^*) = \rho. \tag{5.27}$$

Under the assumption that agents are boundedly rational and believe that the price p_t follows a univariate AR(1) process, the implied actual law of motion for prices is

$$\begin{cases} p_t = \dfrac{1}{R}[\alpha + \beta^2(p_{t-1} - \alpha) + a + \rho y_t], \\ y_t = a + \rho y_{t-1} + \varepsilon_t. \end{cases} \tag{5.28}$$

A straightforward computation shows that the corresponding first-order autocorrelation coefficient $F(\beta)$ of the ALM (5.28) is given by

$$F(\beta) = \frac{\beta^2 + R\rho}{\rho\beta^2 + R}. \tag{5.29}$$

Hommes and Zhu (2014) show that in the asset pricing model (5.28), the BLE (α^*, β^*) is unique, $\alpha^* = \frac{\bar{y}}{R-1} = \overline{p^*}$ and $\beta^* > \rho$. This means that along the BLE the forecast is on average *unbiased* and prices exhibit *persistence amplification*, that is, the persistence β^* is larger than the persistence ρ under RE. Furthermore, the BLE is stable under SAC-learning.

Figure 5.3a illustrates the existence of a unique BLE $(\alpha^*, \beta^*) = (1, 0.997)$. The time series of fundamental prices and market prices along the BLE $(\alpha^*, \beta^*) = (1, 0.997)$ are shown in Fig. 5.3b, illustrating that the BLE exhibits *excess volatility*

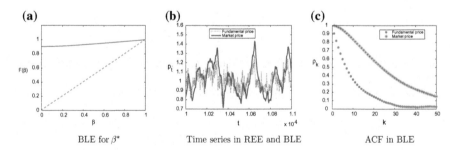

Fig. 5.3 a BLE $\beta^*(= 0.997)$ is the intersection point of the first-order autocorrelation coefficient $F(\beta) = \frac{\beta^2 + R\rho}{\rho\beta^2 + R}$ (*bold curve*) with the perceived first-order autocorrelation β (*dotted line*); **b** RE fundamental prices (*dotted curve*) and market prices (*bold curve*) along the BLE; **c** Autocorrelation Functions (ACF) of RE fundamental prices (*lower dots*) and market prices (higher stars) along the BLE. Parameters: $R = 1.05$, $\rho = 0.9$, $a = 0.005$, $\varepsilon_t \sim IID\ U(-0.01, 0.01)$ (i.e. uniform distribution on $[-0.01, 0.01]$)

compared to the RE solution. Furthermore, along the BLE the first-order autocorrelation coefficient β^* of market prices is larger than that of the fundamental prices ρ, implying that the market price exhibit *persistence amplification*. The autocorrelation functions of the market prices and the fundamental prices are shown in Fig. 5.3c. Persistence amplification leads to much slower decay of the ACF, and the autocorrelation coefficients of prices along a BLE are substantially higher than those of the RE fundamental price.

Figure 5.4 illustrates how the persistence amplification and excess volatility depend on the autocorrelation coefficient ρ of dividends, which is also the autocorrelation coefficient of the fundamental price. The first-order autocorrelation β^* of market prices is significantly higher than that of fundamental prices, especially for $\rho > 0.4$ (Fig. 5.4a). For $\rho \geq 0.5$ we have $\beta^* > 0.9$, implying that asset prices are close to a random walk and therefore quite unpredictable. Based on empirical findings, e.g. Timmermann (1996) and Branch and Evans (2010), the autoregressive coefficient of dividends ρ is about 0.9, where the corresponding $\beta^* \approx 0.997$, very close to a random walk. In the case $\rho > 0.4$, the corresponding unconditional variance of market prices is larger than that of fundamental prices. As illustrated in Fig. 5.4b, the ratio of the variance of market prices and the variance of fundamental prices is greater than 1 for $0.4 < \rho < 1$, with a peak around 3.5 for $\rho = 0.7$. For $\rho = 0.9$, $\frac{\sigma_p^2}{\sigma_{p^*}^2} \approx 2.5$, that is, excess volatility by a factor of more than two for empirically relevant parameter values.

Figure 5.5 illustrates that the unique BLE (α^*, β^*) is stable under SAC-learning. Figure 5.5a shows that the sample mean of the market prices under SAC-learning, α_t, tends to the mean $\alpha^* = 1$, while Fig. 5.5b shows that the first-order sample autocorrelation coefficient of the market prices under SAC-learning, β_t, tends to the first-order autocorrelation coefficient $\beta^* = 0.997$. Figure 5.5c shows the asset price under SAC-learning, using the same sample path of noise, as the time series of the

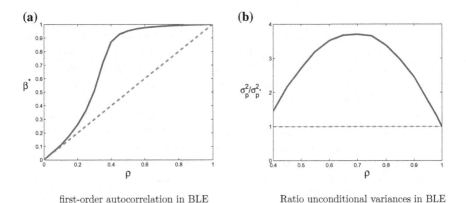

first-order autocorrelation in BLE   Ratio unconditional variances in BLE

Fig. 5.4 **a** First-order BLE β^* with respect to ρ; **b** ratio of unconditional variances of market prices and fundamental prices with respect to ρ, where $R = 1.05$

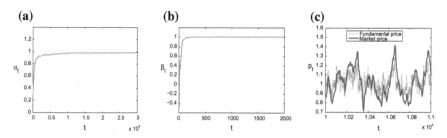

Fig. 5.5 **a** Time series $\alpha_t \rightarrow \alpha^*(1.0)$; **b** time series $\beta_t \rightarrow \beta^*(0.997)$; **c** time series of market prices under SAC-learning and fundamental prices

BLE in Fig. 5.3c. Since the times series are almost the same, SAC-learning converges to the BLE rather quickly.

In summary, the BLE and SAC-learning offer an explanation of high persistence, excess volatility and bubbles and crashes in asset prices within a stationary time series framework.

5.5.2 A New Keynesian Philips Curve

A second application of BLE and SAC-learning uses the New Keynesian macro model (Woodford 2003). In the New Keynesian Philips curve (NKPC) with inflation driven by an exogenous AR(1) process y_t for the firm's real marginal cost or the output gap, inflation and the real marginal cost (output gap) evolve according to

$$\begin{cases} \pi_t = \delta\pi^e_{t+1} + \gamma y_t + u_t, \\ y_t = a + \rho y_{t-1} + \varepsilon_t, \end{cases} \tag{5.30}$$

where π_t is the inflation at time t, π^e_{t+1} is the subjective expected inflation at date $t+1$, y_t is the output gap or real marginal cost, $\delta \in [0, 1)$ is the representative agent's subjective time discount factor, $\gamma > 0$ is related to the degree of price stickiness in the economy and $\rho \in [0, 1)$ describes the persistence of the AR(1) driving process. u_t and ε_t are IID stochastic disturbances with zero mean and finite absolute moments with variances σ^2_u and σ^2_ε, respectively. The most important difference with the asset pricing model is that (5.30) includes two stochastic disturbances, namely the shock ε_t of the AR(1) driving variable and an additional noise term u_t in the New Keynesian Philips curve. We refer to u_t as a supply shock (or markup shock), and to ε_t as a demand shock, that is uncorrelated with the supply shock. We will see that this extra shock allows for the possibility of *multiple equilibria*. Compared with our general framework (5.18), the corresponding parameters are $b_0 = 0$, $b_1 = \delta$ and $b_2 = \gamma$.

Under the assumption that agents are boundedly rational and believe that inflation π_t follows a univariate AR(1) process, the implied actual law of motion becomes[10]

$$\begin{cases} \pi_t = \delta[\alpha + \beta^2(\pi_{t-1} - \alpha)] + \gamma y_t + u_t, \\ y_t = a + \rho y_{t-1} + \varepsilon_t. \end{cases} \tag{5.31}$$

The corresponding first-order autocorrelation coefficient $F(\beta)$ of the implied ALM (5.31) is computed as

$$F(\beta) = \delta\beta^2 + \frac{\gamma^2\rho(1 - \delta^2\beta^4)}{\gamma^2(\delta\beta^2\rho + 1) + (1 - \rho^2)(1 - \delta\beta^2\rho) \cdot \frac{\sigma_u^2}{\sigma_\varepsilon^2}}. \tag{5.32}$$

Hommes and Zhu (2014) show that for $0 < \rho < 1$ and $0 \leq \delta < 1$, there exists at least one nonzero BLE (α^*, β^*) for the New Keynesian Philips curve (5.31) with $\alpha^* = \frac{\gamma a}{(1-\delta)(1-\rho)} = \overline{\pi^*}$. Moreover, a BLE is stable under SAC-learning if $F'(\beta^*) < 1$.

For the New Keyensian Philips curve (5.31), multiple BLE may coexist. In the simulations below, we fix the parameters $\delta = 0.99$, $\gamma = 0.075$, $a = 0.0004$, $\rho = 0.9$, $\sigma_\varepsilon = 0.01$ [$\varepsilon_t \sim N(0, \sigma_\varepsilon^2)$], and $\sigma_u = 0.003162$ [$u_t \sim N(0, \sigma_u^2)$], so that $\frac{\sigma_u^2}{\sigma_\varepsilon^2} = 0.1$. Figure 5.6a illustrates an example where $F(\beta)$ has three fixed points $\beta_1^* \approx 0.3066$, $\beta_2^* \approx 0.7417$ and $\beta_3^* \approx 0.9961$. Hence, we have coexistence of three first-order BLE (α^*, β_j^*), $j = 1, 2, 3$. Figure 5.6b, c illustrate the time series of inflation along the coexisting BLE. Inflation has low persistence along the BLE (α^*, β_1^*), but very high persistence along the BLE (α^*, β_3^*). The time series of inflation along the high persistence BLE in Fig. 5.6c has in fact similar persistence characteristics and amplitude of fluctuation as in empirical inflation data. Furthermore, Fig. 5.6c illustrates that inflation in the high persistence BLE has much stronger persistence than REE inflation, where the first-order autocorrelation coefficient of REE inflation is 0.865, significantly less than $\beta_3^* = 0.9961$.

If multiple BLE coexist, the convergence under SAC-learning depends on the initial state of the system, as illustrated in Fig. 5.7. Since $0 < F'(\beta_j^*) < 1$, for $j = 1$ and $j = 3$, while $F'(\beta_2^*) > 1$, (see Fig. 5.6a), the first-order BLE (α^*, β_1^*) and (α^*, β_3^*) are (locally) stable under SAC-learning, while (α^*, β_2^*) is unstable. For initial state $(\pi_0, y_0) = (0.028, 0.01)$ (Fig. 5.7a, b), the SAC-learning dynamics (α_t, β_t) converges to the stable low-persistence BLE $(\alpha^*, \beta_1^*) = (0.03, 0.3066)$. Figure 5.7b also illustrates that the convergence of the first-order autocorrelation coefficient β_t to the low-persistence first-order autocorrelation coefficient $\beta^* = 0.3066$ is very slow. For a different initial state, $(\pi_0, y_0) = (0.1, 0.15)$, our numerical simulation shows that the sample mean α_t still tends to $\alpha^* = 0.03$, but only slowly (see Fig. 5.7c), while β_t tends to the high persistence BLE $\beta_3^* \approx 0.9961$[11] (see Fig. 5.7d).

---

[10] As in the asset pricing model, we assume that boundedly rational agents do not recognize or do not believe that inflation is driven by output or marginal costs, but simply forecast inflation by an univariate AR(1) rule.

[11] As shown in Fig. 5.6a, $F'(\beta_3^*)$ is close to 1 and, hence, the convergence of SAC-learning is slow.

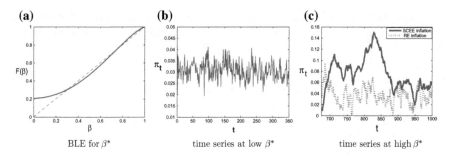

Fig. 5.6 **a** The first-order autocorrelation β^* of the BLE correspond to the three intersection points of $F(\beta)$ in (5.32) (*bold curve*) with the perceived first-order autocorrelation β (*dotted line*); **b** time series of inflation in low-persistence BLE $(\alpha^*, \beta_1^*) = (0.03, 0.3066)$; **c** times series of inflation in high-persistence BLE $(\alpha^*, \beta_3^*) = (0.03, 0.9961)$ (*bold curve*) and time series of REE inflation (*dotted curve*)

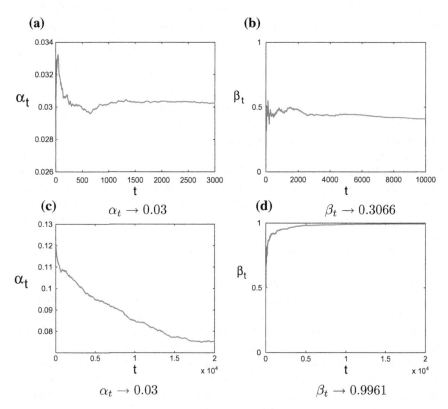

Fig. 5.7 Time series of α_t and β_t under SAC-learning for different initial values. **a-b** For $(\pi_0, y_0) = (0.028, 0.01)$ SAC-learning converges to the low persistence BLE $(\alpha^*, \beta_1^*) = (0.03, 0.3066)$; **c-d** For $(\pi_0, y_0) = (0.1, 0.15)$ SAC-learning converges to the high persistence BLE $(\alpha^*, \beta_1^*) = (0.03, 0.9961)$

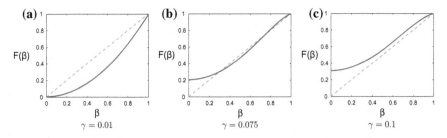

Fig. 5.8 The figure illustrates how the (co-)existence of low and high persistence BLE β^* depends upon the parameter γ, measuring the relative strength of inflation upon the driving variable, the output gap. **a** $\gamma = 0.01$; **b** $\gamma = 0.075$; **c** $\gamma = 0.1$. Other parameters: $\frac{\sigma_u^2}{\sigma_\varepsilon^2} = 0.1, \rho = 0.9$ and $\delta = 0.99$

Numerous simulations (not shown) show that for initial values π_0 of inflation higher than the mean $\alpha^* = 0.03$, the SAC-learning β_t generally enters the high-persistence region. In particular, a large shock to inflation may easily cause a jump of the SAC-learning process into the high-persistence region.[12] In the following we further indicate how high and low persistence BLE depend on different parameters.

5.5.2.1 Multiple Equilibria and Parameter Dependence

Figure 5.8 illustrates how the number of BLE depends on the parameter γ. For sufficiently small $\gamma (< 0.05)$, there exists only one, low persistence BLE β^* (Fig. 5.8a). Moreover, since

$$\frac{\partial F}{\partial \gamma} = \frac{2\rho(1 - \delta^2\beta^4)(1 - \rho^2)(1 - \delta\beta^2\rho)\frac{\sigma_u^2}{\sigma_\varepsilon^2}}{\gamma^3\left[(\delta\beta^2\rho + 1) + (1 - \rho^2)(1 - \delta\beta^2\rho)\frac{1}{\gamma^2}\frac{\sigma_u^2}{\sigma_\varepsilon^2}\right]^2} > 0,$$

the graph of $F(\beta)$ in (5.32) shifts upward as γ increases. At some critical γ-value, a tangent bifurcation occurs. Immediately thereafter, there exist three BLE, β_1^*, β_2^* and β_3^* (see Fig. 5.8b). The low persistence BLE β_1^* and the high persistence BLE β_3^* are stable under SAC-learning, since $0 < F'(\beta_j^*) < 1$, $j = 1$ and $j = 3$, separated by an unstable BLE β_2^*, with $F'(\beta_2^*) > 1$. As γ further increases, another tangent bifurcation occurs and the low persistence BLE disappears. A unique high persistence BLE then remains, which is stable under SAC-learning (Fig. 5.8c).

The dependence of the number of BLE and their persistence upon the parameter γ are quite intuitive. Recall that γ in (5.30) measures the relative strength of the driving variable, the output gap or marginal costs, to inflation. When the driving

---

[12]Hommes and Zhu (2014) also simulate the NKPC under SAC-learning with a constant gain parameter (see the online Supplementary Material) and, similar to Branch and Evans (2010), obtained irregular regime switching between phases of very low persistence and phases of high persistence with near unit root behavior.

Fig. 5.9 First-order
autocorrelation coefficient of
REE inflation (*dotted real
curve*), stable BLE β^* with
respect to ρ (*bold curves*),
unstable BLE β^* (*dotted
curve*), where $\gamma =
0.075, \sigma_u = 0.003162, \sigma_\varepsilon =
0.01, \delta = 0.99$

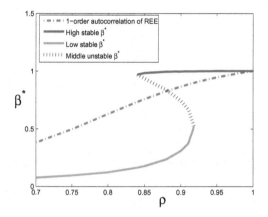

force is relatively weak, a unique, stable low persistence BLE prevails, with much
weaker autocorrelation than in the driving variable. At the other extreme, when the
driving force is sufficiently strong, a unique, stable high persistence BLE prevails,
with significantly stronger autocorrelation and higher persistence than in the driving
variable. In the intermediate case, multiple BLE coexist and the system exhibits path
dependence, where, depending on initial conditions, inflation converges to a low or
a high persistence BLE.

In a similar way, the dependence of the BLE upon the noise ratio $\frac{\sigma_u^2}{\sigma_\varepsilon^2}$ can be
analyzed. $F(\beta)$ in (5.32) can be rewritten as

$$F(\beta) = \delta\beta^2 + \frac{\rho(1 - \delta^2\beta^4)}{(\delta\beta^2\rho + 1) + (1 - \rho^2)(1 - \delta\beta^2\rho) \cdot \frac{\sigma_u^2}{\sigma_\varepsilon^2}\frac{1}{\gamma^2}}.$$

Consequently, the effect of the noise ratio $\frac{\sigma_u^2}{\sigma_\varepsilon^2}$ is inversely related to the effect of γ.
Hence, when the ratio $\frac{\sigma_u^2}{\sigma_\varepsilon^2}$ is high, that is, when the markup shocks to inflation are
high compared to the noise of the driving variable, a unique, stable low persistence
BLE prevails. If on the other hand, the markup shocks to inflation are low compared
to the noise of the driving variable, a unique, stable high persistence BLE prevails.

Furthermore, Fig. 5.9 illustrates how the BLE β^*, together with the first-order autocorrelation coefficient of REE inflation, depends upon the parameter ρ, measuring
the persistence in the driving variable. For intermediate values of $\rho (\in [0.84, 0.918])$,
two stable BLE β^* coexist separated by an unstable BLE. In the high persistence
BLE, β^* is larger than the first-order autocorrelation coefficient of REE inflation,
while in the low persistence BLE β^* is smaller than the first-order autocorrelation
coefficient of REE inflation. For small values of ρ, $\rho < 0.84$, a unique, stable low
persistence BLE prevails, while for large values of ρ, $\rho > 0.918$, a unique, stable
high persistence BLE prevails.

Simulations show that, for plausible values of ρ around 0.9, for a large range of initial values of inflation, the SAC-learning converges to the stable, high persistence BLE β^* with very strong persistence in inflation (see e.g. Fig. 5.7d). This result is consistent with the empirical finding in Adam (2007) that the Restricted Receptions Equilibrium (RPE) describes subjects' inflation expectations surprisingly well and provides a better explanation for the observed persistence of inflation than REE.

In summary, the dependence of the number of equilibria and whether the persistence is high or low are quite intuitive. This intuition essentially follows from the signs of the partial derivatives of the first-order autocorrelation coefficient $F(\beta)$ in (5.32) of the implied ALM (5.31) satisfying:

$$\frac{\partial F}{\partial \gamma} > 0 \qquad \frac{\partial F}{\partial(\frac{\sigma_u^2}{\sigma_\varepsilon^2})} < 0 \qquad \frac{\partial F}{\partial \rho} > 0 \qquad \frac{\partial F}{\partial \delta} > 0. \qquad (5.33)$$

Hence, as in Fig. 5.8, the graph of $F(\beta)$ shifts upwards when γ increases, $\frac{\sigma_u^2}{\sigma_\varepsilon^2}$ decreases, ρ increases or δ increases, and consequently, the equilibria shift from low persistence to high persistence equilibria. When the nonlinearity is strong and F is S-shaped, e.g., as in Fig. 5.6 for empirically relevant parameter values in the NKPC, both the persistence and the number of equilibria shift, and a transition from a unique stable low persistence BLE, through coexisting stable low and high persistence equilibria, to a unique stable high persistence equilibrium occurs. Such a transition from a unique low persistence BLE, through coexisting low and high persistence BLE, toward a unique high persistence BLE occurs when the strength of the AR(1) driving force (the parameter γ) increases, when the ratio of the model noise compared to the noise of the driving force (i.e. $\frac{\sigma_u^2}{\sigma_\varepsilon^2}$) decreases, when the autocorrelation (i.e., the parameter ρ) in the driving force increases, and when the strength of the expectations feedback (i.e., the parameter δ) increases.

5.6   Learning-to-forecast Experiments

In order to study the empirical relevance of different theories of expectations and learning I have been involved in many so-called *learning-to-forecast experiments* (LtFE) in controlled laboratory settings with human subjects. A LtFE, introduced by Marimon and Sunder (1993) and Marimon et al. (1993), is an experiment where a group of subjects repeatedly forecast the price of a good, whose value is endogenously determined by the average group forecast. A LtFE may be seen as an empirical test of the expectations hypothesis of a dynamic economic expectations feedback system, where all other decisions in the economy –consumption, production, trading, etc.– are computerized, consistent with the underlying model assumptions. A LtFE thus becomes an empirical test of the expectations hypothesis, with all other assumptions under the control of the experimenter. See Assenza et al. (2014) for a recent review of

LtFEs and Duffy (2014) for a recent collection of state of the art work in experimental macroeconomics.

To my best knowledge, Jean-Michel Grandmont has never conducted experiments with human subjects himself, but his ideas and suggestions have certainly influenced my own experiments and in particular triggered the positive/negative feedback experimental design in Heemeijer et al. (2009). In a workshop in December 2002, in honor of Volker Böhm, I made the claim that lab experiments in positive feedback environments are *more unstable* than those under negative feedback, based on a comparison between asset pricing experiments (Hommes et al. 2005) and cobweb experiments (Hommes et al. 2007). Jean-Michel—who was in the audience—correctly pointed out that this claim was not warranted, as these experiments differed in multiple dimensions. These suggestions led to the design of a new lab experiment comparing positive versus negative feedback systems in Heemeijer et al. (2009).

Positive expectations feedback is characteristic of speculative asset markets, where an increase of the average price forecast of investors causes the realized market price to rise through higher speculative demand. Negative feedback is more important in producer driven markets of perishable consumption goods, where more optimistic expectations about the price of the good lead to higher production and therefore to lower realized market prices. Heemeijer et al. (2009) investigate how the expectations feedback structure affects individual forecasting behaviour and aggregate market outcomes by considering market environments that *only* differ in the sign of the expectations feedback, but are equivalent along all other dimensions. The realized price is a linear map of the average of the individual price forecasts $p_{i,t}^e$ of six subjects. The (unknown) price generating rules in the *negative* and *positive* feedback systems were respectively:

$$p_t = 60 - \frac{20}{21}[(\sum_{h=1}^{6} \frac{1}{6} p_{ht}^e) - 60] + \epsilon_t, \qquad \text{negative feedback} \quad (5.34)$$

$$p_t = 60 + \frac{20}{21}[(\sum_{h=1}^{6} \frac{1}{6} p_{ht}^e) - 60] + \epsilon_t, \qquad \text{positive feedback} \quad (5.35)$$

where ϵ_t is a (small) exogenous random shock to the pricing rule. The positive and negative feedback systems (5.34) and (5.35) have the same unique RE equilibrium steady state $p^* = 60$ and *only* differ in the sign of the expectations feedback map. Both are linear near-unit-root maps, with slopes $20/21 \approx -0.95$ resp. $+20/21$.[13] Figure 5.10 (top panels) illustrates the difference in the negative and positive expectations feedback maps. Both have the same unique RE fixed point 60. A striking feature of the near-unit-root positive feedback map, is that each point is in fact an *almost self-fulfilling equilibrium*. In near unit root positive feedback systems, agents

---

[13] In both treatments, the absolute value of the slopes is 0.95, implying in both cases that the feedback system is stable under naive expectations.

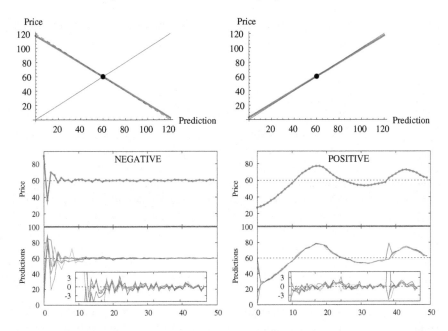

Fig. 5.10 Laboratory experiments of negative (*left panels*) versus positive (*right panels*) feedback systems. Linear feedback maps (*top panels*), realized market prices (*middle panels*), six individual predictions (*bottom panels*) and individual errors (*inside bottom panels*). In the negative expectations feedback market (*left panels*) the realized price quickly converges to the RE benchmark 60. In all positive feedback markets (*right panels*) individuals coordinate on the "wrong" price forecast and as a result the realized market price persistently deviates from the RE benchmark 60

only make small mistakes and these mistakes are almost self-fulfilling. Such near unit root behavior is typical in many macro and financial asset pricing models, where near unit roots arise e.g. due to discount factors close to 1. Will subjects in LtFEs be able to coordinate on the unique RE fundamental price, the only equilibrium that is perfectly self-fulfilling?

Figure 5.10 (bottom panels) shows realized market prices as well as six individual predictions in two typical groups. Aggregate price behavior is very different under positive than under negative feedback. In the negative feedback case, the price settles down to the RE steady state price 60 relatively quickly (within 10 periods), but in the positive feedback treatment the market price does not converge but rather oscillates around its fundamental value. Individual forecasting behavior is also very different: in the case of positive feedback, coordination of individual forecasts occurs extremely fast, within 2–3 periods. The coordination however is on a "wrong", i.e., a non-RE-price. Individual errors are small, but strongly coordinated, leading to large aggregate deviations from the rational fundamental price. In contrast, in the negative feedback case coordination of individual forecasts is slower and takes about 10 periods. More heterogeneity of individual forecasts however ensures that, the realized price quickly converges to the RE benchmark of 60 (within 5–6 periods), after which individual predictions coordinate on the correct RE price.

5.6.1 OLG Experiments

Arifovic et al. (2016) recently performed LtFEs in the overlapping generations modeling framework of Grandmont (1985).[14] In this highly nonlinear OLG economy, infinitely many different periodic and chaotic perfect foresight equilibria may exist. Grandmont (1985) also showed that each periodic or chaotic perfect foresight equilibrium is stable under some suitable adaptive learning algorithm. The purpose of the LtFE is to test empirically on which of these infinitely many different equilibria a group of subjects may coordinate their individual expectations. The expectations feedback system is of the form

$$p_t = G_{\rho_2}(p_{t+1}^e), \qquad (5.36)$$

where p_t is the price of the consumption good, p_{t+1}^e is the average price forecast of young consumers, G is a non-monotonic map and ρ_2 is a parameter (the degree of relative risk aversion of the old consumers). For small values of ρ_2 the dynamics of the map is simple, and the system has a stable steady state. For increasing values of ρ_2 the steady state becomes locally unstable and the dynamics exhibits a period-doubling bifurcation route to chaos and generates infinitely many periodic and chaotic perfect foresight solutions.

Figure 5.11 shows four typical groups of the LtFE of Arifovic et al. (2016). For $\rho_2 = 5$, the map G_5 has an unstable steady state and a stable 2-cycle. In the LtFEs for $\rho_2 = 5$, one group (top left panel) coordinates on a (noisy) 2-cycle, while another group coordinates on a steady state (top right panel). For $\rho_2 = 12$, in the chaotic region of the map, two typical groups are shown, both coordinating on a 2-cycle, one after a long transient and to a somewhat noisy 2-cycle (bottom left panel), and another one converging relatively fast to an almost perfect 2-cycle (bottom right panel). In this LtFE of the highly nonlinear OLG economy, subjects thus *learn-to-believe in a steady state* or *learn-to-believe in a 2-cycle* in a complex chaotic environment.

Figure 5.12 shows that SAC-learning provides a good fit on these laboratory data, and explains convergence to a steady state (top right panel) and to 2-cycles (top left and both bottom panels). These results are consistent with Hommes et al. (2013) (see Sect. 5.4), who showed that SAC-learning in Grandmont's OLG model framework either selects a steady state or a 2-cycle. In Fig. 5.12a constant gain version of the SAC-learning is used, with constant gain parameter $\kappa = 0.2$. The SAC-learning converges rather quickly to the steady state (top right panel) and somewhat slower to the 2-cycle (top left and both bottom panels), but matches both long run outcomes quite nicely. SAC-learning thus explains coordination on simple-steady state and two-

---

[14] See also Heemeijer et al. (2012) for related individual LtFEs in a different OLG economy.

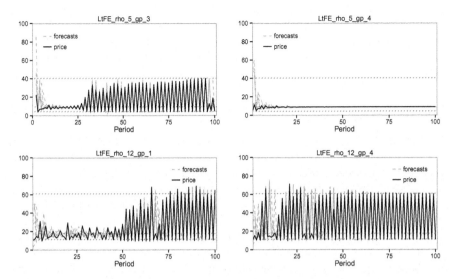

Fig. 5.11 Lab experiments for the OLG-economy of Grandmont (1985). For $\rho_2 = 5$ the map has a stable 2-cycle and the LtFEs converge either to a 2-cycle (*top left*) or to a steady state (*top right*). For $\rho_2 = 12$ the map is chaotic, but both groups in the LtFEs converge to a (noisy) 2-cycle (*bottom panels*)

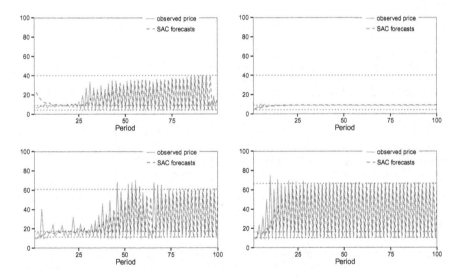

Fig. 5.12 Two-step ahead simulations under SAC learning with gain coefficient $\kappa = 0.2$. SAC-learning explains coordination on a stable steady state or a stable 2-cycle

cycle—equilibria in a complex environment. If agents (wrongly) believe that prices follow a stochastic AR(1) process, this belief becomes self-fulfilling and selects simple equilibria in the complex, nonlinear OLG-economy as the only long run outcomes.

5.7 Concluding Remarks

I am grateful to Jean-Michel Grandmont for much of his inspiration over the years. My work discussed here may be seen as formalizations of his ideas. In particular, the idea of self-fulfilling mistakes (Grandmont 1998), where agents (wrongly) believe that prices follow a (simple) stochastic process, while the true (unknown) law of motion is a nonlinear complex system, and agents are unable to statistically reject these beliefs, has inspired the research program on bounded rationality and learning. Coordination on almost self-fulfilling equilibria in positive feedback systems in laboratory experiments show the empirical relevance of these ideas. Much work on bounded rationality and learning in large socio-economic systems remains to be done, particularly to study how policy should manage self-fulfilling mistakes.

References

Adam, K. (2007). Experimental evidence on the persistence of output and inflation. *The Economic Journal, 117*(520), 603–636.

Arifovic, J., Hommes, C. H., & Salle, I. (2016). Learning to believe in simple equilibria in a complex OLG economy—evidence from the lab. Working Paper, University of Amsterdam.

Assenza, T., Bao, T., Hommes, C. H., & Massaro, D. (2014). Experiments on expectations in macroeconomics and finance, In J. Duffy (Ed.), *Experiments in macroeconomics, research in experimental economics* (Vol. 17). Emerald.

Branch, W. A. (2006). Restricted perceptions equilibria and learning in macroeconomics. In D. Colander (Ed.), *Post walrasian macroeconomics: Beyond the dynamic stochastic general equilibrium model* (pp. 135–160). New York: Cambridge University Press.

Branch, W., & McGough, B. (2005). Consistent expectations and misspecification in stochastic non-linear economies. *Journal of Economic Dynamics and Control, 29*, 659–676.

Branch, W. A., & Evans, G. W. (2010). Asset return dynamics and learning. *Review of Financial Studies, 23*(4), 1651–1680.

Bray, M., & Savin, N. (1986). Rational expectations equilibria, learning and model specification. *Econometrica, 54*, 1129–1160.

Bullard, J. (1994). Learning equilibria. *Journal of Economic Theory, 64*, 468–485.

Bullard, J., Evans, G. W., & Honkapohja, S. (2008). Monetary policy, judgment and near-rational exuberance. *American Economic Review, 98*, 1163–1177.

Bullard, J., Evans, G. W., & Honkapohja, S. (2010). A model of near-rational exuberance. *Macroeconomic Dynamics, 14*, 166–188.

Duffy, J. (Ed.). (2014). *Experiments in macroeconomics, research in experimental economics* (Vol. 17). Emerald.

Evans, G. W., & Honkapohja, S. (2001). *Learning and expectations in macroeconomics*. Princeton: Princeton University Press.

Grandmont, J.-M. (1985). On endogenous competitive business cycles. *Econometrica, 53*, 995–1045.

Grandmont, J.-M. (1998). Expectation formation and stability in large socio-economic systems. *Econometrica, 66*, 741–781.

Hamilton, J. D. (1994). *Time series analysis*. Princeton: Priceton University Press.

Heemeijer, P., Hommes, C. H., Sonnemans, J., & Tuinstra, J. (2009). Price stability and volatility in markets with positive and negative expectations feedback. *Journal of Economic Dynamics & Control, 33*, 1052–1072.

Heemeijer, P., Hommes, C. H., Sonnemans, J., Tuinstra, J. (2012). An experimental study on expectations and learning in overlapping generations models, *Studies in Nonlinear Dynamics & Econometrics, 16*(4), article 1.

Hommes, C. H. (1991). *Chaotic dynamics in economic models*. Wolters-Noordhoff, Groningen: Some simple case studies, Groningen theses in Economics, Management & Organization.

Hommes, C. H. (1998). On the consistency of backward-looking expectations: The case of the cobweb. *Journal of Economic Behavior and Organization, 33*, 333–362.

Hommes, C. H. (2013a). Reflexivity, empirical evidence and laboratory experiments. *Journal of Economic Methodology, 20*, 406–419.

Hommes, C. H. (2013b). *Behavioral rationality and heterogeneous expectations in complex economic systems*. Cambrdige: Cambridge University Press.

Hommes, C. H., & Sorger, G. (1998). Consistent expectations equilibria. *Macroeconomic Dynamics, 2*, 287–321.

Hommes, C. H., & Rosser, J. B. (2001). Consistent expectations equilibria and complex dynamics in renewable resource markets. *Macroeconomic Dynamics, 5*, 180–203.

Hommes, C. H., & Zhu, M. (2014). Behavioral learning equilibria. *Journal of Economic Theory, 150*, 778–814.

Hommes, C. H., Sonnemans, J. H., Tuinstra, J., & van de Velden, H. (2005). Coordination of expectations in asset pricing experiments. *Review of Financial Studies, 18*(3), 955–980.

Hommes, C., Sonnemans, J., Tuinstra, J., & van de Velden, H. (2007). Learning in cobweb experiments. *Macroeconomic Dynamics, 11*(S1), 8–33.

Hommes, C. H., Sorger, G., & Wagener, F. (2013). Consistency of linear forecasts in a nonlinear stochastic economy. In G. I. Bischi, C. Chiarella, & I. Sushko (Eds.), *Global analysis of dynamic models in economics and finance* (pp. 229–287). Berlin: Springer.

Lansing, K. J. (2009). Time-varing U.S. inflation dynamics and the new Keynesian Phillips curve. *Review of Economic Dynamics, 12*, 304–326.

Lansing, K. J. (2010). Rational and near-rational bubbles without drift. *Economic Journal, 120*, 1149–1174.

Marimon, R., & Sunder, S. (1993). Indeterminacy of equilibria in a hyperinflationary world: Experimental evidence. *Econometrica, 61*(5), 1073–1107.

Marimon, R., Spear, S. E., & Sunder, S. (1993). Expectationally driven market volatility: An experimental study. *Journal of Economic Theory, 61*, 74–103.

Sargent, T. J. (1993). *Bounded rationality in macroeconomics*. Oxford: Clarendon Press.

Sargent, T. J. (1999). *The conquest of American inflation*. Princeton, NJ: Princeton University Press.

Schönhofer, M. (1999). Chaotic learning equilibria. *Journal of Economic Theory, 89*, 1–20.

Sögner, L., & Mitlöhner, H. (2002). Consistent expectations equilibria and learning in a stock market. *Journal of Economic Dynamics & Control, 26*, 171–185.

Sorger, G. (1998). Imperfect foresight and chaos: An example of a self-fulfilling mistake. *Journal of Economic Behavior and Organization, 33*, 363–383.

Timmermann, A. (1996). Excess volatility and predictability in stock prices in autoregressive dividend models with learning. *Review Economic Studies, 63*, 523–557.

Tuinstra, J. (2003). Beliefs equilibria in an overlapping generations model. *Journal of Economic Behavior & Organization, 50*, 145–164.

Tuinstra, J., & Wagener, F. O. O. (2007). On learning equilibria. *Economic Theory, 30*, 493–513.

Woodford, M. (2003). *Interest and prices: Foundations of a theory of monetary policy*. Princeton, NJ: Princeton University Press.

Chapter 6
Regime-Switching Sunspot Equilibria in a One-Sector Growth Model with Aggregate Decreasing Returns and Small Externalities

Takashi Kamihigashi

Abstract This paper shows that regime-switching sunspot equilibria easily arise in a one-sector growth model with aggregate decreasing returns and arbitrarily small externalities. We construct a regime-switching sunspot equilibrium under the assumption that the utility function of consumption is linear. We also construct a stochastic optimal growth model whose optimal process turns out to be a regime-switching sunspot equilibrium of the original economy under the assumption that there is no capital externality. We illustrate our results with numerical examples.

6.1 Introduction

In macroeconomics, sunspot equilibria are often associated with local indeterminacy, or the existence of a locally stable steady state. In the context of growth models, the phenomenon of local indeterminacy has been well known since Benhabib and Farmer (1994) and Farmer and Guo (1994). While earlier results required unduly large degrees of increasing returns and externalities,[1] local indeterminacy has been established for various settings under less objectionable assumptions, such as decreasing returns to labor (e.g., Pelloni and Waldmann 1998), moderate externalities (Dufourt et al. 2015), aggregate constant returns to scale (e.g., Benhabib et al. 2000; Mino 2001), and arbitrarily small increasing returns and externalities (e.g., Kamihigashi 2002; Pintus 2006).

This paper seeks to point out the possibility of an additional mechanism that gives rise to sunspot equilibria in an economy with aggregate decreasing returns to scale and arbitrarily small externalities. The combination on which we focus, aggregate decreasing returns to scale in tandem with arbitrarily small externalities, sets a stage

---

[1] See Benhabib and Farmer (1999) for a survey of earlier results. See Grandmont (1989, 1991) for a discussion of the relations between the local stability properties of a steady state and the possibility of sunspot equilibria.

T. Kamihigashi (✉)
Research Institute for Economics and Business Administration (RIEB), Kobe University, Rokkodai, Nada, Kobe 657-8501, Japan
e-mail: tkamihig@rieb.kobe-u.ac.jp

© Springer International Publishing AG 2017
K. Nishimura et al. (eds.), *Sunspots and Non-Linear Dynamics*,
Studies in Economic Theory 31, DOI 10.1007/978-3-319-44076-7_6

almost indistinguishable from the standard neoclassical setting, posing a challenge to proponents of sunspot equilibria.

Instead of small fluctuations around a locally indeterminate steady state, we consider large fluctuations caused by a regime-switching sunspot process. Assuming that the sunspot process is a two-state Markov chain, we construct regime-switching sunspot equilibria in which labor supply is positive in one state and zero in the other.

Although this type of regime-switching sunspot equilibrium is rather extreme and may not match many of the empirical regularities discussed in the local indeterminacy literature, there are merits to studying such equilibria in addition to local sunspot equilibria driven by local indeterminacy. First, local sunspot equilibria can explain only small fluctuations around a steady state, whereas economic events of great magnitude—such as the Great Depression of the early 20th century, Japan's "lost decades" since the early 1990s, and the Global Financial Crisis of 2007–2008—are characterized by large downfalls. Our model can at least generate large and sudden downfalls from a steady state. Second, our analysis suggests that regime-switching sunspot equilibria of the type considered in this paper are widespread in models with externalities, and can even coexist with local sunspot equilibria. Our approach thus complements, rather than substitutes, the more common local indeterminacy approach. It would be possible, in fact, to construct a model in which two types of sunspot shocks are present and both small and large fluctuations are endogenously generated. While the construction of such a model would be beyond the scope of this paper, our analysis can form a basis for further development in this direction.

We present two main results in this paper. First, assuming that the utility function of consumption is linear, we establish the existence of a regime-switching sunspot equilibrium by recursively solving the Euler condition for capital and the first-order condition for labor supply, and then verifying the associated transversality condition. These conditions are easy to verify especially when the utility of consumption is linear, as none of the three aforementioned conditions depend on consumption in such a scenario. Second, assuming an absence of capital externalities, we establish the existence of a regime-switching sunspot equilibrium by constructing a stochastic optimal growth model whose optimal process turns out to be a sunspot equilibrium of the original economy. This latter result is somewhat similar to the observational equivalence result shown by Kamihigashi (1996). In contrast to Kamihigashi (1996), however, we only use the three aforementioned conditions to verify that an optimal process for the stochastic optimal growth model can be interpreted as a sunspot equilibrium. Both results are illustrated with numerical examples.

In addition to the results on local indeterminacy previously mentioned, this paper is also related to results on global indeterminacy (e.g., Drugeon and Venditti 2001; Coury and Wen 2009) and regime-switching sunspot equilibria (e.g., Drugeon and Wigniolle 1996; Dos Santos Ferreira and Lloyd-Braga 2008).[2] While our findings show the existence of regime-switching sunspot equilibria based on global indeterminacy, they differ from the existing literature in that our model deviates only

---

[2]See Clain-Chamosset-Yvrard and Kamihigashi (2015) for an example of a regime-switching sunspot equilibrium in a two-country model with asset bubbles.

slightly from the standard neoclassical setting under our assumptions of aggregate decreasing returns and small externalities.[3] Our model can also be viewed as a variant of the Farmer and Guo (1994) model with aggregate decreasing returns and small externalities.

The remainder of this paper is organized as follows. In the next section we present the model along with basic definitions and assumptions. In Sect. 6.3 we show a standard result that offers a sufficient set of conditions for a feasible process to be an equilibrium. In Sect. 6.4 we present our main results along with numerical examples. In Sect. 6.5 we conclude the paper by discussing possible extensions. Longer proofs are relegated to the appendices.

6.2 The Model

We consider an economy with many agents, each of whom solves the following maximization problem:

$$\max_{\{c_t, n_t, k_{t+1}\}_{t=0}^{\infty}} E \sum_{t=0}^{\infty} \beta^t [u(c_t) - w(n_t)] \tag{6.2.1}$$

$$\text{s.t. } \forall t \in \mathbb{Z}_+, \quad c_t + k_{t+1} = f(k_t, n_t, K_t, N_t) + (1 - \delta)k_t, \tag{6.2.2}$$

$$c_t, k_{t+1} \geqslant 0, \quad n_t \in [0, 1], \tag{6.2.3}$$

where c_t is consumption, n_t is labor supply, k_t is the capital stock at the beginning of period t, N_t is aggregate labor supply, and K_t is the aggregate capital stock. The utility function u of consumption, the disutility function w of labor supply, and the production function f are specified below. The discount factor β and the depreciation rate δ satisfy

$$\beta, \delta \in (0, 1). \tag{6.2.4}$$

In the above maximization problem, the initial capital stock $k_0 > 0$ and the stochastic processes $\{K_t\}_{t=0}^{\infty}$ and $\{N_t\}_{t=0}^{\infty}$ are taken as given. In equilibrium, however, we have

$$\forall t \in \mathbb{Z}_+, \quad K_t = k_t, \quad N_t = n_t. \tag{6.2.5}$$

To formally define an equilibrium of this economy, we first define a *pre-equilibrium* as a five-dimensional stochastic process $\{c_t, n_t, k_t, N_t, K_t\}_{t=0}^{\infty}$ such that $\{c_t, n_t, k_{t+1}\}_{t=0}^{\infty}$ solves the maximization problem (6.2.1)–(6.2.3) given $k_0 > 0$ and $\{N_t, K_t\}_{t=0}^{\infty}$. We define an *equilibrium* as a three-dimensional stochastic process $\{c_t, n_t, k_t\}_{t=0}^{\infty}$ such that the five-dimensional stochastic process $\{c_t, n_t, k_t, n_t, k_t\}_{t=0}^{\infty}$

---

[3]Kamihigashi (2015) shows that multiple steady states are possible even without externalities.

is a pre-equilibrium. We also define a *feasible process* as a three-dimensional stochastic process $\{c_t, n_t, k_t\}_{t=0}^{\infty}$ satisfying (6.2.2), (6.2.3), and (6.2.5).

We specify the functions u, w, and f as follows:

$$u(c) = \frac{c^{1-\sigma} - 1}{1 - \sigma}, \tag{6.2.6}$$

$$w(n) = \eta \frac{n^{\gamma+1}}{\gamma + 1}, \tag{6.2.7}$$

$$f(k, n, K, N) = \theta k^{\alpha} n^{\rho} K^{\overline{\alpha}} N^{\overline{\rho}}. \tag{6.2.8}$$

We impose the following restrictions on the parameters:

$$\sigma \in [0, 1], \tag{6.2.9}$$

$$\theta, \alpha, \rho, \eta > 0, \tag{6.2.10}$$

$$\overline{\alpha}, \overline{\rho}, \gamma \geqslant 0, \tag{6.2.11}$$

$$\overline{\alpha} + \alpha + \overline{\rho} + \rho \leqslant 1. \tag{6.2.12}$$

If $\sigma = 1$, then it is understood that $u(c) = \ln c$. Since $\sigma \in [0, 1]$ by (6.2.9), u is bounded below unless $\sigma = 1$. The inequality in (6.2.12) means that the production function exhibits decreasing returns to scale at the aggregate level.[4] In what follows, we use the nonparametric forms u, w, and f and the parametric forms given by (6.2.6)–(6.2.8) above interchangeably.

Let $\hat{k}$ be the unique strictly positive capital stock $k > 0$ such that $\theta k^{\alpha+\overline{\alpha}} = \delta k$. The capital stock $\hat{k}$ is the maximum sustainable capital stock. It has the property that for any feasible process $\{c_t, n_t, k_t\}_{t=0}^{\infty}$ we have

$$\forall k_t \in \mathbb{Z}_+, \quad k_t \leqslant \max\{k_0, \hat{k}\}. \tag{6.2.13}$$

All equilibria are therefore bounded.

6.3 Sufficient Optimality Conditions

It follows from (6.2.8) that

$$\forall k, n > 0, \quad f_1(k, n, k, 0) = f_2(k, n, k, 0) = 0, \tag{6.3.1}$$

where $f_i(\cdot, \cdot, \cdot, \cdot)$ is the derivative of f with respect to the ith argument. To simplify notation, we define the following for $i = 1, 2$:

---

[4]In the decentralized version of the model, profits are given to consumers, who are the owners of the firms.

$$f_i(k, n) = f_i(k, n, k, n). \tag{6.3.2}$$

For $k, n \geqslant 0$ we also define

$$g(k, n) = f(k, n, k, n) + \zeta k, \tag{6.3.3}$$

where $\zeta = 1 - \delta$.

The first-order condition for labor supply n_t in period t is given by

$$u'(c_t) f_2(k_t, n_t) - w'(n_t) \begin{cases} = 0 & \text{if } n_t \in (0, 1), \\ \geqslant 0 & \text{if } n_t = 1, \\ \leqslant 0 & \text{if } n_t = 0. \end{cases} \tag{6.3.4}$$

Note from (6.3.1) that $n_t = 0$ is always a solution to (6.3.4). This observation forms the basis for our construction of sunspot equilibria. On the other hand, as long as $k_t, K_t, N_t > 0$, it is always optimal to choose a strictly positive labor supply since $f_2(k, 0, K, N) = \infty$ for any $k, K, N > 0$.

The stochastic Euler condition for the capital stock k_{t+1} at the beginning of period $t + 1$ can be written as

$$-u'(c_t) + \beta E_t u'(c_{t+1})[f_1(k_{t+1}, n_{t+1}) + \zeta] \begin{cases} = 0 & \text{if } k_{t+1} \in (0, g(k_t, n_t)), \\ \geqslant & \text{if } k_{t+1} = g(k_t, n_t), \\ \leqslant & \text{if } k_{t+1} = 0. \end{cases} \tag{6.3.5}$$

We also need to consider corner solutions since one of the results shown in the next section assumes that the utility function of consumption is linear. Yet as long as labor supply in period $t + 1$ is strictly positive with strictly positive probability, there is a solution to (6.3.5) with $k_{t+1} > 0$, because $\lim_{k \downarrow 0} f_1(k, n) = \infty$ for any $n > 0$.

The transversality condition is

$$\lim_{T \to \infty} \beta^T E u'(c_T) k_{T+1} = 0. \tag{6.3.6}$$

Kamihigashi (2003, 2005) provides more details on transversality conditions for stochastic problems.

We see from the following result that the above first-order conditions in conjunction with the transversality condition are sufficient for a feasible process to be an equilibrium. The proof is a stochastic version of the standard sufficiency argument; see Brock (1982) for a similar stochastic argument.

Lemma 6.3.1 *A feasible process* $\{c_t, n_t, k_t\}_{t=0}^{\infty}$ *is an equilibrium if it satisfies (6.3.4) and (6.3.5) for all* $t \in \mathbb{Z}_+$ *and (6.3.6).*

Proof See Appendix A. $\qquad\square$

6.4 Sunspot Equilibria

6.4.1 Common Structure

We consider a special type of sunspot equilibrium by taking a regime-switching
sunspot process $\{s_t\}$ as given. In particular, we assume that there are two sunspot
states, 0 and 1, and that $\{s_t\}$ is a two-state Markov chain with transition matrix

$$\begin{bmatrix} p_{00} & p_{01} \\ p_{10} & p_{11} \end{bmatrix}, \tag{6.4.1}$$

where p_{ij} is the probability that $s_{t+1} = j$ given $s_t = i$ for $i, j \in \{0, 1\}$. To simplify
the analysis, we assume that $p_{ij} > 0$ for all $i, j \in \{0, 1\}$. Since (6.4.1) is a transition
matrix, we have

$$p_{00} + p_{01} = p_{10} + p_{11} = 1. \tag{6.4.2}$$

In what follows, all stochastic processes (sequences) are assumed to be adapted to the
σ-field generated by the Markov chain $\{s_t\}_{t=0}^{\infty}$. This simply means that any variable
indexed by t is a function of the history of sunspot states $s_0, s_1, \dots, s_t$ up to period
t. Since $\{s_t\}$ is a sunspot process, it has no direct influence on the fundamentals of
the economy. An equilibrium $\{c_t, n_t, k_t\}$ is a *sunspot equilibrium* if it depends on the
sunspot process $\{s_t\}$ in a nontrivial way.

To see the possibility of a sunspot equilibrium, assume the following in the
maximization problem (6.2.1)–(6.2.3):

$$N_t \begin{cases} > 0 & \text{if } s_t = 1, \\ = 0 & \text{if } s_t = 0. \end{cases} \tag{6.4.3}$$

Then, provided that $k_t = K_t > 0$ for all $t \in \mathbb{Z}_+$, we must have

$$n_t \begin{cases} > 0 & \text{if } s_t = 1, \\ = 0 & \text{if } s_t = 0. \end{cases} \tag{6.4.4}$$

For the rest of the paper, we assume that $k_t = K_t > 0$ for all $t \in \mathbb{Z}_+$, focusing on
regime-switching sunspot equilibria satisfying (6.4.3) and (6.4.4).

Under (6.4.3) the first-order condition (6.3.4) for n_t can be written as

$$s_t = 1 \quad \Rightarrow \quad u'(c_t) f_2(k_t, n_t) - w'(n_t) \begin{cases} = 0 & \text{if } n_t \in (0, 1), \\ \geq 0 & \text{if } n_t = 1, \end{cases} \tag{6.4.5}$$

$$s_t = 0 \quad \Rightarrow \quad n_t = 0. \tag{6.4.6}$$

In (6.4.5) we have no need to consider the case $n_t = 0$ since $f_2(k, 0, k, N) = \infty$ for any $k, N > 0$ by (6.2.8), (6.2.10), and (6.2.12), as mentioned above. If, on the other hand, $s_t = 0$, then $N_t = 0$ by (6.4.3), and $n_t = 0$ since $f_2(k, n, k, 0) = 0$ for any $k > 0$ and $n \geqslant 0$.

6.4.2 Linear Utility of Consumption

One way to show the existence of a sunspot equilibrium is to use Lemma 6.3.1 to explicitly construct a sunspot equilibrium $\{c_t, n_t, k_t\}$ satisfying (6.4.4). To do so, we need to verify the Euler condition (6.3.5) and the transversality condition (6.3.6) in addition to (6.4.5) and (6.4.6). While this is not easy to do in general, we can explicitly construct a sunspot equilibrium using these conditions if we assume that the utility function of consumption is linear, and hence that none of the conditions depends on consumption (except for feasibility). We consider this special case in the following result.

Proposition 6.4.1 *If $\sigma = 0$, then a sunspot equilibrium satisfying (6.4.4) exists.*

Proof See Appendix B. □

The sunspot equilibrium constructed in the proof of Proposition 6.4.1 is generated by the following system of equations:

$$n_t = m(k_t, s_t), \tag{6.4.7}$$
$$k_{t+1} = \min\{q(p_{s_t 1}), g(k_t, n_t)\}, \tag{6.4.8}$$
$$c_t = g(k_t, n_t) - k_{t+1}, \tag{6.4.9}$$

where $m(\cdot, \cdot)$ and $q(\cdot)$ are given by (6.5.21) and (6.5.30), respectively, in Appendix B, and $p_{s_t 1} = p_{01}$ or p_{11} depending on $s_t = 0$ or 1. Given $k_t > 0$ and $s_t \in \{0, 1\}$, n_t is determined by (6.4.7), k_{t+1} is determined by (6.4.8), and c_t is determined by (6.4.9). With a new sunspot variable s_{t+1} drawn according to (6.4.1), n_{t+1} is determined by (6.4.7) again, and so one.

Figure 6.1 depicts the functions in (6.4.7)–(6.4.9) with the following parameter values:

$$\beta = 0.9, \quad \eta = 1, \quad \gamma = 0.1, \quad p_{01} = 0.2, \quad p_{11} = 0.8, \tag{6.4.10}$$
$$\delta = 0.05, \quad \theta = 3, \quad \rho = 0.55, \quad \overline{\rho} = 0.03, \tag{6.4.11}$$
$$\sigma = 0, \quad \alpha = 0.35, \quad \overline{\alpha} = 0.02. \tag{6.4.12}$$

Figure 6.2 shows sample paths for sunspot states, capital, labor, and consumption generated by (6.4.7)–(6.4.9). The sample path for labor supply n_t closely follows the pattern of sunspot states s_t, as expected from (6.4.4). The sample paths for capital and consumption inherit the same pattern to a large extent.

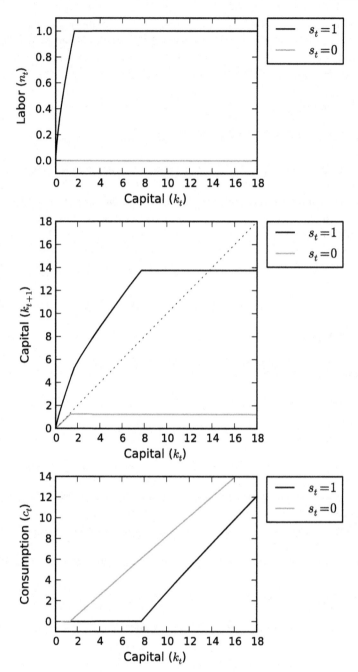

Fig. 6.1 Regime-switching sunspot equilibria under (6.4.7)–(6.4.9) with parameter values given by (6.4.10)–(6.4.12)

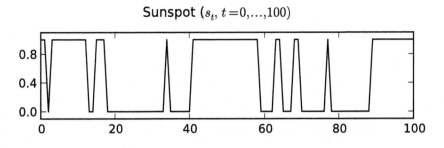

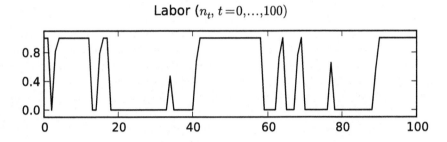

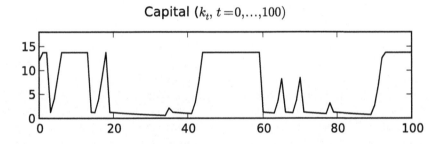

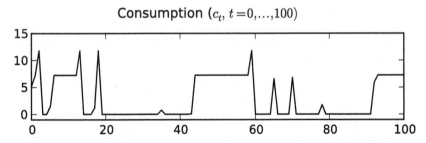

Fig. 6.2 Sample paths for sunspot states, capital, labor, and consumption generated by (6.4.7)–(6.4.9) with parameter values given by (6.4.10)–(6.4.12)

We also find a feature specific to consumption, which rises to its highest level when the sunspot state changes from 1 to 0 after remaining in the state of 1 for a few periods. This is expected from the consumption function in Fig. 6.1. Note that this function is increasing in k_t but unlike the labor and capital functions, decreasing in s_t in the sense that consumption is higher when $s_t = 0$ than when $s_t = 1$.

6.4.3 Stochastic Optimal Growth

One can conjecture that the foregoing feature of the consumption function may not necessarily arise from the presence of externalities, but rather as a consequence of optimal behavior. In this subsection we consider a stochastic optimal growth model without externalities and with regime-switching productivity coefficients. We show that the dynamics of this model are very similar to those of the previous model. The purpose of this subsection is to suggest the presence of a close connection between the two types of models to facilitate the analysis of regime-switching sunspot equilibria in the next subsection.

Consider the following stochastic optimal growth model:

$$\max_{\{c_t,n_t,k_{t+1}\}_{t=0}^{\infty}} E \sum_{t=0}^{\infty} \beta^t [u(c_t) - w(n_t)] \tag{6.4.13}$$

$$\text{s.t. } \forall t \in \mathbb{Z}_+, \quad c_t + k_{t+1} = s_t(k_t)^{\alpha+\overline{\alpha}}(n_t)^{\rho+\overline{\rho}} + (1-\delta)k_t, \tag{6.4.14}$$

$$c_t, k_{t+1} \geqslant 0, \quad n_t \in [0,1], \tag{6.4.15}$$

where $\{s_t\}$ is the same two-state Markov process following (6.4.1). In this subsection, we define s_t not as a sunspot shock, but as a stochastic productivity coefficient that directly affects production. When $s_t = 1$, the aggregate production function in the maximization problem above is unchanged from that in the previous subsection, but the externalities are internalized here. Since output is zero whenever $s_t = 0$, the problem inherits the pattern of (6.4.3).

Figure 6.3 depicts the optimal policy functions for the stochastic optimal growth model (6.4.13)–(6.4.15) under (6.4.10)–(6.4.12). We obtain these functions from (6.4.7)–(6.4.9) by setting

$$\alpha = 0.37, \quad \rho = 0.58, \quad \overline{\alpha} = \overline{\rho} = 0 \tag{6.4.16}$$

in (6.4.7)–(6.4.9).

Note that the consumption function in Fig. 6.3 is decreasing in s_t like the consumption function in Fig. 6.1; in fact, consumption with $s_t = 1$ is even lower than in Fig. 6.1. We can explain this by referring to the capital function in Fig. 6.3, which shows that more capital is accumulated in the stochastic optimal growth model (6.4.13)–(6.4.15) than in the original economy with externalities under (6.4.10)–(6.4.12). The functions in Fig. 6.3 serve as an example of a stochastic optimal growth

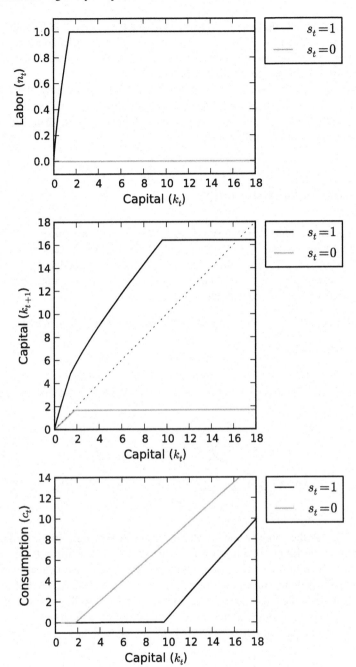

Fig. 6.3 Optimal policy functions for (6.4.13)–(6.4.15)

model in which consumption is *decreasing* in productivity, while capital and labor are increasing in productivity. This contrasts sharply with the Brock and Mirman (1972) model with i.i.d. productivity shocks, where consumption is always increasing in productivity; see Kamihigashi (2008, Theorem 2.1).

Figure 6.4 shows sample paths generated by these optimal policy functions with productivity states identical to the sunspot states in Fig. 6.2. The capital and consumption paths are also similar to those in Fig. 6.2, but the capital path and peaks in consumption are both overall higher.

6.4.4 No Capital Externality

Our analysis in the previous two subsections suggests that the sunspot equilibria of the original economy are closely connected to the optimal process of a stochastic optimal growth model. While this connection between the models is not trivial to show in general, we can establish it fairly easily in the absence of capital externalities. The proof of the following result utilizes the connection.

Proposition 6.4.2 *If $\overline{\alpha} = 0$, then a sunspot equilibrium satisfying (6.4.4) exists.*

Proof See Appendix C.   □

The condition $\overline{\alpha} = 0$ in the above proposition means that there is no capital externality. In the proof of Proposition 6.4.2, we consider the following stochastic optimal growth model:

$$\max_{\{c_t, n_t, k_{t+1}\}_{t=0}^{\infty}} E \sum_{t=0}^{\infty} \beta^t \left[u(c_t) - \frac{\rho + \overline{\rho}}{\rho} w(n_t) \right] \tag{6.4.17}$$

$$\text{s.t. } \forall t \geq 0, \quad c_t + k_{t+1} = s_t \theta (k_t)^{\alpha} (n_t)^{\rho+\overline{\rho}} + \zeta k_t, \tag{6.4.18}$$

$$c_t, k_{t+1} \geq 0, \quad n_t \in [0, 1], \tag{6.4.19}$$

where $\{s_t\}_{t=0}^{\infty}$ is the same two-state Markov chain following (6.4.1). In the proof, we show that the Euler condition for k_{t+1}, the first order condition for n_t, and the transversality condition are necessary for optimality, and are equivalent to the sufficient optimality conditions for the original economy (6.2.1)–(6.2.3) with $\overline{\alpha} = 0$. We can thus establish the existence of a sunspot equilibrium by showing the existence of an optimal process for the above stochastic optimal growth model.

The Bellman equation for (6.4.17)–(6.4.19) can be written as

$$v(k_t) = \max_{c_t, n_t, k_{t+1}} \left\{ u(c_t) - \frac{\rho + \overline{\rho}}{\rho} w(n_t) + \beta E_t v(k_{t+1}) \right\} \tag{6.4.20}$$

$$\text{s.t. } c_t + k_{t+1} = s_t \theta (k_t)^{\alpha} (n_t)^{\rho+\overline{\rho}} + \zeta k_t, \tag{6.4.21}$$

$$c_t, k_{t+1} \geq 0, \quad n_t \in [0, 1]. \tag{6.4.22}$$

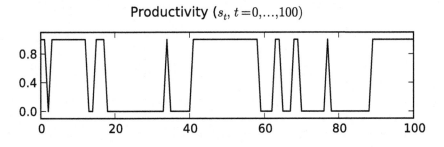

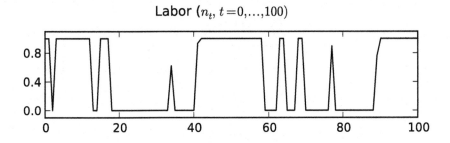

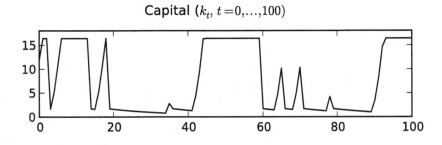

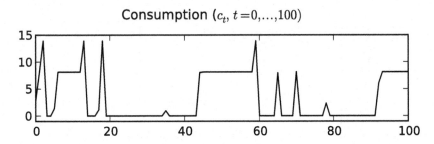

Fig. 6.4 Sample paths for productivity states, capital, labor, and consumption

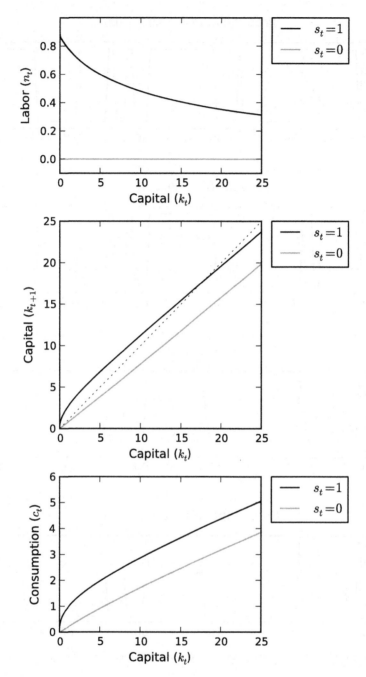

Fig. 6.5 Optimal policy functions for (6.4.20)–(6.4.22) and regime-switching sunspot equilibria for (6.2.1)–(6.2.3) under (6.4.10), (6.4.11), and (6.4.23)

Sunspot (s_t, $t=0,\ldots,100$)

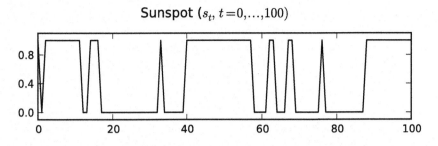

Labor (n_t, $t=0,\ldots,100$)

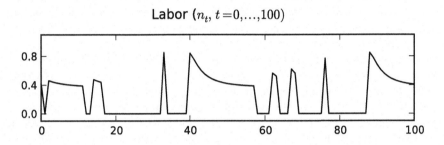

Capital (k_t, $t=0,\ldots,100$)

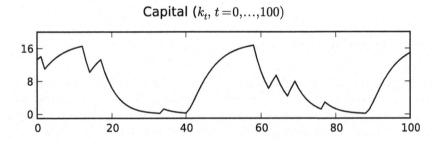

Consumption (c_t, $t=0,\ldots,100$)

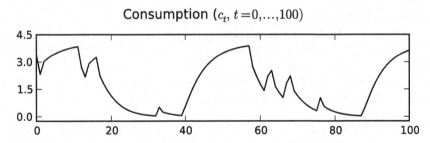

Fig. 6.6 Sample paths for sunspot states, capital, labor, and consumption under (6.4.10), (6.4.11), and (6.4.23)

From the proof of Proposition 6.4.2, we can interpret the optimal policy functions for
the above Bellman equation as constituting a regime-switching sunspot equilibrium.
To illustrate these functions, we use the same parameter values used in (6.4.10) and
(6.4.11), but replace the values of σ, α, and $\overline{\alpha}$ as follows:

$$\sigma = 0.99, \quad \alpha = 0.37, \quad \overline{\alpha} = 0. \tag{6.4.23}$$

With the above parameter values, the utility function u is almost logarithmic, and the
aggregate production function $f(k, n, k, n)$ remains unchanged from that in (6.4.11).

Figure 6.5 shows the optimal policy functions for the Bellman equation (6.4.20)–
(6.4.22) under the above parameter values. These functions are computed by numerically solving the Bellman equation using modified policy iteration (e.g., Puterman
2005) with 5,000 equally spaced grid points.

Figure 6.6 shows sample paths for sunspot states, labor, capital, and consumption
generated by the functions in Fig. 6.5. Compared to those in Fig. 6.2, the sample paths
in Fig. 6.6 appear less extreme due to the concavity of the utility function u. Capital
accumulates while the sunspot state is 1, and decumulates while it is 0. Consumption
follows almost exactly the same pattern as capital, while labor supply moves in
the opposite directions when $s_t = 1$, as expected from the labor and consumption
functions in Fig. 6.5.

6.5   Concluding Comments

In this paper we have shown that regime-switching sunspot equilibria easily arise in
a one-sector growth model with aggregate decreasing returns and arbitrarily small
externalities. We have explicitly constructed a regime-switching sunspot equilibrium
under the assumption that the utility function of consumption is linear. We have also
constructed a stochastic optimal growth model whose optimal process turns out to be a
regime-switching sunspot equilibrium of the original economy under the assumption
that there is no capital externality.

Although our results assume aggregate decreasing returns to scale, the existence
of regime-switching sunspot equilibria can easily be shown for models with increasing returns and large externalities, at least when the utility function is linear, as in
Sect. 6.4.2. On the other hand, the proof of Proposition 6.4.2 relies on the assumption of decreasing returns; it could be non-trivial to extend the proof to the case of
increasing returns.

To conclude this paper, we discuss possible ways to extend our analysis. First,
we can use a similar approach to construct more realistic sunspot equilibria than the
rather extreme equilibria considered in this paper, where labor supply and output
are both zero when the sunspot state is zero. Consider, for example, a one-sector
growth model with externalities in which the first-order condition for labor supply
has multiple solutions. Such a model can easily be produced if we are allowed to

assume externalities of a general form. With such a model, we can construct a sunspot equilibrium that switches between the multiple solutions of the first-order condition for labor supply according to the sunspot state.

Second, although we have focused on sunspot equilibria, we can also construct deterministic equilibria that exhibit chaotic dynamics. We can take, for example, a deterministic sequence of states s_t each assigned a value of either 0 or 1, and solve the deterministic version of the maximization problem (6.4.17)–(6.4.19). The resulting optimal path then follows the pattern of the sequence $\{s_t\}$. We can think of this as an example of symbolic dynamics; see Kamihigashi (1999) for economic applications of symbolic dynamics.

Finally, in Proposition 6.4.2 we have only considered the case without capital externality. While it seems difficult to extend the same approach to models with capital externalities (as long as the capital depreciation rate is less than one), there is a way to deal with such models. In particular, if we allow for nonlinear discounting along the lines of Kamihigashi (2002), we can construct a stochastic optimal growth model whose optimal process turns out to be a sunspot equilibrium of the original economy.

Appendix A Proof of Lemma 6.3.1

Let $\{c_t^*, n_t^*, k_t^*\}_{t=0}^\infty$ be a feasible process satisfying (6.3.4)–(6.3.6) (with c_t^*, n_t^*, k_t^* replacing c_t, n_t, k_t). To simplify notation, for $t \in \mathbb{Z}_+$ and $i = 1, 2$ we define

$$f(t) = f(k_t^*, n_t^*, k_t^*, n_t^*), \tag{6.5.1}$$

$$f_i(t) = f_i(k_t^*, n_t^*, k_t^*, n_t^*). \tag{6.5.2}$$

We are to show that for any feasible process $\{c_t, n_t, k_t\}_{t=0}^\infty$, we have

$$E \sum_{t=0}^\infty \beta^t [u(c_t) - w(n_t)] - E \sum_{t=0}^\infty \beta^t [u(c_t^*) - w(n_t^*)] \leqslant 0. \tag{6.5.3}$$

To this end, let $\{c_t, n_t, k_t\}_{t=0}^\infty$ be a feasible process. Fix $T \in \mathbb{N}_+$ for the moment. Let

$$\Delta_T = E \sum_{t=0}^T \beta^t [u(c_t) - w(n_t)] - E \sum_{t=0}^T \beta^t [u(c_t^*) - w(n_t^*)] \tag{6.5.4}$$

$$\leqslant E \sum_{t=0}^T \beta^t \{u'(c_t^*)(c_t - c_t^*) - w'(n_t^*)(n_t - n_t^*)\}, \tag{6.5.5}$$

where $u'(c_t^*)$ is the right derivative of u at 0 if $c_t^* = 0$, and similarly for $w'(n_t^*)$. We have

$$\Delta_T \leqslant E \sum_{t=0}^{T} \beta^t \{u'(c_t^*)[f(k_t, n_t, k_t^*, n_t^*) - f(t)$$

$$+ \zeta(k_t - k_t^*) - (k_{t+1} - k_{t+1}^*)] - w'(n_t^*)(n_t - n_t^*)\} \tag{6.5.6}$$

$$\leqslant E \sum_{t=0}^{T} \beta^t [u'(c_t^*)(f_1(t) + \zeta)(k_t - k_t^*)$$

$$+ \{u'(c_t^*)f_2(t) - w'(n_t^*)\}(n_t - n_t^*) - u'(c_t^*)(k_{t+1} - k_{t+1}^*)]. \tag{6.5.7}$$

Recalling the first-order condition (6.3.4) for n_t, we see that for all $t \in \mathbb{Z}_+$,

$$\{u'(c_t^*)f_2(t) - w'(n_t^*)\}(n_t - n_t^*) \leqslant 0. \tag{6.5.8}$$

Substituting into (6.5.7) we obtain

$$\Delta_T \leqslant E \sum_{t=0}^{T} \beta^t [u'(c_t^*)(f_1(t) + \zeta)(k_t - k_t^*) - u'(c_t^*)(k_{t+1} - k_{t+1}^*)] \tag{6.5.9}$$

$$= E \sum_{t=0}^{T-1} \beta^t [-u'(c_t^*) + \beta u'(c_{t+1}^*)(f_1(t+1) + \zeta)](k_{t+1} - k_{t+1}^*) \tag{6.5.10}$$

$$- \beta^T E u'(c_T^*)(k_{T+1} - k_{T+1}^*) \tag{6.5.11}$$

$$= E \sum_{t=0}^{T-1} \beta^t [-u'(c_t^*) + \beta E_t u'(c_{t+1}^*)(f_1(t+1) + \zeta)](k_{t+1} - k_{t+1}^*) \tag{6.5.12}$$

$$- \beta^T E u'(c_T^*)(k_{T+1} - k_{T+1}^*), \tag{6.5.13}$$

where the last equality holds by the law of iterated expectations. Recalling the Euler condition (6.3.5) for k_{t+1}, we see that for all $t \in \mathbb{Z}_+$,

$$[-u'(c_t^*) + \beta E_t u'(c_{t+1}^*)(f_1(t+1) + \zeta)](k_{t+1} - k_{t+1}^*) \leqslant 0. \tag{6.5.14}$$

Substituting into (6.5.13) we obtain

$$\Delta_T \leqslant -\beta^T E u'(c_T^*)(k_{T+1} - k_{T+1}^*) \tag{6.5.15}$$

$$\leqslant \beta^T E u'(c_T^*)k_{T+1}^* \to 0, \tag{6.5.16}$$

where the second inequality holds since $k_{T+1} \geqslant 0$, and the convergence holds by the transversality condition (6.3.6). This completes the proof of Lemma 6.3.1.

Appendix B Proof of Proposition 6.4.1

Suppose that $\sigma = 0$. Then conditions (6.4.5) and (6.4.6) can be written as

$$s_t = 1 \quad \Rightarrow \quad \rho\theta(k_t)^{\alpha+\bar{\alpha}}(n_t)^{\rho+\bar{\rho}-1} - \eta(n_t)^{\gamma} \begin{cases} = 0 & \text{if } n_t \in (0,1), \\ \geqslant 0 & \text{if } n_t = 1, \end{cases} \quad (6.5.17)$$

$$s_t = 0 \quad \Rightarrow \quad n_t = 0. \quad (6.5.18)$$

The Euler condition for k_{t+1}, (6.3.5), can be written as

$$\beta E_t[\alpha\theta(k_{t+1})^{\alpha+\bar{\alpha}-1}(n_{t+1})^{\rho+\bar{\rho}} + \zeta] \begin{cases} = 1 & \text{if } k_{t+1} \in (0, g(k_t, n_t)), \\ \geqslant 1 & \text{if } k_{t+1} = g(k_t, n_t), \\ \leqslant 1 & \text{if } k_{t+1} = 0. \end{cases} \quad (6.5.19)$$

The transversality condition (6.3.6) reduces to

$$\lim_{T \to \infty} \beta^T E k_{T+1} = 0. \quad (6.5.20)$$

Note that (6.5.17) and (6.5.18) can be combined into

$$n_t = m(k_t, s_t) \equiv s_t \min \left\{ \left[\frac{\rho\theta}{\eta}(k_t)^{\alpha+\bar{\alpha}} \right]^{\frac{1}{\gamma+1-\rho-\bar{\rho}}}, 1 \right\}. \quad (6.5.21)$$

Substituting into the left-hand side of (6.5.19) we obtain

$$E_t[\alpha\theta(k_{t+1})^{\alpha+\bar{\alpha}-1}(n_{t+1})^{\rho+\bar{\rho}} + \zeta] \quad (6.5.22)$$

$$= E_t[\alpha\theta(k_{t+1})^{\alpha+\bar{\alpha}-1}m(k_{t+1}, s_{t+1})^{\rho+\bar{\rho}} + \zeta] \quad (6.5.23)$$

$$= \begin{cases} p_{01}h(k_{t+1}) + \zeta & \text{if } s_t = 0, \\ p_{11}h(k_{t+1}) + \zeta & \text{if } s_t = 1, \end{cases} \quad (6.5.24)$$

where

$$h(k) = \alpha\theta k^{\alpha+\bar{\alpha}-1}m(k,1)^{\rho+\bar{\rho}} \quad (6.5.25)$$

$$= \min \left\{ \alpha\theta \left[\frac{\rho\theta}{\eta} \right]^{\frac{\rho+\bar{\rho}}{\gamma+1-\rho-\bar{\rho}}} k^{\frac{(\alpha+\bar{\alpha}-1)(\gamma+1)+\rho+\bar{\rho}}{\gamma+1-\rho-\bar{\rho}}}, \alpha\theta k^{\alpha+\bar{\alpha}-1} \right\}. \quad (6.5.26)$$

Both expressions in the curly brackets are strictly decreasing in k by (6.2.10) and (6.2.12) (note that $(\alpha + \bar{\alpha} - 1)(\gamma + 1) + \rho + \bar{\rho} < \alpha + \bar{\alpha} - 1 + \rho + \bar{\rho} \leqslant 1$ by (6.2.12)). Thus $h(\cdot)$ is strictly decreasing, which implies that the inverse $h^{-1}(\cdot)$ exists. Indeed, for $z > 0$ we have

$$h^{-1}(z) = \min \left\{ \left[\left[\frac{z}{\alpha\theta} \right] \left[\frac{\eta}{\rho\theta} \right]^{\frac{\rho+\bar{\rho}}{\gamma+1-\rho-\bar{\rho}}} \right]^{\frac{\gamma+1-\rho-\bar{\rho}}{(\alpha+\bar{\alpha}-1)(\gamma+1)+\rho+\bar{\rho}}} , \left[\frac{z}{\alpha\theta} \right]^{\frac{1}{\alpha+\bar{\alpha}-1}} \right\}. \qquad (6.5.27)$$

Note that

$$\lim_{k\downarrow 0} h(k) = \infty. \qquad (6.5.28)$$

Substituting (6.5.22)–(6.5.24) into (6.5.19) we obtain

$$\beta[p_{s_t 1} h(k_{t+1}) + \zeta] \begin{cases} = 1 & \text{if } k_{t+1} \in (0, g(k_t, n_t)), \\ \geqslant 1 & \text{if } k_{t+1} = g(k_t, n_t), \\ \leqslant 1 & \text{if } k_{t+1} = 0, \end{cases} \qquad (6.5.29)$$

where $p_{s_t 1} = p_{01}$ or p_{11} depending on $s_t = 0$ or 1. For $p > 0$ define

$$q(p) = h^{-1}\left(\frac{1 - \beta\zeta}{\beta p} \right). \qquad (6.5.30)$$

Note from (6.5.28) that we can rule out the case $k_{t+1} = 0$ in (6.5.29). Hence we can write (6.5.29) as

$$k_{t+1} = \min\{q(p_{s_t 1}), g(k_t, n_t)\}. \qquad (6.5.31)$$

We construct a process $\{c_t, n_t, k_t\}_{t=0}^{\infty}$ recursively as follows: given $k_t > 0$ and $s_t \in \{0, 1\}$, let

$$n_t = m(k_t, s_t). \qquad (6.5.32)$$

Determine k_{t+1} by (6.5.31). Let

$$c_t = g(k_t, n_t) - k_{t+1}. \qquad (6.5.33)$$

Draw s_{t+1} according to (6.4.1). Determine n_{t+1} by (6.5.32), and so on. By construction, this process is feasible and satisfies (6.5.17)–(6.5.19). It also satisfies (6.5.20) by (6.2.13). Thus it is a sunspot equilibrium. The conclusion of the proposition now follows.

C Proof of Proposition 6.4.2

Suppose that $\bar{\alpha} = 0$. Consider the stochastic optimal growth model (6.4.17)–(6.4.19). The Euler condition for k_{t+1} is written as

$$- u'(c_t) + \beta E_t u'(c_{t+1})[s_{t+1}\alpha\theta(k_{t+1})^{\alpha-1}(n_{t+1})^{\rho+\overline{\rho}} + \zeta]$$

$$\begin{cases} = 0 & \text{if } k_{t+1} \in (0, g(k_t, n_t)), \\ \geqslant & \text{if } k_{t+1} = g(k_t, n_t), \\ \leqslant & \text{if } k_{t+1} = 0. \end{cases} \tag{6.5.34}$$

This is equivalent to the equilibrium Euler condition (6.3.5) for k_{t+1} for the original economy (6.2.1)–(6.2.3) with $\overline{\alpha} = 0$ and (6.4.3). The first-order condition for n_t for the above stochastic optimal growth model is given by

$$u'(c_t)s_t(\rho + \overline{\rho})\theta(k_t)^\alpha(n_t)^{\rho+\overline{\rho}-1} - \frac{\rho+\overline{\rho}}{\rho}w'(n_t) \begin{cases} = 0 & \text{if } n_t \in (0, 1), \\ \geqslant 0 & \text{if } n_t = 1, \\ \leqslant 0 & \text{if } n_t = 0, \end{cases} \tag{6.5.35}$$

which simplifies to

$$u'(c_t)s_t\rho\theta(k_t)^\alpha(n_t)^{\rho+\overline{\rho}-1} - w'(n_t) \begin{cases} = 0 & \text{if } n_t \in (0, 1), \\ \geqslant 0 & \text{if } n_t = 1, \\ \leqslant 0 & \text{if } n_t = 0. \end{cases} \tag{6.5.36}$$

This is equivalent to (6.3.4) with $\overline{\alpha} = 0$ and (6.4.3). The transversality condition for the above problem is identical to (6.3.6).

Conditions (6.5.34) and (6.5.36) are necessary for optimality by standard arguments. The transversality condition (6.3.6) is also necessary by the argument of Kamihigashi (2005, Sect. 6).[5] Given that the sunspot variable s_t is discrete, we can easily establish the existence of an optimal process for the optimal stochastic growth model (6.4.17)–(6.4.19) by a standard argument (e.g., Ekeland and Scheinkman 1986). Let $\{c_t, n_t, k_t\}_{t=0}^\infty$ be an optimal process for (6.4.17)–(6.4.19). Then by the above argument, the process satisfies (6.3.4)–(6.3.6). Thus by Lemma 6.3.1, the process is an equilibrium of the original economy (6.2.1)–(6.2.3). Since it depends on s_t in a nontrivial way, it is a sunspot equilibrium.

References

Benhabib, J., & Farmer, R. E. A. (1994). Indeterminacy and increasing returns. *Journal of Economic Theory, 63*, 19–41.

Benhabib, J., & Farmer, R. E. A. (1999) Chapter 6: Indeterminacy and sunspots in macroeconomics. Handbook of Macroeconomics, 1A, 387–448.

Benhabib, J., Meng, Q., & Nishimura, K. (2000). Indeterminacy under constant returns to scale in multisector economies. *Econometrica, 68*, 1541–1548.

---

[5]The condition $\sigma \in [0, 1]$ is needed here.

Brock, W. A. (1982). Asset prices in a production economy. In J. J. McCall (Ed.), *The Economics of Information and Uncertainty* (pp. 1–46). Press: University of Chicago.

Brock, W. A., & Mirman, L. J. (1972). Optimal economic growth and uncertainty: The discounted case. *Journal of Economic Theory, 4*, 479–513.

Clain-Chamosset-Yvrard, L., & Kamihigashi, T. (2015). International transmission of bubble crashes: Stationary sunspot equilibria in a two-country overlapping generations model, RIEB Discussion Paper DP2015-21, Kobe University.

Coury, T., & Wen, Y. (2009). Global indeterminacy in locally determinate real business cycle models. *International Journal of Economic Theory, 5*, 49–60.

Dos Santos Ferreira, R., & Lloyd-Braga, T. (2008). Business cycles with free entry ruled by animal spirits. *Journal of Economic Dynamics and Control, 32*, 3502–3519.

Drugeon, J.-P., & Wigniolle, B. (1996). Continuous-time sunspot equilibria and dynamics in a model of growth. *Journal of Economic Growth, 69*, 24–52.

Drugeon, J.-P., & Venditti, A. (2001). Intersectoral external effects, multiplicities & indeterminacies. *Journal of Economic Dynamics and Control, 25*, 765–787.

Dufoourt, F., Nishimura, K., & Vendittie, A. (2015). Indeterminacy and sunspots in two-sector RBC models with generalized no-income-effect preferences. *Journal of Economic Theory, 157*, 1056–1080.

Ekeland, I., & Scheinkman, S. A. (1986). Transversality conditions for some infinite horizon discrete time optimization problems. *Mathematics of Operations Research, 11*, 216–229.

Farmer, R. E. A., & Guo, J.-T. (1994). Real business cycles and the animal spirits hypothesis. *Journal of Economic Theory, 63*, 42–72.

Grandmont, J.-M. (1989). *Local bifurcation and stationary sunspots. Economic Complexity: Chaos, Sunspots, Bubbles, and Nonlinearity* (pp. 45–60).

Grandmont, J.-M. (1991). Expectations driven business cycles. *European Economic Review, 35*, 293–299.

Kamihigashi, T. (1996). Real business cycles and sunspot fluctuations are observationally equivalent. *Journal of Monetary Economics, 37*, 105–117.

Kamihigashi, T. (1999). Chaotic dynamics in quasi-static systems: Theory and applications. *Journal of Mathematical Economics, 31*, 183–214.

Kamihigashi, T. (2002). Externalities and nonlinear discounting: Indeterminacy. *Journal of Economic Dynamics and Control, 26*, 141–169.

Kamihigashi, T. (2003). Necessity of transversality conditions for stochastic problems. *Journal of Economic Theory, 109*, 140–149.

Kamihigashi, T. (2005). Necessity of the transversality condition for stochastic models with bounded or CRRA utility. *Journal of Economic Dynamics and Control, 29*, 1313–1329.

Kamihigashi, T. (2008). Stochastic optimal growth with bounded or unbounded utility and with bounded or unbounded shocks. *Journal of Mathematical Economics, 43*, 477–500.

Kamihigashi, T. (2015). Multiple interior steady states in the Ramsey model with elastic labor supply. *International Journal of Economic Theory, 11*, 25–37.

Mino, K. (2001). Indeterminacy and endogenous growth social constant returns. *Journal of Economic Theory, 97*, 203–222.

Pelloni, A., & Waldmann, R. (1998). Stability properties of a growth model. *Economics Letters, 61*, 55–60.

Pintus, P. A. (2006). Indeterminacy with almost constant returns to scale: Capital-labor substitution matters. *Economic Theory, 28*, 633–649.

Puterman, M. L. (2005). *Markov Decision Processes: Discrete Stationary Dynamic Programming*. Hoboken, NJ: John Wiley & Sons.

Chapter 7
Equilibrium Dynamics in a Two-Sector OLG Model with Liquidity Constraint

Antoine Le Riche and Francesco Magris

Abstract We study a two-sector *OLG* economy in which a share of old age consumption expenditures must be paid out of money balances and we appraise its dynamic features. We first show that competitive equilibrium is dynamically efficient if and only if the share of capital on total income is large enough while a steady state capital per capita above its Golden Rule level is not consistent with a binding liquidity constraint. We thus focus on the gross substitutability in consumption and on dynamic efficiency assumptions and show that, gathered together, they ensure the local determinacy of equilibrium and, as a consequence, rule out sunspot fluctuations. In addition, we prove that the unique steady state may change its stability from a saddle configuration to a source one (undergoing a flip bifurcation) for a capital intensive investment good as well as for a capital intensive consumption good, when the elasticity of the interest rate is set low enough. However, when the investment good is not too capital intensive, the flip bifurcation turns out to be compatible with high elasticities of the interest rate too. Analogous results within dynamic efficiency are found in the non-monetary model, the existence of a flip bifurcation requiring now a capital intensive investment good. Eventually, under dynamic inefficiency, in the non-monetary economy local indeterminacy may instead appear, either through a Hopf bifurcation or through a flip one, and its scope improves as soon as the consumption good becomes more and more capital intensive.

We thank an anonymous referee for his precious remarks and suggestions. We would like to thank Robert Becker, Stefano Bosi, Rodolphe Dos Santos Ferreira, Jean-Pierre Drugeon, Jean-Michel Grandmont, Teresa Lloyd-Braga, Eleonor Modesto, Kazuo Nishimura, Carine Nourry, Thomas Seegmuller, Alain Venditti and participants to the conference "Financial and Real Interdependencies: Volatility, Inequalities and Economic Policies", Católica Lisbon School of Business & Economics, Lisbon, May 2015. Any errors are our own.

A. Le Riche (✉)
School of Economics, Sichuan University, No. 24, South Section 1, Yihuan Road,
Chengdu, P. R. China
e-mail: antoinelerichee@yahoo.fr

F. Magris
LEO, University "François Rabelais" of Tours, 50, Avenue Jean Portalis,
37206 Tours Cedex 03, Bureau B246, France
e-mail: francesco.magris@univ-tours.fr

© Springer International Publishing AG 2017
K. Nishimura et al. (eds.), *Sunspots and Non-Linear Dynamics*,
Studies in Economic Theory 31, DOI 10.1007/978-3-319-44076-7_7

Keywords Overlapping generations · Two-sector · Money demand · Dynamic efficiency · Equilibrium dynamics

JEL Classification D24 · E30 · E32 · E41 · E50

7.1 Introduction

In this paper we study a two-sector *OLG* model with capital accumulation, in which agents supply labor elastically when young and consume when old. However, an exogenously given share of old age consumption must be financed out of outside *fiat* money, accumulated therefore by young people beside physical capital. We carry out a complete qualitative analysis of the local stability and appraise the conditions under which endogenous fluctuations and sunspots equilibria may or may not arise. We also provide an accurate bifurcation analysis in order to appreciate the changes in stability of the steady state giving rise, nearby, to close orbits as well as deterministic cycles, whose stability, in turn, depends upon the direction of the bifurcation studied.

Analyzing a two-sector OLG model with cash-in-advance on consumption expenditures is by far more than a mere theoretical curiosity. It is not a mystery, indeed, that the production of consumption goods is usually made possible by means of technologies which are rather different with respect to the ones employed to produce the investment goods, as many empirical studies suggest. As an example, according to Takashi et al. (2012) and Baxter (1996) empirical estimates, the consumption sector appears often to be more capital intensive that the investment one. In addition, the choice of focusing on a two-sector model allows to enrich the equilibrium dynamics: some phenomena are indeed exclusive pertinence of such class of models as, for example, the cyclical behavior arising in the optimal infinite horizon models studied, among the others, in Benhabib and Nishimura (1985) and Venditti (2005). By extending the hypothesis of different technologies to an OLG model, one is able to further enrich the dynamic features since, in addition to the mechanism relying on the different factor intensities, one faces the typical instability linked to the limited market participation as stressed, e.g., by Gale (1973), Azariadis (1981), Azariadis and Guesnerie (1982), Grandmont (1985) and Reichlin (1986).

The choice of focusing on a two-sector OLG model is even more motivated, in addition to standard arguments for valuing positively *fiat* money, in the light of the specific assumption of a cash-in-advance constraint on consumption expenditures. As pointed out by Bosi et al. (2005), indeed, within a one-sector model, and thus a unique representative firm, the requirement of a fraction of the consumption good to be paid cash could be easily avoided (and along with it the loss represented by the nominal interest rate) since customers could always formulate their demand in terms of the investment good even though in concrete, after the purchase, they consume it. Actually, in view of the perfect substitutability of the two goods, the firm is unable to discern to which end the amount of the good purchased is employed. Conversely, under the assumption of two distinct representative firms, each of them producing and selling only one good and thus with its own market, the distinction between the consumption good (subject to the CIA constraint) and the investment one (which could be bought "credit") does not give rise to any kind of ambiguity and the CIA constraint cannot thus be avoided.

Since the seminal studies by, among the others, Gale (1973), Azariadis (1981) Azariadis and Guesnerie (1982), Grandmont (1985), it is well established that the limited market participation characterizing *OLG* models may be a source of expectations-driven fluctuations in a framework of competitive economy where prices flexibility ensures equilibrium simultaneously in all markets. Such fluctuations occur even in the absence of any exogenous shock affecting economic fundamentals (tastes, endowments, technology) but are generated by the volatility of agents' state of expectations and/or by the non-linearities of the dynamics describing intertemporal equilibrium. Such kinds of fluctuations are often labelled as "sunspot" equilibria, in homage to the early neoclassical economist Stanley Jevons who postulated the existence of a close relationship between the occurrence of sunspots and the harvest outcome. As proved, among others, in Woodford (1986), Grandmont et al. (1988) and Bloise (2001), a sufficient condition in order to get sunspot equilibria is to be found in the existence of an indeterminate steady state of the intertemporal equilibrium, i.e. a steady state reached in correspondence to an infinite set of choices for the non-predetermined variables describing the economy. In fact, in such a case, in each period agents are faced with quite a large number of choices, all of them compatible with long-run equilibrium (namely the limit conditions): it follows that the unique available selective device rests upon individual expectations—no matter how they are formed, but simply reflecting some shared beliefs—of the future "state of affairs" of the whole dynamic process.

The aforementioned contributions, however, rest upon the assumption of pure-exchange or productive economies without capital accumulation in which the unique asset to invest savings in is to be found in an exogenously given amount of *fiat* money. Under such an hypothesis, local indeterminacy and sunspot fluctuations (and also deterministic cycles and maybe chaotic dynamics, in the spirit of Ruelle 1989) require strong enough income effects, namely a saving function reacting negatively with respect to its rate of return which, in a purely monetary economy, boils down to the deflation rate. The requirement of strong income effects for the emergence of sunspot fluctuations has been the object of several criticisms based on empirical grounds: for example, Eichenbaum et al. (1988) and Hall (1988) find that the estimated value of the elasticity of intertemporal substitution in consumption falls within the -0.0–10-range, which means that the saving rate may easily be positively related to its return. Such a criticism, it is worthwhile noticing, can be extended to infinite horizon models too, with cash-in-advance constraint on consumption expenditures in which the steady state is locally indeterminate only under the hypothesis of a strong complementariness in intertemporal consumption (Bloise et al. 2000). Things change dramatically when one assumes a fractional liquidity constraint: if the share of consumption to be paid in cash is set low enough, as shown somewhat paradox-ically in Bosi et al. (2005) in an economy with productive capital, indeterminacy is bound to prevail for whatever parameters configuration.

Reichlin (1986) repairs the lack of empirical evidence of the early *OLG* models by accounting for a non-monetary one-sector economy with capital accumulation and gross substitutability in intertemporal consumption. He proves that indeterminacy requires dynamic inefficiency and strong inputs complementariness. Cazzavillan and Pintus (2007) show that the co-existence of dynamic efficiency and local indetermi-nacy is not robust to the consideration of any positive elasticity of capital-labor-substitution. Benhabib and Laroque (1988) show in an analogous economy that if one introduces money as a bubble, cyclical equilibria require both the quantity of money to be negative at the Golden Rule and inputs to be complementary. Cazzav-illan and Pintus (2005) prove, on the other hand, that when capital externalities are introduced into the Benhabib and Laroque (1988) model, stationary sunspots may occur when the quantity of money is positive and inputs are substitutable enough. Rochon and Polemarchakis (2006) extend the Benhabib and Laroque (1988) model by considering an *OLG* economy with cash-in-advance constraint and government bonds. The coexistence of dynamic efficiency and local indeterminacy in two-sector models are studied by Drugeon et al. (2010), Nourry and Venditti (2011, 2012) and Le Riche et al. (2013). Drugeon et al. (2010) and Nourry and Venditti (2011) prove, in a two-sector model with one consumption good and one investment good, that local indeterminacy is ruled out when the steady state is dynamically efficient, provided the sectoral technologies are not too close to the Leontief production function. On the other hand, Nourry and Venditti (2012) and Le Riche et al. (2013) show, in a two-sector model with one pure consumption good and one consumable capital good, that dynamic efficiency together with local indeterminacy is compatible with stan-dard sectorial technologies if the share of the pure consumption good is low enough. Meanwhile *OLG* economies with *CIA* constraint on consumption expenditures are the object of the studies of Crettez et al. (1999, 2002) and Michel and Wigliolle (2005). However they are mostly concerned with the effects of the monetary policy on aggregate welfare properties. Grandmont et al. (1988) within an infinite horizon model with heterogeneous agents and cash-in-advance constraint find a picture for the local dynamics very close to that analyzed by Reichlin (1986): sunspot fluctua-tions occur when inputs are complementary enough. Cazzavillan et al. (1998) extend such a model by introducing aggregate externalities in production and show that local indeterminacy reappears for inputs high enough substitutable.

In our study we depart from Reichlin (1986), Crettez et al. (1999) and Drugeon et al. (2010), by accounting at the same time for different sectoral factor intensities in the production of the consumption good and of the investment good and for a fractional liquidity constraint on consumption expenditures. By restricting attention to the case in which the gross substitutability assumption holds, our first task is to ensure the dynamic efficiency of the economy at the Golden Rule level: only under such a feature is money dominated by capital in terms of returns and is the cash-in-advance constraint thus binding. We show that dynamic efficiency requires, at the unique steady state, a share of capital in total income jointly with a share of consumption to be paid cash not too low, features easily falling within standard empirical estimates.

We show that under dynamic efficiency and gross substitutability, local determinacy is bound to prevail and thus sunspot equilibria are ruled out. However, there is still room for a flip bifurcation, even though some additional requirements are needed in terms of the capital intensity in the consumption sector that must be either sufficiently high or low enough, jointly with an elasticity of the real interest rate and an elasticity of the offer curve high enough. As is well known both in the literature on infinite horizon models as well as on *OLG* ones (Benhabib and Nishimura 1985; Galor 1992; Venditti 2005; Nourry and Venditti 2011) a capital intensive investment good favors the occurrence of endogenous fluctuations, while a capital intensive consumption good seems to reduce the scope for such phenomena. These results are confirmed in our study, both in the non-monetary economy as well as in the monetary one.

In the non-monetary economy, which extends the Reichlin (1986) model and is characterized by the absence of the financial constraint, under dynamic efficiency and gross substitutability, local indeterminacy and sunspot fluctuations are ruled out. Still, there is room for a flip bifurcation, provided the investment sector is rather capital intensive and the elasticity of the real interest rate large enough. Conversely, local indeterminacy and sunspot fluctuations may occur under capital over-accumulation. Namely, the scope of such phenomena improves as soon as the investment good is made more and more capital intensive: we obtain that the range of the (high) values for the elasticity of the interest rate compatible with a stable stationary solution improves as soon as the relative capital intensive in the sector producing the investment good becomes larger and larger.

In the monetary economy, dynamic efficiency is required to ensure a binding cash-in-advance constraint and local determinacy is then bound to prevail. When the consumption good is capital intensive, however, and differently from the non-monetary economy, deterministic cycles may arise along a flip bifurcation (obtained by increasing continuously the elasticity of the offer curve), provided the elasticity of the interest rate is set low enough. When the investment good is to be capital intensive, the local dynamics becomes richer and different pictures are obtained by varying the parameters configuration. To synthesize the main results, we obtain, for a not very capital intensive investment good, a flip bifurcation in correspondence, first, to low elasticities of the interest rate and, afterwards, to sufficiently large elasticities of this type. On the other hand, in correspondence to a strongly capital intensive investment good, the flip bifurcation is compatible only under the assumption of an elasticity of the real interest rate low enough.

The remainder of the paper is organized as follows. In Sect. 7.2 we present the agents' behavior, the technology, and we provide the definition of intertemporal equilibrium. We also calibrate a particular stationary solution. Section 7.3 is devoted to the analysis of the dynamic efficiency while the main results of the paper, in terms of the local dynamics features, are left to Sect. 7.4. Section 7.5 contains the concluding remarks. Some proofs are left to the Appendix.

7.2 The Model

7.2.1 Technology

We consider a competitive economy in which there are two sectors producing, respectively, a pure consumption good and a pure investment good. In each sector operates a representative firm. We denote the consumption good, produced in period t, $Y_{0,t}$, and the investment good Y_t. The consumption good is taken as the *numéraire*. Each sector uses two factors, physical capital K_t and labor L_t, and both factors are perfectly mobile across sectors. Capital fully depreciates from one period to another[1] and therefore one has $K_{t+1} = Y_t$, with K_{t+1} being the total amount of capital available in period $t + 1$. A constant returns to scale technology is used in each sector and the two goods are produced according to the technological relationships $Y_{0,t} = F^0\left(K_t^0, L_t^0\right)$ and $Y_t = F^1\left(K_t^1, L_t^1\right)$, with $K_t^0 + K_t^1 \leq K_t$ and $L_t^0 + L_t^1 \leq L_t$, where K_t^j and L_t^j, $j = 0, 1$, denote, respectively, the amount of capital and labor utilized in the sector j and K_t and L_t the total amount of capital and labor available in the economy. The production functions satisfy the following properties:

Assumption 1 The production function $F^j : \mathbb{R}_+^2 \rightarrow \mathbb{R}_+^2$, $j = 0, 1$ is C^2, increasing, concave, homogeneous of degree one and satisfies the Inada conditions such that, for any $\mu > 0$, $F_1^j(0, \mu) = F_2^j(\mu, 0) = \infty$, $F_1^j(\infty, \mu) = F_2^j(\mu, \infty) = 0$.[2]

The optimal allocation of factors between sectors is defined by the social production function $T(K_t, Y_t, L_t)$:

$$T(K_t, Y_t, L_t) = \max_{K_t^j, L_t^j} \left\{ Y_{0,t} : | : Y_t \leq F^1(K_t^1, L_t^1), : K_t^0 + K_t^1 \leq K_t, : L_t^0 + L_t^1 \leq L_t \right\}. \quad (7.1)$$

Under Assumption 1, the function $T(K_t, Y_t, L_t)$ is homogeneous of degree one, concave and twice continuously differentiable.[3] Let us denote r_t the rental rate of capital, p_t the price of the investment good and w_t the wage rate, all in terms of the price of the consumption good. Using the envelope theorem we obtain the following three relationships:

$$r(K_t, Y_t, L_t) = T_1(K_t, Y_t, L_t), p(K_t, Y_t, L_t) = -T_2(K_t, Y_t, L_t), w(K_t, Y_t, L_t) = T_3(K_t, Y_t, L_t). \quad (7.2)$$

The relative capital intensity difference b is derived from the factor-price frontier:

$$b = \frac{L^1}{Y}\left(\frac{K^1}{L^1} - \frac{K^0}{L^0}\right). \quad (7.3)$$

---

[1] In a two-period OLG model, full depreciation of capital is justified by the fact that the length of the period is about 30 years.

[2] $F_1^j(K^j, L^j)$ and $F_2^j(K^j, L^j)$ denote, respectively, the derivatives $\partial F^j(K^j, L^j)/\partial K^j$ and $\partial F^j(K^j, L^j)/\partial L^j$.

[3] See Benhabib and Nishimura (1981).

The sign of b is positive (resp. negative) if and only if the consumption good is labor (resp. capital) intensive. The Stolper-Samuelson effect $(dr/dp, dw/dp)$ and the Rybczynski effect $(dY^0/dK, dY/dK)$ are determined, respectively, by the factor-price frontier and the full employment condition, and are given by:

$$\frac{dr}{dp} = \frac{dY}{dK} = b^{-1}, \frac{dw}{dp} = \frac{dY^0}{dK} = -ab^{-1}. \tag{7.4}$$

Under a consumption good labor intensive, the Stolper-Samuelson effect states that an increase of the relative price of the investment good decreases the rental rate of capital and increases the wage rate whereas the Rybczynski effect specifies that an increase of the capital-labor ratio decreases the production of the consumption good and increases the production of the investment good. Furthermore, from the GDP function $T(K_t, Y_t, L_t) + p_t Y_t = w_t L_t + r_t K_t$, we get the share s of capital on total income:

$$s(K_t, Y_t, L_t) = \frac{r_t K_t}{T(K_t, Y_t, L_t) + p_t Y_t} \in (0, 1). \tag{7.5}$$

7.2.2 Preferences

We assume an infinite horizon discrete-time economy populated by overlapping generations of agents living for two periods: in the first one they are young, in the second old. There is no population growth and the total size of the population is normalized to one. In the first period, young agents supply elastically L_t units of labor, with $L_t \in (0, \mathscr{L})$, and receive a wage income. They invest this income in capital K_{t+1} and money balances M_{t+1}. In the second period, old agents are retired and purchase the consumption good C_{t+1} out of capital and money income. We assume that agents are subject to a cash-in-advance (CIA) constraint on old age consumption purchases, following analogous lines as in Hahn and Solow (1955): $\chi p_{t+1}^C C_{t+1} \leq M_{t+1}$. Such a constraint claims that at least a share $\chi \in (0, 1)$ of old age consumption expenditures C_{t+1} must be financed out of money balances M_{t+1} saved in the first period of life. Agents have preferences defined over old age consumption C_{t+1} and young age labor L_t resumed by the following utility function:

$$u(C_{t+1}) - Bv(L_t).$$

We make the following standard Assumption on the preferences:

Assumption 2 The functions $u(C)$ and $v(L)$ are defined and continuous for all $c \geq 0$ and $0 \leq L \leq \mathscr{L}$, respectively. Moreover $u(C)$ and $v(L)$ are C^r, for r large enough, with $u'(C) > 0$, $u''(C) < 0$, $v'(L) > 0$ $v''(L) > 0$, for all $C > 0$ and $0 \leq L \leq \mathscr{L}$, with $\lim_{L \to \mathscr{L}} v'(L) = +\infty$.

A young agent born at period t solves the following dynamic program:

$$\max_{C_{t+1}, L_t, K_{t+1}, M_{t+1}} \quad u\left(C_{t+1}\right) - Bv\left(L_t\right)$$

$$s.t. \quad M_{t+1} + p_t^I K_{t+1} = \omega_t L_t$$
$$p_{t+1}^C C_{t+1} = \rho_{t+1} K_{t+1} + M_{t+1}$$
$$\chi p_{t+1}^C C_{t+1} \leq M_{t+1} \tag{7.6}$$
$$C_{t+1}, L_t, K_{t+1}, M_{t+1} \geq 0$$
$$L_t \in (0, \mathscr{L})$$

where p_{t+1}^C is the price of the consumption good, p_t^I the price of the investment good, ρ_{t+1} the nominal rental rate of capital and ω_t the nominal wage. Choosing the consumption good as the *numéraire* gives the following alternative formulation for the optimization problem:

$$\max_{C_{t+1}, L_t, K_{t+1}, M_{t+1}} \quad u\left(C_{t+1}\right) - Bv\left(L_t\right)$$

$$s.t. \quad q_t M_{t+1} + p_t K_{t+1} = w_t L_t$$
$$C_{t+1} = r_{t+1} K_{t+1} + q_{t+1} M_{t+1}$$
$$\chi C_{t+1} \leq q_{t+1} M_{t+1} \tag{7.7}$$
$$C_{t+1}, L_t, K_{t+1}, M_{t+1} \geq 0$$
$$L_t \in (0, \mathscr{L})$$

where $p_t = p_t^I / p_t^C$, $q_t = 1/p_t^C$, $r_{t+1} = R_{t+1}/p_{t+1}^C$ and the real balances are given by $q_{t+1} M_{t+1}$. We focus on the case where

$$\frac{r_{t+1}}{p_t} > \frac{q_{t+1}}{q_t} \tag{7.8}$$

holds at all dates, which means that the gross rate of return on capital is higher than the profitability of money holding. Under this Assumption, the CIA constraint in (7.7) binds and we obtain the following arbitrage equation (for the first-order conditions see Appendix "First-order Conditions of the Maximization Program of the Consumer"):

$$u'\left(C_{t+1}\right) - \frac{Bv'(L_t)}{w_t}\left[\frac{\chi q_t}{q_{t+1}} + \frac{(1-\chi)p_t}{r_{t+1}}\right] = 0. \tag{7.9}$$

Equation (7.9) states that by increasing of one unit the labor supply (with the associated increase of the effort disutiliy), the corresponding increase of the consumption utility is a weighted average of the capital real return and of the deflation rate.

7.2.3 Equilibrium

At intertemporal equilibrium all markets clear in each period. Since there are four markets, respectively the investment good one, the consumption good one, the labor one, and the money one, by exploiting the Walras law, intertemporal equilibrium, beside the fact that the utility maximization problem is to be solved, must satisfy:

Definition 1

i⌋ Capital accumulation is determined by $p_{t+1}K_{t+1} = w_t L_t - q_t M_{t+1}$;

ii⌋ The consumption good satisfies $C_{t+1} = T(K_{t+1}, Y_{t+1}, L_{t+1})$;

iii⌋ Money dynamics respects $\overline{M} = M_t$ for all t.

From old age budget constraint one derives $C_{t+1} = r_{t+1}K_{t+1} + q_{t+1}M_{t+1}$ and from the binding CIA constraint $\chi C_{t+1} = q_{t+1}M_{t+1}$. Therefore we have $r_{t+1}K_{t+1} - (1-\chi)C_{t+1} = 0$. By exploiting the consumption good market clearing condition $C_{t+1} = T(K_{t+1}, Y_{t+1}, L_{t+1})$, we obtain:

$$T_1(K_{t+1}, Y_{t+1}, L_{t+1})K_{t+1} - (1-\chi)T(K_{t+1}, Y_{t+1}, L_{t+1}) = 0. \qquad (7.10)$$

Applying the Implicit Function Theorem to the static relation (7.10), we are able to solve locally for the labor supply in order to obtain a smooth function $L_{t+1} = L(K_{t+1}, K_{t+2})$. By differentiating the latter, we get:

$$dL_{t+1} = -\frac{T_1\chi + T_{11}K}{T_{13}K - (1-\chi)T_3}dK_{t+1} - \frac{T_{12}K - (1-\chi)T_2}{T_{13}K - (1-\chi)T_3}dK_{t+2}, \forall t. \quad (7.11)$$

Using the trade-off between consumption and leisure (7.9), the static relation (7.10) and the equilibrium condition in the money $q_t/q_{t+1} = C_t/C_{t+1}$, we derive the intertemporal equilibrium with perfect-foresight:

Definition 2 An intertemporal equilibrium with perfect-foresight is a sequence $\{K_t, L_t\}_{t=0}^{\infty}$, with $K_{t=0}$ given, satisfying the following difference equation:

$$u'\left[T(K_{t+1}, K_{t+2}, L_{t+1})\right] - \frac{Bv'(L_t)}{T_3(K_t, K_{t+1}, L_t)}\left[\frac{\chi T(K_t, K_{t+1}, L_t)}{T(K_{t+1}, K_{t+2}, L_{t+1})} - \frac{(1-\chi)T_2(K_t, K_{t+1}, L_t)}{T_1(K_{t+1}, K_{t+2}, L_{t+1})}\right] = 0$$
$$(7.12)$$

with $L_t = L(K_t, K_{t+1})$ and $L_{t+1} = L(K_{t+1}, K_{t+2})$.

Let us define $U(C_{t+1}) = u'(C_{t+1})C_{t+1}$ and $V(L_t) = v'(L_t)L_t$. We have therefore $U'(C) = (1-\varepsilon_u)u'(C)$, with $\varepsilon_u = -u''(C)C/u'(C)$, and $V'(L) = (1+\varepsilon_v)v'(L)$, with $\varepsilon_v = v''(L)L/v'(L)$.

Notice that

$$\left(\frac{V'L}{V}\right)\left(\frac{U}{U'C}\right) = \frac{1+\varepsilon_v}{1-\varepsilon_u} = 1 + \frac{1}{\varepsilon_{uv}}$$

and

$$\varepsilon_{uv} = \frac{1 - \varepsilon_u}{\varepsilon_v + \varepsilon_u}$$

where ε_{uv} is the average wage elasticity of labor supply (or the interest factor elasticity of saving) that we will call in the reminder of the paper as the elasticity of the offer curve, ε_v the elasticity of labor supply and ε_u the elasticity of intertemporal substitution in consumption. Notice that under gross substitutability we have $\varepsilon_{uv} \in (0, +\infty)$.

7.2.4   The Normalized Steady State

A steady state is defined as a sequence $\{K_t, L_t\}_{t=0}^{\infty} = (K^*, L^*)$ for all t. We now show that it is possible to calibrate a particular stationary solution of the dynamic system defined by (7.10) and (7.12) by choosing appropriately the scaling parameter B. To this and, let us fix $K^* = 1$ and let us analyze Eq. (7.10). One immediately verifies that

$$\lim_{L^* \to 0} T_1(1, 1, L^*) - (1 - \chi)T(1, 1, L^*) = +\infty$$

and

$$\lim_{L^* \to +\infty} T_1(1, 1, L^*) - (1 - \chi)T(1, 1, L^*) = -\infty.$$

It follows that there exists a unique positive $L^*(1)$ solving

$$T_1(1, 1, L^*) - (1 - \chi)T(1, 1, L^*) = 0.$$

Therefore, the pair $(1, L^*(1))$ is an interior stationary solution of the system defined by (7.10) and (7.12) if and only if the scaling parameter $B = B^*(1, L^*(1))$ is set such that

$$B = B^*(1, L(1)) = \frac{U'[T(1, 1, L^*(1))] T_3(1, 1, L^*(1))}{V'(L(1)) \left[\frac{\chi T(1,1,L^*(1))}{T(1,1,L^*(1))} - \frac{(1-\chi)T_2(1,1,L^*(1))}{T_1(1,1,L^*(1))} \right]}. \quad (7.13)$$

In the remainder of the paper we make the following Assumption in order to ensure the existence of a normalized steady state (NSS):

Assumption 3 $B = B^*(1, L^*(1))$.

7.3   Dynamic Efficiency

In this Section we analyze the dynamic efficiency properties of the competitive equilibrium around the NSS. We know that in a one-sector OLG model, competitive equilibrium can be not Pareto optimal (Diamond 1965) since intertemporal exchanges are restricted in view of agents' limited planning horizon (two periods). As a matter of fact, if too much capital is accumulated, the economy turns out to be dynamically inefficient. This occurs when the population growth factor (1) exceeds the steady state marginal product of capital (r/p) and the capital-labor ratio exceeds the Golden Rule level. We first characterize the Golden Rule level, i.e. the steady state allocation chosen by a central planner that maximizes the utility of each individual at the steady state. The highest utility is defined as the maximum of the utility function $u(C) - Bv(L)$ subject to the total stationary consumption $C = T(K, K, L)$. The central planner must select non-negative values for capital, labor and consumption in order to solve the following optimization program:

$$\max_{\hat{C}, \hat{K}, \hat{L}} \quad u(C) - Bv(L)$$
$$s.t. \quad C = T(K, K, L). \tag{7.14}$$

In the Appendix "First-order Conditions for Dynamic Efficiency" we provide the expressions for the Lagrangian. The first-order conditions are:

$$u'(C) = \frac{Bv'(L)}{T_3}, \, R\left(\hat{K}, \hat{K}, \hat{L}\right) = -\frac{T_1\left(\hat{K}, \hat{K}, \hat{L}\right)}{T_2\left(\hat{K}, \hat{K}, \hat{L}\right)} = 1. \tag{7.15}$$

As in the traditional one-sector OLG model (Diamond 1965), the Golden Rule level does not depend upon the intertemporal allocation of consumption. We have thus the following Proposition:

Proposition 1 *Under Assumptions 1–3, there exits a unique optimal stationary path* $(\hat{K}, \hat{L})$ *which is characterized by the following conditions:*

$$R\left(\hat{K}, \hat{K}, \hat{L}\right) = 1, \hat{C} = T(\hat{K}, \hat{K}, \hat{L}), u'\left(\hat{C}\right) = \frac{Bv'\left(\hat{L}\right)}{T_3(\hat{K}, \hat{K}, \hat{L})}$$

with $\hat{K}/\hat{L}$ *the Golden Rule capital-labor ratio.*

Proof See Appendix "Proof of Proposition 1".   □

From Proposition 1, the dynamic efficiency properties of the equilibrium paths are appraised through the comparison of the NSS with respect to the Golden Rule level. The concept of feasible path is therefore defined as:

Definition 3 A sequence of capital stock $\{K_t\}_{t=0}^{\infty}$ is a feasible path if, for all $t \geq 0$, the associated total level of consumption is non-negative.

From Definition 3, we can introduce the property of efficiency of a feasible path:

Definition 4 A feasible sequence of capital stock $\{K_t\}_{t=0}^{\infty}$ is efficient if it is not possible to increase the total consumption at one date without decreasing total consumption at another date, i.e. if there does not exist another feasible path $\{K'_t\}_{t=0}^{\infty}$ with $K'_0 = K_0$, such that:

i] $T(K'_t, K'_{t+1}, L_t) \geq T(K_t, K_{t+1}, L_t) \ \forall \ t$;

ii] $T(K'_t, K_{t+1}', L_t) > T(K_t, K_{t+1}, L_t)$ for some $t \geq 0$.

Let us consider the stationary gross rate of return $(R = r/p)$. Using the binding CIA constraint $qM = T\chi$, the budget constraint $qM + pK = wL$ and the fact that $wL/rK = (1 - s)/s$, we determine the stationary gross rate of return R evaluated at the NSS:

$$R = \frac{(1 - \chi)s}{1 - \chi - s}. \tag{7.16}$$

We assume through the paper that the following Assumption holds:

Assumption 4 $\chi < 1 - s \equiv \overline{\chi}$.

Then we are able to provide a condition on the share of capital in the economy to get a NSS (K^*, L^*) lower than the Golden-Rule level $(R > 1)$ and therefore ensuring the dynamic efficiency of the intertemporal equilibrium. Following the proof of Proposition 3 in Drugeon et al. (2010), the following Proposition is immediately proved:

Proposition 2 *Under Assumptions 1–4, let $\underline{s} = (1 - \chi)/(2 - \chi)$. Then, the NSS (K^*, L^*) is characterized by an under-accumulation of capital if and only if $s > \underline{s}$.*

Under-accumulation of capital can be attained provided the share of capital in the economy is large enough, namely $s > \underline{s}$.

7.4 Local Dynamics

In the system describing intertemporal equilibrium, there is one pre-determinate variable, the initial stock of capital, and one forward-looking variables, the labor supply. In such a configuration, the existence of local indeterminacy requires that the two characteristic roots associated with the linearization of the dynamic system (7.12) around the normalized steady state have modulus less than one. In the opposite case the steady state is locally determinate. It is useful to introduce here the elasticity ε_{rk} of the rental rate of capital evaluated at the normalized steady state:

$$\varepsilon_{rk} = -\frac{T_{11}(K^*, K^*, L^*)K^*}{T_1(K^*, K^*, L^*)} \in (0, +\infty). \tag{7.17}$$

Drugeon (2004) points out that the elasticity of the rental rate of capital is negatively linked to the elasticities of capital-labor substitution:

$$\Sigma = \frac{(Y_0 + pY)\left(pYK^0L^0\sigma_0 + Y_0K^1L^1\sigma_1\right)}{pYKY_0}, \quad \varepsilon_{rk} = \left(\frac{L^0}{Y_0}\right)^2 \frac{w\,(Y_0 + pY)}{\Sigma} \tag{7.18}$$

with $\sigma_0 \in (0, +\infty)$ and $\sigma_1 \in (0, +\infty)$ being the sectorial elasticities of inputs substitution.

Then the following proposition holds:

Proposition 3 *Under Assumptions 1–3, the characteristic polynomial is defined by* $\mathscr{P}(\lambda) = \lambda^2 - \lambda\mathscr{T} + \mathscr{D}$ *where:*

$$\mathscr{T}(\varepsilon_{uv}) = \frac{1 - \chi - s - s\chi\varepsilon_{uv} + \varepsilon_{rk}\left[bs + \varepsilon_{uv}(bs\chi + 1 - \chi)\right]}{(bs\chi + 1 - \chi - s)\,\varepsilon_{rk}\varepsilon_{uv}} \tag{7.19}$$

$$\mathscr{D}(\varepsilon_{uv}) = \frac{s(\varepsilon_{rk} - \chi)(1 + \varepsilon_{uv})}{(bs\chi + 1 - \chi - s)\,\varepsilon_{rk}\varepsilon_{uv}} \tag{7.20}$$

Proof See Appendix Proof of Proposition 3. □

In view of the complicated form of the above expressions, it may seem that the study of the local dynamics of system (7.12) requires long and tedious computations. However, by applying the geometrical method adopted in Grandmont et al. (1988) and Cazzavillan et al. (1998), it is possible to analyze qualitatively the (in)stability of the characteristic roots of the Jacobian evaluated at the steady state of system defined by (7.12) and their bifurcations (changes in stability) by locating the point $(\mathscr{T}, \mathscr{D})$ in the plane and studying how $(\mathscr{T}, \mathscr{D})$ varies when the value of some parameter changes continuously. In our case, the bifurcation parameter is $\varepsilon_{uv} \in (0, +\infty)$. In the Appendix "The Geometrical Method", we present the geometrical approach adopted to study the local stability.

7.4.1 The Non-monetary Economy

Before analyzing the monetary economy, let us consider first the case without cash-in-advance constraint $\chi = 0$. Under such an hypothesis, we are able to appraise the role played on local dynamics by the two technological parameters of the model: b and ε_{rk}. Within such an hypothesis, the Trace $\mathscr{T}$, the Determinant $\mathscr{D}$ and the slope $\mathscr{S}$ defined in (7.19), (7.20) and (7.33) boil down to, respectively:

$$\mathscr{T} = \frac{1 - s + \varepsilon_{rk}(bs + \varepsilon_{uv})}{\varepsilon_{rk}\varepsilon_{uv}(1 - s)}; \quad \mathscr{D} = \frac{s(1 + \varepsilon_{uv})}{\varepsilon_{uv}(1 - s)}; \quad \mathscr{S} = \frac{s\varepsilon_{rk}}{\varepsilon_{rk}bs + 1 - s} \quad (7.21)$$

To analyze the local dynamics of the economy, we consider the properties of the starting point $(\mathscr{T}_\infty, \mathscr{D}_\infty)$ and of the slope $\mathscr{S}$ as a function of the two parameters $b \in (-\infty, 1)$ and $\varepsilon_{rk} \in (0, +\infty)$. Using Eq. (7.21), we find that the starting point, obtained setting $\varepsilon_{uv} = +\infty$, has coordinates:

$$\mathscr{D}_\infty = \frac{s}{1 - s}; \quad \mathscr{T}_\infty = \mathscr{D}_\infty + 1 \quad (7.22)$$

Notice that the location of the starting point does not depend upon b and ε_{rk}. Moreover, the starting point lies always on the $\mathscr{D} = \mathscr{T} - 1$ line. In the following, we will consider two cases; the first corresponding to the over-accumulation of capital, i.e. $s < 1/2$, and the other to the under-accumulation of capital, i.e. $s > 1/2$, since both configurations are compatible with the non-monetary economy. When $s < 1/2$ and then the economy is characterized by capital over-accumulation, one has $\mathscr{D}_\infty < 1$ and $\mathscr{T}_\infty < 2$ and thus the starting point lies always below the point $\mathscr{C}$ depicted in Fig. 7.1.

To appraise the local stability properties and the occurrence of bifurcations, we must analyze how does the slope $\mathscr{S}$ move as soon as b and ε_{rk} are made to vary. Notice that, for $\varepsilon_{rk} = +\infty$, the slope is $1/b$. When b increases from $-\infty$ to 1, the half-line Δ undergoes a clockwise rotation. Let us now define $b_1 = (2s - 3)/(1 - 2s)$ the relative critical capital intensity difference such that the half-line Δ goes through the point $\mathscr{B}$ and $b_2 = -1$ the relative critical capital intensity difference such that the slope of the half-line Δ is -1.

After having fixed b, let us vary ε_{rk}. Consider first the case $b < b_1$. Here we can define three critical values for ε_{rk}: ε_{rk}^1 such that the half-line Δ goes through the point $\mathscr{B}$; ε_{rk}^2 such that the slope $\mathscr{S}$ of the half-line Δ is equal to -1 and ε_{rk}^3 such

Fig. 7.1 Hopf and flip bifurcations under over-accumulation of capital

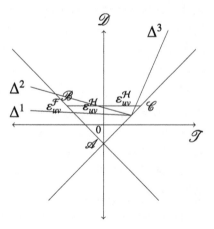

that the slope $\mathscr{S}$ is equal to one. The critical values for ε_{rk} can be summarized with the help of the following notation:

$$\Delta \text{ goes through } \mathscr{B} \iff \varepsilon_{rk}^1 = -\frac{(1-s)(1-2s)}{s[3-2s+b(1-2s)]};$$

$$\mathscr{S} = -1 \iff \varepsilon_{rk}^2 = -\frac{1-s}{s(1+b)};$$

$$\mathscr{S} = 1 \iff \varepsilon_{rk}^3 = \frac{1-s}{s(1-b)}.$$

Notice that ε_{rk}^1 exists if and only if $b < b_1$, ε_{rk}^2 requires $b < -1$ and ε_{rk}^3 is well defined for any values of b. It is immediately verifiable that inequalities $\varepsilon_{rk}^1 > \varepsilon_{rk}^2 > \varepsilon_{rk}^3 > 0$ do hold. Let now set $b < b_1$. It follows that when $\varepsilon_{rk} < \varepsilon_{rk}^3$ the steady state is always a saddle. When $\varepsilon_{rk} \in (\varepsilon_{rk}^3, \varepsilon_{rk}^2)$, by relaxing continuously ε_{uv} in the $(0, +\infty)$ interval, we obtain first a source configuration and then, through a Hopf bifurcation, a sink one. The half-line Δ^3 in Fig. 7.1 represents this case. When $\varepsilon_{rk} \in (\varepsilon_{rk}^2, \varepsilon_{rk}^1)$, by increasing ε_{uv}, we have a steady state which is first a saddle, then, through a flip bifurcation, it becomes a source and eventually, through a Hopf bifurcation, a sink. The half-line Δ^2 in Fig. 7.1 corresponds to this case. If $\varepsilon_{rk} > \varepsilon_{rk}^1$, by relaxing continuously ε_{uv}, one obtains first a saddle configuration and then, through a flip bifurcation, a sink one, as depicted in Fig. 7.1.

Let us now consider the case $b \in (b_1, b_2)$. Notice that here ε_{rk}^1 does not exist anymore since $b > b_1$. Then, when $\varepsilon_{rk} < \varepsilon_{rk}^3$, the steady state is a saddle. When $\varepsilon_{rk} \in (\varepsilon_{rk}^3, \varepsilon_{rk}^2)$, by relaxing continuously ε_{uv}, we obtain first a source configuration and then, through a Hopf bifurcation, a sink one. When $\varepsilon_{rk} > \varepsilon_{rk}^2$ we have a steady state which is first a saddle and then, through a flip bifurcation, it becomes a source and eventually, through a Hopf bifurcation, a sink.

Finally, let us consider the case $b > b_2$. It follows that ε_{rk}^1 and ε_{rk}^2 do not exist anymore since $b > b_2$. Then, when $\varepsilon_{rk} < \varepsilon_{rk}^3$, the steady state is bound to be a saddle; on the other hand, if $\varepsilon_{rk} > \varepsilon_{rk}^3$, we obtain first a source configuration and then, through a Hopf bifurcation, a sink one.

All these results are summarized in the following Proposition.

Proposition 4 *Under Assumptions 1–4, let assume $s < 1/2$. Then there exist $b_1 < b_2$, $\varepsilon_{rk}^1 > \varepsilon_{rk}^2 > \varepsilon_{rk}^3 > 0$, $\varepsilon_{uv}^{\mathscr{F}} > 0$ and $\varepsilon_{uv}^{\mathscr{H}} > 0$ such that the following results hold: i] Let $b < b_1$. If $\varepsilon_{rk} < \varepsilon_{rk}^3$, the steady state is a saddle, i.e. locally determinate. If $\varepsilon_{rk} \in (\varepsilon_{rk}^3, \varepsilon_{rk}^2)$, the steady state is a source, i.e. locally determinate, for $\varepsilon_{uv} < \varepsilon_{uv}^{\mathscr{H}}$, and a sink, i.e. locally indeterminate, for $\varepsilon_{uv} > \varepsilon_{uv}^{\mathscr{H}}$. If $\varepsilon_{rk} \in (\varepsilon_{rk}^2, \varepsilon_{rk}^1)$, the steady state is a saddle, i.e. locally determinate, for $\varepsilon_{uv} < \varepsilon_{uv}^{\mathscr{F}}$, a source, i.e. locally determinate, for $\varepsilon_{uv} \in (\varepsilon_{uv}^{\mathscr{F}}, \varepsilon_{uv}^{\mathscr{H}})$, and a sink, i.e. locally indeterminate, for $\varepsilon_{uv} > \varepsilon_{uv}^{\mathscr{H}}$. If $\varepsilon_{rk} > \varepsilon_{rk}^1$, the steady state is a saddle, i.e. locally determinate, for $\varepsilon_{uv} < \varepsilon_{uv}^{\mathscr{F}}$ and a sink, i.e. locally indeterminate, for $\varepsilon_{uv} > \varepsilon_{uv}^{\mathscr{F}}$. ii] Let $b \in (b_1, b_2)$. If $\varepsilon_{rk} < \varepsilon_{rk}^3$, the steady state is a saddle, i.e. locally determinate. If $\varepsilon_{rk} \in (\varepsilon_{rk}^3, \varepsilon_{rk}^2)$, the steady state is a source, i.e. locally determinate, for $\varepsilon_{uv} < \varepsilon_{uv}^{\mathscr{H}}$*

and a sink, i.e. locally indeterminate, for $\varepsilon_{uv} > \varepsilon^{\mathcal{H}}_{uv}$. *If* $\varepsilon_{rk} > \varepsilon^2_{rk}$, *the steady state is a saddle, i.e. locally determinate, for* $\varepsilon_{uv} < \varepsilon^{\mathcal{F}}_{uv}$, *a source, i.e. locally determinate, for* $\varepsilon_{uv} \in (\varepsilon^{\mathcal{F}}_{uv}, \varepsilon^{\mathcal{H}}_{uv})$, *and a sink, i.e. locally indeterminate, for* $\varepsilon_{uv} > \varepsilon^{\mathcal{H}}_{uv}$.

iii] Let $b > b_2$. *If* $\varepsilon_{rk} < \varepsilon^3_{rk}$, *the steady state is a saddle, i.e. locally determinate. If* $\varepsilon_{rk} > \varepsilon^3_{rk}$, *the steady state is a source, i.e. locally determinate, for* $\varepsilon_{uv} < \varepsilon^{\mathcal{H}}_{uv}$, *and a sink, i.e. locally indeterminate, for* $\varepsilon_{uv} > \varepsilon^{\mathcal{H}}_{uv}$.

When ε_{uv} *goes through* $\varepsilon^{\mathcal{F}}_{uv}$ *and* $\varepsilon^{\mathcal{H}}_{uv}$ *the steady state undergoes, respectively, a flip and a Hopf bifurcation.*

Let us now consider the case of under-accumulation of capital, i.e. $s > 1/2$. From Eq. (7.21), we derive that the starting point lies always above the point $\mathscr{C}$ in Fig. 7.2 and thus the NSS is always locally determinate. Then, in the light of the above considerations with the help of Fig. 7.2, the following Proposition is immediately proved:

Proposition 5 *Under Assumptions 1–4, let assume* $s > 1/2$. *Then there exist* $b_2 < 0$, $\varepsilon^1_{rk} > \varepsilon^2_{rk} > \varepsilon^3_{rk} > 0$, $\varepsilon^{\mathcal{F}}_{uv} > 0$ *and* $\varepsilon^{\mathcal{H}}_{uv} > 0$ *such that the following results hold:*
i] let $b < b_2$. *If* $\varepsilon_{rk} < \varepsilon^3_{rk}$, *the steady state is a saddle, i.e. a locally determinate. If* $\varepsilon_{rk} \in (\varepsilon^3_{rk}, \varepsilon^2_{rk})$, *the steady state is a source, i.e. locally determinate. If* $\varepsilon_{rk} > \varepsilon^2_{rk}$, *the steady state is a saddle, i.e. locally determinate, for* $\varepsilon_{uv} < \varepsilon^{\mathcal{F}}_{uv}$, *and a source, i.e. locally determinate, for* $\varepsilon_{uv} > \varepsilon^{\mathcal{F}}_{uv}$.
ii] let $b > b_2$. *When* $\varepsilon_{rk} < \varepsilon^3_{rk}$, *the steady state is a saddle, i.e. locally determinate. If* $\varepsilon_{rk} > \varepsilon^3_{rk}$, *the steady state is a source, i.e. locally determinate.*

When ε_{uv} *goes through* $\varepsilon^{\mathcal{F}}_{uv}$, *the steady state undergoes a flip bifurcation.*

It could be interesting, at this point, to compare our results in terms of dynamics features with those found in Drugeon et al. (2010) and Nourry and Venditti (2011) since their OLG models are similar to our own. In these two papers labor is assumed to be supplied inelastically and agents to consume when young and when old meanwhile, in our framework, labor supply is endogenous and agents consume only when

Fig. 7.2 Flip bifurcation with under-accumulation of capital

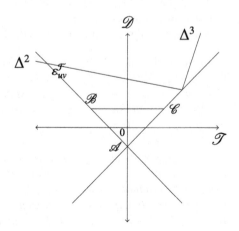

old. Nevertheless, in both models, the dynamical system characterizing the equilibrium is two-dimensional and therefore we can carry out a comparaison of the two models. Under the gross substitutability assumption, Drugeon et al. (2010) and Nourry and Venditti (2011) show that sunspot fluctuations are ruled out when the steady state is dynamical efficient provided that the sectoral elasticities of capital-labor substitution are not too low. More in details, Drugeon et al. (2010) show that, when the steady state is dynamically inefficient, sunspot fluctuations are likely to occur whatever the sectoral elasticities of capital-labor substitution is. On the contrary, in the non-monetary version of our model, under dynamic efficiency, local indeterminacy is bound to be ruled out for whatever parameters configuration. On the other hand, under dynamic efficiency, we obtain local indeterminacy too, even thought our bifurcation parameter is the elasticity of the offer curve meanwhile in Drugeon et al. (2010) and Nourry and Venditti (2011), it is the elasticity of intertemporal substitution in consumption. It follows that a complete comparison between the two models is rather difficult to carry out although their features are mutually consistent.

7.4.2 The Monetary Economy

We now focus on the monetary economy obtained by setting $\chi > 0$ under the hypothesis of a binding *CIA* constraint, condition requiring a dynamically efficient *NSS*. We will consider first the case of a capital intensive investment good, i.e. $b > 0$, and then the case of a capital intensive consumption good, i.e. $b < 0$.

7.4.2.1 A Capital Intensive Investment Good

Let $b > 0$. From the homogeneity of the social production function $T(K_t, K_{t+1}, L_t)$ and the first-order conditions of the producer (7.2), we have that, at the *NSS* (K^*, L^*), $b < 1$. Since $b > 0$, it follows from Eqs. (7.33) to (7.34) that the properties of the starting point $(\mathcal{T}_\infty, \mathcal{D}_\infty)$ and of the slope $\mathcal{S}$ depend upon ε_{rk}. Moreover, from (7.34), one has that $\mathcal{D}_\infty = \mathcal{T}_\infty - 1$. From (7.34), one has that $\mathcal{D}_\infty$ is greater than one if $\varepsilon_{rk} > (1 - \chi)(1 - s)/s(1 - b\chi) \equiv \varepsilon_{rk}^{5\chi}$. It follows that, when $\varepsilon_{rk} > \varepsilon_{rk}^{5\chi}$, the starting point lies above the point $\mathscr{C}$ depicted in Fig. 7.3.

Let fix $\varepsilon_{rk} = +\infty$. We then obtain:

$$\lim_{\varepsilon_{rk} \to +\infty} \mathcal{D}_\infty(\varepsilon_{rk}) \equiv \mathcal{D}_\infty^\infty = \frac{s}{bs\chi + 1 - \chi - s} > 1; \lim_{\varepsilon_{rk} \to +\infty} \mathcal{T}_\infty(\varepsilon_{rk}) \equiv \mathcal{T}_\infty^\infty = \mathcal{D}_\infty^\infty + 1 < 2; \mathcal{S}^\infty = b^{-1}$$

(7.23)

Since $\mathcal{D}_\infty^\infty$ is greater than one and the starting point lies on the $\mathcal{D} = \mathcal{T} - 1$ line, the pair $(\mathcal{T}_\infty^\infty, \mathcal{D}_\infty^\infty)$ lies above the point $\mathscr{C}$. The steady state is a then source if the half-line Δ has slope greater than one. When $\varepsilon_{rk} = +\infty$, this is true when $b < 1$. By a direct inspection of (7.33), one has that $\mathcal{S} > 1$ if and only if $\varepsilon_{rk} > (1 - \chi)$

Fig. 7.3 Flip bifurcation

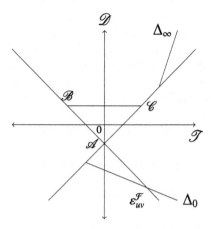

$(1 - s)/s(1 - b) \equiv \varepsilon_{rk}^{3\chi}$. In addition, in view of (7.20), we have $D'(\varepsilon_{uv}) > 0$. Thus, for $\varepsilon_{rk} \in (\varepsilon_{rk}^{3\chi}, +\infty)$, the steady state is always a source. The corresponding half-line Δ_∞ is depicted in Fig. 7.3.

Let us now decrease ε_{rk} so that the starting point lies now on $\mathscr{A}$. This is the case if $\varepsilon_{rk} = (s\chi)/(1 - \chi + bs\chi) \equiv \varepsilon_{rk}^{4\chi}$. Let finally decrease ε_{rk} up to zero. We have:

$$\lim_{\varepsilon_{rk} \to 0} \mathscr{D}_\infty(\varepsilon_{rk}) \equiv \mathscr{D}_\infty^0 = -\infty; \lim_{\varepsilon_{rk} \to 0} \mathscr{T}_\infty(\varepsilon_{rk}) \equiv \mathscr{T}_\infty^0 = \mathscr{D}_\infty^0 + 1 = -\infty; \mathscr{S}^0 = \frac{-s\chi}{1 - \chi - s} \tag{7.24}$$

Notice that, under Assumption 4, the slope is negative and always greater than -1. In addition, from the expression for the Determinant (7.20), we easily derive $D'(\varepsilon_{uv}) < 0$. Then, for $\varepsilon_{rk} \in (\varepsilon_{rk}^{4\chi}, \varepsilon_{rk}^{3\chi})$, the steady state is a saddle. When ε_{rk} is further reduced below $\varepsilon_{rk}^{4\chi}$, by relaxing continuously ε_{uv} from 0 to $+\infty$, we obtain that the steady state is first a saddle and then, through a flip bifurcation, it becomes a source. In Fig. 7.3 we have depicted the half-line Δ_0 corresponding to the case in which the steady state undergoes a flip bifurcation.

These results are summarized in the following Proposition:

Proposition 6 *Under Assumptions 1–4, there exist $\varepsilon_{rk}^{3\chi} > \varepsilon_{rk}^{4\chi} > 0$ and $\varepsilon_{uv}^{\mathscr{F}} > 0$ such that for $b \in (0, 1)$, the following results hold:*
When $\varepsilon_{rk} > \varepsilon_{rk}^{3\chi}$, the steady state is a source, i.e. locally determinate. When $\varepsilon_{rk} \in (\varepsilon_{rk}^{4\chi}, \varepsilon_{rk}^{3\chi})$, the steady state is a saddle, i.e. locally determinate. When $\varepsilon_{rk} < \varepsilon_{rk}^{4\chi}$, the steady state is a saddle, i.e. locally determinate, for $\varepsilon_{uv} \in (0, \varepsilon_{uv}^{\mathscr{F}})$, and a source, i.e. locally determinate, for $\varepsilon_{uv} > \varepsilon_{uv}^{\mathscr{F}}$.
In addition, when ε_{uv} goes through $\varepsilon_{uv}^{\mathscr{F}}$, the steady state undergoes a flip bifurcation.

Fig. 7.4 Flip bifurcation

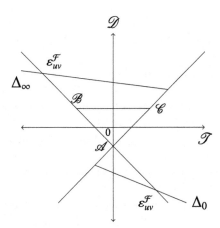

7.4.2.2   A Capital Intensive Consumption Good

Let us now consider the case of a capital intensive consumption good, i.e. $b < 0$, depicted in Fig. 7.4.

Since $b < 0$, it follows from Eqs. (7.33) to (7.34) that the properties of the starting point $(\mathcal{T}_\infty, \mathcal{D}_\infty)$ and of the slope $\mathcal{S}$ depend upon ε_{rk} and b. From (7.34), we have $\mathcal{D}_\infty = \mathcal{T}_\infty - 1$. From (7.34), one has that $\mathcal{D}_\infty$ is greater than one if $\varepsilon_{rk} > (1 - \chi)(1 - s)/s(1 - b\chi) \equiv \varepsilon_{rk}^{5\chi}$. It follows that, when $\varepsilon_{rk} > \varepsilon_{rk}^{5\chi}$, the starting point lies above the point $\mathcal{C}$. Let us now define $b_3 = -(1 - \chi - s)/s\chi$ the relative critical capital intensity difference such that the denominator of $\mathcal{D}_\infty$ changes of sign.

Consider first the case $b > b_3$. Here we must introduce three critical values for ε_{rk}: $\varepsilon_{rk}^{2\chi}$ such that the slope $\mathcal{S}$ is equal to -1; $\varepsilon_{rk}^{3\chi}$ such that the slope $\mathcal{S}$ is equal to one; $\varepsilon_{rk}^{4\chi}$ such that the half-line Δ goes through the point $\mathcal{B}$. The expressions for the critical values for ε_{rk} are the following:

$$\mathcal{S} = -1 \iff \varepsilon_{rk}^{2\chi} = -\frac{1-s}{s(1+b)};$$

$$\mathcal{S} = 1 \iff \varepsilon_{rk}^{3\chi} = \frac{1-s}{s(1-b)};$$

$$\Delta \text{ goes through } \mathcal{A} \iff \varepsilon_{rk}^{4\chi} = (s\chi)/(1 - \chi + bs\chi).$$

Notice that $\varepsilon_{rk}^{4\chi}$ exists if and only if $b > -(1 - \chi)/s\chi \equiv b_4$ which, as it is immediately verifiable, satisfies inequality $b_4 < b_3$. It is also easy to prove that the inequalities $\varepsilon_{rk}^{2\chi} > \varepsilon_{rk}^{3\chi} > \varepsilon_{rk}^{5} > \chi > \varepsilon_{rk}^{4} > 0$ do hold. Let us now set $b > b_3$. When $\varepsilon_{rk} < \varepsilon_{rk}^{4\chi}$ by relaxing continuously ε_{uv}, we obtain that the steady state is first a saddle and then, through a flip bifurcation, it becomes a source. On the other hand, when $\varepsilon_{rk} \in (\varepsilon_{rk}^{4\chi}, \varepsilon_{rk}^{3\chi})$, the steady state is bound to be a saddle. At the same time, it is immediately verifiable that, when $\varepsilon_{rk} \in (\varepsilon_{rk}^{3\chi}, \varepsilon_{rk}^{2\chi})$, the steady state is a source. Eventually, when $\varepsilon_{rk} > \varepsilon_{rk}^{2\chi}$, by increasing continuously ε_{uv}, one obtains first a saddle configuration and then, through a flip bifurcation, a source one.

Let us now set $b \in (b_4, b_3)$. It is easily verifiable that the inequalities $\varepsilon_{rk}^{2\chi} < \varepsilon_{rk}^{3\chi} < \varepsilon_{rk}^{5\chi} < \chi < \varepsilon_{rk}^{4\chi} > 0$ do hold. It follows that, when $\varepsilon_{rk} < \varepsilon_{rk}^{2\chi}$, by relaxing continuously ε_{uv}, one obtains first a saddle configuration for the steady state and then, through a flip bifurcation, a source one. At the same time, it is immediate to see that, when $\varepsilon_{rk} \in (\varepsilon_{rk}^{2}, \varepsilon_{rk}^{3\chi})$, the steady state is bound to be a source and that, when $\varepsilon_{rk} \in (\varepsilon_{rk}^{3\chi}, \varepsilon_{rk}^{4\chi})$, it is a saddle. Eventually, for $\varepsilon_{rk} > \varepsilon_{rk}^{4\chi}$, by increasing continuously ε_{uv}, one obtains a steady state that is first a saddle and then, through a flip bifurcation, becomes a source.

Let us now suppose $b < b_4$. It follows that the critical value $\varepsilon_{rk}^{4\chi}$ does not more exist. It is easy to prove that, when $\varepsilon_{rk} < \varepsilon_{rk}^{2\chi}$, by increasing ε_{uv}, one obtains first a saddle configuration for the steady state and then, through a flip bifurcation, a source one. It is then immediate to see that, when $\varepsilon_{rk} \in (\varepsilon_{rk}^{2}, \varepsilon_{rk}^{3\chi})$, the steady state is a source and, when $\varepsilon_{rk} > \varepsilon_{rk}^{3\chi}$, a saddle.

All the results are summarized in the following Proposition.

Proposition 7 *Under Assumptions 1–4, there exist $\varepsilon_{rk}^{2\chi} > \varepsilon_{rk}^{3\chi} > \varepsilon_{rk}^{4\chi} > 0$, $b_4 < b_3 < 0$ and $\varepsilon_{uv}^{\mathcal{F}} > 0$ such that the following results hold:*

i⌋ *Let $b > b_3$. If $\varepsilon_{rk} < \varepsilon_{rk}^{4\chi}$, the steady state is a saddle, i.e. a locally determinate, for $\varepsilon_{uv} \in (0, \varepsilon_{uv}^{\mathcal{F}})$, and a source, i.e. locally determinate, for $\varepsilon_{uv} > \varepsilon_{uv}^{\mathcal{F}}$. If $\varepsilon_{rk} \in (\varepsilon_{rk}^{4\chi}, \varepsilon_{rk}^{3\chi})$, the steady state is a saddle, i.e. locally determinate. If $\varepsilon_{rk} \in (\varepsilon_{rk}^{3\chi}, \varepsilon_{rk}^{2\chi})$, the steady state is a source, i.e. locally determinate. If $\varepsilon_{rk} > \varepsilon_{rk}^{2\chi}$, the steady state is a saddle, i.e. locally determinate, for $\varepsilon_{uv} \in (0, \varepsilon_{uv}^{\mathcal{F}})$, and a source, i.e. locally determinate, for $\varepsilon_{uv} > \varepsilon_{uv}^{\mathcal{F}}$.*

ii⌋ *let $b \in (b_4, b_3)$. If $\varepsilon_{rk} < \varepsilon_{rk}^{2\chi}$, the steady state is a saddle, i.e. locally determinate, for $\varepsilon_{uv} \in (0, \varepsilon_{uv}^{\mathcal{F}})$, and a source, i.e. locally determinate, for $\varepsilon_{uv} > \varepsilon_{uv}^{\mathcal{F}}$. If $\varepsilon_{rk} \in (\varepsilon_{rk}^{2\chi}, \varepsilon_{rk}^{3\chi})$, the steady state is a source, i.e. locally determinate. If $\varepsilon_{rk} \in (\varepsilon_{rk}^{3\chi}, \varepsilon_{rk}^{4\chi})$, the steady state is a saddle, i.e. locally determinate. If $\varepsilon_{rk} > \varepsilon_{rk}^{4\chi}$, the steady state is a saddle, i.e. locally determinate, for $\varepsilon_{uv} \in (0, \varepsilon_{uv}^{\mathcal{F}})$, and a source, i.e. locally determinate, for $\varepsilon_{uv} > \varepsilon_{uv}^{\mathcal{F}}$;*

iii⌋ *let $b < b_4$. If $\varepsilon_{rk} < \varepsilon_{rk}^{2\chi}$, the steady state is a saddle, i.e. locally determinate, for $\varepsilon_{uv} \in (0, \varepsilon_{uv}^{\mathcal{F}})$ and a source, i.e. locally determinate, for $\varepsilon_{uv} > \varepsilon_{uv}^{\mathcal{F}}$. If $\varepsilon_{rk} \in (\varepsilon_{rk}^{2\chi}, \varepsilon_{rk}^{3\chi})$, the steady state is a source, i.e. locally determinate. If $\varepsilon_{rk} > \varepsilon_{rk}^{3\chi}$, the steady state is a saddle, i.e. locally determinate.*

In addition, when ε_{uv} goes through $\varepsilon_{uv}^{\mathcal{F}}$, the steady state undergoes a flip bifurcation.

7.4.2.3   Interpretation of the Results

We have seen that endogenous fluctuations under the hypothesis of dynamic efficiency occur in both non-monetary model and the monetary one when the consumption good is capital intensive, i.e. $b < 0$. As a matter of fact, in such a case persistent cycles arise through a flip bifurcation. The underlying mechanism is to be found in the combinaison of the Rybzcinsky effect and of the Stolper-Samuelson one. Such a mechanism is described, among the others, in Venditti (2005).

The intuition is the following. Let us suppose that the capital stock at period t increases. Since the consumption good is the most capital intensive good, there is a raise in the production of the consumption good. Meanwhile, this increase of the capital stock implies, through the Rybzsinsky, a decrease of the output of the investment good. This tends to reduce the capital stock at period $t + 1$. In period $t + 1$, the decrease of the capital stock implies a raise in the investment good. Therefore, this tends to increase the investment in period $t + 1$ and thus the capital stock in period $t + 2$. Notice also that the increase of the production of the investment good in period $t + 1$ implies a decrease of the rental rate of capital in period $t + 1$ and, by the Stolper-Samuelson effect, an increase the relative price of investment in period $t + 1$.

What is new in our paper is the finding that deterministic cycles in the monetary economy occur also when the consumption good is labor intensive, i.e. $b > 0$, meanwhile, in the non-monetary model, persistent fluctuations require a capital intensive consumption good. In the non-monetary case, the intuition is the following. Let us suppose that the capital stock at period t increases. Since the investment good is the most capital intensive good, there is a raise in the production of the investment good. Meanwhile, this increase of the capital stock implies, through the Rybzsinsky effect, an increase of the output of the investment good. This tends to raise the capital stock at period $t + 1$. In period $t + 1$, the increase of the capital stock implies a raise in the investment good. In turn, this tends to increase the investment in period $t + 1$ and thus the capital stock in period $t + 2$. Therefore, oscillations are ruled out.

Consider now the monetary case. Let us suppose that the capital stock at period t increases. In principle, this should be accompanied by an increase in the investment good through the Rybzcinscky effect along with an increase in consumption in period $t + 1$. However, the raise in consumption in period $t + 1$ implies, in view of the money market clearing condition, a decrease in the inflation rate which makes money holding less expensive. It follows that, in period t, a larger share of the savings will be devoted to money holdings. Such an increase is large enough to off-set the initial increase of physical capital in period $t + 1$. For the same reason, investment in period $t + 1$, and thus capital stock in period $t + 2$, will be larger. This, together with the hypothesis of gross substitutability (and thus a saving function positively correlated with its returns), explains the occurrence of persistent fluctuations.

7.5 Concluding Remarks

In this paper we have considered a two-sector *OLG* economy with partial cash-in-advance constraint applying on old age consumption expenditures, while young agents supply labor elastically and save their income in physical capital and money balances. We first showed that the capital-labor ratio is above the Golden Rule stationary equilibrium if and only if the share of capital on total income is large enough. As a consequence, we established a general result that shed additional light on the topic: dynamic efficiency under the gross substitutability assumption in consumption, without any additional requirement, is sufficient to rule out local indeterminacy and thus sunspot fluctuations. Such a finding is even more worth emphasizing once one observes that dynamic efficiency must be assumed in order to guarantee that money is dominated by capital in terms of returns and thus the liquidity constraint is binding. Were this not the case, money would have to be interpreted as a bubble whose rate of return (deflation) would adjust in each period to equalize the real interest rate, as in Tirole (1985) and Benhabib and Laroque (1988). If, on the one hand, local determinacy is bound to prevail under dynamic efficiency, on the other, the latter is nevertheless compatible with deterministic cycles arising along a flip bifurcation, obtained by varying continuously the elasticity of the offer curve. Such a bifurcation occurs under whatever assumption concerned with the relative sectoral capital intensity: it is just worth noticing that, under a capital intensive consumption good, in order to get a flip bifurcation one needs low enough elasticities of the real interest rate, while under the hypothesis of a capital intensive investment good, such an occurrence is compatible with arbitrarily large elasticities of the interest rate too.

Analogous results are found, under dynamic efficiency, in the non-monetary case. However, here one can put forward the hypothesis of dynamic inefficiency and thus obtain a richer picture of the local dynamics which includes now also the occurrence of indeterminacy. Specifically, when the consumption good is capital intensive, local indeterminacy arises through a Hopf bifurcation and for high enough elasticities of the interest rate. On the other hand, within a capital intensive investment good, local indeterminacy occurs through either a Hopf or a flip bifurcation, according to the magnitude of the relative capital intensity.

A fruitful extension of the model could take into account the presence of externalities which, as proved in Cazzavillan and Pintus (2005), could restore the compatibility between dynamic efficiency and local indeterminacy. Another line of research might take advantage from the study of Benhabib and Laroque (1988) and provide the analysis of a similar economy with rational bubbles within a two-sector technology. Eventually, it should be worthwhile to extend the Rochon and Polemarchakis (2006) model with capital, government bonds and liquidity constraint, to a two-sector framework.

Appendix A

A.1 First-order Conditions of the Maximization Program of the Consumer

The associated Lagrangian of (7.7) is:

$$L = u\left(C_{t+1}\right) - Bv\left(L_t\right) + \lambda_{0,t}\left[w_t L_t - q_t M_{t+1} - p_t K_{t+1}\right] + \lambda_{1,t}\left[q_{t+1} M_{t+1} + r_{t+1} K_{t+1} - C_{t+1}\right] + \lambda_{2,t}\left[q_{t+1} M_{t+1} - \chi C_{t+1}\right]$$

The associated first-order conditions are:

$$u'\left(C_{t+1}\right) = \lambda_{1,t} + \chi \lambda_{2,t}, \tag{7.25}$$

$$\frac{Bv'\left(L_t\right)}{w_t} = \lambda_{0,t}, \tag{7.26}$$

$$\frac{r_{t+1}}{p_t} = \frac{\lambda_{0,t}}{\lambda_{1,t}}, \tag{7.27}$$

and

$$\frac{q_{t+1}}{q_t} = \frac{\lambda_{0,t}}{\lambda_{1,t} + \lambda_{2,t}}. \tag{7.28}$$

A.2 First-order Conditions for Dynamic Efficiency

The associated Lagrangian of (7.14) is:

$$\mathscr{L} = u\left(C\right) - Bv\left(L\right) + \lambda\left[T\left(K, K, L\right) - C\right].$$

The first-order conditions are easily obtained:

$$u'\left(C\right) = \lambda, \quad \frac{Bv'\left(L\right)}{T_3} = \lambda, \quad T_1\left(K, K, L\right) + T_2\left(K, K, L\right) = 0. \tag{7.29}$$

A.3 Proof of Proposition 1

Since $R(K, K, L) = -T_1/T_2$, we have that

$$R'(K, K, L) = -\frac{T_{11}}{T_2}(1 - b)(1 - Rb).$$

From the homogeneity of degree one of the social production function $T(K_t, K_{t+1}, L_t)$ and from the first-order conditions of the producer problem (7.2), one obtains that, at the NSS (K^*, L^*), $b < 1$. Let us define the factor price frontier:

$$\begin{pmatrix} a^{0l} & a^{0k} \\ a^{1l} & a^{1k} \end{pmatrix} \begin{pmatrix} w \\ r \end{pmatrix} = \begin{pmatrix} 1 \\ p \end{pmatrix} \tag{7.30}$$

where $a^{0l} = \frac{L^0}{Y_0}$, $a^{1l} = \frac{L^1}{Y}$, $a^{0k} = \frac{K^0}{Y_0}$ and $a^{1k} = \frac{K^1}{Y}$ indicate the amount of capital and labor used in each sector. From the definition of the relative capital intensity $b3$ we obtain:

$$1 - Rb = -\frac{T_{11}(1 - b)(1 - Rb)}{T_2}$$

and $R'(K, K, L) < 0$. Let us consider now the first order condition of the consumption maximization problem at the steady state with respect to K: $-T_1(K, K, L)/T_2(K, K, L) = 1$. This is equivalent to the equation defining the stationary capital and labor quantities in a two-sector optimal growth model. Since $R'(K, K, L) < 0$, the proof of Theorem 3.1 in Becker and Tsyganov (2002) here applies and there exists a unique solution $\hat{K}$ of (7.12). Along a stationary path of capital stocks, the highest utility is finally defined as the maximum of $U(C) - BV(L)$ subject to $C = T(K, K, L)$.

A.4 Proof of Proposition 3

Under Assumption 1, the first order conditions of firm's profit maximization problem (7.1) yield

$$\begin{aligned} T_{12} &= -T_{11}b = -\frac{\partial p}{\partial k}\frac{\partial r}{\partial k}, & T_{22} &= T_{11}b^2 = -\frac{\partial p}{\partial y}, \\ T_{31} &= -T_{11}a = \frac{\partial w}{\partial p}\frac{\partial p}{\partial k}, & T_{32} &= T_{11}ab = \frac{\partial w}{\partial p}\frac{\partial p}{\partial y} \end{aligned} \tag{7.31}$$

where $a \equiv K^0/L^0 > 0$, b is defined by (7.3) and $T_{11} < 0$.[4] Consider now the expressions for ε_{rk}, ε_v, ε_u and ε_{uv} together with $T_1 K^*/T_3 L^* = s/(1 - s)$, $T = T_1 K^*/(1 - \chi)$, $Bv'/T_3 = u's/(1 - s)(1 - \chi)$ and $-T_1/T_2 = (1 - \chi)s/(1 - \chi - s)$. Keeping in mind that the homogeneity of $T(K, Y, L)$ implies $a = (1 - b) K^*/L^*$, one has that the total differentiation of (7.12) using (7.10) and (7.11) evaluated at the NSS gives the characteristic polynomial $\mathscr{P}(\lambda) = \lambda^2 - \lambda\mathscr{T} + \mathscr{D}$ where $\mathscr{T}$ is the Trace and $\mathscr{D}$ the Determinant.   ∎

---

[4] See Benhabib and Nishimura (1985), Bosi et al. (2005) and Venditti (2005).

A.5 The Geometrical Method

If $\mathcal{T}$ and $\mathcal{D}$ lie in the interior of the triangle $\mathcal{ABC}$ depicted in Fig. 7.5, the stationary solution is a sink, hence locally indeterminate. In the opposite case, it is locally determinate: it is either a saddle when $|\mathcal{T}| > |1 + \mathcal{D}|$, or a source in the opposite case. If we fix all the parameters of the model with exception of ε_{uv} (which we let vary from zero to $+\infty$) we obtain a parametrized curve $\{\mathcal{T}(\varepsilon_{uv}), \mathcal{D}(\varepsilon_{uv})\}$ that describes a half-line Δ starting from the point $(\mathcal{T}_0, \mathcal{D}_0)$ when ε_{uv} is close to zero. The linearity of such locus can be verified by direct inspection of the expressions for $\mathcal{T}$ and $\mathcal{D}$ and from the fact they share the same denominator. This geometrical method makes it possible also to characterize the different bifurcations that may arise when ε_{uv} moves from zero to $+\infty$. In particular, as shown in Fig. 7.5, when the half-line Δ intersects the line $\mathcal{D} = \mathcal{T} - 1$ (at $\varepsilon_{uv} = \varepsilon_{uv}^{T}$), one eigenvalue goes through unity and a saddle-node bifurcation generically occurs; accordingly, we should expect a change in the number and in the stability of the steady states. When Δ goes through the line $\mathcal{D} = -\mathcal{T} - 1$ (at $\varepsilon_{uv} = \varepsilon_{uv}^{F}$), one eigenvalue is equal to -1 and we expect a flip bifurcation: it follows that there will arise nearby two-period cycles, stable or unstable, according to the direction of the bifurcation. Eventually, when Δ intersects the interior of the segment $\mathcal{BC}$ (at $\varepsilon_{uv} = \varepsilon_{uv}^{H}$), the modulus of the complex conjugate eigenvalues is one and the system undergoes, generically, a Hopf bifurcation. Therefore, around the stationary solution, there will emerge a family of closed orbits, stable or unstable, depending on the nature of the bifurcation (supercritical or subcritical).

Following Grandmont et al. (1988) and Cazzavillan et al. (1998), this analysis is also powerful enough to chracterize the occurrence of sunspot equilibria around an indeterminate stationary solution of system (7.12) as well as along flip and Hopf bifurcations.[5] Actually, as is the case in Grandmont et al. (1988) and Cazzavillan et al. (1998), system (7.12) has at each period t one predetermined variable, the initial stock of capital, and two forward-looking variables, the capital stocks of the two consecutive periods. In such a configuration, the existence of local indeterminacy requires that the two characteristic roots associated with the linearization of the dynamic system (7.12) around the normalized steady state have modulus less than one. In the opposite case, the steady state is locally determinate. Accordingly, multiple equilibria and sunspot fluctuations occur when the modulus of both eigenvalues is lower than unity, i.e. the steady state is located in the interior of the triangle $\mathcal{ABC}$ or along a supercritical flip bifurcation or a supercritical Hopf bifurcation.

The bifurcation parameter we will adopt through our analysis is the elasticity of intertemporal substitution in consumption ε_{uv}. Then the variation of the Trace $\mathcal{T}$ and of the Determinant $\mathcal{D}$ in the $(\mathcal{T}, \mathcal{D})$ plane will be studied as ε_{uv} is made to vary continuously within the $(1, +\infty)$ interval. The relationship between $\mathcal{T}$ and $\mathcal{D}$

---

[5]In the case of supercritical flip bifurcation and s supercritical Hopf bifurcation, sunspot remain in a compact set containing in its interior, respectively, the stable two-period cycle and the stable closed orbit. Unstable cycles and closed orbits emerge in the opposite case of subcritical bifurcations.

Fig. 7.5 Stability triangle and $\Delta(\mathscr{T})$ segment

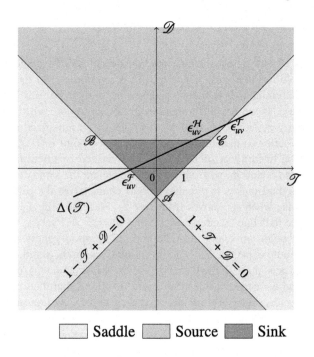

☐ Saddle   ☐ Source   ■ Sink

is given by a half-line $\Delta(\mathscr{T})$ (Fig. 7.5). $\Delta(\mathscr{T})$ is obtained from (7.19) to (7.20) and yields to the following linear relationship:

$$\mathscr{D} = \Delta(\mathscr{T}) = \mathscr{S}\mathscr{T} + \mathscr{Z} \tag{7.32}$$

where $\mathscr{Z}$ is a constant term. The slope of $\Delta(\mathscr{T})$ is given by:

$$\mathscr{S} = \frac{\mathscr{D}'(\varepsilon_{uv})}{\mathscr{T}'(\varepsilon_{uv})} = \frac{s(\varepsilon_{rk} - \chi)}{\varepsilon_{rk}bs + 1 - \chi - s}. \tag{7.33}$$

For a given value $B = B^*$, as ε_{uv} is made to vary in $(0, +\infty)$, $\mathscr{T}(\varepsilon_{uv})$ and $\mathscr{D}(\varepsilon_{uv})$ move linearly along the line $\Delta(\mathscr{T})$. As $\varepsilon_{uv} \in (0, +\infty)$, the properties of the line $\Delta(\mathscr{T})$ are derived from the consideration of its extremities. Actually, the starting point is the couple $(\lim_{\varepsilon_{uv} \to +\infty} \mathscr{T} \equiv \mathscr{T}_{\infty}, \lim_{\varepsilon_{uv} \to +\infty} \mathscr{D} \equiv \mathscr{D}_{\infty})$:

$$\mathscr{T}_{\infty}(\varepsilon_{rk}) = \frac{\varepsilon_{rk}(bs\chi + 1 - \chi) - s\chi}{\varepsilon_{rk}(bs\chi + 1 - \chi - s)}; \quad \mathscr{D}_{\infty}(\varepsilon_{rk}) = \frac{s(\varepsilon_{rk} - \chi)}{\varepsilon_{rk}(bs\chi + 1 - \chi - s)}. \tag{7.34}$$

Using the expressions of $\mathscr{T}_{\infty}$ and $\mathscr{D}_{\infty}$, one is able to show that $\mathscr{D}_{\infty} = \mathscr{T}_{\infty} - 1$. Finally, the half-line $\Delta(\mathscr{T})$ is pointing upward or downward depending on the sign of $\mathscr{D}'(\varepsilon_{uv})$:

$$\mathscr{D}'(\varepsilon_{uv}) = -\frac{s(\varepsilon_{rk} - \chi)}{(bs\chi + 1 - \chi - s)\varepsilon_{rk}\varepsilon_{uv}^2} \tag{7.35}$$

References

Azariadis, C. (1981). Self-fulfilling prophecies. *Journal of Economic Theory, 25*, 380–396.

Azariadis, C., & Guesnerie, R. (1982). Sunspots and cycles. *Review of Economic Studies, 53*(5), 725–737.

Baxter, M. (1996). Are consumer durables important for business cycles? *Review of Economics and Statistics, 78*, 147–155.

Becker, R., & Tsyganov, E. (2002). Ramsey equilibrium in a two-sector model with heterogeneous households. *Journal of Economic Theory, 105*, 188–225.

Benhabib, J., & Laroque, G. (1988). On competitive cycles in productive economies. *Journal of Economic Theory, 45*, 145–170.

Benhabib, J., & Nishimura, K. (1981). Stability of equilibrium in dynamic models of capital theory. *International Economic Review, 22*, 275–293.

Benhabib, J., & Nishimura, K. (1985). Competitive equilibrium cycles. *Journal of Economic Theory, 35*, 284–306.

Bloise, G. (2001). A geometric approach to sunspot equilibria. *Journal of Economic Theory, 101*(2), 519–539.

Bloise, G., Bosi, S., & Magris, F. (2000). Indeterminacy and cycles in a cash-in-advance economy with production. *Rivista Internazionale di Scienze Sociali CVIII, 3*, 263–275.

Bosi, S., Magris, F., & Venditti, A. (2005). Competitive equilibrium cycles with endogenous labor. *Journal of Mathematical Economics, 41*, 325–349.

Bosi, S., Magris, F., & Venditti, A. (2005). Multiple equilibria in a cash-in-advance two-sector economy. *International Journal of Economic Theory, 1*, 131–149.

Cazzavillan, G., & Pintus, P. A. (2005). On competitive cycles and sunspots in productive economies with a positive money stock. *Research in Economics, 59*, 137–147.

Cazzavillan, G., & Pintus, P. A. (2007). Dynamic inefficiency in an overlapping generations economy with production. *Journal of Economic Theory, 5137*, 754–759.

Cazzavillan, G., Lloyd-Braga, T., & Pintus, P. A. (1998). Multiple steady states and endogenous fluctuations with increasing returns to scale in production. *Journal of Economic Theory, 80*(1), 60–107.

Crettez, B., Michel, P., & Wigniolle, B. (1999). Cash-in-advance constraints in the diamond overlapping generations model: Neutrality and optimality of monetary policy. *Oxford Economic Papers, 51*, 431–452.

Crettez, B., Michel, P., & Wigniolle, B. (2002). Seigniorage and public good in an OLG model with cash-in-advance constraints. *Research in Economics, 56*, 333–364.

Diamond, P. A. (1965). National debt in a neoclassical growth model. *American Economic Review, 55*, 1126–1150.

Drugeon, J.-P. (2004). On consumptions, inputs and outputs substituabilities and the evanescence of optimal cycles. *Journal of Difference Equations and Applications, 10*, 473–487.

Drugeon, J.-P., Nourry, C., & Venditti, A. (2010). On efficiency and local uniqueness in two-sector OLG economies. *Mathematical Social Sciences, 59*, 120–144.

Eichenbaum, M. S., Hansen, P., & Singleton, K. J. (1988). A time series analysis of representative agent models of consumption and labor choice under uncertainty. *The Quarterly Journal of Economics, 103*, 51–78.

Gale, D. (1973). Pure exchange equilibrium of dynamic economic models. *Journal of Economic Theory, 6*, 12–36.

Galor, O. (1992). A two-sector overlapping-generations model: A global characterization of the dynamical system. *Econometrica, 60*(6), 1351–1386.

Grandmont, J.-M. (1985). On endogenous competitive business cycles. *Econometrica, 53*(5), 995–1045.

Grandmont, J.-M., Pintus, P., & de Vilder, R. (1998). Capital-labor substitution and competitive nonlinear endogenous business cycles. *Journal of Economic Theory, 80*, 14–59.

Hahn, F., & Solow, R. (1995). *A critical essay on modern macroeconomic theory.* Basil Blackwell.

Hall, R. H. (1988). Intertemporal substitution in consumption. *Journal of Political Economy*, *96*(2), 339–357.

Le Riche, A., Nourry, C., & Venditti, A. (2012). Efficient endogenous fluctuations in two-sector overlapping generations model. *AMSE Working Paper*.

Michel, P., & Wigniolle, B. (2005). Cash-in-advance constraints, bubbles and monetary policy. *Macroeconomic Dynamics*, *9*, 28–56.

Nourry, C., & Venditti, A. (2011). Local indeterminacy under dynamic efficiency in a two-sector overlapping generations economy. *Journal of Mathematical Economics*, *47*, 164–169.

Nourry, C., & Venditti, A. (2012). Endogenous business cycles in OLG economies with multiple consumption goods. *Macroeconomic Dynamics*, *16*, 86–102.

Reichlin, P. (1986). Equilibrium cycles in overlapping generations economy with production. *Journal of Economic Theory*, *40*, 89–102.

Rochon, C., & Polemarchakis, H. M. (2006). Debt, liquidity and cycles. *Economic Theory*, *27*(1), 179–211.

Ruelle, D. (1989). *Elements of differentiable dynamics and bifurcation theory*. San Diego: Academic Press.

Takahashi, H., Mashiyama, K., & Sakagami, R. (2012). Does the capital intensity matter? Evidence from the Postwar Japanese economy and other OECD countries. *Macroeconomic Dynamics*, *16*, 103–116.

Tirole, J. (1985). Asset bubbles and overlapping generations. *Econometrica*, *53*(6), 1499–1528.

Venditti, A. (2005). The two-sector overlapping generations model: A simple formulation. *Research in Economics*, *59*, 164–188.

Woodford, M. (1986). Stationary sunspot equilibria in a finance constrained economy. *Journal of Economic Theory*, *40*, 128–137.

Chapter 8
Homoclinic Orbit and Stationary Sunspot Equilibrium in a Three-Dimensional Continuous-Time Model with a Predetermined Variable

Hiromi Murakami, Kazuo Nishimura and Tadashi Shigoka

Abstract We treat a three-dimensional continuous-time model that includes one predetermined variable and two non-predetermined variables. We assume (1) that the model has a two-dimensional well-located invariant manifold and (2) that the manifold includes a one-dimensional closed curve that could be either a homoclinic orbit or a closed orbit. We construct a stationary sunspot equilibrium in this three-dimensional model by means of generalizing the methods due to Nishimura and Shigoka (2006) and Benhabib et al. (2008). By appealing to the same argument as in Nishimura and Shigoka (2006) we can apply our result to some variants of the Lucas (1988) model the transitional dynamics of which is three-dimensional and undergoes homoclinic bifurcation.

8.1 Introduction

Nishimura and Shigoka (2006) shows that transitional dynamics of the Lucas (1988) model and a variant of the Romer (1990) model due to Benhabib et al. (1994), which includes one predetermined variable and two non-predetermined variables, undergoes *supercritical* Hopf bifurcation for some parameter values and undergoes *subcritical* Hopf bifurcation for other parameter values in such a way that a closed

We are deeply grateful to an anonymous referee, Alain Venditti and Ken Urai for their invaluable advice.

H. Murakami
Graduate School of Economics, Osaka University, Osaka, Japan
e-mail: pge027mh@student.econ.osaka-u.ac.jp

K. Nishimura
Research Institute for Economics and Business Administration,
Kobe University, Kobe, Japan
e-mail: nishimura@rieb.kobe-u.ac.jp

T. Shigoka (✉)
Institute of Economic Research, Kyoto University, Kyoto, Japan
e-mail: sigoka@kier.kyoto-u.ac.jp

© Springer International Publishing AG 2017
K. Nishimura et al. (eds.), *Sunspots and Non-Linear Dynamics*,
Studies in Economic Theory 31, DOI 10.1007/978-3-319-44076-7_8

orbit bifurcates on a two-dimensional invariant manifold that is well-located in a three-dimensional ambient space. Equilibrium is *globally indeterminate* on a subset of the two-dimensional invariant manifold enclosed by the closed orbit. For the case of supercritical Hopf bifurcation, Nishimura and Shigoka (2006) provides the method of constructing a stationary sunspot equilibrium in the three-dimensional transitional dynamics. But one can not apply this method to the case of subcritical Hopf bifurcation.

Mattana et al. (2009) and Bella et al. (2009) construct continuous-time two-sector endogenous growth models that are variants of the Lucas (1988) model. The deterministic equilibrium dynamics in these models is composed of (1) three-dimensional transitional dynamics whose steady state corresponds to a balanced growth path BGP, and (2) a log-linear component characterizing a trend component. Transitional dynamics includes one predetermined variable and two non-predetermined variables, and for some parameter values, the transitional dynamics undergoes *homoclinic bifurcation* in such a way that either a homoclinic orbit or a closed orbit together with two steady states bifurcates on a two-dimensional invariant manifold that is well-located in a three-dimensional ambient space. Equilibrium is *globally indeterminate* on a subset of the two-dimensional invariant manifold enclosed by either the homoclinic orbit or the closed orbit. Benhabib et al. (2008) treats a two-dimensional continuous-time model with one predetermined variable and one non-predetermined variable that includes either a homoclinic orbit or a closed orbit. Equilibrium is *globally indeterminate* on a subset of the ambient space enclosed by either the homoclinic orbit or the closed orbit. Benhabib et al. (2008) constructs a sunspot equilibrium in this model. The method due to Benhabib et al. (2008) is applicable only to the case where both the ambient space and the set on which equilibrium is indeterminate have the same dimension, which is two. Thus one can not directly apply this method to the models due to Mattana et al. (2009) and Bella et al. (2009), because in the transitional dynamics of these models the dimension of the ambient space, which is three, is greater than the dimension of the subset on which equilibrium is indeterminate, which is two.

In the present study, by means of generalizing the methods due to Nishimura and Shigoka (2006) and Benhabib et al. (2008), we show how to construct a stationary sunspot equilibrium in a three-dimensional continuous-time model that includes one predetermined variable and two non-predetermined variables and that includes a two-dimensional invariant manifold which is well-located in an ambient space and on which there exists either a homoclinic orbit or a closed orbit. Our result is applicable to the case of subcritical Hopf bifurcation as well. And although Benhabib et al. (2008) lacks the proof of stationarity, our result includes this proof. By appealing to the same argument as in Nishimura and Shigoka (2006), we can apply the result of the present study to the endogenous growth models due to Lucas (1988), Benhabib et al. (1994), Mattana et al. (2009), and Bella et al. (2009). The rest of the present paper is composed of four sections. In Sect. 8.2 we specify an underlying deterministic model. In Sect. 8.3 we specify a stochastic process that generates sunspot variables. In Sect. 8.4 we state our result formally and explain the idea of a proof intuitively. In Sect. 8.5 we provide the statement with a formal proof.

8.2 Deterministic Equilibrium Dynamics

In the present section we specify a three-dimensional continuous-time deterministic model that includes one predetermined variable and two non-predetermined variables. We will assume that the model has a two-dimensional invariant manifold well-located in a three-dimensional ambient space and that the manifold includes a one-dimensional invariant closed curve. The invariant closed curve might be a homoclinic orbit or a closed orbit of the three-dimensional deterministic model.

Let V be a non-empty open subset of $\mathbb{R}^2$ homeomorphic to some convex set. Let I be a non-empty open connected subset of $\mathbb{R}$. Let W be defined as $W :=V \times I$. Let $f_i : W \to \mathbb{R}, i = 1, 2, 3$, be a continuously differentiable function, i.e., a C^1-function, and let $F : W \to \mathbb{R}^3$ be a C^1-function defined as

$$F(X, u, Q) := \begin{bmatrix} f_1(X, u, Q) \\ f_2(X, u, Q) \\ f_3(X, u, Q) \end{bmatrix}$$

for $(X, u, Q) \in W$. We assume that X is a predetermined variable and that u and Q are non-predetermined variables. This notation as to the one predetermined variable and the two non-predetermined variables is the same as that in Nishimura and Shigoka (2006) that is conformable to the notation in Benhabib and Perli (1994) that analyzes transitional dynamics of the Lucas (1988) model. This is also the same as the notation in Mattana et al. (2009) and Bella et al. (2009) both of which analyze transitional dynamics of variants of the Lucas (1988) model. We use this notation, because we have in mind the application of our main result in Sect. 8.4 to these endogenous growth models.

Since the ordinary differential equation $[\dot{X}, \dot{u}, \dot{Q}]^{\mathrm{T}} = F(X, u, Q)$ has a solution for a given initial value $(X, u, Q) \in W$, there exist an open subset $\hat{N}$ of $\mathbb{R} \times W$ and a continuous function $\hat{\phi} : \hat{N} \to W$ such that $\{0\} \times W \subset \hat{N} \subset \mathbb{R} \times W$ and such that the following relations hold.

(1) For each $(t, X, u, Q) \in \hat{N}$, $\hat{\phi}(t, X, u, Q)$ is a C^1-function of t with $\hat{\phi}(0, X, u, Q) = (X, u, Q) \in W$ and with $\hat{\phi}(t, X, u, Q) \subset W$.

(2) For each $(t, X, u, Q) \in \hat{N}$,

$$\lim_{h \to 0} \frac{\hat{\phi}(t + h, X, u, Q) - \hat{\phi}(t, X, u, Q)}{h} = F(\hat{\phi}(t, X, u, Q)).$$

We define a homoclinic orbit and a periodic orbit of $[\dot{X}, \dot{u}, \dot{Q}]^{\mathrm{T}} = F(X, u, Q)$ in the following way.

Definition 1 Suppose that there exists a one-dimensional closed curve $\hat{\gamma}$ in W that satisfies $\mathbb{R} \times \hat{\gamma} \subset \hat{N}$ and $\hat{\phi}(\mathbb{R} \times \hat{\gamma}) = \hat{\gamma}$.

(1) If there exists a point $(\bar{X}, \bar{u}, \bar{Q})$ in $\hat{\gamma}$ such that $F(\bar{X}, \bar{u}, \bar{Q}) = 0$ and such that for each $(X, u, Q) \in \hat{\gamma} \setminus \{(\bar{X}, \bar{u}, \bar{Q})\}$, $\lim_{t \to +\infty} \hat{\phi}(t, X, u, Q) = (\bar{X}, \bar{u}, \bar{Q}) \wedge$ $\lim_{t \to -\infty} \hat{\phi}(t, X, u) = (\bar{X}, \bar{u}, \bar{Q})$, then $\hat{\gamma}$ is called a homoclinic orbit of $[\dot{X}, \dot{u}, \dot{Q}]^{\mathrm{T}} = F(X, u, Q)$.

(2) If there exists a positive constant $T > 0$ such that $\hat{\phi}(t + T, X, u, Q) = \hat{\phi}(t, X, u, Q)$ for each $(t, X, u, Q) \in \mathbb{R} \times \gamma$ and such that if $0 < s < T$, $\hat{\phi}(t + s, X, u, Q) \neq \hat{\phi}(t, X, u, Q)$, then $\hat{\gamma}$ is called a period-T closed orbit of $[\dot{X}, \dot{u}, \dot{Q}]^{\mathrm{T}} = F(X, u, Q)$.

We make the following assumption that is due to Nishimura and Shigoka (2006).

Assumption 1 There exists a C^1-function $\varphi : V \to I$ such that for $(X, u) \in V$,

$$f_3(X, u, \varphi(X, u)) = \frac{\partial \varphi}{\partial X}(X, u) f_1(X, u, \varphi(X, u)) + \frac{\partial \varphi}{\partial u}(X, u) f_2(X, u, \varphi(X, u)).$$

Under Assumption 1 for each $(X, u) \in V$ we have $(0, X, u, \varphi(X, u)) \subset \hat{N}$, and for each $(t, X, u, \varphi(X, u)) \subset \hat{N}$ we have $\hat{\phi}(t, X, u, \varphi(X, u)) \subset \{(X, u, Q) \in W : Q = \varphi(X, u)\}$. Therefore under Assumption 1 $\{(X, u, Q) \in W : Q = \varphi(X, u)\}$ constitutes a two-dimensional manifold that is invariant under the action of $[\dot{X}, \dot{u}, \dot{Q}]^{\mathrm{T}} = F(X, u, Q)$.

Let $G : V \to \mathbb{R}^2$ be a C^1-function defined as

$$G(X, u) := \begin{bmatrix} f_1(X, u, \varphi(X, u)) \\ f_2(X, u, \varphi(X, u)) \end{bmatrix}$$

for $(X, u) \in V$. Since the ordinary differential equation $[\dot{X}, \dot{u}]^{\mathrm{T}} = G(X, u)$ has a solution for a given initial value $(X, u) \in V$, by Assumption 1 there exists an open subset N of $\mathbb{R} \times V$ and a continuous function $\phi : N \to V$ such that $\{0\} \times V \subset N \subset \mathbb{R} \times V$ and such that the following relations hold.

(1) For each $(t, X, u) \in N$, $\phi(t, X, u)$ is a C^1-function of t with $\phi(0, X, u) = (X, u) \in V$ and with $\phi(t, X, u) \subset V$.

(2) For each $(t, X, u) \in N$,

$$\lim_{h \to 0} \frac{\phi(t + h, X, u) - \phi(t, X, u)}{h} = G(\phi(t, X, u)).$$

(3) For each $(t, X, u) \in N$, $(\phi(t, X, u), \varphi(\phi(t, X, u))) = \hat{\phi}(t, X, u, \varphi(X, u))$. We define a homoclinic orbit and a periodic orbit of $[\dot{X}, \dot{u}]^{\mathrm{T}} = G(X, u)$ in the following way.

Definition 2 Suppose that there exists a one-dimensional closed curve γ in V that satisfies $\mathbb{R} \times \gamma \subset N$ and $\phi(\mathbb{R} \times \gamma) = \gamma$.

(1) If there exists a point $(\bar{X}, \bar{u})$ in γ such that $G(\bar{X}, \bar{u}) = 0$ and such that for each $(X, u) \in \gamma \setminus \{(\bar{X}, \bar{u})\}$, $\lim_{t \to +\infty} \phi(t, X, u) = (\bar{X}, \bar{u}) \wedge \lim_{t \to -\infty} \phi(t, X, u) = (\bar{X}, \bar{u})$, then γ is called a homoclinic orbit of $[\dot{X}, \dot{u}]^{\mathrm{T}} = G(X, u)$.

(2) If there exists a positive constant $T > 0$ such that $\phi(t + T, X, u) = \phi(t, X, u)$ for each $(t, X, u) \in \mathbb{R} \times \gamma$ and such that if $0 < s < T, \phi(t + s, X, u) \neq \phi(t, X, u)$, then γ is called a period-T closed orbit of $[\dot{X}, \dot{u}]^{\mathrm{T}} = G(X, u)$.

By Assumption 1, if a one-dimensional closed curve γ in V is a homoclinic orbit of $[\dot{X}, \dot{u}]^{\mathrm{T}} = G(X, u)$, then $\{(X, u, Q) \in W : (X, u) \in \gamma \wedge Q = \varphi(X, u)\}$ is a homoclinic orbit of $[\dot{X}, \dot{u}, \dot{Q}]^{\mathrm{T}} = F(X, u, Q)$. And if a one-dimensional closed curve γ in V is a period-T closed orbit of $[\dot{X}, \dot{u}]^{\mathrm{T}} = G(X, u)$, then $\{(X, u, Q) \in W : (X, u) \in \gamma \wedge Q = \varphi(X, u)\}$ is a period-T closed orbit of $[\dot{X}, \dot{u}, \dot{Q}]^{\mathrm{T}} = F(X, u, Q)$. If Assumption 1 holds and if $[\dot{X}, \dot{u}]^{\mathrm{T}} = G(X, u)$ has either a homoclinic orbit or a closed orbit, then the following assumption holds.

Assumption 2 There exists a closed subset D of V homeomorphic to the two-dimensional closed unit disk $\{(x, y) \in \mathbb{R}^2 : x^2 + y^2 \leq 1\}$ that satisfies the following conditions.

(1) $\phi(\mathbb{R} \times D) = D$.

(2) Let γ be the boundary of D and let U be the set of all interior points of D. Then $\phi(\mathbb{R} \times \gamma) = \gamma$, and $\phi(\mathbb{R} \times U) = U$.

In the rest of the paper we also make Assumption 2 that is due to Benhabib et al. (2008). Figure 8.1 depicts the case where Assumptions 1 and 2 hold. Figure 8.2a depicts the case where γ is a stable closed orbit with a unique steady state. Figure 8.2b depicts the case where γ is an unstable closed orbit with a unique steady state. Figure 8.2c depicts the case where γ is a homoclinic orbit with two steady states. Figure 8.2d depicts the case where γ is a closed orbit with two steady states such that one steady state belongs to U and such that the other steady state is outside D. See Nishimura and Shigoka (2006) for the case amenable to Fig. 8.2a, b, and see Mattana et al. (2009) for the case amenable to Fig. 8.2c, d.

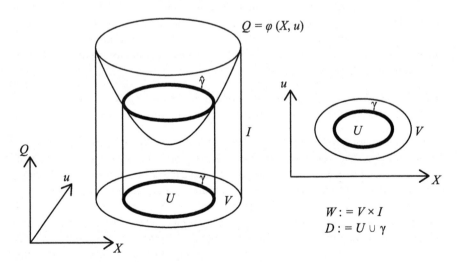

Fig. 8.1 Homoclinic orbit and stationary sunspot equilibrium

(a)   **(b)**

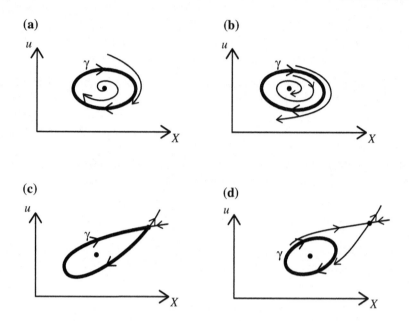

(c)   **(d)**

Fig. 8.2 Homoclinic orbit and stationary sunspot equilibrium

Let X_a and X_b be defined as

$$X_a := \arg \min X \text{ subject to } (X, u) \in \gamma,$$
$$X_b := \arg \max X \text{ subject to } (X, u) \in \gamma.$$

Since γ is a compact subset of $\mathbb{R}^2$ and since the bounded interior region U of D enclosed by γ is non-empty, X_a and X_b are well defined and $X_a < X_b$. Let $\mathbb{T}$ be the set of all non-negative real number, i.e., $\mathbb{T} := \mathbb{R}_+$. For a given initial condition X_0 that belongs to the closed interval $[X_a, X_b]$, we define a perfect foresight equilibrium in the following way.

Definition 3 For a given initial condition $X_0 \in [X_a, X_b]$, $\{(\phi(t, X_0, u), \varphi(\phi(t, X_0, u))\}_{t\in\mathbb{T}}$ constitutes a perfect foresight equilibrium, if $(X_0, u) \in D$.

If an initial condition X_0 belongs to the open interval (X_a, X_b), the set $\{(X, u) \in D : X = X_0\}$ constitutes a one-dimensional manifold whose cardinality is that of *continuum*. See Fig. 8.3. Under Assumptions 1 and 2 the perfect foresight equilibrium is *indeterminate*, because there exists a continuum of perfect foresight equilibria.

Suppose that $(\bar{X}, \bar{u}, \bar{Q}) \in W$ is a steady state of $[\dot{X}, \dot{u}, \dot{Q}]^T = F(X, u, Q)$ and consider the Jacobian of $F(X, u, Q)$ evaluated at $(\bar{X}, \bar{u}, \bar{Q})$. If this Jacobian is hyperbolic and if the number of stable roots is greater than one, we might say that equilibrium is *locally indeterminate* near the steady state $(\bar{X}, \bar{u}, \bar{Q}) \in W$. If equilibrium is indeterminate for a reason different from the reason why equilibrium is

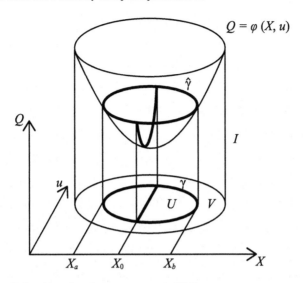

$Q = \varphi(X, u)$

Fig. 8.3 Homoclinic orbit and stationary sunspot equilibrium

locally indeterminate, we might say that equilibrium is *globally indeterminate*. See Definition 1 in Mattana et al. (2009) p. 30. Under Assumptions 1 and 2 the perfect foresight equilibrium is globally indeterminate in this sense.

8.3 Sunspot Variables

In the present section we specify a continuous-time stochastic process that generates sunspot variables. We assume that the stochastic process is subject to a *separable two-state Markov process with stationary transition matrices*.[1] A sample function of a sunspot variable subject to this process generates a sequence of discontinuous points such that each discontinuous point is, of itself, a random variable. We will utilize a sequence of discontinuous points in the sample function of a sunspot variable in order to construct a sunspot equilibrium.

We have set $\mathbb{T} := \mathbb{R}_+$. We denote the set of all functions from $\mathbb{T}$ to $\{1, 2\}$ by $\{1, 2\}^{\mathbb{T}}$. Let B be some subset of $\{1, 2\}^{\mathbb{T}}$. Let $\varepsilon(t, b)$ denote the t-th coordinate of $b \in B$. Let $\mathcal{B}(B)$ be some σ-field in B such that $\varepsilon(t, b)$ is a measurable function of b for each $t \in \mathbb{T}$. And let $\hat{\mathbf{P}}_1 : \mathcal{B}(B) \to [0, 1]$ be some probability measure defined on $\mathcal{B}(B)$. $\varepsilon(t, b)$ that is considered as a function of $t \in \mathbb{T}$ is called a sample function of $b \in B$. Let $\mathcal{L} = \mathcal{L}(\mathbb{R}^2)$ be the set of all 2×2 real square matrices. Let $\lambda > 0$ be

---

[1] In order to generate sunspot variables, Shigoka (1994) uses a continuous-time finite-state Markov process with stationary transition matrices, where the number of states could be two, and Drugeon and Wigniolle (1996), Nishimura and Shigoka (2006) and Benhabib et al. (2008) use a continuous-time two-state Markov process with stationary transition matrices.

a given positive constant, and let $\mathbf{\Lambda} \in \mathcal{L}(\mathbb{R}^2)$ be given by

$$\mathbf{\Lambda} := \begin{bmatrix} -\lambda & \lambda \\ \lambda & -\lambda \end{bmatrix}.$$

Let $\hat{\mathbf{Q}} : \mathbb{T} \to \mathcal{L}(\mathbb{R}^2)$ be defined as

$$\hat{\mathbf{Q}}(h) := \begin{bmatrix} 1 & 0 \\ 0 & 1 \end{bmatrix} + \sum_{k=1}^{\infty} \frac{h^k}{k!} \mathbf{\Lambda}^k$$

for each $h \in \mathbb{T}$. Let $\hat{B}(\{1,2\}^{\mathbb{T}})$ be the set of all functions b in $\{1,2\}^{\mathbb{T}}$ such that b is a piecewise-continuous function of $t \in \mathbb{T}$ and such that b is continuous on the right at each discontinuous point in $\mathbb{T}$. Let $\mathbb{N}$ be the set of all positive integers, i.e., $\mathbb{N} := \{1, 2, \ldots\}$. For $b \in B \cap \hat{B}(\{1,2\}^{\mathbb{T}})$ and for $m \in \mathbb{N}$, let $\hat{t}(m, b) \in \mathbb{T}$ be the m-th discontinuous point of the sample function $\varepsilon(t, b)$ of b if the m-th discontinuous point exists. Then we have the following proposition, where $B \subset \hat{B}(\{1,2\}^{\mathbb{T}})$ means that $(B, \mathcal{B}(B), \hat{\mathbf{P}}_1)$ constitutes a *separable stochastic process*. See Doob (1953) pp. 51–52 for the concept of a separable stochastic process.

Proposition 1 *There exists a continuous-time two-state Markov process $(B, \mathcal{B}(B), \hat{\mathbf{P}}_1)$ with stationary transition probabilities that satisfies the following conditions.*

(1) The initial probability is given by

$$\hat{\mathbf{P}}_1\{b \in B : \varepsilon(0, b) = 1\} = \frac{1}{2} \wedge \hat{\mathbf{P}}_1\{b \in B : \varepsilon(0, b) = 2\} = \frac{1}{2}.$$

(2) A family of the stationary transition probabilities is given by a family of the matrices $\hat{\mathbf{Q}}(h)$ with $h \in \mathbb{T}$.

(3) $B \subset \hat{B}(\{1,2\}^{\mathbb{T}})$.

(4) For each $b \in B$ and for each $m \in \mathbb{N}$ there exists the m-th discontinuous point $\hat{t}(m, b)$ in the sample function $\varepsilon(t, b)$ of b, and $\lim_{m \to \infty} \hat{t}(m, b) = \infty$.

Proof Since the initial probability in (1) and a family of the transition probabilities $\hat{\mathbf{Q}}(h)$, $h \in \mathbb{T}$, in (2) provide *a consistent family of finite dimensional distributions* for a continuous-time two-state Markov process, there exists a stochastic process that satisfies (1) and (2), by *Kolmogorov consistency theorem*. Thus by Theorem II.2.4 in Doob (1953) there exists a stochastic process that satisfies (1)–(3). Doob (1953) p. 249 shows that there exists a subset B' of B such that $\hat{\mathbf{P}}_1(B') = 1$ and such that for each $b \in B'$, (4) holds. Thus if we restrict B to B', and redefine B' as B, then (4) holds. □

The stochastic process $(B, \mathcal{B}(B), \hat{\mathbf{P}}_1)$ is stationary. We will later use $b \in B$ as a sunspot variable. Then $\varepsilon(t, b)$ is a sample function of a sunspot variable $b \in B$, and $\{\hat{t}(m, b)\}_{m \geq 1}$ constitutes a sequence of discontinuous points in the sample function of a sunspot variable $b \in B$. For each $b \in B$, $\varepsilon(t, b)$ is continuous at $t = 0$ and $\hat{t}(1, b) > 0$.

Let $\mathbb{N}_0$ denote the set of all non-negative integers, i.e., $\mathbb{N}_0 := \{0\} \cup \mathbb{N}$. Let $\tau : \mathbb{N}_0 \times B \to \mathbb{T}$ be a function constructed in the following way. For each $b \in B$, set $\tau(0, b) = 0$. For each $m \in \mathbb{N}$ and for each $b \in B$, set $\tau(m, b) = \hat{t}(m, b)$. Then by Proposition 1 $\tau(m, b) \in \mathbb{T}$ is well-defined for all $(m, b) \in \mathbb{N}_0 \times B$. By construction and since $B \subset \hat{B}(\{1, 2\}^{\mathsf{T}})$, $\tau(0, b) = 0$ for each $b \in B$, and $\tau(m + 1, b) - \tau(m, b) > 0$ for each $(m, b) \in \mathbb{N}_0 \times B$. Let $\mathcal{B}_t(B)$ be *the smallest σ-field* with respect to which $(\varepsilon(s, b), 0 \leq s \leq t)$ is a family of measurable functions of $b \in B$. We have the following proposition.

Proposition 2 *(1) For each $m \in \mathbb{N}_0$, if $t \geq \tau(m, b)$, $\tau(m, b)$ is a $\mathcal{B}_t(B)$-measurable function of $b \in B$ and if $\tau(m, b) > t \geq 0$, $\tau(m, b)$ is not $\mathcal{B}_t(B)$-measurable function of $b \in B$. (2) If $\{m_i\}_{i=0}^N$ is a given set of points in $\mathbb{N}_0$ with $N \geq 1$ such that $m_0 < m_1 < \cdots < m_N$, then $N + 1$ random variables $\tau(m_0 + 1, b) - \tau(m_0, b)$, $\tau(m_1 + 1, b) - \tau(m_1, b)$, and $\tau(m_N + 1, b) - \tau(m_N, b)$ are mutually independent and identically distributed.*

(3) For a given $m \in \mathbb{N}_0$ and for a given $T \in \mathbb{T}$,

$$\hat{\mathbf{P}}_1\{b \in B : \tau(m + 1, b) - \tau(m, b) \leq T\} = 1 - e^{-\lambda T}.$$

Proof Note that for each $(t, b) \in \mathbb{T} \times B$ there exists a unique element m in $\mathbb{N}_0$ such that $\tau(m, b) \leq t < \tau(m + 1, b)$. Let $\xi : \mathbb{T} \times B \to \mathbb{N}_0$ be defined as follows. For each $(t, b) \in \mathbb{T} \times B$, if $\tau(m, b) \leq t < \tau(m + 1, b)$, set $\xi(t, b) := m$. Since $\frac{d}{dh}\hat{\mathbf{Q}}(0) = \begin{bmatrix} -\lambda & \lambda \\ \lambda & -\lambda \end{bmatrix}$, we can use Theorems V1.1.2–VI.1.4 in Doob (1953) to show that $\{\xi(t, b)\}_{t \in \mathbb{T}}$ thus constructed is subject to a *separable Poisson process with a parameter $\lambda > 0$ and with a jump of magnitude equal to 1*. Therefore we can use the arguments in Doob (1953) pp. 399–404 to show Proposition 2. $\square$

8.4 Stationary Sunspot Equilibrium

In the present section we assume that the underlying deterministic dynamics $[\dot{X}, \dot{u}, \dot{Q}]^{\mathsf{T}} = F(X, u, Q)$ satisfies Assumptions 1 and 2. Under these assumptions, we define stationary sunspot equilibrium *formally*, and state that there exists a stationary sunspot equilibrium thus defined. We also explain intuitively how to prove the statement. We will provide the statement with a formal proof in the next section.

Let $\hat{D} \subset W$ be defined as $\hat{D} := \{(X, u, Q) \in W : (X, u) \in D \wedge Q = \varphi(X, u)\}$, where φ and D are specified as in Assumptions 1 and 2, respectively. Let $\mathcal{B}(\hat{D})$ be the Borel σ-field in $\hat{D}$, and let $\hat{\mathbf{P}}_0 : \mathcal{B}(\hat{D}) \to [0, 1]$ be some probability measure on $\mathcal{B}(\hat{D})$. Let $(B, \mathcal{B}(B), \hat{\mathbf{P}}_1)$ be the probability space the existence of which is assured by Proposition 1, and let $\Omega := \hat{D} \times B$ and let $\mathcal{B}_\Omega = \mathcal{B}(\Omega)$ be the product σ-field of $\mathcal{B}(\hat{D})$ and $\mathcal{B}(B)$, i.e., $\mathcal{B}(\Omega) := \mathcal{B}(\hat{D}) \otimes \mathcal{B}(B)$. Let $\pi_0 : \Omega \to \hat{D}$ be the projection of

$\hat{D} \times B$ onto $\hat{D}$. Let $\pi_1 : \Omega \to B$ be the projection of $\hat{D} \times B$ onto B. By construction $\pi_0(\omega) = (X, u, Q)$ with $Q = \varphi(X, u)$, and $\varepsilon(t, \pi_1(\omega)) = \varepsilon(t, b)$. We use $b \in B$ as a sunspot variable. We have denoted the t-th coordinate of $b \in B$ as $\varepsilon(t, b)$. Let $\mathcal{B}_t(\Omega)$ be *the smallest σ-field* with respect to which $(\pi_0(\omega), \varepsilon(t, \pi_1(\omega)), 0 \le s \le t)$ is a family of measurable functions of $\omega \in \Omega$. Let $\mathbf{P}$ be the product measure of $\hat{\mathbf{P}}_0$ and $\hat{\mathbf{P}}_1$, i.e., $\mathbf{P} := \hat{\mathbf{P}}_0 \otimes \hat{\mathbf{P}}_1$, and let $\mathbf{E}_t[\cdot]$ be the conditional expectation operator relative to $\mathcal{B}_t(\Omega)$. We denote the set of all functions from $\mathbb{T}$ to $\hat{D}$ by $\hat{D}^{\mathbb{T}}$. Let $\hat{l} : \Omega \to \hat{D}^{\mathbb{T}}$ be a function such that the t-th coordinate of $\hat{l}(\omega) \in \hat{D}^{\mathbb{T}}$ is a $\mathcal{B}_t(\Omega)$-measurable function of $\omega \in \Omega$ for each $t \in \mathbb{T}$. Let $(X(t, \omega), u(t, \omega), Q(t, \omega))$ denote the t-th coordinate of $\hat{l}(\omega) \in \hat{D}^{\mathbb{T}}$. Let $\mathcal{B}(\hat{D}^{\mathbb{T}})$ be some σ-field in $\hat{D}^{\mathbb{T}}$, and let $\hat{\mathbf{P}} : \mathcal{B}(\hat{D}^{\mathbb{T}}) \to [0, 1]$ be some probability measure defined on $\mathcal{B}(\hat{D}^{\mathbb{T}})$. We define *stationary sunspot equilibrium* in the following way.

Definition 4 If the probability measure $\mathbf{P} : \mathcal{B}(\Omega) \to [0, 1]$ satisfies the following conditions, then function $\hat{l} : \Omega \to \hat{D}^{\mathbb{T}}$ constitutes a stationary sunspot equilibrium.

(1) For each $t \in \mathbb{T}$, $(X(t, \omega), u(t, \omega), Q(t, \omega)) \in \hat{D}$ is a $\mathcal{B}_t(\Omega)$-measurable function of $\omega \in \Omega$.

(2) The distribution of $(X(0, \omega), u(0, \omega), Q(0, \omega))$ is given by $(\hat{D}, \mathcal{B}(\hat{D}), \hat{\mathbf{P}}_0)$.

(3) There exists a stochastic process $(\hat{D}^{\mathbb{T}}, \mathcal{B}(\hat{D}^{\mathbb{T}}), \hat{\mathbf{P}})$ on $\hat{D}^{\mathbb{T}}$ such that if $\{t_i\}_{i=1}^{N}$ is a given set of points in $\mathbb{T}$ with $N \ge 1$ and if $\hat{Y}$ is a given Borel subset of $\hat{D}^N$, then

$$\mathbf{P}\{\omega \in \Omega : (\hat{l}(t_1, \omega), \dots, \hat{l}(t_N, \omega)) \in \hat{Y}\}$$
$$= \hat{\mathbf{P}}\{\hat{d} \in \hat{D}^{\mathbb{T}} : (\hat{d}(t_1), \dots, \hat{d}(t_N)) \in \hat{Y}\},$$

where $\hat{l}(t, \omega) := (X(t, \omega), u(t, \omega), Q(t, \omega)) \in \hat{D}$ and $\hat{d}(t)$ denotes the t-th coordinate of $\hat{d} \in \hat{D}^{\mathbb{T}}$.

(4) For each $\omega \in \Omega$, $X(t, \omega)$ is a continuous function of $t \in \mathbb{T}$, and for each $t > 0$,

$$\mathbf{P}\{\omega \in \Omega : \lim_{h \to 0} \frac{X(t + h, \omega) - X(t, \omega)}{h} = f_1(X(t, \omega), u(t, \omega), Q(t, \omega))\} = 1.$$

(5) For each $\omega \in \Omega$, $(u(t, \omega), Q(t, \omega))$ is a piecewise-continuous function of $t \in \mathbb{T}$ and continuous on the right at each discontinuous point in $\mathbb{T}$, and for each $t \in \mathbb{T}$,

$$\begin{bmatrix} \lim_{h \to +0} \frac{X(t+h,\omega)-X(t,\omega)}{h} \\ \mathbf{E}_t[\lim_{h \to +0} \frac{u(t+h,\omega)-u(t,\omega)}{h}] \\ \mathbf{E}_t[\lim_{h \to +0} \frac{Q(t+h,\omega)-Q(t,\omega)}{h}] \end{bmatrix} = \begin{bmatrix} f_1(X(t, \omega), u(t, \omega), Q(t, \omega)) \\ f_2(X(t, \omega), u(t, \omega), Q(t, \omega)) \\ f_3(X(t, \omega), u(t, \omega), Q(t, \omega)) \end{bmatrix}.$$

(6) For each $t > s \ge 0$, $(X(t, \omega), u(t, \omega), Q(t, \omega))$ is not $\mathcal{B}_s(\Omega)$-measurable function of $\omega \in \Omega$.

(7) If $\{t_i\}_{i=1}^{N}$ is a given set of points in $\mathbb{T}$ with $N \ge 1$ and if $\hat{Y}$ is a given Borel subset of $\hat{D}^N$, for any $h \in \mathbb{R}$ such that $\{t_i + h\}_{i=1}^{N} \subset \mathbb{T}$,

$$\mathbf{P}\{\omega \in \Omega : (\hat{l}(t_1, \omega), \ldots, \hat{l}(t_N, \omega)) \in \hat{Y}\}$$
$$= \mathbf{P}\{\omega \in \Omega : (\hat{l}(t_1 + h, \omega), \ldots, \hat{l}(t_N + h, \omega)) \in \hat{Y}\},$$

where $\hat{l}(t, \omega) := (X(t, \omega), u(t, \omega), Q(t, \omega)) \in \hat{D}$.

By (1) $(X(t, \omega), u(t, \omega), Q(t, \omega))$ is a random variable whose information structure is given by $(\Omega, \mathcal{B}_t(\Omega))$. Since $\mathbf{P} := \hat{\mathbf{P}}_0 \otimes \hat{\mathbf{P}}_1$, the distribution of $(X(t, \omega), u(t, \omega), Q(t, \omega))$ at $t = 0$ is independent of the distribution of a sunspot variable $b \in B$ by (2). This implies among other things that a sunspot variable $b \in B$ does not affect the distribution of a value taken by the predetermined variable $X(t, \omega)$ at $t = 0$. Since $\hat{D}$ is a closed subset of $\mathbb{R}^3$, $\hat{D}$ is a *complete and separable metric space*, and $\hat{D}^{\mathbb{T}}$ is a *product space composed of a family of complete and separable metric spaces*. We will use this fact to show the existence of a stochastic process required by (3). $(\hat{D}^{\mathbb{T}}, \mathcal{B}(\hat{D}^{\mathbb{T}}), \hat{\mathbf{P}})$ constitutes the product space representation of the set of random variables $\{\hat{l}(t, \omega)\}_{t \in \mathbb{T}}$ by (3). Since $X(t, \omega)$ is a predetermined variable, $X(t, \omega)$ should be a continuous function of $t \in \mathbb{T}$ as required by (4). This stochastic process constitutes an *equilibrium* stochastic process by (1), (2), (4) and (5). The equilibrium stochastic process is a *non-trivial stochastic process* by (6). In this sense $\hat{l} : \Omega \to \hat{D}^{\mathbb{T}}$ constitutes a *sunspot equilibrium*. This sunspot equilibrium is *stationary* by (7). We have the following result.

Theorem 1 *Suppose that the underlying deterministic dynamics* $[\dot{X}, \dot{u}, \dot{Q}]^{\mathrm{T}} = F(X, u, Q)$ *satisfies Assumptions 1 and 2. Then there exists a stationary sunspot equilibrium.*

By appealing to the same argument as in Nishimura and Shigoka (2006) that treats the Lucas (1988) model and a variant of the Romer (1990) model, we can apply Theorem 1 to two variants of the Lucas (1988) model due to Mattana et al. (2009) and Bella et al. (2009) for the case of homoclinic bifurcation as well as the Lucas (1988) model and the variant of the Romer (1990) model for the case of Hopf bifurcation whether supercritical or subcritical. Growth rates of endogenous variables such as human capital, physical capital, and consumption are subject to a stationary stochastic process in a sunspot equilibrium thus constructed in each of these models. We provide Theorem 1 with a formal proof in the next section. Before leaving this section, we explain the idea of the proof intuitively based on Fig. 8.4 that is adapted from Fig. 3 in Benhabib et al. (2008). The following argument should provide an intuition for the construction of a set of random variables $\{(X(t, \omega), u(t, \omega), Q(t, \omega))\}_{t \in \mathbb{T}}$ in Sect. 8.5.1

Figure 8.4a depicts a typical sample function $\varepsilon_t = \varepsilon(t)$ of a sunspot variable $b \in B$. For $m \geq 1$, $\tau(m)$ denotes the m-th discontinuous point of the sample function. Figure 8.4b depicts the case where $[\dot{X}, \dot{u}]^{\mathrm{T}} = G(X, u)$ has a homoclinic orbit γ with two steady states (X^*, u^*) and (X^{**}, u^{**}) such that $(X^*, u^*) \in U \wedge (X^{**}, u^{**}) \in \gamma$. As shown in the next section, there exists a continuous function $d : D \to \mathbb{R}_+$ such that for each $(X, u) \in U, d(X, u) > 0 \wedge (X, u + \frac{1}{2}d(X, u)) \in U$. Choose some point (X_0, u_0) from U. Set $(X(t), u(t)) = \phi(t, X_0, u_0)$ for $t \in [0, \tau(1))$. Set $(X(t), u(t)) = \phi(t - \tau(1), \phi(\tau(1), X_0, u_0) + (0, \frac{1}{2}d(\phi(\tau(1), X_0, u_0))))$

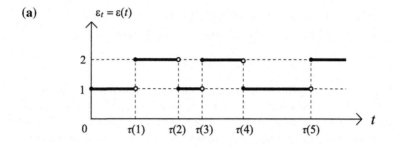

(a)

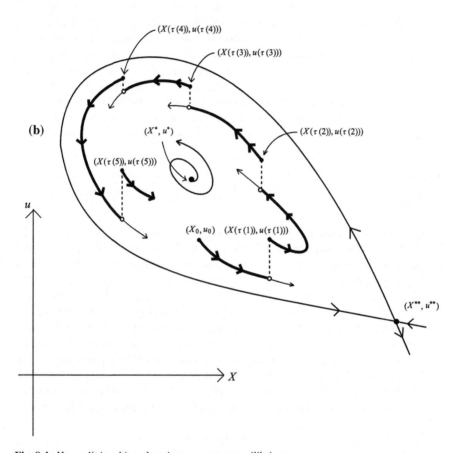

(b)

Fig. 8.4 Homoclinic orbit and stationary sunspot equilibrium

for $t \in [\tau(1), \tau(2))$. Set $(X(t), u(t)) = \phi(t - \tau(2), \phi(\tau(2) - \tau(1), X(\tau(1)), u(\tau(1))))$
$+ (0, \frac{1}{2}d(\phi(\tau(2) - \tau(1), X(\tau(1)), u(\tau(1)))))$ for $t \in [\tau(2), \tau(3))$. For each $m \geq$
2, set $(X(t), u(t)) = \phi(t - \tau_m, \phi(\tau_m - \tau_{m-1}, X(\tau_{m-1}), u(\tau_{m-1}))) + (0, \frac{1}{2}d(\phi(\tau_m - \tau_{m-1}, X(\tau_{m-1}), u(\tau_{m-1}))))$ for $t \in [\tau(m), \tau(m+1))$, where $\tau_m := \tau(m) \wedge \tau_{m-1} := \tau(m-1)$. See Fig. 8.4b. Then by construction $X(t)$ is a continuous function of
$t \in \mathbb{T}$, and $\dot{X}(t) = f_1(X(t), u(t), \varphi(X(t), u(t)))$ if $\varepsilon_t = \varepsilon(t)$ is continuous at $t >$
0. By Proposition 2 we have $\hat{\mathbf{P}}_1\{b \in B : \lim_{s \to t} \varepsilon(s, b) = \varepsilon(t, b)\} = 1$ for each
$t \in \mathbb{T}$. See also Theorem VI.1.2 in Doob (1953). Thus for each $t > 0$, $\dot{X}(t) = f_1(X(t), u(t), \varphi(X(t), u(t)))$ with probability 1. And by construction for each $t \in \mathbb{T}$
we also have

$$[\frac{dX}{dt}(t), \frac{du}{dt}(t)]^{\mathrm{T}} = G(X(t), u(t)),$$

where $\frac{dX}{dt}(t) := \lim_{h \to +0} \frac{X(t+h) - X(t)}{h} \wedge \frac{du}{dt}(t) := \lim_{h \to +0} \frac{u(t+h) - u(t)}{h}$. This is the
method due to Benhabib et al. (2008).

Next set $Q(t) = \varphi(X(t), u(t))$. Then we have

$$\frac{dQ}{dt}(t) = f_3(X(t), u(t), \varphi(X(t), u(t))),$$

where $\frac{dQ}{dt}(t) := \lim_{h \to +0} \frac{Q(t+h) - Q(t)}{h}$, because by Assumption 1

$$\begin{aligned}
\frac{dQ}{dt}(t) &= \frac{\partial \varphi}{\partial X}(X(t), u(t)) \frac{dX}{dt}(t) + \frac{\partial \varphi}{\partial u}(X(t), u(t)) \frac{du}{dt}(t) \\
&= \frac{\partial \varphi}{\partial X}(X(t), u(t)) f_1(X(t), u(t), \varphi(X(t), u(t))) \\
&\quad + \frac{\partial \varphi}{\partial u}(X(t), u(t)) f_2(X(t), u(t), \varphi(X(t), u(t))) \\
&= f_3(X(t), u(t), \varphi(X(t), u(t))).
\end{aligned}$$

Therefore we have

$$[\frac{dX}{dt}(t), \frac{du}{dt}(t), \frac{dQ}{dt}(t)]^{\mathrm{T}} = F(X(t), u(t), Q(t)).$$

This is the method due to Nishimura and Shigoka (2006). Since we can show that
$F(X(t), u(t), Q(t))$ is $\mathcal{B}_t(\Omega)$-measurable so that $\mathbf{E}_t \frac{du}{dt}(t) = \frac{du}{dt}(t) \wedge \mathbf{E}_t \frac{dQ}{dt}(t) = \frac{dQ}{dt}(t)$,
we have

$$[\frac{dX}{dt}(t), \mathbf{E}_t \frac{du}{dt}(t), \mathbf{E}_t \frac{dQ}{dt}(t)]^{\mathrm{T}} = F(X(t), u(t), Q(t)).$$

We can also show that $\{(X(t), u(t), \varepsilon(t))\}_{t \in \mathbb{T}}$ is subject to a Markov process
with stationary transition probabilities on $D \times \{1, 2\}$ and that the associated Markov
operator maps each continuous function on the compact set $D \times \{1, 2\}$ into some
continuous function on $D \times \{1, 2\}$. By using these facts we show that the equilibrium

stochastic process could be stationary. Our proof of stationarity under Assumption 2 is new. The argument in Sect. 8.5 is more rigorous than that in Benhabib et al. (2008), because we need a rigorous argument in order to prove the stationarity of an equilibrium stochastic process under this assumption.

8.5   Proof of Theorem 1

In the present section we assume that Assumptions 1 and 2 are satisfied, and we formally prove Theorem 1 in the following way. In Sect. 8.5.1 we construct a set of random variables $\{(X(t, \omega), u(t, \omega), Q(t, \omega))\}_{t \in \mathbb{T}}$ by following the argument based on Fig. 8.4 in the previous section. In Sect. 8.5.2 we show that the set of random variables thus constructed satisfies the conditions (1)–(6) in Definition 4, which implies the existence of a non-trivial equilibrium stochastic process. In Sect. 8.5.3 we show that the equilibrium stochastic process could be stationary and thus satisfy the condition (7) in Definition 4.

8.5.1   *Construction of Random Variables*

In the present section we construct a set of random variables $\{(X(t, \omega), u(t, \omega), Q(t, \omega))\}_{t \in \mathbb{T}}$ in a recursive way by following the argument in Sect. 8.4. Let $d : D \to \mathbb{R}_+$ be defined as

$$d(X, u) := \min \sqrt{(X - x_1)^2 + (u - x_2)^2} \text{ subject to } (x_1, x_2) \in \gamma.$$

Since γ is a compact set, $d = d(X, u)$ is well-defined and since $d = d(X, u)$ is a distance between a point in D and set γ, $d = d(X, u)$ is a continuous function of $(X, u) \in D$. Since γ is the boundary of D that is homeomorphic to the closed unit disk and since $U \subset \mathbb{R}^2$ is a set of all interior points of D, for any $(X, u) \in U$, $d = d(X, u) > 0$, and $(X, u + \frac{1}{2}d(X, u)) \in U$.

We have set $\hat{D} := \{(X, u, Q) \in W : (X, u) \in D \wedge Q = \varphi(X, u)\}$. Let $\hat{\mathbf{P}}_0 : \mathcal{B}(\hat{D}) \to [0, 1]$ be some probability measure that satisfies

$$\hat{\mathbf{P}}_0\{(X, u, Q) \in \hat{D} : (X, u) \in U\} = 1.$$

One can safely assume that there exists *at least one* such probability measure, because for example one has only to choose some specific point $(X^{'}, u^{'})$ from U in advance and assign probability 1 to the point $(X^{'}, u^{'}, \varphi(X^{'}, u^{'}))$ in $\hat{D}$ in order to obtain such a measure. In Sect. 8.5.3 we will show that there exists so well-behaved a measure $\hat{\mathbf{P}}_0 : \mathcal{B}(\hat{D}) \to [0, 1]$ that the resulting product measure $\hat{\mathbf{P}}_0 \otimes \hat{\mathbf{P}}_1$ satisfies the condition (7) in Definition 4, where $\hat{\mathbf{P}}_1 : \mathcal{B}(B) \to [0, 1]$ is the probability measure in Proposition 1.

We have defined $\mathbf{P} : \mathcal{B}(\Omega) \to [0, 1]$ as the product measure of $\hat{\mathbf{P}}_0$ and $\hat{\mathbf{P}}_1$, i.e., $\mathbf{P} := \hat{\mathbf{P}}_0 \otimes \hat{\mathbf{P}}_1$. Let $\hat{\pi} : \hat{D} \to D$ be the projection of $\hat{D}$ onto D so that $\hat{\pi}(X, u, Q) = (X, u)$ for $(X, u, Q) \in \hat{D}$. Since $\pi_0(\omega) = (X, u, Q)$ for $\omega = (X, u, Q, b) \in \hat{D} \times B = \Omega$, $\hat{\pi}(\pi_0(\omega)) = (X, u)$. Thus we have

$$\mathbf{P}\{\omega \in \Omega : \hat{\pi}(\pi_0(\omega)) \in U\} = 1,$$

because

$$\begin{aligned} \mathbf{P}\{\omega \in \Omega : \hat{\pi}(\pi_0(\omega)) \in U\} \\ = \mathbf{P}\{\omega \in \Omega : (X, u) \in U\} \\ = \mathbf{P}\{(X, u, Q, b) \in \hat{D} \times B : (X, u) \in U\} \\ = \hat{\mathbf{P}}_0\{(X, u, Q) \in \hat{D} : (X, u) \in U\} = 1. \end{aligned}$$

Since $\pi_1(\omega) = b$ for $\omega = (X, u, Q, b) \in \hat{D} \times B = \Omega$, $\varepsilon(t, \pi_1(\omega))$ and $\tau(m, \pi_1(\omega))$ are measurable functions of $\omega \in \Omega$. Since we have set $\tau(0, b) = 0$, $\tau(0, \pi_1(\omega)) = 0$. We have defined $\mathcal{B}_t(\Omega)$ as the smallest σ-field in Ω with respect to which $(\pi_0(\omega), \varepsilon(s, \pi_1(\omega)), 0 \le s \le t)$ is a family of measurable functions of $\omega \in \Omega$. For each $m \in \mathbb{N}_0$, if $t \ge \tau(m, \pi_1(\omega))$, then $\tau(m, \pi_1(\omega))$ is a $\mathcal{B}_t(\Omega)$-measurable function of $\omega \in \Omega$ by Proposition 2.

Since $\phi(\mathbb{T}, U) = U$ and since $\tau(0, \pi_1(\omega)) = 0 \wedge \tau(m + 1, \pi_1(\omega)) - \tau(m, \pi_1(\omega)) > 0 \wedge \lim_{m \to \infty} \tau(m, \pi_1(\omega)) = \infty$ for each $(m, \omega) \in \mathbb{N}_0 \times \Omega$, the following constructions are well-defined. Let $f : \mathbb{N}_0 \times \Omega \to U$ and $g : \mathbb{N}_0 \times \Omega \to U$ be defined as follows. For $\omega \in \Omega$, if $\hat{\pi}(\pi_0(\omega)) = (X, u) \in U$, let $f(0, \omega)$ and $g(0, \omega)$ be given by

$$\begin{aligned} f(0, \omega) : = (\hat{\pi}(\pi_0(\omega))), \\ g(0, \omega) : = f(0, \omega). \end{aligned}$$

Choose some specific point (X', u') from U in advance, and for $\omega \in \Omega$, if $\hat{\pi}(\pi_0(\omega)) = (X, u) \in \gamma$, let $f(0, \omega)$ and $g(0, \omega)$ be given by

$$\begin{aligned} f(0, \omega) : = (X', u'), \\ g(0, \omega) : = f(0, \omega). \end{aligned}$$

For $(m, \omega) \in \mathbb{N}_0 \times \Omega$ and for given $f(m, \omega)$ and $g(m, \omega)$, let $f(m + 1, \omega)$ and $g(m + 1, \omega)$ be given by

$$\begin{aligned} f(m + 1, \omega) : = \phi(\tau(m + 1, \pi_1(\omega)) - \tau(m, \pi_1(\omega)), g(m, \omega)), \\ g(m + 1, \omega) : = f(m + 1, \omega) + (0, \frac{1}{2}d(f(m + 1, \omega))). \end{aligned}$$

Then $f(m, \omega) \subset U \wedge g(m, \omega) \subset U$ for each $(m, \omega) \in \mathbb{N}_0 \times \Omega$. For each $m \in \mathbb{N}_0$, if $t \geq \tau(m, \pi_1(\omega))$, $f(m, \omega)$ and $g(m, \omega)$ are $\mathcal{B}_t(\Omega)$-measurable functions of $\omega \in \Omega$.

Note that for each $(t, \omega) \in \mathbb{T} \times \Omega$ there exists a unique element m in $\mathbb{N}_0$ such that $\tau(m, \pi_1(\omega)) \leq t < \tau(m + 1, \pi_1(\omega))$. Let $\theta : \mathbb{T} \times \Omega \to U$ be defined as follows. For each $(t, \omega) \in \mathbb{T} \times \Omega$, if $\tau(m, \pi_1(\omega)) \leq t < \tau(m + 1, \pi_1(\omega))$, let $\theta(t, \omega)$ be given by

$$\theta(t, \omega) := \phi(t - \tau(m, \pi_1(\omega)), g(m, \omega)).$$

Then for each $(t, \omega) \in \mathbb{T} \times \Omega$, $\theta(t, \omega) \in U$ and for each $t \in \mathbb{T}$, $\theta(t, \omega)$ is a $\mathcal{B}_t(\Omega)$-measurable function of $\omega \in \Omega$. For each $(t, \omega) \in \mathbb{T} \times \Omega$, let $(X(t, \omega), u(t, \omega), Q(t, \omega))$ be defined as

$$(X(t, \omega), u(t, \omega), Q(t, \omega)) := (\theta(t, \omega), \varphi(\theta(t, \omega))).$$

By construction for each $(t, \omega) \in \mathbb{T} \times \Omega$, $(X(t, \omega), u(t, \omega)) \in U \wedge (X(t, \omega), u(t, \omega), Q(t, \omega)) \in \hat{D}$.

8.5.2 Existence of a Non-trivial Equilibrium Stochastic Process

In the present section we will refer to (1)–(6) in Definition 4 as the conditions (1)–(6) respectively. We show that these conditions are satisfied, which implies the existence of a nontrivial equilibrium stochastic process.

Since $\theta(t, \omega) \in U$ is a $\mathcal{B}_t(\Omega)$-measurable function of $\omega \in \Omega$, $(X(t, \omega), u(t, \omega), Q(t, \omega))$ is a $\mathcal{B}_t(\Omega)$-measurable function of $\omega \in \Omega$. Therefore the condition (1) is satisfied.

For E and F in $\mathcal{B}(\Omega)$, if $\omega \in E$ implies $\omega \in F$, then $\mathbf{P}\{E\} \leq \mathbf{P}\{F\}$, because $E \subset F$. And we have set $\mathbf{P} := \hat{\mathbf{P}}_0 \otimes \hat{\mathbf{P}}_1$. We will use these facts repeatedly in the following arguments. By construction for each $\omega \in \Omega$ we have $(X(0, \omega), u(0, \omega)) \in U$ and thus

$$\mathbf{P}\{\omega \in \Omega : (X(0, \omega), u(0, \omega)) \in U\} = 1.$$

If $(X, u) \in U$, $g(0, \omega) = (X, u) \wedge (X, u) \in U$ by construction. Thus since $(X(0, \omega), u(0, \omega)) = g(0, \omega) \wedge \hat{\pi}(\pi_0(\omega)) = (X, u)$ by construction, $(X, u) \in U$ implies $(X(0, \omega), u(0, \omega)) = \hat{\pi}(\pi_0(\omega)) \wedge (X, u) \in U$. Therefore we have

$$\mathbf{P}\{\omega \in \Omega : (X(0, \omega), u(0, \omega)) = (X, u)\} = 1,$$

because

$$\mathbf{P}\{\omega \in \Omega : (X(0, \omega), u(0, \omega)) = (X, u)\}$$
$$= \mathbf{P}\{\omega \in \Omega : (X(0, \omega), u(0, \omega)) = \hat{\pi}(\pi_0(\omega))\}$$
$$\geq \mathbf{P}\{\omega \in \Omega : (X(0, \omega), u(0, \omega)) = \hat{\pi}(\pi_0(\omega)) \wedge (X, u) \in U\}$$
$$\geq \mathbf{P}\{\omega \in \Omega : (X, u) \in U\}$$
$$= \mathbf{P}\{(X, u, Q, b) \in \hat{D} \times B : (X, u) \in U\}$$
$$= \hat{\mathbf{P}}_0\{(X, u, Q) \in \hat{D} : (X, u) \in U\} = 1.$$

And we have constructed $Q(0, \omega)$ in such a way that $Q(0, \omega) = \varphi(X(0, \omega), u(0, \omega))$. Let $\hat{l}(0, \omega) := (X(0, \omega), u(0, \omega), Q(0, \omega))$. Then since $\mathbf{P}\{\omega \in \Omega : (X(0, \omega), u(0, \omega)) = (X, u)\} = 1$ implies that

$$\mathbf{P}\{\omega \in \Omega : (X(0, \omega), u(0, \omega)) = (X, u) \wedge Q = \varphi(X, u)\}$$
$$= \mathbf{P}\{\omega \in \Omega : Q = \varphi(X, u)\},$$

we have

$$\mathbf{P}\{\omega \in \Omega : \hat{l}(0, \omega) = \pi_0(\omega)\} = 1,$$

because

$$\mathbf{P}\{\omega \in \Omega : \hat{l}(0, \omega) = \pi_0(\omega)\} = \mathbf{P}\{\omega \in \Omega : \hat{l}(0, \omega) = (X, u, Q)\}$$
$$= \mathbf{P}\{\omega \in \Omega : (X(0, \omega), u(0, \omega)) = (X, u) \wedge Q = \varphi(X, u)\}$$
$$= \mathbf{P}\{\omega \in \Omega : Q = \varphi(X, u)\} = \mathbf{P}\{\omega \in \Omega : (X, u, Q) \in \hat{D}\}$$
$$= \mathbf{P}\{(X, u, Q, b) \in \hat{D} \times B : (X, u, Q) \in \hat{D}\} = \hat{\mathbf{P}}_0\{\hat{D}\} = 1.$$

And for any $\hat{A} \in \mathcal{B}(\hat{D})$,

$$\mathbf{P}\{\omega \in \Omega : \hat{l}(0, \omega) \in \hat{A}\} = \hat{\mathbf{P}}_0\{(X, u, Q) \in \hat{D} : (X, u, Q) \in \hat{A}\},$$

because $\mathbf{P}\{\omega \in \Omega : \hat{l}(0, \omega) = \pi_0(\omega)\} = 1 \wedge \pi_0(\omega) = (X, u, Q)$ so that

$$\mathbf{P}\{\omega \in \Omega : \hat{l}(0, \omega) \in \hat{A}\}$$
$$= \mathbf{P}\{\omega \in \Omega : \hat{l}(0, \omega) = \pi_0(\omega) \wedge \pi_0(\omega) \in \hat{A}\}$$
$$= \mathbf{P}\{\omega \in \Omega : \pi_0(\omega) \in \hat{A}\}$$
$$= \mathbf{P}\{\omega \in \Omega : (X, u, Q) \in \hat{A}\}$$
$$= \mathbf{P}\{(X, u, Q, b) \in \hat{D} \times B : (X, u, Q) \in \hat{A}\}$$
$$= \hat{\mathbf{P}}_0\{(X, u, Q) \in \hat{D} : (X, u, Q) \in \hat{A}\}.$$

Therefore the condition (2) is satisfied.

If $A_0 \subset D \wedge \hat{\pi}^{-1}(A_0) \in \mathcal{B}(\hat{D})$ and if $A_1 \in \mathcal{B}(B)$, we have

$$\mathbf{P}\{(X, u, Q, b) \in \Omega : (X, u, b) \in A_0 \times A_1\}$$
$$= \hat{\mathbf{P}}_0\{(X, u, Q) \in \hat{D} : (X, u) \in A_0\} \times \hat{\mathbf{P}}_1\{b \in B : b \in A_1\}$$
$$= \hat{\mathbf{P}}_0\{(X, u, Q) \in \hat{D} : (X, u, Q) \in \hat{\pi}^{-1}(A_0)\} \times \hat{\mathbf{P}}_1\{A_1\}$$
$$= \hat{\mathbf{P}}_0\{\hat{\pi}^{-1}(A_0)\} \times \hat{\mathbf{P}}_1\{A_1\}.$$

Therefore for each $A_0 \in D$ such that $\hat{\pi}^{-1}(A_0) \in \mathcal{B}(\hat{D})$ and for each $A_1 \in \mathcal{B}(B)$, the probability $\mathbf{P}\{(X, u, Q, b) \in \Omega : (X, u, b) \in A_0 \times A_1\}$ is well-defined. Let $\hat{l}(t, \omega) := (X(t, \omega), u(t, \omega), Q(t, \omega))$ so that $\hat{l}(t, \omega) = (\theta(t, \omega), \varphi(\theta(t, \omega)))$. For a given $(t, X, u, b) \in \mathbb{T} \times D \times B$ such that $\omega = (\hat{\pi}^{-1}(X, u), b) \in \Omega$, $\hat{l}(t, \omega)$ is uniquely determined and $\hat{l}(t, \omega) \in \hat{D}$ from the construction of $f(m, \omega), g(m, \omega)$, and $\theta(t, \omega)$. For a given $t \in \mathbb{T}$, $\hat{l}(t, \omega)$ is a $\mathcal{B}_t(\Omega)$-measurable and thus $\mathcal{B}(\Omega)$-measurable function of $\omega \in \Omega$. Therefore if $\{t_i\}_{i=1}^N$ is a given set of points in $\mathbb{T}$ with $N \geq 1$, and if $\hat{Y}$ is a given Borel subset of $\hat{D}^N$, then the probability that $(\hat{l}(t_1, \omega), \ldots, \hat{l}(t_N, \omega)) \in \hat{Y}$ is well defined. Therefore $\{\hat{l}(t, \omega)\}_{t \in \mathbb{T}}$ is a set of random variables on the probability space $(\Omega, \mathcal{B}(\Omega), \mathbf{P})$ and has *a consistent family of finite dimensional distributions*. Since $\hat{D}$ is a closed subset of $\mathbb{R}^3$ so that $\hat{D}$ is a *complete and separable metric space*, the condition (3) is satisfied by *Kolmogorov consistency theorem*.

$(X(t, \omega), u(t, \omega), Q(t, \omega)) := (\theta(t, \omega), \varphi(\theta(t, \omega)))$ satisfies the conditions (4) and (5) from the construction of $f(m, \omega), g(m, \omega)$, and $\theta(t, \omega)$. See the argument in Sect. 8.4 that explains the idea of the proof intuitively based on Fig. 8.4.

For each $m \in \mathbb{N}_0$, and for each $t \in [\tau(m, \pi_1(\omega)), \tau(m + 1, \pi_1(\omega)))$,

$$(X(t, \omega), u(t, \omega)) = \phi(t - \tau(m, \pi_1(\omega)), g(m, \omega)),$$
$$Q(t, \omega) = \varphi(X(t, \omega), u(t, \omega)),$$

$$f(m + 1, \omega) = \phi(\tau(m + 1, \pi_1(\omega)) - \tau(m, \pi_1(\omega)), g(m, \omega)),$$
$$g(m + 1, \omega) = f(m + 1, \omega) + (0, \frac{1}{2}d(f(m + 1, \omega))),$$

$$(X(\tau(m + 1, \pi_1(\omega)), \omega), u(\tau(m + 1, \pi_1(\omega)), \omega)) = g(m + 1, \omega).$$

Since $f(m + 1, \omega) \in U$, $\frac{1}{2}d(f(m + 1, \omega)) \neq 0$. If $t > s \geq 0$,

$$\mathbf{P}\{\omega \in \Omega : \exists t' \in \mathbb{T} : t \geq t' > s \wedge \varepsilon(t', \pi_1(\omega)) \neq \varepsilon(s, \pi_1(\omega))\} = 1 - e^{-\lambda(t-s)} > 0$$

by Proposition 2. Thus for $t > s \geq 0$, $(X(t, \omega), u(t, \omega), Q(t, \omega))$ is not $\mathcal{B}_s(\Omega)$-measurable function of $\omega \in \Omega$. Therefore the conditions (6) is satisfied.

8.5.3 Stationarity of the Equilibrium Stochastic Process

In the present section we show the stationarity of an equilibrium stochastic process in the following way. Let $\mathcal{B}(D \times \{1, 2\})$ be the Borel σ-field in $D \times \{1, 2\}$, and let $\mathcal{B}(U \times \{1, 2\})$ be the Borel σ-field in $U \times \{1, 2\}$. Let $z(t, \omega) := \varepsilon(t, \pi_1(\omega))$. We first show that $\{(\theta(t, \omega), z(t, \omega))\}_{t \in \mathbb{T}}$ is subject to a Markov process with stationary transitional probabilities. Next we construct a family of some transitional probabilities $\mathbf{Q}(\cdot : \cdot | \cdot) : \mathbb{T} \times \mathcal{B}(D \times \{1, 2\}) \times D \times \{1, 2\} \to [0, 1]$ such that the restriction of $\mathbf{Q}(\cdot : \cdot | \cdot)$ to $\mathbb{T} \times \mathcal{B}(U \times \{1, 2\}) \times U \times \{1, 2\}$ provides a family of transitional probabilities for $\{(\theta(t, \omega), z(t, \omega))\}_{t \in \mathbb{T}}$. Next we show that the Markov operator associated with $\mathbf{Q}(\cdot : \cdot | \cdot)$ maps each continuous function on the compact set $D \times \{1, 2\}$ into some continuous function on $D \times \{1, 2\}$. Next by using these facts we show that there exists an invariant measure $\mathbf{p} : \mathcal{B}(D \times \{1, 2\}) \to [0, 1]$ for $\mathbf{Q}(\cdot : \cdot | \cdot)$ such that $\mathbf{p}\{U \times \{1, 2\}\} = 1$. Finally by using this invariant measure $\mathbf{p}$ we construct $\hat{\mathbf{P}}_0 : \mathcal{B}(\hat{D}) \to [0, 1]$ in such a way that the resulting product measure $\hat{\mathbf{P}}_0 \otimes \hat{\mathbf{P}}_1$ provides a stationary distribution for $\{(\theta(t, \omega), \varphi(\theta(t, \omega)))\}_{t \in \mathbb{T}}$ and satisfies the condition (7) in Definition 4.

Let $\Delta \mathbb{T}$ be defined as

$$\Delta \mathbb{T} := \{(T_1, T_2) \in \mathbb{T}^2 : T_1 < T_2\}.$$

For $(T_1, T_2) \in \Delta \mathbb{T}$, let $\{1, 2\}^{[T_1, T_2]}$ denote the set of all functions from $[T_1, T_2]$ to $\{1, 2\}$, and for $b^{[T_1, T_2]} \in \{1, 2\}^{[T_1, T_2]}$ and for $t \in [T_1, T_2]$, let $b^{[T_1, T_2]}(t)$ denote the t-th coordinate of $b^{[T_1, T_2]}$. For $(T_1, T_2) \in \Delta \mathbb{T}$, let $\pi^{[T_1, T_2]} : \{1, 2\}^{\mathbb{T}} \to \{1, 2\}^{[T_1, T_2]}$ be the projection of $\{1, 2\}^{\mathbb{T}}$ onto $\{1, 2\}^{[T_1, T_2]}$. For $(T_1, T_2) \in \Delta \mathbb{T}$, let $B^{[T_1, T_2]} \subset \{1, 2\}^{[T_1, T_2]}$ be defined as

$$B^{[T_1, T_2]} := \{b^{[T_1, T_2]} \in \{1, 2\}^{[T_1, T_2]} : \exists b \in B : \pi^{[T_1, T_2]}(b) = b^{[T_1, T_2]}\}.$$

For $b^{[T_1, T_2]} \in B^{[T_1, T_2]}$ and for $t \in [T_1, T_2]$, $b^{[T_1, T_2]}(t) = \varepsilon(t, b)$, because $\varepsilon(t, b)$ is the t-th coordinate of $b \in B$. Let $\mathcal{B}(B^{[T_1, T_2]})$ be the smallest σ-field in $B^{[T_1, T_2]}$ with respect to which $(b^{[T_1, T_2]}(t), T_1 \leq t \leq T_2)$ is a family of measurable functions of $b^{[T_1, T_2]} \in B^{[T_1, T_2]}$. For each $A^{[T_1, T_2]} \in \mathcal{B}(B^{[T_1, T_2]})$, we have $\{b \in B : \pi^{[T_1, T_2]}(b) \in A^{[T_1, T_2]}\} \in \mathcal{B}(B)$, and let $\hat{\mathbf{P}}_1^{[T_1, T_2]}\{A^{[T_1, T_2]}\}$ be defined as

$$\hat{\mathbf{P}}_1^{[T_1, T_2]}\{A^{[T_1, T_2]}\} := \hat{\mathbf{P}}_1\{b \in B : \pi^{[T_1, T_2]}(b) \in A^{[T_1, T_2]}\}.$$

Then by Proposition 1 $(B^{[T_1, T_2]}, \mathcal{B}(B^{[T_1, T_2]}), \hat{\mathbf{P}}_1^{[T_1, T_2]})$ constitutes a Markov process.

Since $\hat{\mathbf{P}}_1$ is stationary by Proposition 1, $\hat{\mathbf{P}}_1^{[T_1, T_2]}\{b^{[T_1, T_2]} \in B^{[T_1, T_2]} : b^{[T_1, T_2]}(T_1) = z\} = \hat{\mathbf{P}}_1\{b \in B : \varepsilon(T_1, b) = z\} = \frac{1}{2}$ for each $z \in \{1, 2\}$. Thus for a given $(T_1, T_2, A^{[T_1, T_2]}, z) \in \Delta \mathbb{T} \times \mathcal{B}(B^{[T_1, T_2]}) \times \{1, 2\}$, if we set

$$\hat{\mathbf{P}}_1^{[T_1,T_2]}(A^{[T_1,T_2]}|z) := \frac{\hat{\mathbf{P}}_1^{[T_1,T_2]}\{b^{[T_1,T_2]} \in B^{[T_1,T_2]} : b^{[T_1,T_2]} \in A^{[T_1,T_2]} \wedge b^{[T_1,T_2]}(T_1) = z\}}{\hat{\mathbf{P}}_1^{[T_1,T_2]}\{b^{[T_1,T_2]} \in B^{[T_1,T_2]} : b^{[T_1,T_2]}(T_1) = z\}},$$

then $\hat{\mathbf{P}}_1^{[T_1,T_2]}(A^{[T_1,T_2]}|z)$ is well-defined. By the stationarity of $\hat{\mathbf{P}}_1$ for each $b^{[T_1,T_2]} \in B^{[T_1,T_2]}$ there exists $b^{[0,T_2-T_1]} \in B^{[0,T_2-T_1]}$ such that $b^{[T_1,T_2]}(t) = b^{[0,T_2-T_1]}(t-T_1)$ for each $t \in [T_1, T_2]$. For $A^{[T_1,T_2]} \in \mathcal{B}(B^{[T_1,T_2]})$ and for $A^{[0,T_2-T_1]} \subset B^{[0,T_2-T_1]}$, if we have $A^{[0,T_2-T_1]} = \{b^{[0,T_2-T_1]} \in B^{[0,T_2-T_1]} : \exists b^{[T_1,T_2]} \in A^{[T_1,T_2]} : \forall t \in [0, T_2 - T_1] : b^{[0,T_2-T_1]}(t) = b^{[T_1,T_2]}(T_1 + t)\}$, then we have $A^{[0,T_2-T_1]} \in \mathcal{B}(B^{[0,T_2-T_1]})$ and $\hat{\mathbf{P}}_1^{[0,T_2-T_1]}(A^{[0,T_2-T_1]}|z) = \hat{\mathbf{P}}_1^{[T_1,T_2]}(A^{[T_1,T_2]}|z)$.

Let $\mathcal{B}(U)$ be the Borel σ-field in U, and let $\mathcal{B}(U \times B^{[T_1,T_2]})$ be the product σ-field of $\mathcal{B}(U)$ and $\mathcal{B}(B^{[T_1,T_2]})$. For each $t \in \mathbb{T}$, $\phi(t, v)$ is a uniformly continuous function of $v \in D$, because D is compact and because for each $t \in \mathbb{T}$, $\phi(t, v)$ is a continuous function of $v \in D$. Therefore, from the construction of $f(m, \omega)$, $g(m, \omega)$, and $\theta(t, \omega)$, for each $T > 0$ there exists a function $\psi_T : U \times B^{[0,T]} \to U$ such that if $\theta(0, \omega) = v \in U$, then $\theta(T, \omega) = \psi_T(v, b^{[0,T]})$ and such that $\psi_T(v, b^{[0,T]})$ is *uniformly continuous* in $v \in U$ and $\mathcal{B}(B^{[0,T]})$-measurable in $b^{[0,T]} \in B^{[0,T]}$, which implies in turn that $\psi_T(v, b^{[0,T]})$ is $\mathcal{B}(U \times B^{[0,T]})$-measurable in $(v, b^{[0,T]}) \in U \times B^{[0,T]}$. As mentioned above, for each $b^{[T_1,T_2]} \in B^{[T_1,T_2]}$ there exists $b^{[0,T_2-T_1]} \in B^{[0,T_2-T_1]}$ such that $b^{[T_1,T_2]}(t) = b^{[0,T_2-T_1]}(t - T_1)$ for each $t \in [T_1, T_2]$. Therefore for $(T_1, T_2) \in \Delta\mathbb{T}$, if we construct a function $\psi^{[T_1,T_2]} : U \times B^{[T_1,T_2]} \to U$ from $\psi_{T_2-T_1} : U \times B^{[0,T_2-T_1]} \to U$ in such a way that

$$\forall t \in [T_1, T_2] : b^{[T_1,T_2]}(t) = b^{[0,T_2-T_1]}(t - T_1)$$
$$\to \forall v \in U : \psi^{[T_1,T_2]}(v, b^{[T_1,T_2]}) := \psi_{T_2-T_1}(v, b^{[0,T_2-T_1]}),$$

then $\psi^{[T_1,T_2]} = \psi^{[T_1,T_2]}(v, b^{[T_1,T_2]})$ is well-defined. For $(T_1, T_2) \in \Delta\mathbb{T}$, if $\theta(T_1, \omega) = v \in U$, then by construction $\theta(T_2, \omega) = \psi^{[T_1,T_2]}(v, b^{[T_1,T_2]})$. And by construction $\psi^{[T_1,T_2]}(v, b^{[T_1,T_2]})$ is uniformly continuous in $v \in U$ and $\mathcal{B}(B^{[T_1,T_2]})$-measurable in $b^{[T_1,T_2]} \in B^{[T_1,T_2]}$, which implies in turn that $\psi^{[T_1,T_2]}(v, b^{[T_1,T_2]})$ is $\mathcal{B}(U \times B^{[T_1,T_2]})$-measurable in $(v, b^{[T_1,T_2]}) \in U \times B^{[T_1,T_2]}$.

As mentioned above, $(B^{[T_1,T_2]}, \mathcal{B}(B^{[T_1,T_2]}), \hat{\mathbf{P}}_1^{[T_1,T_2]})$ constitutes a Markov process. Since we have set $z(t, \omega) := \varepsilon(t, \pi_1(\omega))$, $z(t, \omega) = b^{[T_1,T_2]}(t)$ for $b^{[T_1,T_2]} \in B^{[T_1,T_2]}$ and for $t \in [T_1, T_2]$. Therefore $\{(\theta(t, \omega), z(t, \omega))\}_{t \in \mathbb{T}}$ is subject to a Markov process. By construction, if $b^{[T_1,T_2]}$ and $b^{[0,T_2-T_1]}$ satisfy $b^{[T_1,T_2]}(t) = b^{[0,T_2-T_1]}(t - T_1)$ for each $t \in [T_1, T_2]$, then $\psi^{[T_1,T_2]}(v, b^{[T_1,T_2]}) = \psi^{[0,T_2-T_1]}(v, b^{[0,T_2-T_1]})$ for each $v \in U$. And as mentioned above, for each $(T_1, T_2, A^{[T_1,T_2]}, z) \in \Delta\mathbb{T} \times \mathcal{B}(B^{[T_1,T_2]}) \times \{1,2\}$, if $A^{[0,T_2-T_1]} \subset B^{[0,T_2-T_1]}$ satisfies $A^{[0,T_2-T_1]} = \{b^{[0,T_2-T_1]} \in B^{[0,T_2-T_1]} : \exists b^{[T_1,T_2]} \in A^{[T_1,T_2]} : \forall t \in [0, T_2 - T_1] : b^{[0,T_2-T_1]}(t) = b^{[T_1,T_2]}(T_1 + t)\}$, then $A^{[0,T_2-T_1]} \in \mathcal{B}(B^{[0,T_2-T_1]})$ and $\hat{\mathbf{P}}_1^{[0,T_2-T_1]}(A^{[0,T_2-T_1]}|z) = \hat{\mathbf{P}}_1^{[T_1,T_2]}(A^{[T_1,T_2]}|z)$. Therefore transitional probabilities for $\{(\theta(t, \omega), z(t, \omega))\}_{t \in \mathbb{T}}$ are stationary. Thus $\{(\theta(t, \omega), z(t, \omega))\}_{t \in \mathbb{T}}$ is subject to a Markov process with stationary transitional probabilities.

Let $\mathcal{B}(D)$ be the Borel σ-field in D, and let $\mathcal{B}(D \times B^{[T_1,T_2]})$ be the product σ-field of $\mathcal{B}(D)$ and $\mathcal{B}(B^{[T_1,T_2]})$. Since D is the compact closure of U and since for each $b^{[T_1,T_2]} \in B^{[T_1,T_2]}$, $\psi^{[T_1,T_2]}(v, b^{[T_1,T_2]})$ is uniformly continuous in

$v \in U$ with $\psi^{[T_1,T_2]}(U, B^{[T_1,T_2]}) \subset U$, there exists a $\mathcal{B}(D \times B^{[T_1,T_2]})$-measurable function $\bar{\psi}^{[T_1,T_2]} : D \times B^{[T_1,T_2]} \to D$ such that $\bar{\psi}^{[T_1,T_2]}(v, b^{[T_1,T_2]})$ is *continuous* in $v \in D$ and such that $\bar{\psi}^{[T_1,T_2]}(v, b^{[T_1,T_2]}) = \psi^{[T_1,T_2]}(v, b^{[T_1,T_2]})$ for each $(v, b^{[T_1,T_2]}) \in U \times B^{[T_1,T_2]}$. If $b^{[T_1,T_2]}$ and $b^{[0,T_2-T_1]}$ satisfy $b^{[T_1,T_2]}(t) = b^{[0,T_2-T_1]}(t - T_1)$ for each $t \in [T_1, T_2]$ so that $\psi^{[T_1,T_2]}(v, b^{[T_1,T_2]}) = \psi^{[0,T_2-T_1]}(v, b^{[0,T_2-T_1]})$ for each $v \in U$, then $\bar{\psi}^{[T_1,T_2]}(v, b^{[T_1,T_2]}) = \bar{\psi}^{[0,T_2-T_1]}(v, b^{[0,T_2-T_1]})$ for each $v \in D$, because the extension of $\psi^{[T_1,T_2]}$ on $U \times B^{[T_1,T_2]}$ to $\bar{\psi}^{[T_1,T_2]}$ on $D \times B^{[T_1,T_2]}$ is unique.

Let $\chi_E : U \times \{1, 2\} \to \{0, 1\}$ denote the characteristic function for $E \in \mathcal{B}(U \times \{1, 2\})$. $\psi^{[T_1,T_2]}$ is $\mathcal{B}(U \times B^{[T_1,T_2]})$-measurable in $(v, b^{[T_1,T_2]}) \in U \times B^{[T_1,T_2]}$ and $b^{[T_1,T_2]}(T_2)$ is $\mathcal{B}(B^{[T_1,T_2]})$-measurable in $b^{[T_1,T_2]} \in B^{[T_1,T_2]}$. Therefore for each $E \in \mathcal{B}(U \times \{1, 2\})$, $\chi_E(\psi^{[T_1,T_2]}(v, b^{[T_1,T_2]}), b^{[T_1,T_2]}(T_2))$ is $\mathcal{B}(U \times B^{[T_1,T_2]})$-measurable in $(v, b^{[T_1,T_2]}) \in U \times B^{[T_1,T_2]}$. Thus for each $(T_1, T_2, E, v, z) \in \Delta\mathbb{T} \times \mathcal{B}(U \times \{1, 2\}) \times U \times \{1, 2\}$, if we set

$$Q(T_1, T_2 : E|v, z) := \int_{B^{[T_1,T_2]}} \chi_E(\psi^{[T_1,T_2]}(v, b^{[T_1,T_2]}), b^{[T_1,T_2]}(T_2)) \hat{\mathbf{P}}_1^{[T_1,T_2]}(db^{[T_1,T_2]}|z),$$

then $Q(T_1, T_2 : E|v, z)$ is well-defined. Let m be the Lebesgue measure on $(U, \mathcal{B}(U))$. $\chi_E(\psi^{[T_1,T_2]}(v, b^{[T_1,T_2]}), b^{[T_1,T_2]}(T_2))$ is $\mathcal{B}(U \times B^{[T_1,T_2]})$-measurable in $(v, b^{[T_1,T_2]}) \in U \times B^{[T_1,T_2]}$ and integrable with respect to the product measure $m \otimes \hat{\mathbf{P}}_1^{[T_1,T_2]}(\cdot|z)$ of m and $\hat{\mathbf{P}}_1^{[T_1,T_2]}(\cdot|z)$. Thus by Fubini's theorem $Q(T_1, T_2, E|v, z)$ is $\mathcal{B}(U)$-measurable in $v \in U$. By construction, for each $(T_1, T_2, v, z) \in \Delta\mathbb{T} \times U \times \{1, 2\}$, $Q(T_1, T_2 : U \times \{1, 2\}|v, z) = 1$. By construction, $Q(T_1, T_2 : E|v, z) = Q(0, T_2 - T_1 : E|v, z)$. And from the construction of $\psi^{[T_1,T_2]}$, $Q(T_1, T_2, E|v, z)$ provides the transitional probability that $(\theta(T_2, \omega), z(T_2, \omega)) \in E$ under the condition $(\theta(T_1, \omega), z(T_1, \omega)) = (v, z)$.

Let $\chi_E : D \times \{1, 2\} \to \{0, 1\}$ denote the characteristic function for $E \in \mathcal{B}(D \times \{1, 2\})$. $\bar{\psi}^{[T_1,T_2]}$ is $\mathcal{B}(D \times B^{[T_1,T_2]})$-measurable in $(v, b^{[T_1,T_2]}) \in D \times B^{[T_1,T_2]}$ and $b^{[T_1,T_2]}(T_2)$ is $\mathcal{B}(B^{[T_1,T_2]})$-measurable function in $b^{[T_1,T_2]} \in B^{[T_1,T_2]}$. Therefore for each $E \in \mathcal{B}(D \times \{1, 2\})$, $\chi_E(\bar{\psi}^{[T_1,T_2]}(v, b^{[T_1,T_2]}), b^{[T_1,T_2]}(T_2))$ is $\mathcal{B}(D \times B^{[T_1,T_2]})$-measurable in $(v, b^{[T_1,T_2]}) \in D \times B^{[T_1,T_2]}$. Thus for each $(T_1, T_2, E, v, z) \in \Delta\mathbb{T} \times \mathcal{B}(D \times \{1, 2\}) \times D \times \{1, 2\}$, if we set

$$\bar{Q}(T_1, T_2 : E|v, z) := \int_{B^{[T_1,T_2]}} \chi_E(\bar{\psi}^{[T_1,T_2]}(v, b^{[T_1,T_2]}), b^{[T_1,T_2]}(T_2)) \hat{\mathbf{P}}_1^{[T_1,T_2]}(db^{[T_1,T_2]}|z),$$

then $\bar{Q}(T_1, T_2 : E|v, z)$ is well-defined. Let $\bar{m}$ be the Lebesgue measure on $(D, \mathcal{B}(D))$. $\chi_E(\bar{\psi}^{[T_1,T_2]}(v, b^{[T_1,T_2]}), b^{[T_1,T_2]}(T_2))$ is $\mathcal{B}(D \times B^{[T_1,T_2]})$-measurable in $(v, b^{[T_1,T_2]}) \in D \times B^{[T_1,T_2]}$ and integrable with respect to the product measure $\bar{m} \otimes \hat{\mathbf{P}}_1^{[T_1,T_2]}(\cdot|z)$ of $\bar{m}$ and $\hat{\mathbf{P}}_1^{[T_1,T_2]}(\cdot|z)$. Thus by Fubini's theorem $\bar{Q}(T_1, T_2, E|v, z)$ is $\mathcal{B}(D)$-measurable in $v \in D$. By construction for $(E, v) \in \mathcal{B}(U \times \{1, 2\}) \times U$ and for $(T_1, T_2, z) \in \Delta\mathbb{T} \times \{1, 2\}$, $\bar{Q}(T_1, T_2 : E|v, z) = Q(T_1, T_2, E|v, z)$. For each $E \in \mathcal{B}(D \times \{1, 2\})$ and for each $\{E_n\}_{n \geq 1}$ such that for all n, $E_n \in \mathcal{B}(D \times \{1, 2\})$ and such that $E = \sum_{n=1}^{\infty} E_n$, we have

$$\bar{Q}(T_1, T_2 : E|v, z) = \sum_{n=1}^{\infty} \bar{Q}(T_1, T_2 : E_n|v, z),$$

because

$$\bar{Q}(T_1, T_2 : E|v, z) = \int_{B^{[T_1, T_2]}} \chi_E(\bar{\psi}^{[T_1, T_2]}(v, b^{[T_1, T_2]}), b^{[T_1, T_2]}(T_2)) \hat{\mathbf{P}}_1^{[T_1, T_2]}(db^{[T_1, T_2]}|z)$$

$$= \int_{B^{[T_1, T_2]}} \sum_{n=1}^{\infty} \chi_{E_n}(\bar{\psi}^{[T_1, T_2]}(v, b^{[T_1, T_2]}), b^{[T_1, T_2]}(T_2)) \hat{\mathbf{P}}_1^{[T_1, T_2]}(db^{[T_1, T_2]}|z)$$

$$= \sum_{n=1}^{\infty} \int_{B^{[T_1, T_2]}} \chi_{E_n}(\bar{\psi}^{[T_1, T_2]}(v, b^{[T_1, T_2]}), b^{[T_1, T_2]}(T_2)) \hat{\mathbf{P}}_1^{[T_1, T_2]}(db^{[T_1, T_2]}|z)$$

$$= \sum_{n=1}^{\infty} \bar{Q}(T_1, T_2 : E_n|v, z).$$

And by construction $\bar{Q}(T_1, T_2 : \emptyset|v, z) = 0 \wedge \bar{Q}(T_1, T_2 : D \times \{1, 2\}|v, z) = 1$ for $(T_1, T_2, v, z) \in \Delta\mathbb{T} \times D \times \{1, 2\}$. Therefore $\bar{Q}(T_1, T_2 : \cdot|v, z) : \mathcal{B}(D \times \{1, 2\}) \to [0, 1]$ constitutes a probability measure for $(T_1, T_2, v, z) \in \Delta\mathbb{T} \times D \times \{1, 2\}$, and $\bar{Q}(T_1, T_2 : E|\cdot) : D \times \{1, 2\} \to [0, 1]$ is a measurable function of $(v, z) \in D \times \{1, 2\}$ for $(T_1, T_2, E) \in \Delta\mathbb{T} \times \mathcal{B}(D \times \{1, 2\})$. By construction $\bar{Q}(T_1, T_2 : E|v, z) = \bar{Q}(0, T_2 - T_1 : E|v, z)$. By construction for $v \in U$ and for $(T_1, T_2, z) \in \Delta\mathbb{T} \times \{1, 2\}$, $\bar{Q}(T_1, T_2 : U \times \{1, 2\}|v, z) = 1$.

For $(h, E, v, z) \in \mathbb{T} \times \mathcal{B}(D \times \{1, 2\}) \times D \times \{1, 2\}$, let $\mathbf{Q}(h : E|v, z)$ be defined as follows.

$$\mathbf{Q}(0 : E|v, z) = 1, \text{ if } (v, z) \in E, \text{ and } \mathbf{Q}(0 : E|v, z) = 0, \text{ if } (v, z) \notin E.$$

$$\mathbf{Q}(h : E|v, z) := \bar{Q}(0, h : E|v, z), \text{ for } h > 0.$$

Then for $(h, v, z) \in \mathbb{T} \times D \times \{1, 2\}$, $\mathbf{Q}(h : \cdot|v, z) : \mathcal{B}(D \times \{1, 2\}) \to [0, 1]$ constitutes a probability measure, and for $(h, E) \in \mathbb{T} \times \mathcal{B}(D \times \{1, 2\})$, $\mathbf{Q}(h : E|\cdot) : D \times \{1, 2\} \to [0, 1]$ is a measurable function of $(v, z) \in D \times \{1, 2\}$. Therefore, we might consider $\mathbf{Q}(\cdot : \cdot|\cdot)$ as a family of stationary transitional probabilities for some Markov process. By construction for $v \in U$ and for each $(h, z) \in \mathbb{T} \times \{1, 2\}$, $\mathbf{Q}(h : U \times \{1, 2\}|v, z) = 1$. The restriction of $\mathbf{Q}(\cdot : \cdot|\cdot)$ to $\mathbb{T} \times \mathcal{B}(U \times \{1, 2\}) \times U \times \{1, 2\}$ provides a family of transitional probabilities for $\{(\theta(t, \omega), z(t, \omega))\}_{t \in \mathbb{T}}$, because $\mathbf{Q}(h : E|v, z) = Q(t, t + h, E|v, z)$ for each $t \in \mathbb{T}$, for each $h > 0$ and for each $(E, v, z) \in \mathcal{B}(U \times \{1, 2\}) \times U \times \{1, 2\}$.

Let $C(D \times \{1, 2\})$ be the set of all continuous functions from the compact set $D \times \{1, 2\}$ to $\mathbb{R}$. For each $g(\cdot) \in C(D \times \{1, 2\})$, for each $h > 0$ and for each $(v, z) \in D \times \{1, 2\}$, we have

$$\int_{\hat{D} \times \{1, 2\}} g(v', z') d\mathbf{Q}(h : v', z'|v, z)$$

$$= \int_{\hat{D} \times \{1, 2\}} g(\bar{\psi}^{[0, h]}(v, b^{[0, h]}), b^{[0, h]}(h)) \hat{\mathbf{P}}_1^{[0, h]}(db^{[0, h]}|z).$$

Since $\bar{\psi}^{[T_1, T_2]}(v, b^{[T_1, T_2]})$ is a continuous function of $v \in D$ and since $g = g(v', z')$ is a bounded continuous function of $(v', z') \in D \times \{1, 2\}$, by the Lebesgue dominated convergence theorem the right hand side of the above equality is a continuous function of $v \in D$ and thus a continuous function of $(v, z) \in D \times \{1, 2\}$. Therefore the Markov operator associated with the family of transitional probabilities $\mathbf{Q}(\cdot : \cdot | \cdot)$ maps $C(D \times \{1, 2\})$ into $C(D \times \{1, 2\})$. Since $D \times \{1, 2\}$ is *compact*, this implies that the Markov process associated with $\mathbf{Q}(\cdot : \cdot | \cdot)$ *does not include any dissipative part in the sense of* Yosida (1980) p. 395 in $D \times \{1, 2\}$. And by construction we have $\mathbf{Q}(h : U | v, z) = 1$ for each $h \in \mathbb{T}$ and for each $(v, z) \in U \times \{1, 2\}$. Based on these facts we can prove the following claim by the argument for the proof of Theorem 1 of XIII.4 in Yosida (1980). We will provide the proof for this claim at the end of this section.

Claim There exists a probability measure $\mathbf{p} : \mathcal{B}(D \times \{1, 2\}) \to [0, 1]$ such that $\mathbf{p}\{U \times \{1, 2\}\} = 1$ and that if we assign $\mathbf{p}$ to a family of transitional probabilities $\mathbf{Q}(\cdot : \cdot | \cdot)$ as an initial probability measure, then the resulting Markov process is stationary.

Since $\hat{\mathbf{P}}_1$ has already been stationary, we should already have had for each $i \in \{1, 2\}$

$$\hat{\mathbf{P}}_1\{b \in B : \varepsilon(0, b) = i\} = \mathbf{p}\{(v, z) \in D \times \{1, 2\} : z = i\}.$$

For each $\hat{A} \in \mathcal{B}(\hat{D})$, which in turn implies $\hat{\pi}(\hat{A}) \in \mathcal{B}(D)$, let $\hat{\mathbf{P}}_0\{\hat{A}\}$ be defined as

$$\hat{\mathbf{P}}_0\{\hat{A}\} := \mathbf{p}\{(v, z) \in D \times \{1, 2\} : v \in \hat{\pi}(\hat{A})\}.$$

Since $\mathbf{p}(U \times \{1, 2\}) = 1$, the condition $\hat{\mathbf{P}}_0\{(X, u, Q) \in \hat{D} : (X, u) \in U\} = 1$ that has been required in Sect. 8.5.1 is satisfied. As mentioned above, the restriction of $\mathbf{Q}(\cdot : \cdot | \cdot)$ to $\mathbb{T} \times \mathcal{B}(U \times \{1, 2\}) \times U \times \{1, 2\}$ provides a family of transitional probabilities for $\{(\theta(t, \omega), z(t, \omega))\}_{t \in \mathbb{T}}$. Therefore the resulting product measure $\hat{\mathbf{P}}_0 \otimes \hat{\mathbf{P}}_1$ provides a stationary distribution for $\{(\theta(t, \omega), \varphi(\theta(t, \omega)))\}_{t \in \mathbb{T}}$ and satisfies the condition (7) in Definition 4.

Proof of Claim. Let $\rho : D \times \{1, 2\} \to \mathbb{R}_+$ be defined as

$$\rho(y) := \min d(y, y') \text{ subject to } y' \in \gamma \times \{1, 2\},$$

where $d(y, y')$ denotes the distance between two points y and y' in $D \times \{1, 2\} \subset \mathbb{R}^3$. Since $\gamma \times \{1, 2\}$ is compact, $\rho = \rho(y)$ is well-defined and continuous in $y \in D \times \{1, 2\}$. For $n \in \mathbb{N}$, let $f_n : D \times \{1, 2\} \to [0, 1]$ be defined as

$$f_n(y) := \min\{n \cdot \rho(y), 1\}.$$

Then $\{f_n(\cdot)\}_{n \in \mathbb{N}}$ is an increasing sequence of non-negative functions such that for each $n \in \mathbb{N}$, $f_n(\cdot) \in C(D \times \{1, 2\})$. By construction, $\rho(y) = 0$ for $y \in \gamma \times \{1, 2\}$, and $\rho(y) > 0$ for $y \in U \times \{1, 2\}$. Therefore we have

$$\lim_{n\to\infty} f_n(y) = \chi_{U\times\{1,2\}}(y),$$

where $\chi_{U\times\{1,2\}}(\cdot)$ denotes the characteristic function of $U \times \{1, 2\}$.

By construction for each $(v, z) \in U \times \{1, 2\}$ we have $\mathbf{Q}(1 : U|v, z) = 1$. For each $\bar{v} \in D$ there exists a sequence $\{v_m\}_{m\in\mathbb{N}}$ such that $v_m \in U$ for each $m \in \mathbb{N}$ and such that $\lim_{m\to\infty} v_m = \bar{v}$. And by the Lebesgue dominated convergence theorem we have

$$\mathbf{Q}(1 : U|\bar{v}, z) = \int_{D\times\{1,2\}} \chi_{U\times\{1,2\}}(\bar{\psi}^{[0,1]}(\bar{v}, b^{[0,1]}), b^{[0,1]}(1))\hat{\mathbf{P}}_1^{[0,1]}(db^{[0,1]}|z)$$

$$= \lim_{n\to\infty} \int_{D\times\{1,2\}} f_n(\bar{\psi}^{[0,1]}(\bar{v}, b^{[0,1]}), b^{[0,1]}(1))\hat{\mathbf{P}}_1^{[0,1]}(db^{[0,1]}|z)$$

$$= \lim_{n\to\infty} \int_{D\times\{1,2\}} \lim_{m\to\infty} f_n(\bar{\psi}^{[0,1]}(v_m, b^{[0,1]}), b^{[0,1]}(1))\hat{\mathbf{P}}_1^{[0,h]}(db^{[0,1]}|z)$$

$$= \lim_{n\to\infty} \lim_{m\to\infty} \int_{D\times\{1,2\}} f_n(\bar{\psi}^{[0,1]}(v_m, b^{[0,1]}), b^{[0,1]}(1))\hat{\mathbf{P}}_1^{[0,1]}(db^{[0,1]}|z)$$

$$= \lim_{m\to\infty} \lim_{n\to\infty} \int_{D\times\{1,2\}} f_n(\bar{\psi}^{[0,1]}(v_m, b^{[0,1]}), b^{[0,1]}(1))\hat{\mathbf{P}}_1^{[0,1]}(db^{[0,1]}|z)$$

$$= \lim_{m\to\infty} \int_{D\times\{1,2\}} \chi_{U\times\{1,2\}}(\bar{\psi}^{[0,1]}(v_m, b^{[0,1]}), b^{[0,1]}(1))\hat{\mathbf{P}}_1^{[0,1]}(db^{[0,1]}|z)$$

$$= \lim_{m\to\infty} \mathbf{Q}(1 : U|v_m, z) = \lim_{m\to\infty} 1 = 1.$$

Therefore for each $\bar{v} \in D$ we have $\mathbf{Q}(1 : U|\bar{v}, z) = 1$. Thus for each $y \in D \times \{1, 2\}$ we have $\mathbf{Q}(1 : U|y) = 1$.

Because the Markov process associated with $\mathbf{Q}(\cdot : \cdot|\cdot)$ *does not include any dissipative part in the sense of* Yosida (1980) p. 395 in $D \times \{1, 2\}$ as mentioned in the main text, by the argument for the proof of Theorem 1 of XIII.4 (pp. 394–395, especially, by the equation in the second line of p. 395) in Yosida (1980), for each point $x \in D \times \{1, 2\}$ there exists a probability measure $\varphi_x : \mathcal{B}(D \times \{1, 2\}) \to [0, 1]$ such that for each $h > 0$ and for each $E \in \mathcal{B}(D \times \{1, 2\})$,

$$\varphi_x\{E\} = \int_{y\in D\times\{1,2\}} \mathbf{Q}(h : E|y)\varphi_x(dy),$$

and such that for each $g(\cdot) \in C(D \times \{1, 2\})$,

$$\int_{y\in D\times\{1,2\}} g(y)\varphi_x(dy) = \int_{y\in D\times\{1,2\}} [\int_{y'\in D\times\{1,2\}} g(y')\mathbf{Q}(1 : dy'|y)]\varphi_x(dy).$$

Thus for each $x \in D \times \{1, 2\}$ we have

$$\int_{y\in D\times\{1,2\}} f_n(y)\varphi_x(dy) = \int_{y\in D\times\{1,2\}} [\int_{y'\in D\times\{1,2\}} f_n(y')\mathbf{Q}(1 : dy'|y)]\varphi_x(dy).$$

Thus by the Lebesgue dominated convergence theorem,

$$
\begin{aligned}
\varphi_x\{U \times \{1,2\}\} &= \int_{y \in D \times \{1,2\}} \chi_{U \times \{1,2\}}(y)\varphi_x(dy) \\
&= \lim_{n \to \infty} \int_{y \in D \times \{1,2\}} f_n(y)\varphi_x(dy) \\
&= \lim_{n \to \infty} \int_{y \in D \times \{1,2\}} [\int_{y' \in D \times \{1,2\}} f_n(y')\mathbf{Q}(1 : dy'|y)]\varphi_x(dy) \\
&= \int_{y \in D \times \{1,2\}} [\lim_{n \to \infty} \int_{y' \in D \times \{1,2\}} f_n(y')\mathbf{Q}(1 : dy'|y)]\varphi_x(dy) \\
&= \int_{y \in D \times \{1,2\}} [\int_{y' \in D \times \{1,2\}} \chi_{U \times \{1,2\}}(y')\mathbf{Q}(1 : dy'|y)]\varphi_x(dy) \\
&= \int_{y \in D \times \{1,2\}} \mathbf{Q}(1 : U \times \{1,2\}|y)\varphi_x(dy) \\
&= \varphi_x\{D \times \{1,2\}\} = 1,
\end{aligned}
$$

where we have used the following relation in the last equality.

$$
\mathbf{Q}(1 : U|y) = 1 \text{ for each } y \in D \times \{1,2\}.
$$

Therefore for each $x \in D \times \{1,2\}$ we have $\varphi_x\{U \times \{1,2\}\} = 1$. Thus we have only to set $\mathbf{p} = \varphi_{x'}$ for some $x' \in D \times \{1,2\}$. $\square$

References

Bella, G., Mattana, P., Nishimura, K., & Shigoka, T. (2009). *Homoclinic bifurcations and global indeterminacy in Chamley's model of endogenous growth, mimeo.* (A revised version in preparation.)

Benhabib, J., & Perli, R. (1994). Uniqueness and indeterminacy: On the dynamics of endogenous growth. *Journal of Economic Theory, 63*, 113–142.

Benhabib, J., Perli, R., & Xie, D. (1994). Monopolistic competition, indeterminacy, and growth. *Ricerche Economiche, 48*, 279–298.

Benhabib, J., Nishimura, K., & Shigoka, T. (2008). Bifurcation and sunspots in the continuous time equilibrium model with capacity utilization. *International Journal of Economic Theory, 4*, 337–355.

Doob, J. L. (1953). *Stochastic processes.* New York: Wiley.

Drugeon, J. P., & Wigniolle, B. (1996). Continuous-time sunspot equilibria and dynamics in a model of growth. *Journal of Economic Theory, 69*, 24–52.

Lucas, R. E. (1988). On the mechanics of economic development. *Journal of Monetary Economics, 22*, 3–42.

Mattana, P., Nishimura, K., & Shigoka, T. (2009). Homoclinic bifurcation and global indeterminacy of equilibrium in a two-sector endogenous growth model. *International Journal of Economic Theory, 5*, 25–47.

Nishimura, K., & Shigoka, T. (2006). Sunspots and Hopf bifurcations in continuous time endogenous growth models. *International Journal of Economic Theory*, *2*, 199–216.

Romer, P. (1990). Endogenous technological change. *Journal of Political Economy*, *98*, S71–S102.

Shigoka, T. (1994). A note on Woodford's conjecture: constructing stationary sunspot equilibria in a continuous time model. *Journal of Economic Theory*, *64*, 531–540.

Yosida, K. (1980). *Functional analysis* (6th ed.). New York: Springer.

Part II
Bubbles and Stabilizing Policy

Chapter 9
Rational Land and Housing Bubbles in Infinite-Horizon Economies

Stefano Bosi, Cuong Le Van and Ngoc-Sang Pham

Abstract This paper considers rational land and housing bubbles in an infinite-horizon general equilibrium model. Their demands rest on two different grounds: the land is an input to produce while the house may be consumed. Our work differs from the existing literature in two respects. First, dividends on both these long-lived assets are endogenous and their sequences are computed. Second, we introduce and study different concepts of bubbles, including individual and strong bubbles.

Keywords Infinite horizon · General equilibrium · Land bubble · Housing bubble

JEL classification   C62 · D51 · D9 · G13

9.1   Introduction

The existence of rational bubbles in general equilibrium model is a challenging issue since the early Eighties. Thinking bubbles in (dynamic) general equilibrium becomes indispensable to understand the real effects of financial crises. A general equilibrium approach captures the interplay between (financial and real) markets. If the issue of

We are very grateful to the Referee for her/his remarks, comments, suggestions and questions. They have helped us to substantially improve our previous version. Ngoc-Sang Pham acknowledges the financial support of the LabEx MME-DII (ANR11-LBX-0023-01).

S. Bosi
EPEE, University of Evry, Evry, France
e-mail: stefano.bosi@univ-evry.fr

C. Le Van
IPAG, CNRS, and Paris School of Economics, Paris, France
e-mail: Cuong.Le-Van@univ-paris1.fr

N.-S. Pham (✉)
LEM, University of Lille 3, Lille, France
e-mail: pns.pham@gmail.com

© Springer International Publishing AG 2017   203
K. Nishimura et al. (eds.), *Sunspots and Non-Linear Dynamics*,
Studies in Economic Theory 31, DOI 10.1007/978-3-319-44076-7_9

rational bubbles was initially raised in the OLG models, today it is mostly addressed in infinite-horizon intertemporal models.

Our paper is part of this recent literature and focuses on rational bubbles of specific long-lived assets, land and houses.

The conventional definition of asset bubble is the positive difference between equilibrium price and fundamental value. In general equilibrium models, price and fundamental value are endogenous, and the bubble as well. The equilibrium price is observed and determined by market clearing conditions, while the fundamental value depends on the definition of discounting and returns. Different definitions of fundamental value exist in literature.

The novelty of our paper is twofold.

(1) Land and houses are emblematic assets representing two alternative fundamental pricing mechanisms (land is used to produce good, and house is consumed and gives utility). The proofs of equilibrium existence are routine. Nevertheless, while most of the infinite-horizon models introduce exogenous sequence of dividends (for mathematical tractability), we compute instead endogenous dividends.

(2) We provide three different notions of bubble: the conventional one, of course, but also new definitions (individual and strong). We compare them to shed light on the financial underworld (assets) and its consequences on the real world (goods).

We focus first on a long-lived productive asset. Agents buy land today to produce a consumption good and resell it tomorrow. Any agent is consumer and producer at the same time and technology is supposed to be landowner-specific. One unit of land produces a no longer exogenous amount of consumption good. Our contribution rests on the concept of *dividend of land* denoted by d_t and endogenously determined by the following asset-pricing equation:

$$\frac{q_t}{p_t} = \gamma_{t+1}\left(\frac{q_{t+1}}{p_{t+1}} + d_{t+1}\right)$$

where p_t and q_t are the prices of consumption good and land at date t and γ_{t+1} is the endogenous discount factor of the economy from date t to date $t+1$.

We show that the land dividend lies between the lowest and the highest marginal productivities. The fundamental value of land at date 0 denoted by FV_0 is defined as a sum of discounted values of land dividends: $FV_0 := \sum_{t=1}^{\infty} Q_t d_t$ where $Q_t := \gamma_1 \ldots \gamma_t$ is the discount factor of the economy from the initial date to date t. We will show that the price of land consists of two parts. First, the reselling price of one unit of land. Second, the fundamental value, that is what one unit of land delivers in terms of dividend. A land bubble is said to exist when the equilibrium price of land (in consumption good units) exceeds its fundamental value: $q_0/p_0 > FV_0$.

A particular case of our productive asset is when every agents share the same and linear technology. In this case, one unit purchased at each date delivers an exogenous amount of consumption good (a real dividend) at the next date. This asset is studied by Tirole (1982), Kocherlakota (1992) and Santos and Woodford (1997). The definition of fundamental value of the asset with exogenous dividends is just the sum (over time) of discounted values of dividends. When dividends are nil, the fundamental value of the asset is zero; this zero-dividend asset is studied by Tirole (1985) in OLG frameworks and by Aoki et al. (2014), Hirano and Yanagawa (2013) in infinite-horizon models.

Revisiting the notion of discounting, we introduce a new concept of bubble based on individual preferences. Formally, an i-*bubble* rests on the *individual asset-pricing equation*:

$$\frac{q_t}{p_t} = \gamma_{i,t+1} \left(\frac{q_{t+1}}{p_{t+1}} + d_{i,t+1} \right)$$

where $\gamma_{i,t+1}$ is the endogenous discount factor of agent i from date t to date $t+1$, and $d_{i,t+1}$ is her individual dividend at date $t+1$. In this case, the individual fundamental value of land expected by agent i is $FV_i := \sum_{t=1}^{\infty} Q_{i,t} d_{i,t}$ where $Q_{i,t} := \gamma_{i,1} \cdots \gamma_{i,t}$. As above, an i-bubble exists if $q_0/p_0 > FV_i$. We also say that *strong bubble* exists if $q_0/p_0 > \max_i FV_i$ meaning that the land price is strictly higher than every individual fundamental value.

From these definitions, number of results come. First, there exists an agent i such that her expected fundamental value of land equals its equilibrium price: $FV_i = q_0/p_0$, that is, the i-bubble is ruled out. Indeed, when an agent expects a fundamental value of land less than its price, she does not buy land in the long run. At equilibrium, this is not the case for any agent, otherwise the land market clearing condition fails. Most importantly, this result implies the impossibility of strong bubbles, i.e. $q_0/p_0 = \max_i FV_i$. Second, the ratio $Q_t/Q_{i,t}$ is uniformly bounded from above for any i, then, there is no room for land bubbles. In particular, if the agents' discount factors are identical, bubbles are ruled out; by the way, we recover Le Van and Pham (2014). Third, the agents' expected fundamental values are not less than the fundamental value of land. Moreover, when any individual value of land coincides with its fundamental value, both bubbles and i-bubbles are ruled out. To illustrate this theory and the occurrence of land bubbles, we give an example where endowments fluctuate and agents' TFPs converge to zero. Indeed, agents concerned by a drop in endowments tomorrow buy land today at a higher price in order to smooth their consumption over time. This price is independent on their technologies. We show that when the agent's TFP goes to zero, the fundamental value of land tends to zero as well. By consequence, land bubbles arise.

While land assets and bubbles are valued on the production side, other assets and bubbles may be valued on the consumption side. In the second part of the article, we focus on a leading example: the *housing bubbles*. Houses are goods that yield utility. An agent buys a house today to enjoy its services, and may resell it tomorrow. Thus, differently from land, the house enters the utility function of its owner. Its marginal utility represents a sort of *housing dividend*. Most of the results for land bubbles still hold for housing bubbles. A constructive example complements the theory: as above, fluctuations in agents' endowments jointly with a housing preference rate which tends to zero, promote a bubbly equilibrium.

Our paper contribute to the existing literature on asset price bubble, in particular on bubbles of assets delivering endogenous dividends.[1] Among others, two approaches deserve mention.

(1) Miao and Wang (2012, 2015) consider bubbles of firm values with endogenous dividends. They divide in two parts the value $V(K)$ of a firm endowed with K units of initial capital: $V(K) = QK + B$, where Q represents an endogenous Tobin's Q (marginal). They interpret QK and B as fundamental value (of the firm) and bubble, respectively.

(2) Becker et al. (2015) introduce instead the concept of *physical capital bubble*. In this case, the fundamental value of physical capital is the sum of discounted values of capital returns (after depreciation) and physical capital bubbles occurs when the equilibrium price of physical capital exceeds this fundamental value. In our model, land behaves as physical capital and is used to produce a consumption good. However, in Becker et al. (2015), the good is produced by a representative firm, while, in our model, any agent can be viewed as a producer and, therefore, there are as many producers as agents.

Bosi et al. (2015a) introduce concepts of bubbles in aggregate and capital goods. They then show how these two kinds of goods generate different bubbles: a bubble in aggregate good may exist even if (1) the present value of output is finite, (2) all consumers are identical, (3) borrowing constraints of consumers are never binding. By contrast, bubbles in capital good[2] are ruled out if one of three conditions is violated.

The rest of the paper is organized as follows. Section 9.2 introduces the model and provides some preliminary equilibrium properties. Sections 9.3 and 9.4 study land and housing bubbles respectively. Section 9.5 presents an alternative unified model. Section 9.6 concludes. All the formal proofs are gathered in Appendices 9.7 and 9.8.

9.2 Fundamentals and Equilibrium Existence

We consider a discrete-time general equilibrium model without uncertainty. The economy starts at time $t = 0$ and goes on forever. There are two markets, one for consumption good and the other for land. A set I of m infinite-lived heterogeneous agents own the land, produce, exchange and consume.

Consumption good. At any date t, the ith agent is endowed with $e_{i,t}$ of consumption good. Facing a price p_t, she decides to consume $c_{i,t}$ units of this good.

---

[1] The reader is referred to Miao (2014) for an introduction to bubbles in infinite-horizon models. Brunnermeier and Oehmke (2012) is a good survey on bubbles in OLG models with asymmetric information or heterogeneous belief.

[2] A particular case of bubbles in capital good is that of bubbles in asset with exogenous dividends as in Kocherlakota (1992), Santos and Woodford (1997), Huang and Werner (2000) and Le Van and Pham (2014).

Land. The aggregate endowment of land is constant over time and equal to L. At the date t, the ith agent buys $l_{i,t+1}$ units of land at a price q_t in order to produce $F_{i,t}(l_{i,t+1})$ units of consumption good. Technology is non-stationary and $F_{i,t}$ is a time-dependent agent-specific production function. When the production is over, the same agent may resell this amount of land at a price q_{t+1}. Land allows agents to transfer their wealth over time.

At date 0, taking the sequence of prices $(p, q) = (p_t, q_t)_{t=0}^{\infty}$ as given, the infinite-lived agent chooses the sequence of consumption and land $(c_i, l_i) := (c_{i,t}, l_{i,t+1})_{t=0}^{\infty}$ to maximize her intertemporal utility. The program writes:

$$P_i(p,q): \quad \max \sum_{t=0}^{\infty} \beta_i^t u_i(c_{i,t}) \tag{9.1}$$

$$\text{subject to}: \quad l_{i,t+1} \geq 0$$

$$p_t c_{i,t} + q_t l_{i,t+1} \leq p_t e_{i,t} + q_t l_{i,t} + p_t F_{i,t}(l_{i,t})$$

for any $t \geq 0$. The initial endowment of land $l_{i,0} > 0$ is given.

There are no financial markets. This assumption prevents agents from borrowing.

Definition 1 A sequence of prices and quantities $\left(\bar{p}_t, \bar{q}_t, (\bar{c}_{i,t}, \bar{l}_{i,t+1})_{i=1}^{m}\right)_{t=0}^{\infty}$ is an equilibrium of the economy without financial market if the following conditions are satisfied.

(i) Price positivity: $\bar{p}_t, \bar{q}_t > 0$ for any $t \geq 0$.
(ii) Market clearing conditions:

$$\sum_{i=1}^{m} \bar{c}_{i,t} = \sum_{i=1}^{m} \left[e_{i,t} + F_{i,t}(\bar{l}_{i,t})\right] \text{ and } \sum_{i=1}^{m} \bar{l}_{i,t} = L$$

for any t.
(iii) Agents' optimality: $(\bar{c}_{i,t}, \bar{l}_{i,t+1})_{t=0}^{\infty}$ is a solution of the problem $P_i(\bar{p}, \bar{q})$ for any i.

In the sequel, we denote for simplicity the equilibrium sequence by $(p, q, (c_i, l_i)_{i=1}^{m})$. The economy, denoted by $\mathcal{E}$, is identified by a list of fundamentals: $\mathcal{E} := (F_i, u_i, \beta_i, e_i, l_{i,0})$.

As seen above, unavailability of financial assets prevents agents from borrowing. Other authors allow agents to borrow: Bosi et al. (2015b) is an example of economy with land bubbles where imperfect financial markets take place.

If $F_i(x) = \xi_t x$ for every i, where (ξ_t) is an exogenous sequence of returns, land becomes an asset as in Kocherlakota (1992), Santos and Woodford (1997) and Huang and Werner (2000), or looks like a Lucas' tree. If $F_i = 0$ for every i, land becomes a pure bubble as in Tirole (1985). In this case, the fundamental value is zero.

The existence of a competitive equilibrium is ensured by some mild assumptions on technology and preferences, namely the positivity of endowments, the concavity of production and utility functions, and the boundedness of intertemporal utility in the set of feasible consumption sequences.

Assumption 1 For any i and t, the function $F_{i,t}$ is concave and continuously differentiable with $F_{i,t}(0) = 0$ and $F'_{i,t} > 0$.

Assumption 2 $l_{i,0} > 0$ for any i and $e_{i,t} > 0$ for any i and t.

Assumption 3 For any i, the function u_i is continuously differentiable and concave with $u'_i(0) = \infty$ and $u'_i(x) > 0$ if $x > 0$.

Assumption 4 $\sum_{t=0}^{\infty} \beta_i^t u_i(W_t) < \infty$ for any i, where $W_t := \sum_{i=1}^{m} \left[e_{i,t} + F_{i,t}(L) \right]$ is the maximum amount of consumption available at time t.

A standard proof of equilibrium existence is given in Appendix 9.8.

Proposition 1 *Under Assumptions 1–4, there exists an equilibrium for the economy $\mathcal{E}$.*

9.3   Land Bubbles

The definition of bubble rests on the notion of fundamental value which depends on discounting in turn. The first-order conditions of programs $P_i(p, q)$ allow us to define the equilibrium discount factors as follows.

Let $\left(p, q, (c_i, l_i)_{i=1}^m \right)$ be an equilibrium for the economy $\mathcal{E}$. Denote by $\lambda_{i,t}$ and $\mu_{i,t+1}$ the ith agent's multipliers with respect to the budget constraint and the borrowing constraint $l_{i,t+1} \geq 0$, respectively. The first-order conditions write:

$$\beta_i^t u'_i(c_{i,t}) = \lambda_{i,t} p_t \tag{9.2}$$
$$\lambda_{i,t} q_t = \lambda_{i,t+1} \left[q_{t+1} + p_{t+1} F'_{i,t+1}(l_{i,t+1}) \right] + \mu_{i,t+1}$$

jointly with the slackness condition $\mu_{i,t+1} l_{i,t+1} = 0$.

Asset pricing rests on no-arbitrage conditions:

$$\frac{q_t}{p_t} = \frac{\beta_i u'_i(c_{i,t+1})}{u'_i(c_{i,t})} \left[\frac{q_{t+1}}{p_{t+1}} + F'_{i,t+1}(l_{i,t+1}) + \frac{\mu_{i,t+1}}{\lambda_{i,t+1} p_{t+1}} \right] \tag{9.3}$$

We introduce the discount factors for individuals and economy.

$$\gamma_{i,t+1} := \frac{\beta_i u'_i(c_{i,t+1})}{u'_i(c_{i,t})} \text{ and } \gamma_{t+1} := \max_{i \in I} \frac{\beta_i u'_i(c_{i,t+1})}{u'_i(c_{i,t})}$$

are, respectively, the individual discount factor and the discount factor of the economy from date t to date $t + 1$. We then define the individual discount factor and the discount factor of the economy from date 0 to date t:

$$Q_{i,t} := \gamma_{i,1} \dots \gamma_{i,t} \text{ and } Q_t := \gamma_1 \dots \gamma_t$$

with $Q_{i,0} := Q_0 := 1$.

We now define the lowest and the highest productivities:

$$\underline{d}_t := \min_{i \in I} F'_{i,t}(l_{i,t}) \text{ and } \bar{d}_t := \max_{i \in I} F'_{i,t}(l_{i,t}).$$

9.3.1 Fundamental Value of Land and Land Bubbles

Asset pricing and productivity bounds are closely related through condition (9.3).

Lemma 1 *The relative price of land is bounded as follows:*

$$\gamma_{t+1} \left(\frac{q_{t+1}}{p_{t+1}} + \underline{d}_{t+1} \right) \leq \frac{q_t}{p_t} \leq \gamma_{t+1} \left(\frac{q_{t+1}}{p_{t+1}} + \bar{d}_{t+1} \right) \tag{9.4}$$

The introduction of land dividends allows us to define the land bubbles.

Definition 2 (*dividends of land*) A land dividend d_{t+1} at date $t + 1$ satisfies a no-arbitrage equation:

$$\frac{q_t}{p_t} = \gamma_{t+1} \left(\frac{q_{t+1}}{p_{t+1}} + d_{t+1} \right) \tag{9.5}$$

The value of a unit of land today in terms of consumption good equals the discounted value of the reselling relative price tomorrow and the amount of consumption good delivered by the unit. Because of condition (9.5), dividends become endogenous. The role of land is captured by the sequence of dividends at the end and Eq. (9.5) can be viewed either as an asset-pricing or a non-arbitrage condition.

Inequalities (9.4) entail that any land dividend is bounded by the lowest and the highest marginal productivities: $d_t \in \left[\underline{d}_t, \bar{d}_t \right]$.

Since $Q_{t+1} = \gamma_{t+1} Q_t$, (9.5) writes

$$Q_t \frac{q_t}{p_t} = Q_{t+1} \left(\frac{q_{t+1}}{p_{t+1}} + d_{t+1} \right)$$

Solving forward, we find an intertemporal asset-pricing equation

$$\frac{q_0}{p_0} = \sum_{t=1}^{T} Q_t d_t + Q_T \frac{q_T}{p_T} \tag{9.6}$$

which holds for any $T \geq 1$. This fundamental equation allows us to introduce the notion of land bubble.

Definition 3 (*land bubble*) The fundamental value of land is given by $FV_0 := \sum_{t=1}^{\infty} Q_t d_t$. A land bubble exists if the relative price of land (in terms of consumption good) strictly exceeds the fundamental value: $q_0/p_0 > FV_0$.

Here the price of land q_0/p_0 is decomposed into two parts: the fundamental value FV_0 and the bubble term $\lim_{t \to \infty} (Q_t q_t/p_t)$ which can be viewed as the reselling price of one unit of land at the infinity.

9.3.1.1   General Results

In the spirit of Montrucchio (2004), Le Van and Pham (2014), we provide a simple characterization of bubble existence.

Proposition 2 *The following statements are equivalent.*

(i) *Land bubbles exist.*
(ii) $\lim_{t \to \infty} (Q_t q_t/p_t) > 0$.
(iii) $\sum_{t=1}^{\infty} (p_t d_t/q_t) < \infty$.

Note that this result only requires the asset pricing Eq. (9.2). Proposition 2 holds for any form of production functions, including non-stationary technologies.

Let us consider a particular case where technologies are stationary ($F_{i,t} = F_i$ for every i and t). In this case, we have $d_t \geq \min_i F_i'(l_{i,t}) \geq \min_i F_i'(L) > 0$ for every t and, by consequence, condition (iii) in Proposition 2 simplifies.

Corollary 1 *If technologies are stationary and a land bubble exists, then $\sum_{t=1}^{\infty} (p_t/q_t) < \infty$.*

In other terms, the existence of land bubbles implies that the relative price of land q_t/p_t tends to infinity. However, this property only holds under stationary technologies. We will readdress this issue in Sect. 9.3.2 through an explicit example.

The transversality condition simplifies under our definition of individual discounting.

Lemma 2 (transversality condition) $\lim_{t \to \infty} (Q_{i,t} l_{i,t+1} q_t/p_t) = 0$ *for any i.*

The structure of discount factors plays also a role in the existence of bubbles.

Proposition 3 *If, for any i, the ratio $Q_t/Q_{i,t}$ is uniformly bounded from above, then there is no bubble.*

By consequence, a land bubble may arise only if the endogenous discount factors $Q_{i,t}$ are different at infinitely many dates. In more standard models of rational bubbles with exogenous dividends such as Kocherlakota (1992), Huang and Werner (2000),

Le Van and Pham (2014), the heterogeneity of endogenous discount factors vanishes if agents' borrowing constraints are no longer binding.

In our paper, the properties of production functions $F_{i,t}$ plays also an important role in the heterogeneity of discount factors. Let agents share the same linear technology, that is $F_{i,t}(X) = \xi_t X$ for any i and t (Lucas' tree). In this case, If $l_{i,t} > 0$ for any i and t, the discount factors turn out to be the same and bubbles are ruled out.

Remark 1 In general, the discount factors may differ even if $l_{i,t} > 0$ for any i and t. Indeed, consider the case where every agent has the same production function but non-linear: $F_i(x) = Ax^\alpha$ with $\alpha \in (0, 1)$. At equilibrium, we have $l_{i,t} > 0$ for any i and t, but the marginal productivities $Al_{i,t}^{\alpha-1}$ and $Al_{j,t}^{\alpha-1}$ may be different. This is because of the absence of financial markets (no agent can borrow).

9.3.1.2 New Concepts of Bubbles

The first-order conditions imply

$$\frac{q_t}{p_t} = \gamma_{i,t+1}\left(\frac{q_{t+1}}{p_{t+1}} + d_{i,t+1}\right) \tag{9.7}$$

where $d_{i,t+1} := F'_{i,t+1}(l_{i,t+1}) + \mu_{i,t+1}/(\lambda_{i,t+1}p_{t+1})$ can be interpreted as an individual land dividend for agent i at date $t + 1$. Since the individual discount factor $\gamma_{i,t+1}$ is less than the factor of the economy γ_{t+1}, the individual dividend $d_{i,t+1}$ expected by agent i is greater than the dividend d_{t+1} of the economy. Moreover, $\min_i d_{i,t+1} = d_{t+1}$.

If we write $\gamma_{i,t+1} = 1/(1 + r_{i,t+1})$ where $r_{i,t+1}$ is the expected interest rate by the ith agent at the date $t + 1$, then the asset-pricing Eq. (9.7) becomes

$$\frac{q_t}{p_t}(1 + r_{i,t+1}) = \frac{q_{t+1}}{p_{t+1}} + d_{i,t+1}$$

In other words, the relative price of land at date t and its interests equals the sum of the same price at date $t + 1$ and individual land dividends.

The asset-pricing Eq. (9.7) shows how the ith agent evaluates the price of land. This agent buys land today and resells it tomorrow at the price q_{t+1} after earning $d_{i,t+1}$ units of consumption good as dividend. She compares the current price with the future gains discounted at the rate $\gamma_{i,t+1}$.

Solving recursively (9.7), we can separate the land price in two parts.

$$\frac{q_0}{p_0} = \sum_{t=1}^{T} Q_{i,t}d_{i,t} + Q_{i,T}\frac{q_T}{p_T} \tag{9.8}$$

(9.8) leads us to define new concepts of bubbles.

Definition 4 (*i-bubble*) The individual value of land with respect to agent i is $FV_i := \sum_{t=1}^{\infty} Q_{i,t} d_{i,t}$. We say that an i-bubble exists if $q_0/p_0 > FV_i$.

Our concept of i-bubble is closely related to that of bubbles of durable goods and collateralized assets considered by Araujo et al. (2011). In their terminology, factors $Q_{i,t}$ are called deflators and multipliers $\mu_{i,t}$ are called non-pecuniary returns.

We apply the same arguments of Proposition 2.

Proposition 4 *The following statements are equivalent.*

 (i) *i-bubbles exist.*
 (ii) $\lim_{t \to \infty} \left(Q_{i,t} q_t / p_t \right) > 0.$
 (iii) $\sum_{t=1}^{\infty} \left(p_t d_{i,t} / q_t \right) < \infty.$

By combining point (ii) of Proposition 4 and Lemma 2 we have the following result.

Lemma 3 *If an i-bubble exists, then $\lim_{t \to \infty} l_{i,t} = 0$.*

In other words, when an agent expects a fundamental value of land strictly less than its price, she does not buy land in the long run. At equilibrium the land market clears $\sum_{i=1}^{m} l_{i,t} = L$. As a consequence, Lemma 3 implies that there exists an agent i such that the individual value of land expected by this agent is equal to the equilibrium price of land.

Proposition 5 *There exists an agent i such that an i-bubble is ruled out.*

Now, let us bridge the two concepts of bubble and i-bubble.

Proposition 6 1. *If an i-land bubble exists for some agent i, then land bubbles exist as well.*
 2. *$FV_0 \leq FV_i$ for any i. Moreover, if $FV_0 = FV_i$ for any i, then $FV_0 = FV_i = q_0/p_0$ for any i: there are neither bubbles nor i-bubbles.*

Note that the converse of point 1 is not true.[3] Point 2 of Proposition 6 is highly intuitive. Since any agent expects an interest rate which is higher than that of economy, the individual value of land expected by any agent will be higher than the fundamental value of land. Interestingly, when any individual value of land coincides with that of economy, both bubble and individual bubbles are ruled out. However when any individual value of land is identical but different from the fundamental value of land, we do not know whether land bubbles are ruled out.[4]

---

[3] See the example in Sect. 9.3.2.
[4] See Remark 4.

Remark 2 Let $B_0 := q_0/p_0 - FV_0$ and $B_i := q_0/p_0 - FV_i$ denote the bubble and i-bubble respectively. We observe that $B_i \leq B_0$ for every i.

Another concept of bubble, more restrictive, rests also on the notions of individual fundamental values.

Definition 5 (*strong bubble*) We say that a strong bubble exists if the asset price exceeds any individual value of land, that is $q_0/p_0 > \max_{i \in I} FV_i$.

The following result is a direct consequence of Proposition 4.

Proposition 7 *The following statements are equivalent.*

(i) *A strong bubble exists.*
(ii) $\min_{i \in I} \lim_{t \to \infty} (Q_{i,t} q_t / p_t) > 0$.
(iii) $\max_{i \in I} \sum_{t=1}^{\infty} (p_t d_{i,t} / q_t) < \infty$.

The existence of a strong land bubble entails the existence of an i-bubble for any i. Nevertheless, Proposition 5 prevents from this possibility and implies a new straightforward result.

Proposition 8 *Strong land bubbles never occur in our model because* $q_0/p_0 = \max_{i \in I} FV_i$.

Remark 3 Strong bubbles are ruled out in our framework because of the transversality conditions: $\lim_{t \to \infty} (Q_{i,t} l_{i,t+1} q_t / p_t) = 0$ for any i. However, in more general frameworks (for example, when uncertainty takes place as in Araujo et al. (2011) or Araujo et al. (2011)), the alternative transversality conditions are quite different and it is not trivial to prove them.

Our concept of strong bubble is related to the notion of speculative bubble in Werner (2014). He considers an asset bringing exogenous dividends in a model with ambiguity and defines a fundamental value based on agent's beliefs as the sum of discounted expected future dividends weighted by these beliefs. Then, he says that a speculative bubble exists if the asset price strictly exceeds any agent's fundamental value.

The readers may wonder why there is room for speculative bubbles à la Werner (2014) but not for strong bubbles in infinite-horizon models. It is hard to compare these two results within different frameworks (with and without ambiguity). The linearity of utility functions also matters. We impose Inada condition for utility functions to ensure a strictly positive individual consumption at equilibrium while Werner (2014) works with linear preferences. With linear utility functions, a simple non-negativity constraint on consumption $c_{i,t} \geq 0$ may be binding. In this case, condition (9.2) no longer holds and the ratio $p_{t+1} \lambda_{i,t+1} / (p_t \lambda_{i,t})$ may become higher than the marginal rate of substitution $\beta_i = \beta_i u_i'(c_{i,t+1}) / u_i'(c_{i,t})$. By consequence, the asset price q_0/p_0 may exceed $\sum_{t=1}^{\infty} \beta_i^t d_{i,t}$, the fundamental value with respect to the ith agent and strong bubbles may exist at the end.

From a theoretical point of view, the existence of bubbles depends on how we define the fundamental value of an asset, a somewhat ambiguous concept.

9.3.2 Example of Land Bubbles

We consider non-stationary production functions and we provide a constructive example of bubbly equilibrium. For simplicity, we normalize the price of consumption good to one: $p_t = 1$ for any t. Focus on an oversimplified economy with two agents A and B and linear technologies: $F_{A,t}(L) = A_t L$ and $F_{B,t}(L) = B_t L$. Preferences are rationalized by a common logarithmic utility function: $u_A(x) = u_B(x) := ln(x)$. Finally, agents' endowments are supposed to be zero any two periods: $e_{A,2t} = e_{B,2t+1} = 0$ for any t, and the supply of land inelastic: $L = 1$.

We need the following conditions to verify the first-order conditions and to identify the discount factors of the economy (γ_t).[5]

$$\beta_A \left(A_{2t} + \frac{\beta_B e_{B,2t}}{1 + \beta_B} \right) \left(A_{2t+1} + \frac{\beta_A e_{A,2t+1}}{1 + \beta_A} \right) \le \beta_B \frac{e_{B,2t}}{1 + \beta_B} \frac{e_{A,2t+1}}{1 + \beta_A} \tag{9.9}$$

$$\beta_B \left(B_{2t} + \frac{\beta_B e_{B,2t}}{1 + \beta} \right) \left(B_{2t-1} + \frac{\beta_A e_{A,2t-1}}{1 + \beta_A} \right) \le \beta_A \frac{e_{B,2t}}{1 + \beta_B} \frac{e_{A,2t-1}}{1 + \beta_A} \tag{9.10}$$

$$\beta_A \left(A_{2t} + \frac{\beta_B e_{B,2t}}{1 + \beta_B} \right) \left(B_{2t+1} + \frac{\beta_A e_{A,2t+1}}{1 + \beta_A} \right) \le \beta_B \frac{e_{B,2t}}{1 + \beta_B} \frac{e_{A,2t+1}}{1 + \beta_A} \tag{9.11}$$

$$\beta_B \left(A_{2t} + \frac{\beta_B e_{B,2t}}{1 + \beta} \right) \left(B_{2t-1} + \frac{\beta_A e_{A,2t-1}}{1 + \beta_A} \right) \le \beta_A \frac{e_{B,2t}}{1 + \beta_B} \frac{e_{A,2t-1}}{1 + \beta_A} \tag{9.12}$$

We will check in Appendix that the sequence of allocations

$$\begin{aligned}
(c_{A,2t}, c_{B,2t}) &= (q_{2t} + A_{2t}, e_{B,2t} - q_{2t}) \\
(c_{A,2t+1}, c_{B,2t+1}) &= (e_{A,2t+1} - q_{2t+1}, q_{2t+1} + B_{2t+1}) \\
(l_{A,2t}, l_{B,2t}) &= (1, 0) \\
(l_{A,2t+1}, l_{B,2t+1}) &= (0, 1)
\end{aligned}$$

and land prices

$$q_{2t} = \frac{\beta_B}{1 + \beta_B} e_{B,2t} \quad \text{and} \quad q_{2t+1} = \frac{\beta_A}{1 + \beta_A} e_{A,2t+1} \tag{9.13}$$

form a general equilibrium $\left(p_t, q_t, (c_{i,t}, l_{i,t+1})_{i \in I} \right)_t$.

The sequence of discount factors is computed:

$$\gamma_{2t} = \frac{\beta_A u_A'(c_{A,2t})}{u_A'(c_{A,2t-1})} \quad \text{and} \quad \gamma_{2t+1} = \frac{\beta_B u_B'(c_{B,2t+1})}{u_B'(c_{B,2t})}.$$

---

[5]These conditions are satisfied if, for instance, $\beta_A = \beta_B := \beta$, $A_{2t}, B_{2t} < e_{B,2t}(1 - \beta)/(1 + \beta)$ and $A_{2t+1}, B_{2t+1} < e_{A,2t+1}(1 - \beta)/(1 + \beta)$ for any t.

It allows us to compute in turn the sequence of dividends:

$$d_{2t} = A_{2t}, \quad d_{2t+1} = B_{2t+1} \tag{9.14}$$

and to write eventually an explicit characterization of bubble existence. Indeed, according to Proposition 2, land bubbles exist if and only if

$$\sum_{t=0}^{\infty} \frac{d_t}{q_t} = \sum_{t=0}^{\infty} \frac{A_{2t}}{e_{B,2t}} + \sum_{t=0}^{\infty} \frac{B_{2t+1}}{e_{A,2t+1}} < \infty$$

In other words, the existence of bubbles requires "low" dividends.

The intuition is straightforward. In the odd periods $(2t + 1)$, agent B has no endowments. She wants to smooth consumption over time according to her logarithm utility (which satisfies the Inada conditions), but she can not transfer her wealth from future to this date because of the borrowing constraint. As a consequence, she accepts to buy land at date $2t$ at a higher price: $q_{2t} \geq e_{B,2t}\beta_B/(1 + \beta_B)$, independent on agents' productivity. A lower productivity implies lower dividends and a lower fundamental value of land. As long as dividends tend to zero, the land price remains higher than this fundamental value.

Let us revisit the characterization of bubble existence under stationary technologies.

Corollary 2 *Assume that* $A_t = A > 0$ *and* $B_t = B > 0$ *for any* t. *Then, land bubbles exist if and only if*

$$\sum_{t=0}^{\infty} \left(\frac{1}{e_{B,2t}} + \frac{1}{e_{A,2t+1}} \right) < \infty$$

In this example, land bubbles exist if and only if $\sum_{t=1}^{\infty} 1/q_t < \infty$. This elegant characterization also illustrates Corollary 1.

Corollary 3 *Assume that* $e_{A,2t+1} = e_{B,2t} = e > 0$ *for any* t. *Then, land bubbles exist if and only if*

$$\sum_{t=0}^{\infty} (A_{2t} + B_{2t+1}) < \infty$$

This result is consistent with the example in Bosi et al. (2015a) where bubbles in capital good arise if the sum of capital good returns is finite.

Remark 4 (bubble vs i-bubble) Since $\lim_{t\to\infty} \left[\beta_i^t u_i'(c_{i,t}) q_t \right] = 0$ for $i = A, B$, there are no i-bubbles. However, a land bubble may occur. In this case, any individual value of land is identical and equals to the equilibrium price but may be strictly higher that the fundamental value of land.

Land bubbles and prices

Corollary 1 points out a necessary condition for land bubbles under stationary technologies: land prices must diverge to infinity. In our example, technologies are nonstationary and the land prices are given by (9.13). Thus, a land bubble may exist with asset prices either increasing or decreasing or fluctuating over time. In this respect, we recover and generalize Weil (1990), an example of bubble with decreasing asset prices. Indeed, his model is a particular case with $A_t = B_t = 0$ for any $t \geq T$ (land will give no fruits from some date on).

Pure bubbles

If $A_t = B_t = 0$ for any t, the fundamental value of land is zero. In this case, a sequence of positive prices ($q_t > 0$ for any t) is called a pure (land) bubble. It is easy to see that our example admits an equilibrium with pure bubbles.

Remark 5 In this example, although agents have linear production functions, these functions are different.

There is a case where the productivity of agent A is higher than that of agent B, i.e., $A_{2t+1} > B_{2t+1}$, but agent A does not produce at date $2t + 1$. For two reasons: (1) agents are prevented from borrowing, (2) agents' endowments change over time. Although A has a higher productivity at date $2t + 1$, she has also a higher endowment at this date, but no endowment at date $2t$. So, she may not need to buy land at date $2t$ to produce and transfer wealth from date $2t$ to date $2t + 1$. Instead, she sells land at date $2t$ to buy and consume consumption good at date $2t$. Therefore, agent A may not produce at date $2t + 1$ even if $A_{2t+1} > B_{2t+1}$.

9.3.3 Example of Individual Land Bubbles

Fundamentals of the economy. Consider the example in Sect. 9.3.2. For the sake of simplicity, we assume that $\beta_A = \beta_B =: \beta$.

We add the third agent: agent D. The utility, the rate of time preference, and the technologies of agent D are

$$u_D(c) = ln(c), \quad \beta_D = \beta, \quad F_{D,t}(L) = D_t L.$$

The endowments $(e_{D,t})_t$ and productivities (D_t) of agents D are defined by

$$\frac{\beta e_{D,t}}{e_{D,t+1}} = \frac{q_t}{q_{t+1} + D_{t+1}} = \gamma_{t+1}$$

where (γ_t) is determined as in Sect. 9.3.2. We see that such sequences $(e_{D,t})_t$ and (D_t) exist. Indeed, for example, we choose $D_t = d_t$ where d_t is determined as in (9.14). Then, we choose $(e_{D,t})_t$ such that $\beta e_{D,t} = \gamma_{t+1} e_{D,t+1}$.

Equilibrium: Prices and allocations of agents A and B are as in Sect. 9.3.2. The allocations of agent D are $c_{D,t} = e_{D,t}$ and $l_{D,t} = 0$ for any t. By using the same argument in Sect. 9.3.2, it is easy to verify that this system of prices and allocations constitutes an equilibrium.

We observe that agent D does not trade and $\gamma_{D,t} = \beta e_{D,t-1}/e_{D,t} = \gamma_t$ for any t. By consequence, $\lim_{t \to \infty} Q_{D,t} q_t = \lim_{t \to \infty} Q_t q_t > 0$. There is a D-bubble, i.e. the equilibrium price of land is strictly higher than the individual value of land with respect to agent D.

Remark 6 In the spirit of this example, we may provide other examples of bubbles with other production functions, for example $F_{i,t}(x) = a_{i,t} \ln(1 + x)$ where $a_{i,t} > 0$.

9.4 Housing Bubbles

Houses are different kind of assets. While land is a production factor, houses are consumption goods. While the former is valued through the production function, the latter are priced through a consumption demand. To formalize and understand asset pricing and bubbles in the case of a housing market, we consider a simple two-good economy with consumption and housing whose prices are p_t and q_t at date t respectively. Household i is endowed with $h_{i,0}$ houses at the initial date and $e_{i,t}$ units of consumption good at any date t. For simplicity, we consider a separable utility function: $u_i(c_{i,t}) + v_{i,t}(h_{i,t})$ where u_i is the consumption utility and $v_{i,t}$ the housing utility function which is assumed to be concave. Without loss of generality, the consumption utility is supposed to be stationary.

Houses are traded every period as follows. The agent i buys $h_{i,t}$ units of house at date $t - 1$. At date t, she enjoys the house services, that is a utility $v_{i,t}(h_{i,t})$, and resells her house at price q_t. Under the assumption of non-stationary housing utility, we obtain more general results on asset pricing and characterization of bubble.

Taking the sequence of prices $(p, q) = (p_t, q_t)_{t=0}^{\infty}$ as given, each household i chooses the sequence of goods $(c_i, h_i) := (c_{i,t}, h_{i,t})_{t=0}^{\infty}$ and solves a program to maximize her intertemporal utility function:

$$R_i(p, q): \quad \max \sum_{t=0}^{\infty} \beta_i^t \left[u_i(c_{i,t}) + v_{i,t}(h_{i,t}) \right]$$

$$\text{subject to}: \quad h_{i,t+1} \geq 0$$

$$p_t c_{i,t} + q_t h_{i,t+1} \leq p_t e_{i,t} + q_t h_{i,t}$$

for any t.

Definition 6 A list of prices and quantities $\left(\bar{p}_t, \bar{q}_t, (\bar{c}_{i,t}, \bar{h}_{i,t+1})_{i=1}^{m} \right)_{t=0}^{\infty}$ is an equilibrium of the economy without financial markets under the following conditions.

(i) Price positivity: $\bar{p}_t, \bar{q}_t > 0$ for any $t \geq 0$.
(ii) Market clearing conditions: $\sum_{i=1}^{m} \bar{c}_{i,t} = \sum_{i=1}^{m} e_{i,t}$ and $\sum_{i=1}^{m} \bar{h}_{i,t} = H$ for any $t \geq 0$.
(iii) Agents' optimality: for any i, $(\bar{c}_{i,t}, \bar{h}_{i,t+1})_{t=0}^{\infty}$ is a solution of program $R_i(\bar{p}, \bar{q})$.

In the sequel, we denote for simplicity the equilibrium sequence by $(p, q, (c_i, h_i)_{i=1}^{m})$. It is easy to prove the existence of reduced multipliers $v_{i,t+1} \geq 0$ such that the slackness condition $v_{i,t+1} h_{i,t+1} = 0$ and the asset-pricing (no-arbitrage) condition

$$\frac{q_t}{p_t} = \frac{\beta_i u_i'(c_{i,t+1})}{u_i'(c_{i,t})} \left[\frac{q_{t+1}}{p_{t+1}} + \frac{v_{i,t}'(h_{i,t+1})}{u_i'(c_{i,t+1})} + v_{i,t+1} \right]$$

hold.

As above, we introduce the transition discount factors (individual and maximal) from time t to time $t + 1$:

$$\gamma_{i,t+1} := \frac{\beta_i u_i'(c_{i,t+1})}{u_i'(c_{i,t})} \quad \text{and} \quad \gamma_{t+1} := \max_{i \in I} \frac{\beta_i u_i'(c_{i,t+1})}{u_i'(c_{i,t})}$$

and the corresponding compound discount factors: $Q_{i,t} := \gamma_{i,1} \ldots \gamma_{i,t}$ and $Q_t := \gamma_1 \ldots \gamma_t$, with $Q_{i,0} := Q_0 := 1$.

As above, the transversality condition involves quantities and prices.

Lemma 4 $\lim_{t \to \infty} \left(Q_{i,t} h_{i,t+1} q_t / p_t \right) = 0$ for any i.

Similarly to Lemma 1, the lowest and the highest marginal rates of substitution between consumption and housing drive the asset pricing:

$$\gamma_{t+1} \left(\frac{q_{t+1}}{p_{t+1}} + \underline{d}_{t+1} \right) \leq \frac{q_t}{p_t} \leq \gamma_{t+1} \left(\frac{q_{t+1}}{p_{t+1}} + \bar{d}_{t+1} \right)$$

where

$$\underline{d}_{t+1} := \min_{i \in I} \frac{v_{i,t+1}'(h_{i,t+1})}{u_i'(c_{i,t+1})} \quad \text{and} \quad \bar{d}_{t+1} := \max_{i \in I} \frac{v_{i,t+1}'(h_{i,t+1})}{u_i'(c_{i,t+1})}$$

We are naturally leaded to introduce the housing dividend d_t and the individual housing dividend $d_{i,t}$ through the asset-pricing equation

$$\frac{q_t}{p_t} = \gamma_{t+1} \left(\frac{q_{t+1}}{p_{t+1}} + d_{t+1} \right)$$

Equivalently, we define the individual housing dividend $d_{i,t}$ of agent i as follows:

$$d_{i,t+1} := \frac{v_{i,t+1}'(h_{i,t+1})}{u_i'(c_{i,t+1})} + v_{i,t+1}.$$

9.4.1 Fundamental Value of Housing and Housing Bubbles

Land and housing are assets of different nature. Land dividends are determined by the marginal productivities while housing dividends are determined by marginal utilities. By the way, our model of housing is related to the money-in-utility-function model by Tirole (1985). These models differ in two main respects: (1) we study infinite-lived agents instead of overlapping generations and (2) Tirole's utility function depends on money prices.

In the spirit of Sect. 9.3, we introduce new concepts of housing bubbles.

Definition 7 (*housing bubble*) The fundamental value of a house is defined by $FV_0 := \sum_{t=1}^{\infty} Q_t d_t$. We say that housing bubbles exist if the market price of houses (in term of consumption good) exceeds the fundamental value, that is $q_0/p_0 > FV_0$.

Definition 8 (*individual housing bubble*) The individual value of a house with respect to the agent i is defined by $FV_i := \sum_{t=1}^{\infty} Q_{i,t} d_{i,t}$.

We say that an i-housing bubble exists if $q_0/p_0 > FV_i$.

We say that a strong housing bubble exists if $q_0/p_0 > \max_i FV_i$.

Results of Propositions 2–8 also hold for housing bubbles.

9.4.2 Example of Housing Bubbles

A constructive example illustrates the emergence of housing bubbles. Consider an economy with two agents A and B whose preferences are logarithmic with respect to consumption: $u_A(x) = u_B(x) := ln(x)$; non-stationary and linear with respect to housing: $v_{A,t}(h) = A_t h$ and $v_{B,t}(h) = B_t h$. For simplicity, the supply of houses is inelastic: $H = 1$. Endowments fluctuate: for any t, $e_{A,2t} = e_{B,2t+1} = 0$ and $e_{A,2t+1}, e_{B,2t} > 0$. We normalize the consumption prices ($p_t = 1$ for any t) and we construct an equilibrium sequence $\left(p_t, q_t, \left(c_{i,t}, h_{i,t+1}\right)_{i \in I}\right)_t$. The sequence of equilibrium allocations is given by

$$\left(c_{A,2t}, c_{B,2t}\right) = \left(q_{2t}, e_{B,2t} - q_{2t}\right)$$
$$\left(c_{A,2t+1}, c_{B,2t+1}\right) = \left(e_{A,2t+1} - q_{2t+1}, q_{2t+1}\right)$$
$$\left(h_{A,2t}, h_{B,2t}\right) = (1, 0)$$
$$\left(h_{A,2t+1}, h_{B,2t+1}\right) = (0, 1)$$

with prices:

$$q_{2t} = \frac{\beta_B(1 + B_{2t+1})}{1 + \beta_B(1 + B_{2t+1})} e_{B,2t} \text{ and } q_{2t-1} = \frac{\beta_A(1 + A_{2t})}{1 + \beta_A(1 + A_{2t})} e_{A,2t-1}$$

The first-order conditions rest on two restrictions: $1 \geq \beta_B B_{2t} + \beta_B^2 (1 + B_{2t+1})$ and $1 \geq \beta_A A_{2t-1} + \beta_A^2 (1 + A_{2t})$. Moreover, inequalities $1 \geq \beta_A^2 (1 + A_{2t})(1 + B_{2t-1})$ and $1 \geq \beta_B^2 (1 + A_{2t})(1 + B_{2t+1})$ imply the inequalities $\gamma_{B,2t+1} \geq \gamma_{A,2t+1}$ and $\gamma_{A,2t} \geq \gamma_{B,2t}$ under which the above sequence of allocations turns out to be an equilibrium sequence. It is easy to compute the discount factors of the economy

$$\gamma_{2t+1} = \frac{\beta_B u_B'(c_{B,2t+1})}{u_B'(c_{B,2t})} \text{ and } \gamma_{2t} = \frac{\beta_A u_A'(c_{A,2t})}{u_A'(c_{A,2t-1})}$$

as well as the equilibrium dividends: $d_{2t} = A_{2t}c_{A,2t}$ and $d_{2t+1} = B_{2t+1}c_{B,2t+1}$.

The characterization of Proposition 2 applies: housing bubbles exist if and only if

$$\sum_{t=0}^{\infty} \frac{d_t}{q_t} = \sum_{t=0}^{\infty} (A_{2t} + B_{2t+1}) < \infty$$

Our example points out the possibility of housing bubble when individual endowments fluctuate over time and the housing preference rates A_t and B_t converges to zero.

9.5 A Unified Framework

So far, we see that both the models presented above share a common approach. We may introduce an asset (with fixed supply G) which is not only used to produce but also gives utility for agents. In this case, taking the sequence of prices $(p, q) = (p_t, q_t)_{t=0}^{\infty}$ as given, each household i chooses the sequence of allocations $(c_i, a_i) := (c_{i,t}, a_{i,t})_{t=0}^{\infty}$ and solves a program to maximize her intertemporal utility function:

$$S_i(p, q): \quad \max \sum_{t=0}^{\infty} \beta_i^t \left[u_i(c_{i,t}) + v_{i,t}(a_{i,t}) \right]$$

$$\text{subject to}: \quad a_{i,t+1} \geq 0$$

$$p_t c_{i,t} + q_t a_{i,t+1} \leq p_t e_{i,t} + q_t a_{i,t} + p_t F_{i,t}(a_{i,t})$$

Definition 9 A list of prices and quantities $\left(\bar{p}_t, \bar{q}_t, (\bar{c}_{i,t}, \bar{a}_{i,t+1})_{i=1}^{m} \right)_{t=0}^{\infty}$ is an equilibrium of the economy without financial markets under the following conditions.

(i) Price positivity: $\bar{p}_t, \bar{q}_t > 0$ for any $t \geq 0$.
(ii) Market clearing conditions: $\sum_{i=1}^{m} \bar{c}_{i,t} = \sum_{i=1}^{m} \left(e_{i,t} + F_{i,t}(\bar{a}_{i,t}) \right)$ and $\sum_{i=1}^{m} \bar{a}_{i,t} = G$ for any $t \geq 0$.
(iii) Agents' optimality: for any i, $(\bar{c}_{i,t}, \bar{a}_{i,t+1})_{t=0}^{\infty}$ is a solution of program $S_i(\bar{p}, \bar{q})$.

In the sequel, we denote for simplicity the equilibrium sequence by $(p, q, (c_i, a_i)_{i=1}^m)$. It is easy to prove the existence of reduced multipliers $x_{i,t+1} \geq 0$ such that the slackness condition $x_{i,t+1} a_{i,t+1} = 0$ and the asset-pricing (no-arbitrage) condition

$$\frac{q_t}{p_t} = \frac{\beta_i u_i'(c_{i,t+1})}{u_i'(c_{i,t})} \left[\frac{q_{t+1}}{p_{t+1}} + \frac{v_{i,t}'(a_{i,t+1})}{u_i'(c_{i,t+1})} + F_{i,t+1}'(a_{i,t+1}) + x_{i,t+1} \right] \qquad (9.15)$$

As above, we introduce the transition discount factors (individual and maximal) from time t to time $t+1$:

$$\gamma_{i,t+1} := \frac{\beta_i u_i'(c_{i,t+1})}{u_i'(c_{i,t})} \text{ and } \gamma_{t+1} := \max_{i \in I} \frac{\beta_i u_i'(c_{i,t+1})}{u_i'(c_{i,t})}$$

and the corresponding compound discount factors: $Q_{i,t} := \gamma_{i,1} \dots \gamma_{i,t}$ and $Q_t := \gamma_1 \dots \gamma_t$, with $Q_{i,0} := Q_0 := 1$.

Transversality conditions still hold.

Lemma 5 $\lim_{t \to \infty} \left(Q_{i,t} a_{i,t+1} q_t / p_t \right) = 0$ for any i.

For each t, individual dividend $d_{i,t}$ of agent i is defined by:

$$d_{i,t} := \frac{v_{i,t}'(a_{i,t})}{u_i'(c_{i,t})} + F_{i,t}'(a_{i,t}) + x_{i,t}$$

We then define the sequence of dividends (d_t) through the asset-pricing equation

$$\frac{q_t}{p_t} = \gamma_{t+1} \left(\frac{q_{t+1}}{p_{t+1}} + d_{t+1} \right)$$

It is easy to see that $d_t \in [\underline{d}_t, \bar{d}_t]$, where

$$\underline{d}_t := \min_{i \in I} \left\{ \frac{v_{i,t}'(a_{i,t})}{u_i'(c_{i,t})} + F_{i,t}'(a_{i,t}) \right\} \text{ and } \bar{d}_t := \max_{i \in I} \left\{ \frac{v_{i,t}'(a_{i,t})}{u_i'(c_{i,t})} + F_{i,t}'(a_{i,t}) \right\}$$

By using the approaches in Sects. 9.3 and 9.4, we can obtain similar results (about bubbles, individual bubbles, and strong bubbles) as the ones with land and house. The difference here is that the endogenous dividends in this general model take into account both roles of the asset: input and utility.

9.6   Conclusion and Further Discussions

We have introduced different concepts of land and housing bubbles, including individual and strong bubbles. While strong bubbles never occur in our model, bubbles may exist but they are ruled out if agents' expected values coincide with the fundamental value of land. Interestingly, there is always an agent whose expected value of land equals the land equilibrium price. Some explicit examples of bubbles are provided to illustrate these theoretical outcomes. Land bubbles may arise when agents experience endowment fluctuations and the ratio between fruits of land and endowments tends to zero. Similarly, housing bubbles may arise if the endowments fluctuate and the housing preference rate converges to zero.

Further discussions (bubbles and efficiency)

Although the results on land and house bubbles are quite similar, we would like to point out that different assets may generate very different bubbles. Readers are referred to Bosi et al. (2015a) where they show that bubbles of capita good are ruled out if borrowing constraints of agents are not binding while bubbles of aggregate good may appear even when the financial market is perfect.

Another interesting issue is the connection between the existence of bubbles and the efficiency.

First, let us discuss about pure bubbles, i.e., bubbles in the asset which pays no dividends. There is a large literature on this kind of bubble and most of papers focus on OLG frameworks. In a standard OLG model of bubbles, Tirole (1985) shows that a pure bubble may occur only if the economy is inefficient.[6] However, Farhi and Tirole (2012) points out that when capital markets are imperfect, the economy can be efficient.

Second, we discuss the connection between physical capital bubble (Becker et al. 2015) and the efficiency of Ramsey equilibrium.[7] In an infinite-horizon model with stationary technologies, Becker et al. (2015) prove that physical capital bubbles are ruled out. In a similar framework, Becker et al. (2014) provide an example where a three-cycle equilibrium is inefficient. One can prove that there is no physical capital bubble in Becker et al. (2014). This shows that an inefficient bubbleless equilibrium may exist.

With non-stationary technologies, Bosi et al. (2015a) give an example where equilibrium is efficient and bubbles occur, and another example where equilibrium is efficient and bubbles are ruled out. These examples suggest that there is no causal relationship between the existence of physical capital bubble and the efficiency of Ramsey equilibrium.

---

[6]According to Tirole (1985), an allocation is efficient if it is not possible to improve the welfare of all generations (and, this, strictly for at least one of them).

[7]A Ramsey equilibrium is efficient if its capital path is efficient in the sense of Malinvaud.

Let us now come back to our framework. The question about the (constrained) efficiency in infinite-horizon general equilibrium models is matter of a long debate. Until now, we didn't provide a general sufficient condition for the inefficiency of equilibria.

If we define the efficiency of equilibrium in terms of equilibrium capital paths, we may use Cass (1972) where he gives a necessary and sufficient condition for the efficiency of capital paths. However, Cass (1972) considers only the capital paths (with a single technology) that are bounded away from zero.

In this respect, we raise two issues and leave them for future research: (1) checking whether any equilibrium in our framework is efficient, (2) bridging the existence of land bubble and the efficiency of equilibrium. In order to solve these fundamental problems, we may use the methods introduced by Cass (1972) and Balasko and Shell (1980).[8]

9.7 Appendix: Proofs for Sections 9.3 and 9.4

Proof of Lemma 2 $\lim_{t \to \infty} \left(Q_{i,t} l_{i,t+1} q_t / p_t \right) = 0$ and $\lim_{t \to \infty} \left[\beta_i^t u_i'(c_{i,t}) l_{i,t+1} q_t / p_t \right] = 0$ are equivalent. We say that l_i is feasible if, for every t, $l_{i,t} \geq 0$ and $q_t l_{i,t+1} \leq p_t e_{i,t} + q_t l_{i,t} + p_t F_i(l_{i,t})$. Note that if l_i is feasible then $(l_{i,0}, l_{i,1}, \ldots, l_{i,t-1}, \lambda l_{i,t}, \lambda l_{i,t+1}, \ldots)$ is also feasible for each $t \geq 1$ and $\lambda \in (0, 1)$. By using the same argument in the proof of Theorem 2.1 by Kamihigashi (2002), we obtain $\lim_{t \to \infty} \left[\beta_i^t u_i'(c_{i,t}) l_{i,t+1} q_t / p_t \right] = 0$. $\qquad \square$

Proof of Proposition 2 According to (6), a bubble exists if and only if $\lim_{T \to \infty} Q_T q_T / p_T > 0$. According to (5), we have

$$Q_t \frac{q_t}{p_t} = Q_{t+1} \frac{q_{t+1}}{p_{t+1}} \left(1 + \frac{p_{t+1}}{q_{t+1}} d_{t+1} \right)$$

for any t. By consequence,

$$\frac{q_0}{p_0} = Q_1 \frac{q_1}{p_1} \left(1 + \frac{p_1}{q_1} d_1 \right) = \cdots = Q_T \frac{q_T}{p_T} \left(1 + \frac{p_T}{q_T} d_T \right) \ldots \left(1 + \frac{p_1}{q_1} d_1 \right)$$

So, $\lim_{T \to \infty} Q_T q_T / p_T > 0$ if and only if

$$\lim_{T \to \infty} \left[\left(1 + \frac{p_T}{q_T} d_T \right) \ldots \left(1 + \frac{p_1}{q_1} d_1 \right) \right] < \infty$$

or, equivalently, $\sum_{t=1}^{\infty} (p_t d_t / q_t) = 0$. $\qquad \square$

---

[8]We observe that (1) Balasko and Shell (1980) consider OLG models and (2) the technique introduced by Cass (1972) doesn't apply directly to a framework with many producers as ours.

Proof of Proposition 3 Assume that $Q_t/Q_{i,t}$ is uniformly bounded from above. According to Lemma 2, we have

$$\lim_{t\to\infty}\left(Q_t\frac{q_t}{p_t}l_{i,t+1}\right)=\lim_{t\to\infty}\left(\frac{Q_t}{Q_{i,t}}Q_{i,t}\frac{q_t}{p_t}l_{i,t+1}\right)=0$$

for any i. Observing that $\sum_{i\in I}l_{i,t+1}=L$ for any t, we obtain $\lim_{t\to\infty}(Q_tq_t/p_t)=0$. □

Proof of Proposition 6 $Q_t\ge Q_{i,t}$ implies $\lim_{t\to\infty}(Q_tq_t/p_t)\ge\lim_{t\to\infty}(Q_{i,t}q_t/p_t)$. However, we have $FV_i+\lim_{t\to\infty}(Q_{i,t}q_t/p_t)=q_0/p_0=FV_0+\lim_{t\to\infty}(Q_tq_t/p_t)$. By consequence, we get $FV_0\le FV_i$ for any i. Therefore, if i-land bubbles exist for some agent i then land bubbles exist.

We now assume that $FV_0=FV_i$ for any i, which implies $\lim_{t\to\infty}(Q_tq_t/p_t)=\lim_{t\to\infty}(Q_{i,t}q_t/p_t)$. If land bubbles exist, then

$$\lim_{t\to\infty}\left(Q_t\frac{q_t}{p_t}\right)=\lim_{t\to\infty}\left(Q_{i,t}\frac{q_t}{p_t}\right)\in(0,\frac{q_0}{p_0})$$

Thus, we find $\lim_{t\to\infty}(Q_{i,t}/Q_t)=1$. We obtain $\lim_{t\to\infty}(Q_tq_t/p_t)=0$ according to Proposition 3. □

Lemma 6 *If a sequence* $(p_t,q_t,(c_{i,t},l_{i,t+1},\mu_{i,t})_{i\in I})_t$ *satisfies:*

(i) *positivity of prices* ($p_t=1$ *and* $q_t>0$ *for any* t) *and non-negativity of allocations and multipliers* ($c_{i,t}>0$, $l_{i,t+1}\ge 0$ *and* $\mu_{i,t}\ge 0$ *for any* i *and* t);
(ii) *asset-pricing conditions*

$$q_t=\frac{\beta_i u_i'(c_{i,t+1})}{u_i'(c_{i,t})}\left[q_{t+1}+F_{i,t}'(l_{i,t+1})\right]+\mu_{i,t+1} \qquad (9.16)$$

and slackness conditions $\mu_{i,t+1}l_{i,t+1}=0$ *for any* i *and* t;
(iii) *transversality conditions:* $\lim_{t\to\infty}\left[\beta_i^t u_i'(c_{i,t})q_tl_{i,t+1}\right]=0$ *for any* i;
(iv) *budget constraints:* $c_{i,t}+q_tl_{i,t+1}=e_{i,t}+q_tl_{i,t}+F_{i,t}(l_{i,t})$ *for any* i *and* t;
(vi) *market clearing:* $L=\sum_{i\in I}l_{i,t}$ *for any* t;

then, the sequence of prices and allocations $(p_t,q_t,(c_{i,t},l_{i,t+1})_{i\in I})_t$ *is an equilibrium for the economy with land.*

Proof of Lemma 6 It is easy to see that market clearing conditions are satisfied. We now prove the optimality of the agents' plan. Let $(c_i',l_i')\ge 0$ be a plan satisfying all the budget constraints and $l_{i,0}'=l_{i,0}$. We have

$$\Delta_T := \sum_{t=0}^{T} \beta_i^t \left[u_i(c_{i,t}) - u_i(c'_{i,t}) \right] \geq \sum_{t=0}^{T} \beta_i^t u_i'(c_{i,t})(c_{i,t} - c'_{i,t})$$

$$= \sum_{t=0}^{T} \beta_i^t u_i'(c_{i,t}) \left[q_t(l_{i,t} - l'_{i,t}) + F_{i,t}(l_{i,t}) - F_{i,t}(l'_{i,t}) - q_t(l_{i,t+1} - l'_{i,t+1}) \right]$$

$$\geq \sum_{t=0}^{T} \beta_i^t u_i'(c_{i,t}) \left[q_t + F_{i,t}'(l_{i,t}) \right] (l_{i,t} - l'_{i,t}) - \sum_{t=0}^{T} \beta_i^t u_i'(c_{i,t}) q_t (l_{i,t+1} - l'_{i,t+1})$$

$$= \sum_{t=1}^{T} \left(\beta_i^t u_i'(c_{i,t}) \left[q_t + F_{i,t}'(l_{i,t}) \right] - \beta_i^{t-1} u_i'(c_{i,t-1}) q_{t-1} \right) (l_{i,t} - l'_{i,t})$$

$$- \beta_i^T u_i'(c_{i,T}) q_T (l_{i,T+1} - l'_{i,T+1}).$$

We obtain $\beta_i^t u_i'(c_{i,t}) \left[q_t + F_{i,t}'(l_{i,t}) \right] - \beta_i^{t-1} u_i'(c_{i,t-1}) q_{t-1} = -\mu_{i,t} \beta_i^{t-1} u_i'(c_{i,t-1})$ according to (9.16). By using this and the fact that $\mu_{i,t} l_{i,t} = 0$ for any t, we have

$$\Delta_T \geq - \sum_{t=1}^{T} \mu_{i,t} \beta_i^{t-1} u_i'(c_{i,t-1})(l_{i,t} - l'_{i,t}) + \beta_i^T u_i'(c_{i,T}) q_T (l'_{i,T+1} - l_{i,T+1})$$

$$= \sum_{t=1}^{T} \beta_i^{t-1} u_i'(c_{i,t-1}) \mu_{i,t} l'_{i,t} + \beta_i^T u_i'(c_{i,T}) q_T (l'_{i,T+1} - l_{i,T+1})$$

$$\geq - \beta_i^T u_i'(c_{i,T}) q_T l_{i,T+1}.$$

$\sum_{t=0}^{\infty} \beta_i^t u_i(c_{i,t}) < \infty$ and $\lim_{t \to \infty} \left[\beta_i^T u_i'(c_{i,T}) q_T l_{i,T+1} \right] = 0$ entail that the sum $\sum_{t=0}^{\infty} \beta_i^t u_i(c'_{i,t})$ is finite. Therefore, we get $\sum_{t=0}^{\infty} \beta_i^t u_i(c_{i,t}) \geq \sum_{t=0}^{\infty} \beta_i^t u_i(c'_{i,t})$. □

Proof for the example in Sect. 9.3.2. It is easy to see that the market clearing conditions are satisfied.
We check the first-order conditions:

$$q_{2t} = \frac{\beta_B u_B'(c_{B,2t+1})}{u_B'(c_{B,2t})} (q_{2t+1} + B_{2t+1}) \geq \frac{\beta_A u_A'(c_{A,2t+1})}{u_A'(c_{A,2t})} (q_{2t+1} + A_{2t+1}) \quad (9.17)$$

$$q_{2t-1} = \frac{\beta_A u_A'(c_{A,2t})}{u_A'(c_{A,2t-1})} (q_{2t} + A_{2t}) \geq \frac{\beta_B u_B'(c_{B,2t})}{u_B'(c_{B,2t-1})} (q_{2t} + B_{2t}) \quad (9.18)$$

The equality in (9.17) holds because

$$\frac{\beta_B u_B'(c_{B,2t+1})}{u_B'(c_{B,2t})} (q_{2t+1} + B_{2t+1}) = \frac{\beta_B(e_{B,2t} - q_{2t})}{q_{2t+1} + B_{2t+1}} (q_{2t+1} + B_{2t+1})$$

$$= \beta_B(e_{B,2t} - q_{2t}) = q_{2t}$$

Now, we prove the inequality in (9.17). We have

$$\frac{\beta_A u'_A(c_{A,2t+1})}{u'_A(c_{A,2t})}(q_{2t+1} + A_{2t+1})$$

$$= \frac{\beta_A(q_{2t} + A_{2t})}{e_{A,2t+1} - q_{2t+1}}(q_{2t+1} + A_{2t+1})$$

$$= \frac{\beta_A(\frac{\beta_B}{1+\beta_B}e_{B,2t} + A_{2t})}{\frac{1}{1+\beta_A}e_{A,2t+1}}\left(\frac{\beta_A}{1+\beta_A}e_{A,2t+1} + A_{2t+1}\right)$$

By consequence, the inequality in (9.17) is equivalent to

$$\beta_A(\frac{\beta_B e_{B,2t}}{1+\beta_B} + A_{2t})(\frac{\beta_A e_{A,2t+1}}{1+\beta_A} + A_{2t+1}) \leq \beta_B \frac{e_{B,2t}}{1+\beta_B}\frac{e_{A,2t+1}}{1+\beta_A}$$

that is (9.9).

We have

$$\frac{\beta_B u'_B(c_{B,2t})}{u'_B(c_{B,2t-1})}(q_{2t} + B_{2t}) = \frac{\beta_B(q_{2t-1} + B_{2t-1})}{e_{B,2t} - q_{2t}}(q_{2t} + B_{2t})$$

$$= \frac{\beta_B(\frac{\beta_A}{1+\beta_A}e_{A,2t-1} + B_{2t-1})}{\frac{1}{1+\beta_B}e_{B,2t}}\left(\frac{\beta_B}{1+\beta_B}e_{B,2t} + B_{2t}\right)$$

By consequence, the inequality in (9.18) is equivalent to

$$\beta_B\left(\frac{\beta_A e_{A,2t-1}}{1+\beta_A} + B_{2t-1}\right)\left(\frac{\beta_B e_{B,2t}}{1+\beta} + B_{2t}\right) \leq \beta_A \frac{e_{B,2t}}{1+\beta_B}\frac{e_{A,2t-1}}{1+\beta_A}$$

that is (9.10).

We now check the transversality conditions:

$$\beta_A^{2t} u'_A(c_{A,2t})q_{2t}l_{A,2t+1} = 0$$

$$\beta_A^{2t-1} u'_A(c_{A,2t-1})q_{2t-1}l_{A,2t} = \frac{\beta_A^{2t-1}}{c_{A,2t-1}}q_{2t-1} = \frac{1}{1+\beta_A}\beta_A^{2t} \to 0$$

Similarly,

$$\beta_B^{2t} u'_B(c_{B,2t})q_{2t}l_{B,2t+1} = \frac{1}{1+\beta_B}\beta_B^{2t+1} \to 0$$

$$\beta_B^{2t-1} u'_B(c_{B,2t-1})q_{2t-1}l_{B,2t} = 0$$

Finally, it is easy to see that conditions (9.11) and (9.12) imply

$$\gamma_{2t} = \frac{\beta_A u'_A(c_{A,2t})}{u'_A(c_{A,2t-1})} \geq \frac{\beta_B u'_B(c_{B,2t})}{u'_B(c_{B,2t-1})}$$

$$\gamma_{2t+1} = \frac{\beta_B u'_B(c_{B,2t+1})}{u'_B(c_{B,2t})} \geq \frac{\beta_A u'_A(c_{A,2t+1})}{u'_A(c_{A,2t})}$$

□

The result of Lemma 6 also holds for housing.

Lemma 7 (sufficient equilibrium condition in housing economies) *If a sequence* $\left(p_t, q_t, (c_{i,t}, h_{i,t+1}, v_{i,t})_{i\in I}\right)_t$ *satisfies:*

(i) *positivity of prices ($p_t = 1$ and $q_t > 0$ for any t) and non-negativity of allocations and multipliers ($c_{i,t} > 0$, $h_{i,t+1} \geq 0$ and $v_{i,t} \geq 0$ for any i and t);*
(ii) *asset-pricing conditions*

$$q_t = \frac{\beta_i u'_i(c_{i,t+1})}{u'_i(c_{i,t})} \left[q_{t+1} + \frac{v'_{i,t+1}(h_{i,t+1})}{u'_i(c_{i,t+1})} \right] + v_{i,t+1}$$

and slackness conditions $v_{i,t+1} h_{i,t+1} = 0$ for any i and t;
(iii) *transversality conditions: $\lim_{t\to\infty} \left[\beta_i^t u'_i(c_{i,t}) q_t h_{i,t+1} \right] = 0$ for any i;*
(iv) *budget constraints: $c_{i,t} + q_t h_{i,t+1} = e_{i,t} + q_t h_{i,t}$ for any i and t;*
(vi) *market clearing: $\sum_{i\in I} H_{i,t} = H$ for any t;*

then, the sequence of prices and allocations $\left(p_t, q_t, (c_{i,t}, h_{i,t+1})_{i\in I}\right)_t$ is an equilibrium for the housing economy.

9.8 Existence of Equilibrium

We adapt the proofs of Becker et al. (2015) and Le Van and Pham (2016). We consider non-stationary production functions $F_{i,t}$ depending on t.

Existence of equilibrium in a truncated economy

We define a T—truncated economy $\mathcal{E}^T$ as $\mathcal{E}$ but without activities from period $T + 1$ on, that is we set $c_{i,t} = l_{i,t} = 0$ for $i = 1, \ldots, m$ and $t \geq T + 1$.

Then, we define a bounded economy $\mathcal{E}_b^T$ as $\mathcal{E}^T$ but sequences of consumption $(c_{i,t})_{t=0}^T$ and land holdings $(l_{i,t})_{t=1}^T$ lie in the following bounded sets $\mathcal{C}_i :=$ $[0, B_c]^{T+1}$ and $\mathcal{L}_i := [0, B_l]^T$ respectively, with $B_c > \max_{t\leq T} \sum_{i=1}^m \left[e_{i,t} + F_{i,t}(B_l) \right]$ and $B_l > L$. Therefore, the economy $\mathcal{E}_b^T$ depends on bounds B_c and B_l. We write $\mathcal{E}_b^T(B_c, B_l)$.

Let us denote $\mathcal{X}_b := \mathcal{C}_i \times \mathcal{L}_i$ and $\mathcal{X} := (\mathcal{X}_b)^{T+1}$, and define

$$\mathcal{P} := \{z_0 = (p, q) : q_T = 0; \ p_t, q_t \geq 0; \ p_t + q_t = 1; \ t = 0, \ldots, T\}$$
$$\Phi := \mathcal{P} \times \mathcal{X}$$

An element $z \in \Phi$ is a list $z = (z_i)_{i=0}^m$ with $z_0 := (p, q)$ and $z_i := (c_i, l_i)$ for each $i = 1, \ldots, m$.

Remark 7 If $z \in \Phi$ is an equilibrium for the economy $\mathcal{E}$ then $c_{i,t} \in [0, B_c)$, $l_{i,t} \in [0, L]$.

Proposition 9 *Under our assumptions, there exists an equilibrium* $(p, q, (c_i, l_i)_{i=1}^m)$ *with* $p_t + q_t = 1$ *for the economy* $\mathcal{E}_b^T(B_c, B_l)$.

Proof We introduce the budget sets:

$$C_i^T(p, q) := \{(c_{i,t}, l_{i,t+1})_{t=0}^T \in \mathcal{X} : l_{i,T+1}, b_{i,T+1} = 0,$$
$$p_t c_{i,t} + q_t l_{i,t+1} \le p_t e_{i,t} + p_t F_{i,t}(l_{i,t}) + q_t l_{i,t} \forall t\}$$
$$B_i^T(p, q) := \{(c_{i,t}, l_{i,t+1})_{t=0}^T \in \mathcal{X} : l_{i,T+1}, b_{i,T+1} = 0,$$
$$p_t c_{i,t} + q_t l_{i,t+1} < p_t e_{i,t} + p_t F_{i,t}(l_{i,t}) + q_t l_{i,t} \forall t\}$$

Since $e_{i,0}, l_{i,0} > 0$ and $(p_0, q_0) \ne (0, 0)$, we always have $p_0 e_{i,0} + p_0 F_{i,0}(l_{i,0}) + q_0 l_{i,0} > 0$. By consequence, $B_i^T(p, q) \ne \emptyset$ and $\bar{B}_i^T(p, q) = C_i^T(p, q)$.

Since $B_i^T(p, q)$ is nonempty and has an open graph, $B_i^T(p, q)$ is lower semi-continuous correspondence on $\mathcal{P}$. And we also have $C_i^T(p, q)$ is continuous on $\mathcal{P}$ with compact convex values.

We introduce the correspondences. We define φ_0 (for additional agent 0) : $\mathcal{X} \to 2^{\mathcal{P}}$:

$$\varphi_0((z_i)_{i=1}^m)$$
$$:= \arg \max_{(p,q) \in \mathcal{P}} \left(\sum_{t=0}^T p_t \sum_{i=1}^m [c_{i,t} - e_{i,t} - F_{i,t}(l_{i,t})] + \sum_{t=0}^{T-1} q_t \sum_{i=1}^m (l_{i,t+1} - l_{i,t}) \right)$$

For $i = 1, \ldots, m$, we define $\varphi_i : \mathcal{P} \to 2^{\mathcal{X}}$:

$$\varphi_i((p, q)) := \arg \max_{(c_i, l_i) \in C_i(p,q)} \sum_{t=0}^T \beta_i^t u_i(c_{i,t})$$

The correspondence φ_i is upper semi-continuous, non-empty, convex and compact-valued for each $i = 0, \ldots, m + 1$. According to the Kakutani Theorem, there exists $(\bar{p}, \bar{q}, (\bar{c}_i, \bar{l}_i)_{i=1}^m)$ such that $(\bar{p}, \bar{q}) \in \varphi_0((\bar{c}_i, \bar{l}_i)_{i=1}^m)$ and $(\bar{c}_i, \bar{l}_i) \in \varphi_i((\bar{p}, \bar{q}))$.

It is easy to prove that $(\bar{p}, \bar{q}, (\bar{c}_i, \bar{l}_i)_{i=1}^m)$ is an equilibrium for the economy $\mathcal{E}_b^T$. □

We apply the argument of Lemma 3 in Le Van and Pham (2016) to prove the existence of equilibrium in unbounded truncated economies.

Proposition 10 *An equilibrium* $(p, q, (c_i, l_i)_{i=1}^m)$, *with* $p_t + q_t = 1$, *of* $\mathcal{E}_b^T$ *is an equilibrium for* $\mathcal{E}^T$.

Existence of equilibrium in an infinite-horizon economy

Proposition 11 *Under Assumptions 1–4, there exists an equilibrium in the economy* $\mathcal{E}$.

Proof We have shown that there exists an equilibrium, say $\left(p^T, q^T, \left(c_i^T, l_i^T\right)_i\right)$, for each truncated economy $\mathcal{E}^T$. Recall that $p_t^T + q_t^T = 1$. It is clear that, for each t, we have $c_{i,t}^T \in (0, W_t)$; $l_{i,t}^T \in [0, L]$; $p_t^T, q_t^T \in [0, 1]$. Therefore, without loss of generality, we can assume that $\left(p^T, q^T, \left(c_i^T, l_i^T\right)_i\right) \overset{T\to\infty}{\longrightarrow} (p, q, (c_i, l_i)_i)$ (for the product topology).

Applying the same argument in the proof of Theorem 1 by Le Van and Pham (2016) or the proof of Theorem 1 by Le Van and Pham (2014), we obtain that $(p, q, (c_i, l_i)_i)$ is an equilibrium for the economy $\mathcal{E}$. □

References

Aoki, K., Nakajima, T., & Nikolov, K. (2014). Safe asset shortages and asset price bubbles. *Journal of Mathematical Economics, 53*, 165–174.

Araujo, A., Novinski, R., & Pascoa, M. R. (2011). General equilibrium, wariness and efficient bubbles. *Journal of Economic Theory, 146*, 785–811.

Araujo, A., Pascoa, M. R., & Torres-Martinez, J. P. (2011). Long-lived collateralized assets and bubbles. *Journal of Mathematical Economics, 47*, 260–271.

Balasko, Y., & Shell, K. (1980). The overlapping generations model I: The case of pure exchange without money. *Journal of Economic Theory, 23*, 281–306.

Becker, R. A., Dubey, R. S., & Mitra, T. (2014). On Ramsey equilibrium: Capital ownership pattern and inefficiency. *Economic Theory, 55*, 565–600.

Becker, R., Bosi, S., Le Van, C., & Seegmuller, T. (2015). On existence and bubbles of Ramsey equilibrium with borrowing constraints. *Economic Theory, 58*, 329–353.

Bosi, S., Le Van, C., & Pham, N. -S. (2015a). *Asset bubbles and efficiency in a generalized two-sector model*, LEM working paper.

Bosi, S., Le Van, C., & Pham, N. -S. (2015b). *Intertemporal equilibrium with heterogeneous agents, endogenous dividends and borrowing constraints*, EPEE working paper.

Brunnermeier M. K., Oehmke, M. (2012). Bubbles, financial crises, and systemic risk. *Handbook of the Economics of Finance, 2*.

Cass, D. (1972). On capital overaccumulation in the aggregative, neoclassical model of economic growth: A complete characterization. *Journal of Economic Theory, 4*, 200–223.

Farhi, E., & Tirole, J. (2012). Bubbly liquidity. *Review of Economic Studies, 79*, 678–706.

Hirano T., & Yanagawa, N. (2013). *Asset bubbles, endogenous growth and financial frictions*. Working Paper, University of Tokyo.

Huang, K. X. D., & Werner, J. (2000). Asset price bubbles in Arrow-Debreu and sequential equilibrium. *Economic Theory, 15*, 253–278.

Kamihigashi, T. (2002). A simple proof of the necessity of the transversality condition. *Economic Theory, 20*, 427–433.

Kocherlakota, N. R. (1992). Bubbles and constraints on debt accumulation. *Journal of Economic Theory, 57*, 245–256.

Le Van, C., & Pham, N. S. (2014). *Financial asset bubble with heterogeneous agents and endogenous borrowing constraints*, Working paper.

Le Van C., & Pham, N. S. (2016). Intertemporal equilibrium with financial asset and physical capital. *Economic Theory, 62*, 155–199.

Miao, J. (2014). Introduction to economic theory of bubbles. *Journal of Mathematical Economics, 53*, 130–136.

Miao, J., & Wang, P. (2012). Bubbles and total factor productivity. *American Economic Review: Papers and Proceedings, 102*, 82–87.

Miao, J., & Wang, P. (2015). *Bubbles and credit constraints*, working paper, Boston University.

Montrucchio, L. (2004). Cass transversality condition and sequential asset bubbles. *Economic Theory, 24*, 645–663.

Santos, M. S., & Woodford, M. (1997). Rational asset pricing bubbles. *Econometrica, 65*, 19–57.

Tirole, J. (1982). On the possibility of speculation under rational expectations. *Econometrica, 50*, 1163–1181.

Tirole, J. (1985). Asset bubbles and overlapping generations. *Econometrica, 53*, 1499–1528.

Weil, P. (1990). On the possibility of price decreasing bubbles. *Econometrica, 58*, 1467–1474.

Werner, J. (2014). *Speculative trade under ambiguity*. Mimeo.

Chapter 10
The Stabilizing Virtues of Monetary Policy on Endogenous Bubble Fluctuations

Lise Clain-Chamosset-Yvrard and Thomas Seegmuller

Abstract We explore the stabilizing role of monetary policy on the existence of endogenous fluctuations when the economy experiences a rational bubble. Considering an overlapping generations model, expectation-driven fluctuations are explained by a portfolio choice between three assets (capital, bonds and money), credit market imperfections and a collateral effect. They occur under a positive bubble on bonds. The key mechanism relies on the existence of gaps between the returns on assets due to financial distortions. Then, we study the stabilizing role of the monetary policy. Such a policy managed by a (standard) Taylor rule has no clear stabilizing virtues.

Keywords Indeterminacy · Rational bubble · Cash-in-advance constraint · Monetary policy · Taylor rule

JEL classification: D91 · E32 · E52

We would like to thank an anonymous referee for his comments. We also thank Stefano Bosi, Marion Davin, Takashi Kamihigashi, Carine Nourry and Alain Venditti for valuable suggestions. We also thank participants to the Conference in honor of Jean-Michel Grandmont held in Aix-Marseille University on June 2013, to the Conference 12th "Journées Louis-André Gérard-Varet", to the Conference PET 2013 and to the 13th Annual SAET Conference. Any remaining errors are our own.

L. Clain-Chamosset-Yvrard
Aix-Marseille University (Aix-Marseille School of Economics), CNRS-GREQAM
and EHESS, Centre de la Vieille Charité, 13236 CEDEX 02, Marseille, France
e-mail: lise.clain-chamosset-yvrard@univ-amu.fr

T. Seegmuller (✉)
Aix-Marseille University (Aix-Marseille School of Economics), CNRS-GREQAM
and EHESS, Centre de la Vieille Charité, 2 Rue de la Charité, CEDEX 02, 13236
Marseille, France
e-mail: thomas.seegmuller@univ-amu.fr

10.1 Introduction

In recent years, asset prices have experienced large fluctuations, and the financial sphere of the economy had strong effects on the real one, as illustrating during the last financial crisis. Some empirical contributions shed light on the excessive asset price volatility, and reveal that asset prices fluctuate more than their fundamental value (see Shiller 1981, 1989, and LeRoy and Porter 1981). One explanation for this excessive volatility is the existence and the fluctuations of asset bubbles.

A large body of theoretical literature explores the role of credit market imperfections in the existence and dynamics of rational bubbles (Farhi and Tirole 2012; Martin and Ventura 2012; Wang and Wen 2012). Despite the fact that most of these contributions deal with credit constraints at the level of entrepreneurs, some empirical studies highlight the existence of credit constraints faced by consumers underlying the role of collateral on their behavior (Campbell and Mankiw 1989; Jappelli 1990; Iacoviello 2004). Such types of credit market imperfections affect the portfolio choices between different existing assets, but also explain the gaps between their returns. We think that such credit market imperfections may be a main transmission channel between the financial and the real spheres. In this paper, we argue that credit constraints faced by consumer play a crucial role to explain expectation-driven fluctuations of speculative bubbles, as illustrated during the recent subprime crisis. This idea already appears in Bosi and Seegmuller (2010) and Clain-Chamosset-Yvrard and Seegmuller (2015).[1] They assume that the share of consumption financed by credit is positively correlated to a collateral. Because of this type of credit market distortions, the portfolio choices are no longer constant through time. A change in agents' expectations generates a new trade-off between asset holdings promoting equilibrium indeterminacy, and thus the occurrence of expectation-driven fluctuations. We enrich these contributions by considering both capital and the stabilizing role of a monetary policy conducted through a Taylor rule.

Indeed, economic fluctuations based on consumer credit constraints open the door to new policy tools for stabilizing issues. A stabilizing policy must dampen or eliminate the mechanism source of indeterminacy. Since our explanation of expectation-driven fluctuations relies on a trade-off between different assets, namely capital, bonds and money, relevant stabilizing policies are those reducing the gaps between their returns. Monetary policy appears to be a natural policy tool, since it affects the opportunity cost of money holdings through the level of the nominal interest rate. In addition, contrary to most of the literature, we analyze the stabilizing role of the monetary policy when bubble fluctuations occur in an economy with both production and a positive bubble (Grandmont 1985, 1986; Bernanke and Woodford 1997; Benhabib et al. 2001; Sorger 2005; and Rochon and Polemarchakis 2006).

---

[1]Only few other contributions have analyzed the existence of bubble fluctuations with an interplay between the real and the financial spheres of the economy (Michel and Wigniolle 2003, 2005; Bosi and Seegmuller 2010; Wigniolle 2014).

We consider a simple overlapping generations (OLG) model with capital accumulation to highlight the role of consumers' credit market imperfections and collateral in an economy characterized by a rational bubble.[2] Households save through bonds, money and capital. Bonds are sold by the monetary authority to supply money. Because of a binding cash-in-advance (CIA) constraint, money is held by households to finance a share of their consumption in the second period of their life. Despite the fact that capital is used for the production, it also serves as a collateral: Holding more capital increases the amount of collateral, and thus allows each household to reduce the share of consumption financed through money. It is important to note that the three assets have different returns. Bonds have larger return than capital because this latter is used as a collateral to relax the consumers' credit constraint, and also a larger return than money because we focus on equilibria with binding constraints. As a direct implication, the Fisher relationship is not satisfied.[3] The violation of this relationship will induce some portfolio choices that promote indeterminacy, and therefore endogenous fluctuations.

We prove the existence of a steady state characterized by a positive rational bubble on bonds (Tirole 1985). In contrast to several existing papers (Farmer 1986; Benhabib and Laroque 1988; Rochon and Polemarchakis 2006), expectation-driven fluctuations occur in the neighborhood of such a steady state with a positive rational asset bubble under gross substitutability and reasonable values of input substitution, without requiring arbitrarily large increasing returns to scale (Cazzavillan and Pintus 2005). This result is obtained because of the role of collateral.

Since expectation-driven fluctuations are mainly driven by the portfolio choices between capital, money and bonds and the violation of the Fisher relationship, a policy may have a stabilizing virtue if it is able to reduce the gaps between the different returns on assets. Therefore, the monetary authority could play an active role in stabilizing the economy by manipulating the nominal interest rate. Following Bernanke (2010) who argues that a rule which responds to expected inflation is relevant to describe the US monetary policy, we consider that the nominal interest rate is determined according to a Taylor rule on expected inflation. We show however that the stabilizing results are mitigated. A weakly active policy can even promote endogenous fluctuations for some relevant parameter configurations. One explanation is that such a rule does not significantly modify the nominal interest rate, and therefore does not alter so much the portfolio choices. More generally, this result provides an adding argument emphasizing that standard policy tools are not so relevant in some circumstances. In our case, these circumstances are the existence of a bubble and consumers' credit constraints affected by a collateral.

This paper is organized as follows. In the next section, we present the model. The intertemporal equilibrium is defined in Sect. 10.3. Section 10.4 is devoted to the

---

[2] Our work is close to the framework developed by Rochon and Polemarchakis (2006). However, our analysis differs in two main points: First, we take into account the role of collateral on the consumption behavior; Second, we analyze a monetary policy that could fit better the practices of central banks, instead of an interest rate pegging.

[3] Recall that the Fisher relationship means that the gross real interest rate is equal to the gross nominal interest rate deflated by the gross inflation rate.

steady state analysis. In Sect. 10.5, we analyze the occurrence of expectation-driven fluctuations when there is a positive bubble and the stabilizing role of monetary policy. Concluding remarks are provided in Sect. 10.6, and all the proofs are gathered in a final Appendix.

10.2 The Model

We consider an OLG model with production in discrete time ($t = 0, 1, ..., +\infty$). This economy consists of identical two period-lived households, firms, a monetary authority and a government.

10.2.1 Households

There is no population growth, and at each date t, a generation of unit size is born and lives for two periods.

An household derives utility from consumption of final good when young (c_t) and old (d_{t+1}). Her preferences are represented by an additively separable life-cycle utility function:

$$u\left(c_t\right) + \beta v\left(d_{t+1}\right) = \frac{c_t^{1-\varepsilon_u}}{1 - \varepsilon_u} + \beta \frac{d_{t+1}^{1-\varepsilon_v}}{1 - \varepsilon_v}, \quad \beta > 0 \tag{10.1}$$

where $\varepsilon_u > 0$ and $\varepsilon_v > 0$ denote respectively the degrees of concavity of $u\left(c_t\right)$ and $v\left(d_{t+1}\right)$. We further note that $\varepsilon_v < 1$ implies gross substitutability meaning that savings are an increasing function of the global return on portfolio.[4]

In her first period of life, the household is young and supplies one unit of labor inelastically remunerated at the wage w_t. With this wage, she can consume an amount c_t of final good at price p_t, and save through a diversified portfolio of nominal

---

[4] As we will see below, the consumer problem has the following structure:

$$max \quad \frac{c_t^{1-\varepsilon_u}}{1 - \varepsilon_u} + \beta \frac{d_{t+1}^{1-\varepsilon_v}}{1 - \varepsilon_v}$$
$$st. \quad c_t + s_t = w_t$$
$$d_{t+1} = \tilde{R}_{t+1} s_t + \Delta_{t+1},$$

where s_t represents global savings of a household, $\tilde{R}_{t+1}$ the global return on her portfolio, w_t her labor income and Δ_{t+1} a monetary transfer. From this problem, we obtain:

$$\frac{ds_t}{d\tilde{R}_{t+1}} \frac{\tilde{R}_{t+1}}{s_t} = \frac{1 - \varepsilon_v \tilde{R}_{t+1} s_t/(\tilde{R}_{t+1} s_t + \Delta_{t+1})}{\varepsilon_u s_t/(w_t - s_t) + \varepsilon_v \tilde{R}_{t+1} s_t/(\tilde{R}_{t+1} s_t + \Delta_{t+1})},$$

which is positive for $\varepsilon_v < 1$.

balances M_{t+1} needed for a transaction motive, productive capital per capita k_{t+1} (with rental factor R_{t+1})[5] and nominal bonds B_{t+1} (with nominal interest rate i_{t+1}). In our framework, bonds denote nominal debts issued by the monetary authority in order to inject money in the economy. In contrast to asset papers with no fundamental value considered as freely disposed of, these bonds can have a negative nominal value ($B_{t+1} < 0$).

In her second period of life, the household is old. She uses her remunerated savings and her monetary transfer Δ_{t+1} received from the monetary authority to purchase an amount d_{t+1} of final good at price p_{t+1}. The first and second-period budget constraints are written as follows:

$$p_t c_t + M_{t+1} + B_{t+1} + p_t k_{t+1} \leq p_t w_t \tag{10.2}$$

$$p_{t+1} d_{t+1} \leq M_{t+1} + (1 + i_{t+1}) B_{t+1}$$
$$+ p_{t+1} R_{t+1} k_{t+1} + \Delta_{t+1} \tag{10.3}$$

The household has to pay cash a part of the second period consumption d_{t+1}: Her money demand is rationalized by a cash-in-advance (CIA) constraint. We use a constraint of the type introduced by Hahn and Solow (1995), i.e. $\gamma p_{t+1} d_{t+1} \leq M_{t+1}$, but we extend it to capture the role of collateral:

$$\gamma(k_{t+1}) p_{t+1} d_{t+1} \leq M_{t+1} \tag{10.4}$$

A binding cash-in-advance constraint means that a share $\gamma(k_{t+1}) \in (0, 1)$ of her second-period consumption has to be paid cash, i.e. with nominal balances M_{t+1}. As underlined in Rochon and Polemarchakis (2006) and Clain-Chamosset-Yvrard and Seegmuller (2015), the household can consume the remaining share $1 - \gamma(k_{t+1})$ of consumption on credit when old. Indeed, because she holds $B_{t+1} + p_{t+1} k_{t+1}$ in her portfolio when young, the household knows that she will have her remunerated savings from bonds and capital $(1 + i_{t+1}) B_{t+1}/p_{t+1} + R_{t+1} k_{t+1}$ at the next period, in addition to the transfer from the monetary authority Δ_{t+1}/p_{t+1}. As a result, she can consume on credit by borrowing from a bank or a financial institution an amount equal to $(1 + i_{t+1}) B_{t+1}/p_{t+1} + R_{t+1} k_{t+1} + \Delta_{t+1}/p_{t+1}$, that she will pay back at the end of her second period of life. In the following, we refer to $1 - \gamma(k_{t+1})$ as the credit share.

Furthermore, we assume that the credit share is increasing with the amount of physical capital held by a household. Through this assumption, we assert that capital acts as a collateral for the household. Since a collateral is by definition an asset that a household offers a bank or a financial institution to secure a loan, we argue that the value of physical capital k_{t+1} can be pledged as a collateral, rather than the capital income $R_{t+1} k_{t+1}$. If the household fails to repay the loan, the financial institution can seize its physical capital to recover its losses, and thus become the owner of this capital. Since the financial institution takes less risk, it would be easier for the

---

[5] We assume a full capital depreciation within a period.

household to obtain credit from the bank or the financial institution by holding more capital in her portfolio, and thus to reduce her need of cash in her second period of life.

This is also in accordance with some empirical studies which, focusing on U.S data, underline the negative correlation between money holdings and wealth (see Wolff 1998). In our framework, k_{t+1} can be seen as a proxy of household's wealth. In any case, it is a simple way to introduce credit market imperfections and to capture the role of collateral on consumption behavior of the household as highlighted by empirical studies (among others, Campbell and Mankiw 1989; Iacoviello 2004).[6]

The following assumption summarizes the properties of the function $\gamma(k)$:

Assumption 1 $\gamma(k) \in (0, 1)$ is a continuous function defined on $[0, +\infty)$, C^2 on $(0, +\infty)$, decreasing $(\gamma'(k) \leq 0)$.

For further references, we define the following elasticities:

$$\eta_1(k) \equiv \frac{[1-\gamma(k)]' k}{1-\gamma(k)} = -\frac{\gamma'(k)k}{1-\gamma(k)} \geq 0, \qquad (10.5)$$

$$\eta_2(k) \equiv -\frac{[1-\gamma(k)]'' k}{[1-\gamma(k)]'} = -\frac{\gamma''(k)k}{\gamma'(k)} \qquad (10.6)$$

Example The following function satisfies these properties:

$$\gamma(k) = 1 - \frac{a+bk^\epsilon}{1+ck^\epsilon}, \qquad (10.7)$$

with $a \in (0, 1)$, $c > 1$, $b \in (ac, c)$ and $\epsilon > 0$. Using this example, $\eta_1(k)$ and $\eta_2(k)$ are given by:

$$\eta_1(k) = \frac{b-ca}{a+bk^\epsilon}\frac{\epsilon k^\epsilon}{1+ck^\epsilon} \geq 0 \quad and \quad \eta_2(k) = 1 + \epsilon\left(\frac{2ck^\epsilon}{1+ck^\epsilon} - 1\right)$$

When collateral does not matter $(\eta_1(k_{t+1}) = 0)$, and γ tends to 0, money is no longer needed and the credit market distortion disappears, whereas when $\gamma > 0$, there is a need of cash. When collateral matters $(\eta_1(k_{t+1}) > 0)$, the households are aware of the credit share function: They are able to relax the CIA constraint by increasing capital holdings.

---

[6]This manner of introducing a collateral effect differs from models with borrowing/collateral constraint *à la* Kiyotaki and Moore (1997). First, borrowing is typically used to finance investment project in these models with collateral constraint, whereas in our paper borrowing finances consumption. Second, our CIA constraint implies a limit on the borrowing's share of total expenditures instead of the borrowing capacity itself. Indeed, using the second-period budget constraint and introducing $A_{t+1} = (1+i_{t+1})B_{t+1}/p_{t+1} + R_{t+1}k_{t+1} + \Delta_{t+1}/p_{t+1}$ as the amount of borrowing, we can rewrite our CIA constraint as follows $A_{t+1}/(p_{t+1}d_{t+1}) \leq 1 - \gamma(k_{t+1})$. Finally, our limit is nonlinear and increasing with collateral, whereas the borrowing limit of a standard collateral constraint is exogenous or linear with collateral.

Using $\pi_{t+1} \equiv p_{t+1}/p_t$ and introducing the real variables $m_{t+1} \equiv M_{t+1}/p_{t+1}$, $b_{t+1} \equiv B_{t+1}/p_{t+1}$ and $\delta_{t+1} \equiv \Delta_{t+1}/p_{t+1}$, the constraints (10.2)–(10.4) can be rewritten as follows:

$$c_t + \pi_{t+1} m_{t+1} + \pi_{t+1} b_{t+1} + k_{t+1} \leq w_t \tag{10.8}$$

$$d_{t+1} \leq m_{t+1} + (1 + i_{t+1}) b_{t+1} + R_{t+1} k_{t+1} + \delta_{t+1} \tag{10.9}$$

$$\gamma (k_{t+1}) d_{t+1} \leq m_{t+1} \tag{10.10}$$

An household derives her optimal consumption choice (c_t, d_{t+1}) and her optimal portfolio choice $(k_{t+1}, m_{t+1}, b_{t+1})$ by maximizing her utility function (10.1) under her budget and cash-in-advance constraints (10.8)–(10.10).

Assumption 2 Let $\tilde{\varepsilon}_u \equiv c \dfrac{1+i}{\pi} \dfrac{i \eta_1 (1 - \gamma)}{\eta_2 d (1 + i\gamma)^2}$.[7] For all $t \geq 0$, we assume $i_t > 0$, $\eta_2(k_t) > 0$ and $\varepsilon_u > \tilde{\varepsilon}_u$.

Since the conditions in Assumption 2 rely on endogenous variables, as $i_t, \pi_t, \gamma(k_t)$, Assumption 2 can seem quite strong. Nevertheless, as we are interested in the occurrence of fluctuations in the vicinity of a steady state, Assumption 2 will be supposed satisfied at the steady state. By continuity, it will also hold in the neighborhood of this steady state. Note that we provide a numerical example satisfying $n_2(k_t) > 0$ at the normalized steady state that we consider for the dynamic analysis. In addition, since ε_u is a free preference parameter, Assumption 2 can always be satisfied.

We can derive the following Lemma[8]:

Lemma 1 *Under Assumptions 1 and 2, constraints (10.8)–(10.10) are binding and the second-order conditions are satisfied.*

Lemma 1 requires that the function of the credit share $1 - \gamma(k_{t+1})$ is concave: Capital holdings increase, at a decreasing rate, the fraction of second-period consumption purchased on credit. Moreover, the CIA constraint is binding if the nominal interest rate i_{t+1} is strictly positive ($i_{t+1} > 0$).

Under Assumptions 1 and 2, the optimal households' behavior is summarized by the following equations:

$$\frac{u'(c_t)}{\beta v'(d_{t+1})} = \frac{1 + i_{t+1}}{\pi_{t+1}} \frac{1}{1 + i_{t+1}\gamma(k_{t+1})} \tag{10.11}$$

$$R_{t+1} = \frac{1 + i_{t+1}}{\pi_{t+1}} - i_{t+1}\eta_1(k_{t+1}) \frac{1 - \gamma(k_{t+1})}{k_{t+1}} d_{t+1} \tag{10.12}$$

---

[7]For simplicity, the arguments of the functions and the time subscripts are omitted.

[8]The proof of Lemma 1 is given in a technical appendix available on https://sites.google.com/site/liseclainchamosset.

When collateral does not matter ($\eta_1(k) = 0$), Eqs. (10.11) and (10.12) rewrite:

$$\frac{u'(c_t)}{\beta v'(d_{t+1})} = \frac{(1+i_{t+1})/\pi_{t+1}}{1+i_{t+1}\gamma} \qquad (10.13a)$$

and

$$R_{t+1} = \frac{1+i_{t+1}}{\pi_{t+1}} \qquad (10.13b)$$

We note that as γ tends to 0, we obtain the intertemporal trade-off found in Diamond (1965) and Tirole (1985), in which there are no credit market distortions in the economy (see Eq. (10.13a)). As $\gamma > 0$, a distortion exists: Old households now have to pay cash γ in order to consume an additional unit of final good, and money entails an opportunity cost. Nevertheless, when collateral does not matter, capital and bonds are perfect substitutes (see Eq. (10.13b)).

When collateral matters ($\eta_1(k) > 0$), capital and bonds are no longer perfect substitutes. Households can now decrease their need of cash by holding more capital in their portfolio. As a consequence, the return on capital is lower than the return on bonds.

The endogeneity of the credit share ensures the portfolio choices to be no longer constant through time. The trade-off between assets is endogenous and depends on the amount of collateral held by the households. A change in expected inflation generates a portfolio effect, i.e. a new trade-off between asset holdings. This portfolio effect is the key mechanism through which expectation-driven fluctuations may occur. Since the portfolio choices are the explanation for fluctuations, we will also focus on a stabilizing policy designed to dampen the portfolio effect by modifying the different returns on assets.

10.2.2 Firms

A representative competitive firm produces the final good using capital and labor under a constant returns to scale technology $f(K/L)L$. Using $k = K/L$, the intensive production function $f(k)$ satisfies:

Assumption 3 $f(k)$ is a continuous function defined on $[0, +\infty)$ and C^2 on $(0, +\infty)$, strictly increasing ($f'(k) > 0$) and strictly concave ($f''(k) < 0$). Defining $\alpha(k) \equiv f'(k)k/f(k) \in (0, 1)$ as the capital share in total income and $\sigma(k) \equiv \left[\frac{f'(k)k}{f(k)} - 1\right]\frac{f'(k)}{kf''(k)} > 0$ as the elasticity of capital-labor substitution, we further assume $f'(1) < 1$, $\lim_{k\to 0^+} f'(k) > 1$ and $\sigma(k) > 1 - \alpha(k)$.

Note that at the end of Sect. 10.4.2, we provide a numerical example that satisfies Assumption 3. The competitive firm takes the prices as given and maximizes the profits $f(K_t/L_t)L_t - w_t L_t - R_t K_t$:

$$R_t = f'(k_t) \equiv R(k_t) \tag{10.14}$$

$$w_t = f(k_t) - k_t f'(k_t) \equiv w(k_t) \tag{10.15}$$

Hence, the interest rate and wage elasticities are respectively equal to $\epsilon_R(k) \equiv R'(k)k/R(k) = -(1 - \alpha(k))/\sigma(k)$ and $\epsilon_w(k) \equiv w'(k)k/w(k) = \alpha(k)/\sigma(k)$. The inequality $\sigma(k) > 1 - \alpha(k)$ involves capital income $R_t k_t$ being increasing with k_t, which is not a restrictive assumption.

10.2.3 Monetary Authority

For implementing monetary policy, the monetary authority (central bank) uses open market operations defined as the purchase or sale of bonds in exchange for nominal balances.[9] At time t, the central bank creates nominal balances M_{t+1}, which offer liquidity at the next period $t + 1$.[10] The money growth factor $\mu_t = M_{t+1}/M_t$ can be written as follows:

$$\mu_t = \pi_{t+1}\frac{m_{t+1}}{m_t} \tag{10.16}$$

In order to supply M_{t+1} in the economy at $t + 1$, the central bank buys bonds from old households, and pays for them in cash through open market operations. The profits made by central bank Δ_t at time t are given by:

$$\Delta_t = B_{t+1} + M_{t+1} - (1 + i_t)B_t - M_t \tag{10.17}$$

These profits are distributed as dividends to the old households at time t. The budget constraint of the monetary authority at time t is written as follows:

$$B_{t+1} + M_{t+1} = (1 + i_t)B_t + M_t + \Delta_t = (1 + i_t)(B_t + M_t) \tag{10.18}$$

or in real terms:

$$\pi_{t+1}(b_{t+1} + m_{t+1}) = (1 + i_t)(b_t + m_t) \tag{10.19}$$

Introducing the variable $\theta_t \equiv (1 + i_t)(b_t + m_t)$, Eq. (10.19) can be rewritten as follows:

$$\pi_{t+1}\theta_{t+1} = (1 + i_{t+1})\theta_t \tag{10.20}$$

---

[9]To study the existence of expectation-driven fluctuations in an OLG model without collateral, Rochon and Polemarchakis (2006) use similar open market operations to issue money in the economy.

[10]Placing a part of their savings in the form of nominal balances in their first period of life, young households have the opportunity to obtain liquidity in their second period of life.

Note that if θ_t is positive at the equilibrium, then a part of bonds, which are purely unbacked public assets (intrinsically useless), has a positive value. Let $b_t = \bar{b}_t + \tilde{b}_t$, where $\bar{b}_t$ denotes the real counterpart of money, and $\tilde{b}_t$ the real value of unbacked public assets. As $\bar{b}_t + m_t = 0$, $\theta_t = (1 + i_t)(b_t + m_t) > 0$ is equivalent to $\tilde{b}_t > 0$. We can argue that there is a bubble on bonds when $\tilde{b}_t > 0$. Therefore, $\theta_t > 0$ pertains to a situation in which a positive bubble on bonds exists.[11] When $\theta_t = 0$, all bonds are the counterpart of money. In this case, all money in the economy corresponds to inside money: No bubbles on bonds exist. When $\theta_t < 0$, there is an excess of households' debt.

In addition, the monetary authority chooses the nominal interest rate i_{t+1} as the monetary instrument, and implements the following interest rate rule:

$$1 + i_{t+1} = \left(1 + i^*\right)\left(\frac{\pi_{t+1}}{\pi^*}\right)^\phi, \tag{10.21}$$

where $\phi \geq 0$ is a measure of monetary policy responses to expected inflation. Furthermore, i^* and π^* are respectively the stationary values of the nominal interest rate and the inflation of an existing stationary equilibrium chosen as the targets by the monetary authority.

When $\phi = 0$, the central bank decides to fix the level of the nominal interest rate at its stationary level i^*. When $\phi > 0$, Eq. (10.21) depicts a Taylor interest rate rule, which responds to expected inflation. Note that according to Bernanke (2010), a rule which responds to expected inflation is more relevant to describe the US monetary policy than a rule responding to observed inflation. For $\phi \in (0, 1)$, the rule weakly reacts to expected inflation. An increase (decrease) in the inflation raises (depresses) the nominal interest rate less than proportionally, involving a decrease (increase) in the real interest rate. For $\phi > 1$, the rule strongly reacts to expected inflation. An increase (decrease) in the inflation raises (depresses) the nominal interest rate more than proportionally, involving an increase (decrease) in the real interest rate. Following Benhabib et al. (2001), we define a rule with $\phi \in (0, 1)$ as a passive one, and a rule with $\phi > 1$ as an active one.

10.3   Intertemporal Equilibrium

At the intertemporal equilibrium, the budget and cash-in-advance constraints of households are given by:

$$c_t + \frac{\pi_{t+1}}{1 + i_{t+1}}\theta_{t+1} + k_{t+1} = w(k_t) \tag{10.22}$$

$$d_{t+1} = \theta_{t+1} + f'(k_{t+1})k_{t+1} \tag{10.23}$$

---

[11] Alternatively, $\theta_t > 0$ corresponds to a situation in which the outside money is positive. A positive outside money indicates that there is fiat money in circulation in the economy. In the literature on rational bubble, the bubble is often considered as being fiat money.

$$\gamma(k_{t+1})d_{t+1} = m_{t+1} \tag{10.24}$$

The budget constraints of the monetary authority is as follows:

$$\pi_{t+1} = (1 + i_{t+1}) \frac{\theta_t}{\theta_{t+1}} \tag{10.25}$$

Substituting Eq. (10.25) into the first-period budget constraint Eq. (10.22), we get:

$$c_t + \theta_t + k_{t+1} = w(k_t) \tag{10.26}$$

Using Eqs. (10.16), (10.24) and (10.25), we deduce the money growth factor:

$$\mu_t = (1 + i_{t+1}) \frac{\theta_t}{\theta_{t+1}} \frac{\gamma(k_{t+1})}{\gamma(k_t)} \frac{\theta_{t+1} + f'(k_{t+1})k_{t+1}}{\theta_t + f'(k_t)k_t} \tag{10.27}$$

Substituting Eqs. (10.22) and (10.26) into Eq. (10.11), and Eqs. (10.23) and (10.25) into Eq. (10.12), the consumers' intertemporal trade-off and the no-arbitrage condition are respectively given by:

$$\begin{cases} \theta_t \dfrac{u'(f(k_t) - f'(k_t)k_t - \theta_t - k_{t+1})}{\beta v'(\theta_{t+1} + f'(k_{t+1})k_{t+1})} = \dfrac{\theta_{t+1}}{1 + i_{t+1}\gamma(k_{t+1})} \\[2ex] \dfrac{\theta_{t+1}}{\theta_t} = \dfrac{1 + i_{t+1}}{\pi_{t+1}} = f'(k_{t+1})H_{t+1}(k_{t+1}, \theta_t), \end{cases} \tag{10.28}$$

$$with \quad H_{t+1}(k_{t+1}, \theta_t) \equiv \frac{1 + i_{t+1}\eta_1(k_{t+1})\left[1 - \gamma(k_{t+1})\right]}{1 - \theta_t i_{t+1}\eta_1(k_{t+1})\left[1 - \gamma(k_{t+1})\right]/k_{t+1}} \tag{10.29}$$

When collateral does not matter ($\eta_1(k) = 0$), we obtain $H_{t+1}(k_{t+1}, \theta_t) = 1$. Therefore, the Fisher equation ($(1 + i_{t+1})/\pi_{t+1} = f'(k_{t+1})$) holds at the intertemporal equilibrium. This means that the return on real asset (capital) is equal to the return on nominal asset (bonds) deflated by the inflation factor. The role of collateral ($\eta_1(k) > 0$) implies the violation of the Fisher equation ($H_{t+1}(k_{t+1}, \theta_t) > 1$). As capital serves as a collateral, its return becomes lower than the real return on bonds ($f'(k_{t+1}) < (1 + i_{t+1})/\pi_{t+1}$). This violation of the Fisher equation will induce some portfolio choices that promote indeterminacy, a source of expectation-driven fluctuations.

Interestingly, the level of nominal interest rate can offset the collateral effect (see Eq. (10.29)). Hence, considering a monetary policy conducted through an usual interest rate rule, like a Taylor one, could a priori be relevant to stabilize macroeconomic fluctuations.

From the budget constraint of the monetary authority Eq. (10.25) and the monetary rule Eq. (10.21), we deduce that:

$$\frac{\pi_{t+1}}{\pi^*} = \left(\frac{\theta_t}{\theta_{t+1}}\right)^{\frac{1}{1-\phi}} \qquad (10.30)$$

Substituting Eq. (10.30) into Eq. (10.21), we obtain the nominal interest rate at the equilibrium:

$$i_{t+1} = (1 + i^*)\left(\frac{\theta_t}{\theta_{t+1}}\right)^{a_\phi} - 1, \quad where \quad a_\phi \equiv \frac{\phi}{1-\phi} \in (-\infty, -1) \cup [0, +\infty) \qquad (10.31)$$

Note that when $a_\phi \in (-\infty, -1)$, the monetary rule is active. When $a_\phi \in (0, +\infty)$, the rule is passive.

Definition 1 Under Assumptions 1–3, an intertemporal equilibrium with perfect foresight is a sequence $(k_t, \theta_t) \in \mathbb{R}_+ \times \mathbb{R}, t = 0, 1, ..., +\infty$, such that the dynamic system (10.28) is satisfied, where $H_{t+1}(k_{t+1}, \theta_t)$ is defined by (10.29), i_{t+1} by Eq. (10.31), and $k_0 > 0$ is given.

Taking into account Eqs. (10.29) and (10.31), we note that k_t is the only predetermined variable of the two-dimensional dynamic system (10.28). The intertemporal sequence of k_t and θ_t enables us to determine all the other variables, namely c_t, d_t, m_t and b_t.

10.4 Steady State Analysis

From the system (10.28), we deduce that two kinds of steady state exist: $\theta = 0$ and $\theta \neq 0$. Since we are interested in fluctuations with a positive bubble, we will focus on steady states with $\theta > 0$. A steady state with a positive bubble is a solution $(k, \theta) \in \mathbb{R}^2_{++}$ that satisfies the following system:

$$\begin{cases} \dfrac{u'(f(k) - f'(k)k - k - \theta)}{\beta v'(\theta + f'(k)k)} = \dfrac{1}{1 + i^*\gamma(k)} \\[4mm] f'(k)H(k, \theta) = 1 \end{cases} \qquad (10.32)$$

$$with \quad H(k, \theta) = \frac{1 + i^*\eta_1(k)\left[1 - \gamma(k)\right]}{1 - \theta i^*\eta_1(k)\left[1 - \gamma(k)\right]/k} \qquad (10.33)$$

Under a constant credit share $(\eta_1(k) = 0)$, we see from the system (10.32) that the steady state is unique, and the monetary policy does not affect the production side. Indeed, the second equation of the system (10.32) reduces to $f'(k) = 1$. Under Assumption 3, this gives a unique stationary solution in k and superneutrality of money holds. When collateral matters $(\eta_1(k) > 0)$, the superneutrality of money is canceled. Because of collateral, the monetary sphere affects the real one.

10.4.1 Existence

From Eq. (10.32), a steady state with $\theta > 0$ is a solution $k \in \mathbb{R}_{++}$ satisfying:

$$
\begin{cases}
\dfrac{u'(c(k))}{\beta v'(d(k))} = \dfrac{1}{1 + i^*\gamma(k)} \\[3mm]
\theta = \dfrac{1 - f'(k)\{1 + i^*\eta_1(k)[1 - \gamma(k)]\}}{i^*\eta_1(k)[1 - \gamma(k)]/k}
\end{cases}
\tag{10.34}
$$

$$
with \quad c(k) = f(k) - k - \frac{k[1 - f'(k)]}{i^*\eta_1(k)[1 - \gamma(k)]} \ and \ d(k) = \frac{k[1 - f'(k)]}{i^*\eta_1(k)[1 - \gamma(k)]}
$$

From these equations, we deduce that $d(k) > 0$ implies $f'(k) < 1$, and from Eqs. (10.25) and (10.27), $1 + i^* = \pi = \mu > 1$.

Assumption 4

$$
\frac{1 - f'(k^*)}{f'(k^*)} > i^*\eta_1(k^*)\left[1 - \gamma(k^*)\right]
\tag{10.35}
$$

Under Assumption 4, any bubble can be positive (see Eq. (10.34)). Note also that this assumption is satisfied by the example provided at the end of Sect. 10.4.2.

Proposition 1 *Let $\bar{k}$ be defined by $c\left(\bar{k}\right) = 0$ and $\underline{k} \in (0, \bar{k})$ by $f'(\underline{k}) = 1$. Under Assumptions 1–4, there exists a steady state characterized by $k^* \in \left(\underline{k}, \bar{k}\right)$ and $0 < \theta^* < f(k^*) - k^* - f'(k^*)k^*$.*

Proof See Appendix "Proof of Proposition 1".

Proposition 1 indicates that a steady state with a positive bubble exists. When collateral does not matter ($\eta_1(k) = 0$), we can see from Eq. (10.33) that the steady state is at the golden rule ($R(k) = 1$). As the well-known result of Tirole (1985), a positive rational asset bubble crowds out capital. When collateral matters ($\eta_1(k) > 0$), this economy experiences an over-accumulation of capital at the steady state ($R(k) < 1$). The existence of collateral incites households to hold more capital in their portfolio in order to relax the cash-in-advance constraint, and therefore, the capital return decreases.

Regarding the monetary policy, we recall that the central bank chooses the value of an existing steady state for its target. Since the steady state k^* always exists, we assume that the central bank selects this steady state as a target, i.e. $\pi^* = 1 + i^*$.[12]

---

[12]Indeed, our analysis does not exclude multiplicity of steady states. See Clain-Chamosset-Yvrard and Seegmuller (2013) for more details.

10.4.2 Normalized Steady State

In order to facilitate the analysis of local dynamics (Sect. 10.5), we establish the existence of a normalized steady state $k^* = 1$ (NSS). We follow the procedure introduced by Cazzavillan et al. (1998), and use the scaling parameter β to give conditions for the existence of such a steady state.

Assumption 5 Let $v(\eta_1) = i^*\eta_1(1)\left[1 - \gamma(1)\right]$, we assume:

$$f(1) > 1 + \frac{1 - f'(1)}{v(\eta_1)}$$

Assumption 5 ensures that the first period consumption at the normalized steady state is positive (i.e. $c(1) > 0$), and it is satisfied when the productivity is sufficiently large. Note that we show that a numerical example satisfies Assumption 5 at the end of Sect. 10.4.2.

Proposition 2 *Under Assumptions 1–5, there exists a unique value $\beta^* > 0$ given by*

$$\beta^* = \frac{u'\left[f(1) - 1 - (1 - f'(1))/v(\eta_1)\right]}{v'\left[(1 - f'(1))/v(\eta_1)\right]}[1 + i^*\gamma(1)]$$

such that $k^* = 1 \in \left(\underline{k}, \overline{k}\right)$ is a steady state of the dynamic system (10.28). Moreover, there is a positive bubble ($\theta^* > 0$) if $1 - f'(1)[1 + v(\eta_1)] > 0$.

Thereafter, we set $\beta = \beta^*$ so that $k^* = 1$. We further note $c^* \equiv c(1)$, $\gamma \equiv \gamma(1)$, $\eta_1 \equiv \eta_1(1)$, $\eta_2 \equiv \eta_2(1)$, $\psi \equiv f'(1)$, $\alpha = \alpha(1)$ and $\sigma = \sigma(1)$.

To convince the reader that all our assumptions leading to our results are compatible, consider the following example:

Example For a non-empty set of parameter values $(a, b, c, \epsilon, A, \alpha, \sigma)$, the function $\gamma(k)$ given by Eq. (10.7) in Sect. 10.2.1 and the production function $f(k) = A\left(\alpha k^{(\sigma-1)/\sigma} + 1 - \alpha\right)^{\sigma/(\sigma-1)}$ evaluated at the normalized steady state $k^* = 1$ fit all the requirements imposed by Assumptions 2–5.

Indeed, there exist critical values $\underline{\sigma}$, $\bar{\alpha}$, $\underline{A}$ and $\bar{A}$ such that for $a < 1$, $c > 1$, $b \in (ac, c)$, $\epsilon > 0$, $\sigma > \underline{\sigma}$, $\alpha \in (1 - \sigma, \bar{\alpha})$ and $A \in (\underline{A}, \bar{A})$, the function $\gamma(k)$ and the production function $f(k)$ evaluated at $k^* = 1$ satisfy Assumptions 2–5.

Proof See Appendix "Proof of Example in Section 10.4.2".

10.5 Expectation-Driven Fluctuations and the Stabilizing Role of Monetary Policy

We now study the emergence of expectation-driven fluctuations with a speculative bubble and an interplay between the financial and the real spheres. We show that when no Taylor rule is implemented ($\phi = 0$), local indeterminacy can occur in the neighborhood of the normalized steady state with a positive bubble under not restrictive conditions, namely gross substitutability and a not too weak input substitution, because of the credit market distortion. The violation of the Fisher relationship and the resulting portfolio choice between bonds, capital and money are the key ingredients to explain these fluctuations. When the Taylor rule is implemented ($\phi > 0$), we analyze the stabilizing role of the monetary policy. We will see that it is quite mitigated.

To study the existence of local indeterminacy, we introduce the following additional assumption:

Assumption 6 Let $\bar{\eta}_1 > 0$ and $\bar{\eta}_2 > 0$.[13] We assume $\eta_1 < \bar{\eta}_1$ and $\eta_2 > \bar{\eta}_2$.

Example Note that $\eta_1 = (b - ca)\epsilon/[(a + b)(1 + c)]$ and $\eta_2 = 1 + \epsilon[2c/(1 + c) - 1]$ at the normalized steady state $k^* = 1$. Recall that $a < 1, c > 1, b \in (ac, c)$ and $\epsilon > 0$.

For b close to ca, η_1 is close to zero, and thus $\eta_1 < \bar{\eta}_1$ is satisfied. As shown in Appendix "Proofs of Proposition 3, Corollaries 1 and 2" (see Eq. (10.54)), $\epsilon > \bar{\epsilon}$ is equivalent to $\eta_2 > \bar{\eta}_2$. Thus, the function $\gamma(k)$ evaluated at $k^* = 1$ satisfies Assumption 6.

To derive our different results, we start by linearizing the dynamic system (10.28) around the normalized steady state $k^* = 1$, and obtain the following lemma[14]:

Lemma 2 *Under Assumptions 1–6, the characteristic polynomial, evaluated at the steady state $k^* = 1$, writes $P(X) \equiv X^2 - TX + D = 0$, where T and D are respectively the trace and the determinant of the associated Jacobian matrix. We have:*

$$1 - T + D = \varepsilon_{dk} \frac{1 - \psi[1 + v(\eta_1)]}{\xi_1} \frac{\varepsilon_v - \varepsilon_v^s}{\varepsilon_v - \bar{\varepsilon}_v} \tag{10.36}$$

$$1 + T + D = \frac{\xi_3}{\xi_1} \frac{\varepsilon_v - \varepsilon_v^f}{\varepsilon_v - \bar{\varepsilon}_v} \tag{10.37}$$

$$1 - D = \frac{\varepsilon_v - \varepsilon_v^h}{\varepsilon_v - \bar{\varepsilon}_v} \tag{10.38}$$

---

[13]The expressions of $\bar{\eta}_1$ and $\bar{\eta}_2$ are given in Appendix "Proofs of Proposition 3, Corollaries 1 and 2".

[14]The proof of Lemma 2 is given in a technical appendix available on https://sites.google.com/site/liseclainchamosset.

where $\varepsilon_v \neq \bar{\varepsilon}_v$, and the expressions of $\xi_1 \equiv \xi_1(a_\phi)$, $\xi_3 \equiv \xi_3(a_\phi)$, $\bar{\varepsilon}_v \equiv \bar{\varepsilon}_v(a_\phi)$, $\varepsilon_v^f \equiv \varepsilon_v(a_\phi)$, $\varepsilon_v^h \equiv \varepsilon_v^h(a_\phi)$, ε_v^s and ε_{dk} are given in Appendix "Proofs of Proposition 3, Corollaries 1 and 2".

We recall that when $1 - T + D = 0$, one eigenvalue is equal to 1. When $1 + T + D = 0$, one eigenvalue is equal to -1. When $1 - T + D > 0$, $1 + T + D > 0$ and $D = 1$, the characteristic roots are complex conjugates with modulus equal to 1. All eigenvalues are inside the unit circle, when the following conditions are satisfied (i) $1 - T + D > 0$, (ii) $1 + T + D > 0$ and (iii) $D < 1$. In other words, when conditions (i)-(iii) are satisfied, the steady state is a sink, i.e. locally indeterminate. The steady state is a saddle point when $1 - T + D < 0$ (resp. > 0) and $1 + T + D > 0$ (resp. < 0). It is a source otherwise.

A (local) bifurcation arises when at least one eigenvalue crosses the unit circle. Therefore, a bifurcation occurs when either (iv) $1 - T + D = 0$, or (v) $1 + T + D = 0$, or (vi) $1 - T + D > 0$, $1 + T + D > 0$ and $D = 1$. According to a continuous change of ε_v, a pitchfork bifurcation emerges when ε_v goes through ε_v^s, defined by $1 - T + D = 0$.[15] A flip bifurcation occurs when ε_v goes through ε_v^f, defined by $1 + T + D = 0$. Finally, a Hopf bifurcation arises, as ε_v goes through ε_v^h, defined by $D = 1$, but we still keep $1 - T + D > 0$ and $1 + T + D > 0$.

The next proposition summarizes the local dynamic properties of the model:

Proposition 3 *Let $\varepsilon_v \neq \bar{\varepsilon}_v$, $\psi < 1/(1 + i^*\gamma)$ and Assumptions 1–6 hold.*

1. When $\varepsilon_u \in (\tilde{\varepsilon}_u, \varepsilon_u^s)$, the following holds:

- *if $a_\phi \in (-\infty, \hat{a}_\phi)$, then the steady state is locally determinate for $\varepsilon_v < max\{\varepsilon_v^h, \varepsilon_v^s\}$, undergoes a pitchfork (resp. Hopf) bifurcation for $\varepsilon_v = \varepsilon_v^s$ (resp. $\varepsilon_v = \varepsilon_v^h$), is locally indeterminate for $\varepsilon_v > max\{\varepsilon_v^h, \varepsilon_v^s\}$.*
- *if $a_\phi \in (\hat{a}_\phi, \bar{a}_\phi)$, then the steady state is locally indeterminate for $\varepsilon_v < \varepsilon_v^f$, undergoes a flip bifurcation for $\varepsilon_v = \varepsilon_v^f$, is locally determinate for $\varepsilon_v^f < \varepsilon_v < max\{\varepsilon_v^h, \varepsilon_v^s\}$, undergoes a pitchfork (resp. Hopf) bifurcation for $\varepsilon_v = \varepsilon_v^s$ (resp. $\varepsilon_v = \varepsilon_v^h$), is locally indeterminate for $\varepsilon_v > max\{\varepsilon_v^h, \varepsilon_v^s\}$.*
- *if $a_\phi \in (\bar{a}_\phi, -1[$, then the steady state is locally indeterminate for $\varepsilon_v < \varepsilon_v^s$, undergoes a pitchfork bifurcation for $\varepsilon_v = \varepsilon_v^s$, is locally determinate for $\varepsilon_v^s < \varepsilon_v < max\{\varepsilon_v^f, \varepsilon_v^h\}$, undergoes a flip (resp. Hopf) bifurcation for $\varepsilon_v = \varepsilon_v^f$ (resp. $\varepsilon_v = \varepsilon_v^h$), is locally indeterminate for $\varepsilon_v > max\{\varepsilon_v^f, \varepsilon_v^h\}$.*
- *if $a_\phi \in [0, \tilde{a}_\phi)$, then the steady state is locally indeterminate for $\varepsilon_v < \varepsilon_v^s$, undergoes a pitchfork bifurcation for $\varepsilon_v = \varepsilon_v^s$, is locally determinate for $\varepsilon_v \in (\varepsilon_v^s, \varepsilon_v^f)$, undergoes a flip bifurcation for $\varepsilon_v = \varepsilon_v^f$, is locally indeterminate for $\varepsilon_v > \varepsilon_v^f$.*
- *if $a_\phi \in (\tilde{a}_\phi, +\infty)$, then the steady state is locally indeterminate for $\varepsilon_v < \varepsilon_v^s$, undergoes a pitchfork bifurcation for $\varepsilon_v = \varepsilon_v^s$, is locally determinate for $\varepsilon_v > \varepsilon_v^s$.*

---

[15]Indeed, we have an odd number of steady states. We prove the existence of at least three steady states in the online technical appendix at https://sites.google.com/site/liseclainchamosset. See also Clain-Chamosset-Yvrard and Seegmuller (2013).

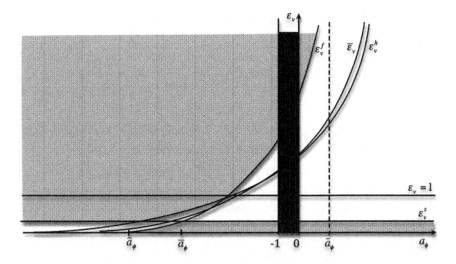

Fig. 10.1 The stabilizing role of monetary policy

2. *When $\varepsilon_u > \varepsilon_u^s$, the steady state is locally determinate for $\varepsilon_v < max\{\varepsilon_v^f, \varepsilon_v^h\}$, undergoes a flip (resp. Hopf) bifurcation for $\varepsilon_v = \varepsilon_v^f$ (resp. $\varepsilon_v = \varepsilon_v^h$), and is locally indeterminate for $\varepsilon_v > max\{\varepsilon_v^f, \varepsilon_v^h\}$.*

Proof See Appendix "Proofs of Proposition 3, Corollaries 1 and 2".

Figure 10.1 provides a qualitative illustration of the local dynamic properties of the model when $\varepsilon_v^s > 0$. Under Assumptions 1–6, $\bar{\varepsilon}_v$, ε_v^f, ε_v^h are increasing functions of a_ϕ, while ε_v^s does not depend on a_ϕ. Furthermore, $\bar{\varepsilon}_v$, ε_v^f, ε_v^h and ε_v^s intersect as indicated in Fig. 10.1 under some parameter conditions. Note that Fig. 10.1 is precisely constructed using Appendix "Proofs of Proposition 3, Corollaries 1 and 2". From Lemma 2 and plotting the different critical and bifurcation values, $\bar{\varepsilon}_v$, ε_v^s, ε_v^f, ε_v^h in the plane (a_ϕ, ε_v), we can determine the local indeterminacy regions corresponding to the grey areas in Fig. 10.1.

Proposition 3 provides general conditions for local dynamics. To clarify their significance, we start by considering the case where a Taylor rule is not implemented $(a_\phi = 0)$. This allows us to understand under which conditions expectation-driven fluctuations occur. In a second step, we will discuss the stabilizing role of a monetary policy managed by a Taylor rule $(a_\phi \in (-\infty, -1) \cup (0, +\infty))$.

From Proposition 3, we can derive the following corollary on the existence of expectation-driven fluctuations when $a_\phi = 0$:

Corollary 1 *Assuming $\varepsilon_v \neq \bar{\varepsilon}_v$ and Assumptions 1–6, the following holds when $a_\phi = 0$:*

1. *When $\varepsilon_u \in \left(\bar{\varepsilon}_u, \varepsilon_u^s\right)$, the steady state is locally indeterminate for $\varepsilon_v < \varepsilon_v^s < 1$, undergoes a pitchfork bifurcation for $\varepsilon_v = \varepsilon_v^s$, is locally determinate for $\varepsilon_v \in \left(\varepsilon_v^s, \varepsilon_v^f\right)$, undergoes a flip bifurcation for $\varepsilon_v = \varepsilon_v^f$, and is locally indeterminate for $\varepsilon_v > \varepsilon_v^f$.*
2. *When $\varepsilon_u > \varepsilon_u^s$, the steady state is locally determinate for $\varepsilon_v < \varepsilon_v^f$, undergoes a flip bifurcation for $\varepsilon_v = \varepsilon_v^f$, and is locally indeterminate for $\varepsilon_v > \varepsilon_v^f$.*

Proof See Appendix "Proofs of Proposition 3, Corollaries 1 and 2".

Corollary 1 shows the occurrence of persistent endogenous fluctuations around the steady state with a positive bubble under gross substitutability and a not too weak capital-labor substitution when no Taylor rule is implemented. Hence, this result extends Bosi and Seegmuller (2010) and Clain-Chamosset-Yvrard and Seegmuller (2015) to a model with both inside money and capital.[16]

When collateral does not matter ($\eta_1 = 0$), the local stability properties of the model correspond to Corollary 1.2. Endogenous fluctuations and two-period cycles occur only for a significant income effect, i.e. a large degree of concavity ε_v.[17] More interestingly, when collateral matters ($\eta_1 > 0$), local indeterminacy also appears under gross substitutability, i.e. for a small degree of concavity $\varepsilon_v < \varepsilon_v^s < 1$.

The basic mechanism for fluctuations under gross substitutability relies on a portfolio trade-off between the three assets. Because of the difference between the returns on physical and monetary assets, a reallocation between the assets takes place following a modification in agents' expectations.

Economic intuition. If households expect an increase in inflation from period t to $t + 1$, the return on bonds becomes less attractive compared to the return on capital. Because of the portfolio effect, households reallocate their savings towards capital. As a consequence, when $\varepsilon_v < \varepsilon_v^s < 1$, the portfolio effect can accelerate capital accumulation. Households consume less by cash (see Eq. (10.24)). The real balances m_{t+1} decrease, entailing a decrease in the return on money. An effective rise in inflation takes place, meaning that the initial expectations are self-fulfilling.

We focus now on the stabilizing role of the monetary policy managed by a Taylor rule. A policy is stabilizing in our framework as soon as it reduces the range of parameter values for which local indeterminacy, and thus expectation-driven fluctuations, emerge. Since the occurrence of fluctuations under gross substitutability is an interesting result, we ensure for the remainder of the paper that $\varepsilon_v^s > 0$, assuming[18]:

---

[16]In these two papers, the stabilizing role of monetary policies is not addressed in the same way as here. Indeed, the monetary authority directly manages the money growth factor, while it fixes the interest rate in our framework.

[17]Since $\varepsilon_v > \varepsilon_v^f > 1$, income effects dominate substitution effects. Hence, global savings ($\theta_t + k_{t+1}$) are a decreasing function of their return.

[18]ε_u^s is given by Eq. (10.45) in Appendix "Proofs of Proposition 3, Corollaries 1 and 2".

Assumption 7 $\varepsilon_u < \varepsilon_u^s$.

Since ε_u is a free preference parameter, Assumption 7 can always be satisfied and is in accordance with the numerical example provided at the end of Sect. 10.4.2.

We recall that since $a_\phi = \phi/(1 - \phi)$, $a_\phi \in (-\infty, -1)$ ($\phi > 1$) means that the monetary rule is active, while $a_\phi \in (0, +\infty)$ ($\phi \in (0, 1)$) means that the rule is passive. Using these notations, we derive the next corollary from Proposition 3 and Fig. 10.1:

Corollary 2 *Assuming $\varepsilon_v \neq \bar{\varepsilon}_v$, $\psi < 1/(1 + i^*\gamma)$ and Assumptions 1–7, the following holds:*

1. *If $a_\phi \in (-\infty, \hat{a}_\phi]$, increasing the responsiveness of the monetary rule ϕ does not affect or reduces the range of parameters for indeterminacy.*
2. *If $a_\phi \in (\hat{a}_\phi, -1)$, increasing the responsiveness of the monetary rule ϕ reduces the range of parameters for local indeterminacy when ε_v is large enough, but raises the range of parameters for local indeterminacy when ε_v is small enough.*
3. *If $a_\phi \in [0, +\infty)$, increasing the responsiveness of the monetary rule ϕ reduces the range of parameters for local indeterminacy when ε_v is large enough, but has no impact on the range of parameters for local indeterminacy when ε_v is small enough.*

Proof See Appendix "Proofs of Proposition 3, Corollaries 1 and 2".

Proposition 3, Corollary 2 and Fig. 10.1 highlight that a monetary policy managed by a Taylor rule has mitigated results concerning the stabilization of expectation-driven fluctuations. Comparing with the results of Corollary 1, a passive rule ($a_\phi \in (0, +\infty)$) stabilizes fluctuations occurring for a large level of ε_v ($\varepsilon_v > \varepsilon_v^f$). However, it has no impact on fluctuations occurring for $\varepsilon_v < \varepsilon_v^f$. On the contrary, an active rule ($a_\phi \in (-\infty, -1)$) could stabilize endogenous fluctuations that occur for $\varepsilon_v < \varepsilon_v^s$, in particular if it is weakly active (a_ϕ sufficiently negative i.e. ϕ close to one). Nevertheless, with respect to the configuration without Taylor rule ($a_\phi = 0$), an active rule destabilizes promoting indeterminacy for new ranges of parameter values. Indeed, under a weakly active rule (a_ϕ sufficiently negative), local indeterminacy occur for all $\varepsilon_v > \varepsilon_v^s$.

To sum, depending on households' preferences (ε_v) and on the responsiveness of the monetary policy with respect to expected inflation (a_ϕ), we may have opposite conclusions concerning the stabilizing role of the monetary policy. This is in contrast with several previous contributions, for instance Bernanke and Woodford (1997), Sorger (2005) or Clain-Chamosset-Yvrard and Seegmuller (2015).

Our explanation of these results is that the monetary authority manipulates the level of the elasticity of the nominal interest rate with respect to the expected inflation (ϕ), which has not a sufficiently significant impact on the interest rate level to dampen or eliminate the impact of the credit market imperfection and the role of collateral. For instance, a weakly active rule has not a huge impact on the nominal interest rate, and therefore does not sufficiently modify the portfolio choices.

10.6   Concluding Remarks

We develop an overlapping generations model with capital accumulation, bonds and money, where the share of consumption purchased on credit depends on a collateral. We show the existence of expectation-driven fluctuations with a positive rational bubble on bonds. In addition, such endogenous fluctuations are in accordance with gross substitutability and a not too weak substitution between inputs. This is explained by a credit market imperfection and the role of a collateral. The basic mechanism for fluctuations relies on a portfolio trade-off between the three assets.

We further analyze the stabilizing role of the monetary policy. To consider an usual policy rule, we focus on a monetary policy fixed according to a Taylor rule on expected inflation. We show that the stabilizing role of such a policy is mitigated. In fact, no clear-cut conclusion on stabilisation is obtained. One reason is that such a policy does not alter sufficiently the portfolio choices. More generally, we think that our results provide an adding argument against the use of standard policy instruments in any circumstances. In this model, these circumstances are the existence of a bubble, a credit market imperfection and the collateral effect.

Appendix

Proof of Proposition 1

A steady state k is a solution of $h\,(k) = j\,(k)$, with:

$$h\,(k) \equiv \frac{u'\,(c(k))}{\beta v'\,(d(k))} \tag{10.39a}$$

$$j\,(k) \equiv \frac{1}{1 + i^* \gamma\,(k)} \tag{10.39b}$$

where $c\,(k) \equiv f(k) - k - d(k)$ and $d\,(k) \equiv \dfrac{k[1 - f'(k)]}{i^* \eta_1(k)[1 - \gamma(k)]}$.

We start by determining the admissible range of values for k. To ensure $d\,(k) > 0$, we get at the steady state $f'\,(k) < 1$. Under Assumption 3, $f'(k)$ is a decreasing function of k. Hence, $k > f'^{-1}\,(1) = \underline{k}$.

Now, we want to determine the range of k such that $c(k) > 0$. The decreasing returns on capital imply $f\,(\underline{k}) > \underline{k}f'(\underline{k})$. Since $f'(\underline{k}) = 1$, $c(\underline{k}) = f\,(\underline{k}) - \underline{k} > 0$. In addition, as $d(k) > 0$, we derive the following inequality:

$$\lim_{k \to +\infty} c\,(k) < \lim_{k \to +\infty} f(k) - k = -\infty$$

because $f'(k) < 1$ for k large enough. As a result, there exists one value $\bar{k}$ such that $\forall k < \bar{k},\, c\left(\bar{k}\right) > 0$. By construction, we have $\underline{k} < \bar{k}$, and therefore $(\underline{k}, \bar{k})$ is a nonempty subset.

To prove the existence of a stationary solution k, we use the continuity of $h(k)$ and $j(k)$. Using Eqs. (10.39a) and (10.39b), we determine the boundary values of $h(k)$ and $j(k)$:

$$\lim_{k \to \underline{k}} h(k) = \frac{u'\left(c\left(\underline{k}\right)\right)}{\beta v'(0)} = 0^+ \qquad \lim_{k \to \bar{k}} h(k) = \frac{u'(c(0))}{\beta v'(d(\bar{k}))} = +\infty$$

$$\lim_{k \to \underline{k}} j(k) = \frac{1}{1 + i(\underline{k})\gamma(\underline{k})} \in (0, 1] \quad \lim_{k \to \bar{k}} j(k) = \frac{1}{1 + i(\bar{k})\gamma(\bar{k})} \le 1$$

We have $\lim_{k \to \underline{k}} h(k) < \lim_{k \to \underline{k}} j(k)$ and $\lim_{k \to \bar{k}} h(k) > \lim_{k \to \bar{k}} j(k)$. Therefore, there exists at least one value $k^* \in (\underline{k}, \bar{k})$ such that $h(k^*) = j(k^*)$. ∎

Proof of Example in Section 10.4.2

Let $\underline{\sigma} \equiv 1 - \frac{1}{2 + v(\eta_1)}$, $\bar{\alpha} \equiv \frac{1}{2 + v(\eta_1)}$, $\underline{A} \equiv \frac{1 + v(\eta_1)}{\alpha + v(\eta_1)}$ and $\bar{A} \equiv \frac{1/\alpha}{1 + v(\eta_1)}$. For $a \in (0, 1)$, $c > 1$, $b \in (ac, c)$ and $\epsilon > 0$, Assumption 2 is satisfied at the normalized steady state. Assumption 3 requires $A < 1/\alpha$ and $\sigma > 1 - \alpha$. For $A > \underline{A}$, Assumption 5 is satisfied. Moreover, the bubble is positive at the normalized steady state when $A < \bar{A}$. As a consequence, the set $(\underline{A}, \bar{A})$ must be non-empty. This is true for $\alpha < \bar{\alpha}$ and $\sigma > \underline{\sigma}$.

Proofs of Proposition 3, Corollaries 1 and 2

Recall that $\psi = f'(1)$, $v(\eta_1) = \eta_1 i^*(1 - \gamma) > 0$, $c^* = f(1) - 1 - \frac{1 - \psi}{v(\eta_1)} > 0$, (10.40)

$$a_\phi = \phi/(1 - \phi) \in (-\infty, -1[\cup [0, +\infty) \text{ and let } \varepsilon_{dk} = \frac{\psi}{1 - \psi}\frac{1 - \alpha}{\sigma} + \eta_2 > 0, \text{ and}$$

$$\eta_1^\theta = \frac{1 - \psi}{\psi}\frac{1}{i^*(1 - \gamma)}.$$

First of all, we suppose for the rest of the proof that $\eta_1 < \eta_1^\theta$, because $\forall \eta_1 < \eta_1^\theta$, $\theta > 0$ (see condtion (10.35)).

From Lemma 2, the expressions of $1 - T + D$, $1 + T + D$ and $1 - D$ can be written as follows:

$$1 - T + D = \varepsilon_{dk} \frac{1 - \psi \left[1 + v(\eta_1)\right]}{\xi_1(a_\phi)} \frac{\varepsilon_v - \varepsilon_v^s}{\varepsilon_v - \bar{\varepsilon}_v} \tag{10.41}$$

$$1 + T + D = \frac{\xi_3(a_\phi)}{\xi_1(a_\phi)} \frac{\varepsilon_v - \varepsilon_v^f}{\varepsilon_v - \bar{\varepsilon}_v} \tag{10.42}$$

$$1 - D = \frac{\varepsilon_v - \varepsilon_v^h}{\varepsilon_v - \bar{\varepsilon}_v} \tag{10.43}$$

where $\varepsilon_v^s = \dfrac{1 - \psi}{v(\eta_1)} \dfrac{v(\eta_1) + \varepsilon_{dk}}{\varepsilon_{dk}} \dfrac{\varepsilon_u^s - \varepsilon_u}{c^*}$, $\quad \varepsilon_v^f = \dfrac{\xi_4(a_\phi)}{\xi_3(a_\phi)} \equiv \varepsilon_v^f(a_\phi)$,

$\varepsilon_v^h = \dfrac{\xi_5(a_\phi)}{\xi_1(a_\phi)} \equiv \varepsilon_v^h(a_\phi)$ and $\bar{\varepsilon}_v = -\dfrac{\xi_2(a_\phi)}{\xi_1(a_\phi)} \equiv \bar{\varepsilon}_v(a_\phi)$ $\tag{10.44}$

with $\varepsilon_u^s = c^* \dfrac{v(\eta_1)}{1 - \psi} \dfrac{1}{v(\eta_1) + \varepsilon_{dk}} \dfrac{v(\eta_1)}{1 + i^*\gamma}$, $\tag{10.45}$

$$\xi_1(a_\phi) = -\psi v(\eta_1) \left(1 - \frac{1-\alpha}{\sigma}\right) \frac{1+i^*}{i^*} a_\phi + \{1 - \psi \left[1 + v(\eta_1)\right]\} \varepsilon_{dk}$$
$$- \frac{\psi}{1-\psi} v(\eta_1) \left(1 - \frac{1-\alpha}{\sigma}\right),$$
$$\tag{10.46}$$

$$\xi_2(a_\phi) = (1 - \psi) \frac{1+i^*}{i^*} \left\{ -\frac{\varepsilon_u}{c^*} + \frac{v(\eta_1)}{1+i^*\gamma} \left[1 + i^*\gamma \frac{\psi}{1-\psi} \left(1 - \frac{1-\alpha}{\sigma}\right)\right] \right.$$
$$\left. - \varepsilon_{dk} \frac{i^*\gamma}{1+i^*\gamma} \right\} a_\phi - \psi \left[1 + v(\eta_1)\right] \frac{\varepsilon_u}{c^*} + v(\eta_1)^2 \frac{\psi}{1+i^*\gamma} + v(\eta_1)\psi$$
$$\left(1 - \frac{1-\alpha}{\sigma} + \frac{1}{1+i^*\gamma}\right) - (1 - \psi) \varepsilon_{dk},$$
$$\tag{10.47}$$

$$\xi_3(a_\phi) = -\psi v(\eta_1) \left(1 - \frac{1-\alpha}{\sigma}\right) \frac{1+i^*}{i^*} a_\phi + \{1 - \psi \left[1 + v(\eta_1)\right]\} \varepsilon_{dk}$$
$$- 2 \frac{\psi}{1-\psi} v(\eta_1) \left(1 - \frac{1-\alpha}{\sigma}\right),$$
$$\tag{10.48}$$

$$\xi_4(a_\phi) = 2(1 - \psi) \frac{1+i^*}{i^*} \left\{ \left(1 + \psi \frac{1-\alpha}{\sigma}\right) \frac{\varepsilon_u}{c^*} - \frac{i^*\gamma v(\eta_1)}{1+i^*\gamma} \left[1 + \frac{\psi}{1-\psi} \left(1 - \frac{1-\alpha}{\sigma}\right)\right] \right.$$
$$\left. + \frac{\varepsilon_{dk} i^*\gamma}{1+i^*\gamma} \right\} a_\phi + \left\{ 2\psi \left[1 + v(\eta_1) + \frac{1-\alpha}{\sigma}\right] + \right.$$
$$\left. \frac{1-\psi}{v(\eta_1)} \left[v(\eta_1) + \varepsilon_{dk}\right] \{1 - \psi \left[1 + v(\eta_1)\right]\} \right\} \frac{\varepsilon_u}{c^*} - \frac{v(\eta_1)^2 \psi}{1+i^*\gamma}$$
$$- 2v(\eta_1)\psi \left(1 - \frac{1-\alpha}{\sigma}\right) - v(\eta_1) \frac{1+\psi}{1+i^*\gamma} + 2(1 - \psi)\varepsilon_{dk}, \tag{10.49}$$

$$\xi_5(a_\phi) = (1 - \psi) \frac{1 + i^*}{i^*} \left\{ \left(1 - \psi \frac{1 - \alpha}{\sigma} \right) \frac{\varepsilon_u}{c^*} - \frac{v(\eta_1)}{1 + i^*\gamma} \right.$$

$$\left. \left[1 + \frac{i^*\gamma\psi}{1 - \psi} \left(1 - \frac{1 - \alpha}{\sigma} \right) \right] + \frac{\varepsilon_{dk} i^*\gamma}{1 + i^*\gamma} \right\} a_\phi + \psi \left[1 + v(\eta_1) - \frac{1 - \alpha}{\sigma} \right] \frac{\varepsilon_u}{c^*} - \frac{v(\eta_1)^2 \psi}{1 + i^*\gamma}$$

$$- v(\eta_1)\psi \left(1 - \frac{1 - \alpha}{\sigma} + \frac{1}{1 + i^*\gamma} \right) + (1 - \psi)\varepsilon_{dk}$$

$$(10.50)$$

We aim to determine the range of parameter values (a_ϕ and ε_v) for which local indeterminacy conditions (i)-(iii) are satisfied. To do this, we must analyze the functions ε_v^f, ε_v^h, $\bar\varepsilon_v$ and ξ_i with $i = \{1, 2, 3, 4, 5\}$, then draw ε_v^f, ε_v^h and $\bar\varepsilon_v$ in the plane (a_ϕ, ε_v).

We observe that ξ_i with $i = \{1, 2, 3, 4, 5\}$ are linear functions of a_ϕ, i.e. $\xi_i^a a_\phi + \xi_i^b$.

Note that $\xi_1^a < 0$ and $\xi_3^a < 0$. Furthermore, there exist $\eta_1^{\xi_1^b} > 0$, $\eta_1^{\xi_2^a} > 0$, $\eta_1^{\xi_2^b} > 0$, $\eta_1^{\xi_3^b} > 0$, $\eta_1^{\xi_4^a} > 0$, $\eta_1^{\xi_4^b} > 0$, $\eta_1^{\xi_5^a} > 0$ and $\eta_1^{\xi_5^b} > 0$ such that $\forall$ $\eta_1 < min\{\eta_1^{\xi_1^b}, \eta_1^{\xi_2^a}, \eta_1^{\xi_2^b}, \eta_1^{\xi_3^b}, \eta_1^{\xi_4^a}, \eta_1^{\xi_4^b}, \eta_1^{\xi_5^a}, \eta_1^{\xi_5^b}\} \equiv \tilde\eta_1$, one has $\xi_1^b > 0$, $\xi_2^a < 0$, $\xi_2^b < 0$, $\xi_3^b > 0$, $\xi_4^a > 0$, $\xi_4^b > 0$, $\xi_5^a > 0$ and $\xi_5^b > 0$. Therefore, we deduce that $\xi_1(a_\phi) \geq 0$ when $a_\phi \leq \frac{\xi_1^b}{\xi_1^a}$ and $\xi_1(a_\phi) < 0$ otherwise. $\xi_2(a_\phi) \geq 0$ when $a_\phi \leq \frac{\xi_2^b}{\xi_2^a} < 0$ and $\xi_2(a_\phi) < 0$ otherwise. $\xi_3(a_\phi) \geq 0$ when $a_\phi \leq \frac{\xi_3^b}{\xi_3^a}$ and $\xi_3(a_\phi) < 0$ otherwise. $\xi_4(a_\phi) \leq 0$ when $a_\phi \leq \frac{\xi_4^b}{\xi_4^a} < 0$ and $\xi_4(a_\phi) > 0$ otherwise. $\xi_5(a_\phi) \leq 0$ when $a_\phi \leq \frac{\xi_5^b}{\xi_5^a} < 0$ and $\xi_5(a_\phi) > 0$ otherwise.

We analyze now ε_v^f, ε_v^h, ε_v^s and $\bar\varepsilon_v$. Suppose that $\eta_1 < min\{\eta_1^\theta, \tilde\eta_1\}$.

First, ε_v^s does not depend on a_ϕ. Second, the different critical and bifurcation values ($\bar\varepsilon_v$, ε_v^f, ε_v^h) are homographic functions of a_ϕ. ε_v^f has a vertical asymptote at $a_\phi = \frac{\xi_3^b}{\xi_3^a} \equiv \tilde a_\phi > 0$. $\bar\varepsilon_v$ and ε_v^h have the same vertical asymptote at $a_\phi = \frac{\xi_1^b}{\xi_1^a} > \tilde a_\phi$. The first derivatives of ε_v^f, ε_v^h and $\bar\varepsilon_v$ with respect to a_ϕ are given by:

$$\frac{\partial \varepsilon_v^f}{\partial a_\phi} = \frac{\xi_4^a \xi_3^b - \xi_4^b \xi_3^a}{\xi_3(a_\phi)^2} > 0, \quad \frac{\partial \varepsilon_v^h}{\partial a_\phi} = \frac{\xi_5^a \xi_1^b - \xi_5^b \xi_1^a}{\xi_1(a_\phi)^2} > 0 \text{ and } \frac{\partial \bar\varepsilon_v}{\partial a_\phi} = -\frac{\xi_2^a \xi_1^b - \xi_2^b \xi_1^a}{\xi_1(a_\phi)^2} > 0.$$

It would be useful to locate the different bifurcation and critical values (ε_v^f, ε_v^h, ε_v^s and $\bar\varepsilon_v$) when $a_\phi = 0$. We can note that $\varepsilon_v^f > 0$, $\varepsilon_v^h > 0$ and $\bar\varepsilon_v > 0$ when $a_\phi = 0$. Moreover, $\varepsilon_v^s > 0$ if and only if $\varepsilon_u \in (\tilde\varepsilon_u, \varepsilon_u^s)$, where $\varepsilon_u > \tilde\varepsilon_u$ is required for the second-order conditions, with:

$$\tilde\varepsilon_u \equiv c^* \frac{v(\eta_1)^2}{\eta_2 (1 - \psi)(1 + i^*\gamma)^2} \qquad (10.51)$$

The set $(\tilde\varepsilon_u, \varepsilon_u^s)$ is nonempty if and only if

$$\eta_2(1 - \psi)i^*\gamma > v(\eta_1)(1 - \psi) + \psi \frac{1 - \alpha}{\sigma} \qquad (10.52)$$

As $v(\eta_1) = i^*\eta_1(1-\gamma)$, the condition (10.52) holds if

$$\eta_1 < \frac{1}{i^*(1-\gamma)}\frac{\eta_2(1-\psi)i^*\gamma - \psi(1-\alpha)/\sigma}{1-\psi} \equiv \bar{\bar{\eta}}_1 \text{ and } \eta_2 > \frac{\psi}{1-\psi}\frac{(1-\alpha)/\sigma}{i^*\gamma} \equiv \bar{\eta}_2 \quad (10.53)$$

Therefore, $\forall \eta_1 < \bar{\bar{\eta}}_1$ and $\eta_2 > \bar{\eta}_2$, we have $\tilde{\varepsilon}_u < \varepsilon_u^s$. We suppose now that $\eta_1 < \min\{\eta_1^\theta, \tilde{\eta}_1, \bar{\bar{\eta}}_1\}$ and $\eta_2 > \bar{\eta}_2$. Note that using the function $\gamma(k)$ given by Eq. (10.7), $\eta_2 > \bar{\eta}_2$ is equivalent to:

$$\epsilon > \frac{1+c}{c-1}\left(\frac{\psi}{1-\psi}\frac{1-\alpha}{\sigma i\gamma} - 1\right) \equiv \bar{\epsilon} \quad (10.54)$$

where $\gamma = 1 - (a+b)/(1+c)$.

We can show that $\varepsilon_v^s < 1$. Since $\varepsilon_{dk} = \frac{\psi}{1-\psi}\frac{1-\alpha}{\sigma} + \eta_2$ and the condition (10.52) is satisfied, we have $v(\eta_1) < (1+i^*\gamma)\varepsilon_{dk}$. Therefore, $\varepsilon_v^s = \frac{v(\eta_1)}{1+i^*\gamma}\frac{1}{\varepsilon_{dk}} - \frac{\varepsilon_u}{c^*}\frac{v(\eta_1)+\varepsilon_{dk}}{\varepsilon_{dk}}$ $\frac{1-\psi}{v(\eta_1)} < 1$.

Furthermore, $\varepsilon_v^h < \bar{\varepsilon}_v$ and $\varepsilon_v^h > 1$ for a sufficiently small η_1. Indeed $\varepsilon_v^h > 1$ is equivalent to $\psi Q(v(\eta_1)) > 0$, where $Q(v(\eta_1))$ is a quadratic polynomial defined on $\mathbb{R}_+$ such that:

$$Q(v(\eta_1)) = -\frac{v(\eta_1)^2}{1+i^*\gamma} + v(\eta_1)\left[\frac{\varepsilon_u}{c^*} - \left(1 - \frac{1-\alpha}{\sigma} + \frac{1}{1+i\gamma}\right)\right.$$
$$\left. + \varepsilon_{dk} + \frac{1-(1-\alpha)/\sigma}{1-\psi}\right] + \left(1 - \frac{1-\alpha}{\sigma}\right)\frac{\varepsilon_u}{c^*}$$

$Q(v(\eta_1))$ is a concave function with $Q(v(0)) > 0$. As a consequence, there is a threshold $\hat{\eta}_1 > 0$ such that $\forall \eta_1 < \hat{\eta}_1, Q(v(\eta_1)) > 0$.

Concerning ε_v^f, we can show that $\bar{\varepsilon}_v < \varepsilon_v^f$ for η_1 small enough. Note that $\varepsilon_v^f > \xi_4^b/\xi_1^b \cdot \xi_4^b/\xi_1^b > \bar{\varepsilon}_v$ is satisfied if $-v(\eta_1)\left[\psi\left(1-\frac{1-\alpha}{\sigma}\right) + \frac{1}{1+i^*\gamma}\right] + (1-\psi)\varepsilon_{dk} > 0$. Therefore, there exists $\underline{\eta}_1 > 0$ such that $\forall \eta_1 < \underline{\eta}_1$, this inequality is satisfied. Hence, $\forall \eta_1 < \underline{\eta}_1, \bar{\varepsilon}_v < \varepsilon_v^f$.

Therefore, $\forall \eta_1 < \min\{\eta_1^\theta, \tilde{\eta}_1, \bar{\bar{\eta}}_1, \hat{\eta}_1, \underline{\eta}_1\}$, one has $\varepsilon_v^s < 1 < \varepsilon_v^h < \bar{\varepsilon}_v < \varepsilon_v^f$ when $a_\phi = 0$. Moreover, we can show that there exists $\eta_1' > 0$ such that $\forall \eta_1 < \eta_1'$, $\varepsilon_v^s < \varepsilon_v^h < \bar{\varepsilon}_v < \varepsilon_v^f$ when $a_\phi = -1$. For the rest of the proof, we assume that $\eta_1 < \min\{\eta_1^\theta, \tilde{\eta}_1, \bar{\bar{\eta}}_1, \hat{\eta}_1, \underline{\eta}_1, \eta_1'\}$.

Recall that a_ϕ is defined on $(-\infty, -1) \cup [0, +\infty)$. At this stage of the proof, we can state by analyzing $1 - T + D$, $1 + T + D$ and $1 - D$ given by Eqs. (10.41)–(10.43) that if $a_\phi \in (-\infty, -1)$, local indeterminacy occurs when $\varepsilon_v < \min\{\bar{\varepsilon}_v, \varepsilon_v^f, \varepsilon_v^h, \varepsilon_v^s\}$ or when $\varepsilon_v > \max\{\bar{\varepsilon}_v, \varepsilon_v^f, \varepsilon_v^h, \varepsilon_v^s\}$. From the previous results, we deduce that if $a_\phi \in [0, \tilde{a}_\phi)$, local indeterminacy occurs when $\varepsilon_v < \varepsilon_v^s$ or when $\varepsilon_v > \varepsilon_v^f$.

Finally, if $a_\phi > \tilde{a}_\phi$, local indeterminacy occurs when $\varepsilon < \varepsilon_v^s$. For the case $a_\phi \in (-\infty, -1)$, we should determine the location of ε_v^f, ε_v^h, ε_v^s and $\bar{\varepsilon}_v$ in the plane (a_ϕ, ε_v).

The functions ε_v^f, ε_v^h and $\bar{\varepsilon}_v$ are continuous and monotone increasing on $a_\phi \in (-\infty, -1)$. We can show that the graph of these functions cross the horizonntal axis on $a_\phi \in (-\infty, -1)$. Let introduce the different following points $a_\phi^{\xi_2}$, $a_\phi^{\xi_4}$ and $a_\phi^{\xi_5}$, which corresponds to the points at which $\bar{\varepsilon}_v$, ε_v^f and ε_v^h cross the horizontal axis. $a_\phi^{\xi_2}$ is defined by $\bar{\varepsilon}_v = 0$ such that $a_\phi^{\xi_2} = -\frac{\xi_2^b}{\xi_2^a} < 0$. $a_\phi^{\xi_4}$ is defined by $\varepsilon_v^f = 0$ such that $a_\phi^{\xi_4} = -\frac{\xi_4^b}{\xi_4^a} < 0$. $a_\phi^{\xi_5}$ is defined by $\varepsilon_v^h = 0$ such that $a_\phi^{\xi_5} = -\frac{\xi_5^b}{\xi_5^a} < 0$.

After some algebra, we can show that since $\psi < 1/(1 + i^*\gamma)$, there exists $\eta_1^a > 0$ such that $\forall \eta_1 < \eta_1^a$, $a_\phi^{\xi_5} < a_\phi^{\xi_2}$. Furthermore, either $a_\phi^{\xi_2} < a_\phi^{\xi_4} \, \forall \eta_1 > 0$ or there exists η_1^b such that $\forall \eta_1 < \eta_1^b > 0$, $a_\phi^{\xi_2} < a_\phi^{\xi_4}$. Hence, if $\eta_1 < min\{\eta_1^a, \eta_1^b, \eta_1^\theta, \tilde{\eta}_1, \overline{\bar{\eta}}_1, \hat{\eta}_1, \underline{\eta}_1, \eta_1'\}$, one has $a_\phi^{\xi_5} < a_\phi^{\xi_2} < a_\phi^{\xi_4} < 0$.

Let $\eta_1'' = min\{\eta_1^a, \eta_1^b, \eta_1^\theta, \tilde{\eta}_1, \overline{\bar{\eta}}_1, \hat{\eta}_1, \underline{\eta}_1, \eta_1'\}$. For the rest of the proof, we suppose that $\eta_1 < \eta_1''$.

Since the functions ε_v^f, ε_v^h and $\bar{\varepsilon}_v$ are continuous and monotone increasing on $(-\infty, -1) \cup [0, \tilde{a}_\phi)$, we can now locate ε_v^f, ε_v^h and $\bar{\varepsilon}_v$ in the plane (a_ϕ, ε_v).

Let $\hat{a}_\phi \equiv a_\phi^{\xi_4}$. Since $\psi < 1/(1 + i^*\gamma)$, $a_\phi^{\xi_5} < a_\phi^{\xi_2} < \hat{a}_\phi$. Using the expressions of $1 - T + D$, $1 + T + D$ and $1 - D$ given by Eqs. (10.41)–(10.43), we state that if $a_\phi \in (-\infty, \hat{a}_\phi)$, local indeterminacy occurs when $\varepsilon_v > max\{\varepsilon_v^f, \varepsilon_v^h, \varepsilon_v^s\}$. If $a_\phi \in [\hat{a}_\phi, -1) \cup [0, \tilde{a}_\phi)$, local indeterminacy occurs when $\varepsilon_v < min\{\varepsilon_v^f, \varepsilon_v^s\}$ or when $\varepsilon_v > max\{\varepsilon_v^f, \varepsilon_v^h, \varepsilon_v^s\}$.

We have shown that ε_v^s can be positive. In such a case, it would be useful to determine when ε_v^f, ε_v^h and $\bar{\varepsilon}_v$ cross ε_v^s. Suppose that $\varepsilon_v^s > 0$. $\varepsilon_v^h = \varepsilon_v^s$ when $a_\phi = -\left(\varepsilon_v^s \xi_1^b - \xi_5^b\right)/(\varepsilon_v^s \xi_1^a - \xi_5^a) \equiv a_\phi^1 < 0$, $\bar{\varepsilon}_v = \varepsilon_v^s$ when $a_\phi = -\left(\varepsilon_v^s \xi_1^b + \xi_2^b\right)/(\varepsilon_v^s \xi_1^a + \xi_2^a) \equiv a_\phi^2 < 0$, and $\varepsilon_v^f = \varepsilon_v^s$ when $a_\phi = -\left(\varepsilon_v^s \xi_3^b - \xi_4^b\right)/(\varepsilon_v^s \xi_3^a - \xi_4^a) \equiv a_\phi^3 < 0$.

Because $\psi < 1/(1 + i^*\gamma)$, we can show after some algebra that either $a_\phi^1 < a_\phi^2 \, \forall \eta_1 > 0$ or there exists $\eta_1^e > 0$ such that $\forall \eta_1 < \eta_1^e$, $a_\phi^1 < a_\phi^2$. It is difficult to determine the location of a_ϕ^3. Nevertheless, if $a_\phi^1 < a_\phi^3 < a_\phi^2 < 0$ or if $a_\phi^3 < a_\phi^1 < a_\phi^2 < 0$, we get $1 - T + D < 0$, $1 + T + D < 0$ and $D > 1$ for some values of ε_v. Since this is not feasible, we can eliminate these configurations. Therefore, if $\eta_1 < min\{\eta_1^e, \eta_1''\}$, one has $a_\phi^1 < a_\phi^2 < a_\phi^3 < 0$.

Let $\bar{\eta}_1 = min\{\eta_1^e, \eta_1''\}$. All conditions on η_1 required in this proof are satisfied when $\eta_1 < \bar{\eta}_1$. We can now derive the dynamic properties of the model. The properties of local dynamics are depicted by Fig. 10.1.[19] Grey areas in Fig. 10.1 correspond to the different regions in which the steady state is a sink, in other words to the indeterminacy regions.

---

[19]Figure 10.1 depicts the dynamic properties of the model with $\varepsilon_u \in (\bar{\varepsilon}_u, \varepsilon_u^s)$. The configuration with $\varepsilon_u > \varepsilon_u^s$ can be easily deduce from the former one.

We deduce Proposition 3 and Corollary 1 from Fig. 10.1. Since $a_\phi = \phi/(1-\phi)$ is increasing with ϕ, we can derive Corollary 2 from Proposition 3 and Fig. 10.1. ∎

References

Benhabib, J., & Laroque, G. (1988). On competitive cycles in productive economies. *Journal of Economic Theory, 45*, 145–170.

Benhabib, J., Schmitt-Grohé, S., & Uribe, M. (2001). The perils of Taylor rules. *Journal of Economic Theory, 96*, 40–69.

Bernanke, B. (2010). Monetary policy and the housing bubble by Ben S. Bernanke Chairman, Speach at the Annual Meeting of the American Economic Association, Atlanta, Georgia, January 3.

Bernanke, B., & Woodford, M. (1997). Inflation forecasts and monetary policy. *Journal of Money, Credit and Banking, 29*, 653–84.

Bosi, S., & Seegmuller, T. (2010). On rational exuberance. *Mathematical Social Sciences, 59*, 249–270.

Campbell, J., & Mankiw, G. (1989). Consumption, income and interest rates: Reinterpreting the time series evidence. In O. Blanchard & S. Fisher (Eds.), *NBER macroeconomics annual* (pp. 185–216). Cambridge MA: MIT Press.

Cazzavillan, G., Lloyd-Braga, T., & Pintus, P. (1998). Multiple steady states and endogenous fluctuations with increasing returns to scale in production. *Journal of Economic Theory, 80*, 60–107.

Cazzavillan, G., & Pintus, P. (2005). On competitive cycles and sunspots in productive economies with a positive money stock. *Research in Economics, 59*, 137–147.

Clain-Chamosset-Yvrard, L., & Seegmuller, T. (2013). *The stabilizing virtues of fiscal vs. monetary policy on endogenous bubble fluctuations*, AMSE Working Paper 2013-43.

Clain-Chamosset-Yvrard, L., & Seegmuller, T. (2015). Rational bubbles and macroeconomic fluctuations: The (de-)stabilizing role of monetary policy policy. *Mathematical Social Sciences, 79*, 1–15.

Diamond, P. A. (1965). National debt in a neoclassical growth model. *American Economic Review, 55*, 1126–1150.

Farhi, E., & Tirole, J. (2012). Bubbly liquidity. *Review of Economic Studies, 79*, 678–706.

Farmer, R. E. A. (1986). Deficits and cycles. *Journal of Economic Theory, 40*, 77–88.

Grandmont, J.-M. (1985). On endogenous competitive business cycles. *Econometrica, 53*, 995–1045.

Grandmont, J.-M. (1986). Stabilizing competitive business cycles. *Journal of Economic Theory, 40*, 57–76.

Hahn, F., & Solow, R. (1995). *A critical essay on modern macroeconomic theory*. Oxford: Basil Blackwell.

Iacoviello, M. (2004). Consumption, house prices, and collateral constraints: A structural econometric analysis. *Journal of Housing Economics, 13*, 304–320.

Jappelli, T. (1990). Who is credit constrained in the US economy? *Quarterly Journal of Economics, 105*, 219–234.

Kiyotaki, N., & Moore, J. (1997). Credit cycles. *Journal of Political Economy, 105*, 211–248.

LeRoy, S., & Porter, R. (1981). The present value relation: Tests based on variance bounds. *Econometrica, 49*, 555–584.

Martin, A., & Ventura, J. (2012). Economic growth with bubbles bubbles. *American Economic Review, 102*, 3033–3058.

Michel, P., & Wigniolle, B. (2003). Temporary bubbles. *Journal of Economic Theory, 112*, 173–183.

Michel, P., & Wigniolle, B. (2005). Cash-in-advance constraints, bubbles and monetary policy. *Macroeconomic Dynamics, 9*, 28–56.

Rochon, C., & Polemarchakis, H. (2006). Debt, liquidity and dynamics. *Economic Theory, 27,* 179–211.

Shiller, R. J. (1981). Do stock prices move too much to be justified by subsequent changes in dividends? *American Economic Review, 71,* 421–436.

Shiller, R. J. (1989). *Market volatility.* Cambridge, MA: MIT Press.

Sorger, G. (2005). Active and passive monetary policy in an overlapping generations model. *Review of Economic Dynamics, 8,* 731–748.

Tirole, J. (1985). Asset bubbles and overlapping generations. *Econometrica, 53,* 1071–1100.

Wang, P., & Wen, Y. (2012). Speculative bubbles and financial crises. *American Economic Journal: Macroeconomics, 43,* 184–221.

Wigniolle, B. (2014). Optimism, pessimism and financial bubbles. *Journal of Economic Dynamics and Control, 41,* 188–208.

Wolff, E. N. (1998). Recent trends in the size distribution of household wealth. *Journal of Economic Perpectives, 12,* 131–150.

Chapter 11
Can Consumption Taxes Stabilize the Economy in the Presence of Consumption Externalities?

Teresa Lloyd-Braga and Leonor Modesto

Abstract We discuss the stabilization role of consumption taxes under a balanced-budget rule in the presence of consumption externalities of the "keeping up with the Joneses" type. We consider a finance constrained economy and depart from a situation where sufficiently strong externalities make the steady state indeterminate, if government intervention is absent. Sufficiently procyclical consumption tax rates are able to ensure local saddle path stability. However, this procyclicality leads to the appearance of another steady state with lower levels of output which is a source or indeterminate. Therefore, government intervention with stabilization purposes may not be successful.

Keywords Indeterminacy · Consumption externalities · Consumption taxation · Steady state multiplicity

JEL classification: E32 · E62

11.1 Introduction

Several works have analyzed the possible effects on economic instability of balanced-budget fiscal policy rules. The typical result, either in standard one-sector infinitely horizon Ramsey models or in the Woodford (1986)/Grandmont et al. (1998) frame-

Financial support from "Fundacão para a Ciência e Tecnologia" under the PTDC/IIM-ECO/4831/2014 is gratefully acknowledged. We thank an anonymous referee for detailed and insightful comments and suggestions. We also wish to thank Jean-Michel Grandmont, Nicolas Abad, Thomas Seegmuller and Alain Venditti for enlightening comments on an earlier version of this paper.

T. Lloyd-Braga (✉)
UCP, Catolica Lisbon School of Business and Economics, Palma de Cima,
1649-023 Lisbon, Portugal
e-mail: tlb@ucp.pt

L. Modesto
UCP, Catolica Lisbon School of Business and Economics and IZA, Lisbon, Portugal

© Springer International Publishing AG 2017
K. Nishimura et al. (eds.), *Sunspots and Non-Linear Dynamics*,
Studies in Economic Theory 31, DOI 10.1007/978-3-319-44076-7_11

work, is that countercyclical tax rates promote indeterminacy, triggering cycles driven by self-fulfilling volatile expectations (sunspots or endogenous fluctuations), that would not exist in the absence of government. In contrast, procyclical tax rates are able to maintain saddle path stability of the unique steady state. See, for instance, the pioneer work of Schmitt-Grohé and Uribe (1997) who considered income (labor and capital) taxes and Nourry et al. (2013) who considered taxes on consumption, both using a Ramsey model.[1] However, these works do not consider the possible stabilizing role of fiscal policy rules when the economy departs from a situation where local indeterminacy would prevail due, for instance, to the existence of externalities.

Although the idea that taxes can be used to correct inefficient market outcomes due to externalities dates back to Pigou, the stabilization role of taxation against endogenous fluctuations induced by externalities has not been extensively studied in the literature.[2] In this paper we address this issue, considering non linear (cyclical) consumption tax rates, a balanced government budget and consumption externalities in preferences. We argue that fiscal policy may not be able to fully stabilize an economy with sufficiently strong consumption externalities of the 'keeping up with the Joneses' type. Indeed, although a sufficient procyclical consumption tax rate is able to ensure local saddle path stability, we show that in this case another steady state, with a lower level of output, may appear.

Giannitsarou (2007) was the first to study the link between consumption taxes and (in)stability, showing that, in a Ramsey model with additively separable utility functions, a Cobb-Douglas technology and a balanced budget rule with constant government spending, indeterminacy does not emerge. In contrast, Lloyd-Braga et al. (2008) have shown that indeterminacy may emerge under countercyclical consumption tax rates in a Woodford (1986)/Grandmont et al. (1998) finance constrained economy with heterogeneous agents. More recently, Nourry et al. (2013), introducing general specifications for preferences, technology and tax rate rules, obtained a similar result in a Ramsey model. Here, considering a Woodford (1986)/Grandmont et al. (1998) framework, instead of focusing on the properties of fiscal rules that may facilitate the emergence of indeterminacy,[3] we focus on the stabilization role of consumption tax rates in the presence of consumption externalities. To our knowledge, this is the first time that the interrelations between 'keeping up with the Joneses' preferences, variable consumption tax schedules and macroeconomic (in)stability are analyzed jointly from a policy point of view.

Our set up is a particular case of the general framework developed in Lloyd-Braga et al. (2014) in order to analyze local stability properties in the presence of market distortions. However, in the current paper we go beyond the mere characterization of local stability properties, discussing also multiplicity of steady states and its rela-

---

[1] See also in a Woodford (1986) framework, Gokan (2006) and Pintus (2004) with income tax rates and Lloyd-Braga et al. (2008) with consumption taxation.

[2] One exception is Guo and Lansing (1998), who studied the stabilization role of (variable) tax rates on income in a one sector Ramsey model with infinitely-lived households, where indeterminacy is created by increasing returns in production.

[3] On this issue see also Gokan (2008), Guo and Harrison (2008) and Kamiguchi and Tamai (2011).

tion to saddle path stability. Several papers have already shown that multiplicity of steady states often appears in the presence of taxation. For instance, Gokan (2006) in a Woodford model with labour income taxes, and Nourry et al. (2013) in a Ramsey model with consumption tax rates, have shown that multiple steady states may exist with countercyclical tax rates. We also find that countercyclicality of the consumption tax rate is a necessary condition for steady state multiplicity if consumption externalities are small or absent. In contrast, when consumption externalities are sufficiently strong this result no longer holds, steady state multiplicity requiring instead procyclical consumption tax rates. Moreover, although multiplicity of equilibria is often related to the existence of a Laffer curve, we show that, in a Woodford model with strong consumption externalities, steady state multiplicity and indeterminacy may emerge even when a Laffer curve does not exist.

We consider that utility is additively separable between leisure and effective consumption (a composite good, including externalities), that technology exhibits constant returns to scale and a general functional form for the consumption tax rate rule, that nests most of the cases considered in the literature. The chosen specifications are general enough to allow for different values and variability of the labour supply elasticity, the degree of consumption externalities, the elasticity of input substitution in production, the tax rate and its degree of cyclicality. We state conditions ensuring existence of a (normalized) steady state, necessary conditions for steady state multiplicity and for the existence of a Laffer curve. Furthermore we show that, under steady state multiplicity, if one steady state is a saddle there is another steady state which is a source or a sink and vice-versa. While discussing local stability properties, we highlight that in the presence of sufficiently important consumption externalities (such that indeterminacy would prevail in the absence of government) a procyclical consumption tax rate is necessary for local saddle path stability and steady state multiplicity, whereas without (or with small) consumption externalities, procyclicality of the tax rate is sufficient for saddle path stability and uniqueness of steady state.

We also show that with isoelastic specifications for the offer curve and effective consumption, steady state uniqueness is guaranteed in the absence of government intervention. However, introducing government intervention in this environment, we prove the existence of steady state multiplicity considering that the consumption tax rate, τ_c, is an isoelastic function of income. Steady state multiplicity emerges in this case because the elasticity of the consumption tax wedge, i.e., the elasticity of $1 + \tau_c$ is not constant. A numerical illustration of this parameterized version confirms our conclusions. Using empirically plausible values for the parameters, we consider sufficiently strong, but realistic consumption externalities, that, in the absence of government intervention, cause indeterminacy of the unique steady state, introducing therefore expectations driven fluctuations (sunspots). In this case, a procyclical consumption tax rate is able to locally eliminate indeterminacy and sunspot fluctuations, rendering the (normalized) steady state saddle stable. However, another steady state with a lower level of activity, which is a sink (indeterminate) appears, so that the economy may end up fluctuating around a lower level of output and employment. We conclude that procyclical tax rates are able to stabilize the economy locally but not globally. These results challenge the view that procyclical tax rates are powerful

stabilization tools, suggesting caution when using variable (consumption) tax rates
for stabilization purposes in the presence of other market distortions.

The rest of the paper is organized as follows. In the next section we present the
model considered and obtain the perfect foresight equilibria. In Sect. 11.3 we prove
steady state existence and discuss multiplicity versus uniqueness of the steady-state.
The local stability properties of the model are analyzed in Sect. 11.4. In Sect. 11.5
we discuss the stabilization role of consumption taxes, providing also a numerical
illustration. Finally some concluding remarks are provided in Sect. 11.6.

11.2 The Model

The model here used extends the Woodford (1986)/Grandmont et al. (1998) framework introducing consumption externalities and taxation. This framework considers
a perfectly competitive monetary economy with discrete time $t = 1, 2, ..., \infty$ and
heterogeneous infinite lived agents of two types: workers and capitalists. Both consume the final good, but only workers supply labor. There is a financial market
imperfection that prevents workers from borrowing against their wage income and
workers are more impatient than capitalists, i.e. they discount the future more than
the latter. Considering that money is a dominated asset and that the workers's borrowing constraint is binding, at equilibrium capitalists hold the whole capital stock
and no money, whereas workers save their wage earnings through money balances
and use them to consume in the following period. The final good is produced by
firms under a technology characterized by constant returns to scale. We introduce
consumption externalities in this framework, i.e., we assume that the individual utility of consumption is affected by the current consumption of others. Finally, we
consider "wasteful" public spending, that is financed by consumption taxes, which
are only used for stabilization purposes. The detailed description of the model is
provided below.

11.2.1 Production

In each period $t = 1, 2, ..., \infty$, both capital $k_{t-1} > 0$ and labor $l_t > 0$ are used to
produce output y_t, under a constant returns to scale technology i.e., output is given
as:

$$y_t = AF(k_{t-1}, l_t) \equiv Al_t f(x_t), \qquad (11.1)$$

where $x_t \equiv k_{t-1}/l_t$ is the capital labor ratio and $A > 0$ is a scaling parameter. We
assume that:

Assumption 1 $f(x)$ is a continuous function for $x \equiv k/l \in \Re_+$, C^r for $x \in \Re_{++}$ with r high enough, with $f'(x) > 0$ and $f''(x) < 0$. The elasticity of $f(x)$ will be denoted by $s(x) \in (0, 1)$, with $s(x) \equiv \frac{f'(x)x}{f(x)}$; and $\sigma(x) > 0$ will denote the elasticity of substitution between capital and labor, i.e. $\frac{1}{\sigma(x)} \equiv -\frac{f(x)f''(x)x}{f'(x)[f(x)-f'(x)x]}$.

From profit maximization, the real interest rate ρ_t and the real wage ω_t are respectively equal to the marginal productivities of capital and labor, i.e.

$$\rho_t = Af'(x_t) \equiv \rho(k_{t-1}/l_t) \tag{11.2}$$
$$\omega_t = A\left[f(x_t) - f'(x_t)x_t\right] \equiv \omega(k_{t-1}/l_t) \tag{11.3}$$

Note that $\rho_t k_{t-1}/y_t = \frac{f'(x_t)x_t}{f(x_t)}$ so that $s(x_t)$ represents also the capital share of output at period t.

11.2.2 The Government

The government balances its budget at each period in time and chooses, for stabilization purposes, the consumption tax rate, $\tau_c(y_t) \geq 0$, which is a function of real income in the economy.[4] Therefore, real public spending in goods and services in period t, $G_t \geq 0$, is given by

$$G_t = h(y_t)c_t - c_t, \text{ with } h(y_t) \equiv 1 + \tau_c(y_t) \tag{11.4}$$

where $c_t = c_t^w + c_t^k$ is total real consumption in the economy, c_t^w denoting consumption of workers and c_t^k consumption of capitalists. We further assume that public expenditures, G_t, neither affect preferences nor the production function.

As in Nourry et al. (2013), the function $h(y)$ is assumed to have the following properties:

Assumption 2 $h(y) : [0, +\infty) \to [1, +\infty)$ is continuous and C^1 on $(0, +\infty)$.

This very general specification allows us to consider in the same framework procyclical and countercyclical consumption tax rate rules. Indeed, let us introduce the following elasticity:

$$\eta(y) \equiv \frac{h'(y)y}{h(y)} \tag{11.5}$$

Denoting by $\phi(y) \equiv \frac{\tau_c'(y)y}{\tau_c(y)}$ the elasticity of the tax rate, we have that $\eta(y) = \phi(y)\frac{\tau_c(y)}{1+\tau_c(y)}$. Therefore, when the tax rate is procyclical, $\phi(y) > 0$, η is positive

---

[4]Since the tax rate policy is only used for stabilization purposes, it makes sense to assume that the consumption tax rate depends on income and not on the tax base.

($h(y)$ increasing) and when that tax rate is countercyclical, $\phi(y) < 0$, η is negative ($h(y)$ decreasing). When $\phi(y) = \eta(y) = 0$ for all y the tax rate is constant. We also obtain $\eta(y) = 0$ in the absence of government where $\tau_c(y) = 0$.

11.2.3 Workers

We introduce consumption externalities in the utility of workers. Consumption externalities correspond to the idea that the individual utility of consumption is affected by the current consumption of others, so that aggregate or average consumption becomes an argument of the utility function.[5] Here we assume that individual workers compare their own consumption, $c^w \geq 0$, to the average consumption of workers, $\overline{c}^w$.

The behavior of the representative worker can be summarized by the maximization of $U(c^w_{t+1}\varphi(\overline{c}^w_{t+1})/B) - V(l_t)$, subject to the budget constraint $h(y_{t+1})p_{t+1}c^w_{t+1} = w_t l_t = m_t$. Here p_{t+1} is the expectation at period t of the price of the final good at $t + 1$, which under perfect foresight is identical to its equilibrium level at $t + 1$, w_t represents the nominal wage at t, l_t are hours worked with $l \in [0, \widetilde{l}]$, where $\widetilde{l}$ is the worker's time endowment, m_t represents money holdings at the beginning of period $t + 1$, $B > 0$ is a scaling parameter, $V(l)$ the desutility of labor, $U(c^w_{t+1}\varphi(\overline{c}^w_{t+1})/B)$ the utility of effective consumption,[6] $\overline{c}^w_{t+1}$ is worker's average consumption and $\varphi(\overline{c}^w_{t+1})$ is the externality function. We assume that:

Assumption 3 $V(l)$ is a continuous function for $l \in [0, \widetilde{l}]$, and C^r, with r high enough, $V'(l) > 0$, $V''(l) > 0$ for $l \in (0, \widetilde{l})$, with $\widetilde{l}$ possibly infinite. Also, $U(q)$ is a continuous function of $q \geq 0$, and C^r, with r high enough, $U'(q) > 0$, $U''(q) \leq 0$ for $q > 0$, and $-qU''(q)/U'(q) < 1$. The function $\varphi(\overline{c}^w)$ is a positively valued continuous non decreasing function for $\overline{c}^w \geq 0$ such that $\varphi(0) = 1$, and C^g for $\overline{c}^w > 0$ with g high enough, with $\chi(\overline{c}^w_{t+1}) \equiv \overline{c}^w_{t+1}\varphi'(\overline{c}^w_{t+1})/\varphi(\overline{c}^w_{t+1}) \geq 0$.

Remark that $\chi(\overline{c}^w_{t+1}) > 0$ corresponds to the "keeping up with the Joneses" case, according to which the marginal utility of individual consumption is increasing in $\overline{c}^w_t$.[7]

Workers take taxes as given when solving their maximization problem.[8] The solution of this problem is given by the intertemporal trade-off between future consumption and leisure:

$$c^w_{t+1}\varphi(\overline{c}^w_{t+1})/B = \gamma(l_t) \tag{11.6}$$

---

[5] See Alonso-Carrera et al. (2008), Gali (1994), Ljungqvist and Uhlig (2000), Wendner (2010).

[6] As in Lloyd-Braga et al. (2014), we denote by effective consumption the argument $c^w_{t+1}\varphi(\overline{c}^w_{t+1})/B$ of the function $U(.)$, which includes consumption externalities.

[7] The desire to 'keep up with the Joneses' is supported by empirical studies, see for example Carlsson et al. (2003), Ferrer-i-Carbonell (2005) and Maurer and Meier (2008).

[8] Since in our framework the tax rate depends on aggregate income, individuals, being atomistic, take tax rates as given.

with $h(y_{t+1})p_{t+1}c_{t+1}^w = w_t l_t$ and where $\gamma(l_t)$ is the usual offer curve, with $\varepsilon_\gamma(l) \equiv \gamma'(l)l/\gamma(l)$ satisfying $\varepsilon_\gamma(l) > 1$ since consumption and leisure are gross substitutes.[9] Accordingly, the labor supply elasticity at the individual level is positive, i.e., $1/(\varepsilon_\gamma(l) - 1) > 0$. Note that $h(y_{t+1})$ represents the consumption tax wedge, i.e., the ratio between gross and net (of taxes) consumption expenditures.

11.2.4 Capitalists

The representative capitalist maximizes the log-linear lifetime utility function $\sum_{t=1}^{\infty} \beta^t \ln c_t^k$ subject to the budget constraint $h(y_t)c_t^k + k_t = (1 - \delta + (r_t/p_t))k_{t-1}$, where $c_t^k \geq 0$ denotes real consumption of the representative capitalist, $\beta \in (0, 1)$ his subjective discount factor, r_t the nominal interest rate and $\delta \in (0, 1)$ the depreciation rate of capital.[10] Capitalists also take the tax rate as given. Solving the capitalist's problem we obtain the capital accumulation equation:

$$k_t = \beta[1 - \delta + (r_t/p_t)]k_{t-1}. \tag{11.7}$$

11.2.5 Equilibrium

Equilibrium on capital and labor markets requires $\rho_t = r_t/p_t$, $\omega_t = w_t/p_t$, with ρ_t and ω_t given respectively by (11.2) and (11.3). Considering that $m > 0$ is the constant money supply, at the monetary equilibrium, where $w_t l_t = m$ in every period t, we have $h(y_{t+1})c_{t+1}^w = \omega_{t+1}l_{t+1}$, where, using (11.1), $h(y_{t+1}) = h\left(Al_{t+1}f\left(\frac{k_t}{l_{t+1}}\right)\right)$. Also, at a symmetric equilibrium we have $\bar{c}_{t+1}^w = c_{t+1}^w$. Therefore from (11.6) and (11.7) we obtain:

Definition 1 A perfect foresight intertemporal equilibrium is a sequence $(k_{t-1}, l_t) \in R_{++}^2, t = 1, 2, ..., \infty$, that, for a given $k_0 > 0$, satisfies

$$\frac{\omega(k_t/l_{t+1})l_{t+1}}{h\left(Al_{t+1}f\left(\frac{k_t}{l_{t+1}}\right)\right)} \varphi\left(\frac{\omega(k_t/l_{t+1})l_{t+1}}{h\left(Al_{t+1}f\left(\frac{k_t}{l_{t+1}}\right)\right)}\right)/B = \gamma(l_t) \tag{11.8}$$

$$k_t = \beta\left[1 - \delta + \rho(k_{t-1}/l_t)\right]k_{t-1}. \tag{11.9}$$

Equations (11.8) and (11.9) represent, respectively, the intertemporal trade-off between consumption and leisure and capital accumulation. They determine the

---

[9]Note that our assumptions on U and V imply that consumption and leisure are gross substitutes.
[10]We do not introduce consumption externalities into capitalists' preferences because, since they have a log-linear utility function, such externalities would not affect the dynamics.

dynamics of this economy through a two-dimensional dynamic system with only one predetermined variable, the capital stock k. On the contrary, employment l, being affected by expectations of future events,[11] is a non predetermined variable.

11.3 Steady State Analysis

A steady state solution $(k_{ss}, l_{ss}) \in \mathfrak{R}^2_{++}$ of the dynamic system (11.8) and (11.9), with $x_{ss} \equiv k_{ss}/l_{ss} > 0$, is a stationary solution $k_t = k_{t-1} = k_{ss} > 0$ and $l_{t+1} = l_t = l_{ss} > 0$ of that system. Define $\theta \equiv 1 - \beta(1 - \delta) \in (0, 1)$ and assume that $\lim_{x \to 0} f'(x) > \theta/A\beta$ and that $\lim_{x \to +\infty} f'(x) < \theta/A\beta$. Then, from (11.2) and (11.9) and under Assumption 1, we obtain a unique steady state solution x_{ss} which satisfies $f'(x_{ss}) = \theta/A\beta$. Substituting in (11.8) k_{ss} by $x_{ss}l_{ss}$ and using (11.3), we then obtain the following definition:

Definition 2 Under Assumption 1, consider a given value for A satisfying $\frac{\theta}{\beta} \frac{1}{\lim_{x \to 0} f'(x)} < A < \frac{\theta}{\beta} \frac{1}{\lim_{x \to +\infty} f'(x)}$ and let $x_{ss} > 0$ be given by the unique solution of $f'(x_{ss}) = \theta/A\beta$. Then, under Assumptions 2 and 3, steady state values $(k_{ss}, l_{ss}) \in \mathfrak{R}^2_{++}$ of the dynamic system (11.8) and (11.9), are solutions of the following equations:

$$Z(l_{ss}) = B, \text{ with} \tag{11.10}$$

$$Z(l) \equiv \frac{Al \left[f(x_{ss}) - f'(x_{ss})x_{ss} \right]}{h(Alf(x_{ss}))} \frac{\varphi \left(\frac{Al[f(x_{ss}) - f'(x_{ss})x_{ss}]}{h(Alf(x_{ss}))} \right)}{\gamma(l)} \tag{11.11}$$

$$k_{ss} = x_{ss}l_{ss} \tag{11.12}$$

It is easy to see that, for a given level of l_{ss}, Eq. (11.12) uniquely determines k_{ss}. However, a steady state solution only exists if there is a value for $l_{ss} > 0$ satisfying Eq. (11.10). We ensure existence of a steady state, namely the normalized steady state k_{nss}, l_{nss} with the corresponding level of output $y_{nss} \equiv 1$, by following the usual procedure of fixing the parameter B at the appropriate level. It is easy to see that (11.10) is satisfied for the normalized steady state if and only if B takes the value $B_{nss} \equiv Z(l_{nss})$. Hence, using also (11.1) and the definition of $s(x)$ given in Assumption 1, we deduce the following proposition.

Proposition 1 *Normalized Steady State: Under Assumptions 1–3, consider a given value for A satisfying $\frac{\theta}{\beta} \frac{1}{\lim_{x \to 0} f'(x)} < A < \frac{\theta}{\beta} \frac{1}{\lim_{x \to +\infty} f'(x)}$ and let $x_{ss} > 0$ be the unique solution of $f'(x_{ss}) = \theta/A\beta$ and*

---

[11] See (11.6) and the budget constraint, according to which current employment depends on future values of consumption and expectations of the future price of output.

$$y_{nss} \equiv 1$$
$$l_{nss} \equiv 1/Af(x_{ss})$$
$$k_{nss} \equiv x_{ss}/Af(x_{ss})$$

Then (k_{nss}, l_{nss}) with the corresponding level of output $y_{nss} = Al_{nss}f(x_{ss}) = 1$ is the (normalized) steady state of the dynamic system (11.8)–(11.9) if and only if

$$B = B_{nss} \equiv \frac{1-s(x_{ss})}{h(1)}\varphi\left(\frac{1-s(x_{ss})}{h(1)}\right)/\gamma(1/Af(x_{ss})) > 0. \qquad (11.13)$$

Moreover, the elasticity of the function $h(y)$, evaluated at the normalized steady state, is given by $\eta(1) \equiv h(1)/h'(1)$.

From now on we will consider that existence of the normalized steady state is persistent. However, when $B = B_{nss}$ nothing guarantees that (k_{nss}, l_{nss}) is the unique steady state solution. The number of steady states is determined by the number of intersections of the curve $Z(l)$ with the horizontal line B_{nss}. As $Z(l)$ is a continuous function of $l \in (0, \tilde{l})$, it is easy to see that a sufficient condition for the existence of at least another steady state solution (besides the normalized one) is that both limits, $\lim_{l \to 0} Z(l)$ and $\lim_{l \to \tilde{l}} Z(l)$, are either greater or smaller than B_{nss}. Accordingly we have the following Proposition.

Proposition 2 *Sufficient Conditions for Steady State Multiplicity: Under Assumptions 1–3, consider that $\frac{\theta}{\beta}\frac{1}{\lim_{x \to 0} f'(x)} < A < \frac{\theta}{\beta}\frac{1}{\lim_{x \to +\infty} f'(x)}$, let $x_{ss} > 0$ be the unique solution of $f'(x_{ss}) = \theta/A\beta$ and $B = B_{nss} > 0$, as under Proposition 1. Then sufficient conditions for the existence of at least two steady states are:*

(i) $\lim_{l \to 0} Z(l) < B_{nss}$ and $\lim_{l \to \tilde{l}} Z(l) < B_{nss}$, or
(ii) $\lim_{l \to 0} Z(l) > B_{nss}$ and $\lim_{l \to \tilde{l}} Z(l) > B_{nss}$.

Obviously, when $Z(l)$ is a monotonic function of l, the conditions of Proposition 2 cannot be satisfied. Therefore, if the function $Z(l)$ is always increasing or always decreasing, uniqueness of the steady state can be guaranteed. It follows that a necessary condition for steady state multiplicity is that $Z(l)$ is non monotonic. To further discuss this issue we will now obtain its elasticity:

$$\epsilon_{Z,l} \equiv \frac{Z'(l)l}{Z(l)} = 1 + \chi(c^w) - \varepsilon_y(l) - (1 + \chi(c^w))\eta(y), \qquad (11.14)$$

with $c^w = Al\left[f(x_{ss}) - f'(x_{ss})x\right]/h(lf(x_{ss}))$ and $y = Alf(x_{ss})$, where x_{ss} satisfies $f'(x_{ss}) = \theta/A\beta$. Since $Z(l) > 0$ and $\varepsilon_y(l) > 1$ for all $l > 0$, we obtain the following Proposition.

Proposition 3 *Necessary Conditions for Steady State Multiplicity: Under Assumptions 1–3, consider a given value for A satisfying $\frac{\theta}{\beta}\frac{1}{\lim_{x \to 0} f'(x)} < A < \frac{\theta}{\beta}\frac{1}{\lim_{x \to +\infty} f'(x)}$ and let $x_{ss} > 0$ be such that $f'(x_{ss}) = \theta/A\beta$, $c^w = Al\left[f(x_{ss}) - f'(xx_{ss})x_{ss}\right]/$*

$h(lf(x_{ss}))$ and $y = Alf(x_{ss})$. Then, a necessary condition for steady state multiplicity is that $Z(l)$ is a non monotonic function of l, with $\epsilon_{Z,l} \gtreqless 0$ if and only

if $\varepsilon_\gamma^*(l) \equiv (1 + \chi(c^w))(1 - \eta(y)) > 1$ and $\varepsilon_\gamma(l) \lesseqgtr \varepsilon_\gamma^*(l)$. On the contrary, if for

every $l \in [0, \tilde{l}]$ we have either $\varepsilon_\gamma(l) < \varepsilon_\gamma^*(l)$ or $\varepsilon_\gamma(l) > \varepsilon_\gamma^*(l)$, there is at most one
steady state.[12]

In the absence of consumption externalities, $\chi(c^w) = 0$, and without government,
$\eta(y) = 0$, the steady state would be unique, since as in this case $\varepsilon_\gamma^*(l) = 1$ we
always have $\varepsilon_\gamma(l) > \varepsilon_\gamma^*(l)$. We can also see that in the absence of government,
$\eta(y) = 0$, steady state multiplicity requires consumption externalities, $\chi(c^w) > 0$,
to guarantee that $\varepsilon_\gamma^*(l) > 1$.[13] Note also that with isoelastic specifications for the
functions $\varphi(c^w)$, $\gamma(l)$ and $h(y)$, so that all the elasticities appearing in the RHS of
(11.14), $\chi(c^w)$, $\varepsilon_\gamma(l)$ and $\eta(y)$ are constant, the steady state will be unique since
$\epsilon_{Z,l} = (1 + \chi)(1 - \eta) - \varepsilon_\gamma$ becomes constant, taking either a negative ($\varepsilon_\gamma > \varepsilon_\gamma^*$) or
a positive value ($\varepsilon_\gamma < \varepsilon_\gamma^*$) for all $l \in [0, \tilde{l}]$, provided that $(1 + \chi)(1 - \eta) - \varepsilon_\gamma \neq 0$.
We can then state the following Corollary.[14]

Corollary 1 *Under Assumptions 1–3, consider a given value for A satisfying*
$\frac{\theta}{\beta} \frac{1}{\lim_{x \to 0} f'(x)} < A < \frac{\theta}{\beta} \frac{1}{\lim_{x \to +\infty} f'(x)}$ *and let* $x_{ss} > 0$ *be such that* $f'(x_{ss}) = \theta/A\beta$,
$c^w = Al\left[f(x_{ss}) - f'(xx_{ss})x_{ss}\right]/h(lf(x_{ss}))$ *and* $y = Alf(x_{ss})$. *Consider also isoe-
lastic specifications for the externality function* $\varphi(c^w)$ *and for the offer curve* $\gamma(l)$,
so that $\chi(c^w)$ *and* $\varepsilon_\gamma(l)$ *are given by constants, respectively* χ *and* ε_γ. *Then, there
exists at most one steady state, under the following conditions:*

(i) *in the absence of government or with a constant consumption tax rate,* $\eta(y) = 0$
for all $y > 0$, *provided* $1 + \chi - \varepsilon_\gamma \neq 0$.
(ii) *or with a variable consumption tax rate, provided the consumption tax wedge,
$h(y)$, is an isoelastic function of income with a constant elasticity* $\eta \neq (1 + \chi - \varepsilon_\gamma)/(1 + \chi)$.

Note that with isoelastic specifications for the functions $\varphi(c^w)$ and $\gamma(l)$, even
if we consider also an isoelastic specification for the tax rate $\tau_c(y_t)$, we can still
obtain steady state multiplicity, since obviously in this case $\eta(y)$, the elasticity of
$h(y) = 1 + \tau_c(y_t)$, will not be constant.

---

[12]In Sect. 11.4, we will show that when $\varepsilon_\gamma(l)$ crosses the critical value $\varepsilon_\gamma^*(l)$ a transcritical bifurcation generically occurs. Note that we disregard the case where $\varepsilon_\gamma(l) = \varepsilon_\gamma^*(l)$ for all l. In this case
$Z(l)$ becomes constant describing an horizontal line in the space $(l, Z(l))$. Hence, either $Z(l) \neq B$
for all l and there would be no steady state, or $Z(l) = B$ for all l and we would obtain a continuum
of steady states.

[13]The same result was obtained by Alonso-Carrera et al. (2008), in a growth model with endogenous
labor supply.

[14]This result, which links constant elasticities with steady state uniqueness is not specific to our
framework. See, for example, Guo and Lansing (1998).

Using Proposition 3, we now discuss which type of tax rate cyclicality promotes steady state multiplicity. From (11.14), we can see that if consumption externalities are sufficiently weak so that $1 + \chi(c^w) - \varepsilon_y(l)$ is always negative, then $Z(l)$ is a decreasing function when $\eta(y) \geq 0$. Also, if $1 + \chi(c^w) - \varepsilon_y(l)$ is always positive then $Z(l)$ is an increasing function when $\eta(y) \leq 0$. Accordingly we obtain the following Corollary.

Corollary 2 *Under Assumptions 1–3, consider a given value for A satisfying* $\frac{\theta}{\beta} \frac{1}{\lim_{x \to 0} f'(x)} < A < \frac{\theta}{\beta} \frac{1}{\lim_{x \to +\infty} f'(x)}$ *and let* $x_{ss} > 0$ *be such that* $f'(x_{ss}) = \theta/A\beta$ *and* $c^w = Al [f(x_{ss}) - f'(xx_{ss})x_{ss}]/h(lf(x_{ss}))$ *and* $y = Alf(x_{ss})$.

1. *Consider* $(1 + \chi(c^w) - \varepsilon_y(l)) < 0$ *for every* $l > 0$. *Then, if the consumption tax rate is procyclical, i.e.* $\eta(y) \geq 0$ *for all* $y > 0$, *there exists at most one steady state. A necessary condition for multiplicity is that* $\eta(y) < 0$, *at least for some values of* $y > 0$.
2. *Consider* $(1 + \chi(c^w) - \varepsilon_y(l)) > 0$ *for every* $l > 0$. *Then, if the consumption tax rate is countercyclical, i.e.* $\eta(y) \leq 0$ *for all* $y > 0$, *there exists at most one steady state. A necessary condition for multiplicity is that* $\eta(y) > 0$, *at least for some values of* $y > 0$.

Corollary 2.1 corresponds to the case of not too strong consumption externalities. We can see that in this case, as in Nourry et al. (2013) where consumption externalities are not considered, procyclical consumption tax rates with $\eta(y) > 0$ ensure uniqueness of the steady state. However, the result that procyclical tax rates ensure steady state uniqueness is reversed in the presence of sufficiently important consumption externalities. In this case, uniqueness is ensured instead by countercyclical or constant consumption tax rates. See Corollary 2.2.

Steady state multiplicity has been related to the existence of a Laffer curve. However, taxation is the only distortion considered in those works. See, for example Nourry et al. (2013).[15] We show below that in a Woodford model with sufficiently strong consumption externalities, the link between the existence of a Laffer curve and steady state multiplicity does not necessarily hold.

Using (11.2), (11.3) and (11.4) and the budget constraints of workers and capitalists, a steady state solution for τ can be obtained from Definition 2 as the solution of the following equation:

$$M(\tau) = H(\tau)$$

with[16]

---

[15] See also Schmitt-Grohe and Uribe (1997) for a seminal contribution on the relation between the Laffer curve and equilibrium indeterminacy.

[16] Note that $M(\tau) \equiv \tau [c^w(\tau) + c^k(\tau)]$ and $H(\tau) \equiv h(Al(\tau)f(x_{ss}))[c^w(\tau) + c^k(\tau)] - [c^w(\tau) + c^k(\tau)]$ with $c^w(\tau) = \frac{1}{1+\tau} Al[f(x_{ss}) - f'(x_{ss})x_{ss}]$ and $c^k(\tau) = \frac{1}{1+\tau} \{k[1 - \delta + Af'(x_{ss})] - k\}$.

$$M(\tau) \equiv \frac{\tau}{1+\tau} l(\tau) [Af(x_{ss}) - \delta x_{ss}] \tag{11.15}$$

$$H(\tau) \equiv \frac{l(\tau) [Af(x_{ss}) - \delta x_{ss}]}{1+\tau} [h(Al(\tau)f(x_{ss})) - 1], \tag{11.16}$$

where $l(\tau)$ is implicitly defined by the following equation:

$$\frac{Al[f(x_{ss}) - f'(x_{ss})x_{ss}]}{1+\tau} \frac{\varphi\left(\frac{Al[f(x_{ss})-f'(x_{ss})x_{ss}]}{1+\tau}\right)}{\gamma(l)} = B,$$

provided that $1 + \chi\left(\frac{Al[f(x_{ss})-f'(x_{ss})x_{ss}]}{1+\tau}\right) - \varepsilon_\gamma(l) \neq 0$. Let $\chi_{(\tau)} \equiv \chi\left(\frac{Al(\tau)[f(x_{ss})-f'(x_{ss})x_{ss}]}{1+\tau}\right)$, $\eta_{(\tau)} \equiv \eta(Al(\tau)f(x_{ss}))$, $\varepsilon_{\gamma(\tau)} \equiv \varepsilon(l(\tau))$ and assume that $1 + \chi_{(\tau)} - \varepsilon_{\gamma(\tau)} \neq 0$ for any $\tau \geq 0$. Then, using (11.15) and (11.16) and defining $\epsilon_{M,\tau} \equiv M'(\tau)\tau/M(\tau)$, $\epsilon_{H,\tau} \equiv H'(\tau)\tau/H(\tau)$, we obtain:

$$\epsilon_{M,\tau} = \frac{1}{1+\tau}\left[1 + \frac{\tau(1+\chi_{(\tau)})}{1+\chi_{(\tau)}-\varepsilon_{\gamma(\tau)}}\right] \qquad \epsilon_{H,\tau} = \frac{(1+\tau)(1+\chi_{(\tau)})\eta_{(\tau)}+\tau\varepsilon_{\gamma(\tau)}}{(1+\chi_{(\tau)}-\varepsilon_{\gamma(\tau)})(1+\tau)} \tag{11.17}$$

A Laffer curve is obtained if $M(\tau)$ is a non monotonic function of $\tau \in (0, \infty)$, i.e. if $\epsilon_{M,\tau}$ changes sign. From (11.17) and defining $\tau^* = \frac{\varepsilon_{\gamma(\tau)}-(1+\chi_{(\tau)})}{1+\chi_{(\tau)}}$, a hump shaped Laffer curve exists if and only if $\tau^* > 0$, i.e. $1 + \chi_{(\tau)} - \varepsilon_{\gamma(\tau)} < 0$, with $\epsilon_{M,\tau} \overset{\leq}{\underset{>}{=}} 0$ if and only if $\tau \overset{\leq}{\underset{>}{=}} \tau^*$. Therefore we can state the following.

Proposition 4 *Under Assumptions 1–3, if* $1 + \chi_{(\tau)} - \varepsilon_{\gamma(\tau)} > 0$ *for all values of* $\tau > 0$, *a Laffer curve does not exist.*

This Proposition states that in the presence of sufficiently important consumption externalities such that $\chi_{(\tau)} > \varepsilon_{\gamma(\tau)} - 1 > 0$ for all $\tau > 0$, a Laffer curve is never obtained. Combining now Corollary 2.2 and Proposition 4, we deduce that steady state multiplicity may emerge, provided tax rates are procyclical, even when a Laffer curve does not exist. In fact the number of steady states is determined by the number of intersections of the curves $M(\tau)$ and $H(\tau)$. Hence steady state multiplicity requires that the difference $\epsilon_{M,\tau} - \epsilon_{H,\tau}$ changes sign at least once.[17] However this last condition does not require a change of sign of $\epsilon_{M,\tau}$, i.e., the existence of a Laffer curve. In fact, in Sect. 11.5, we provide a numerical example where, in the absence of a Laffer curve, steady state multiplicity emerges.

---

[17]Note that $\epsilon_{M,\tau} - \epsilon_{H,\tau} = \epsilon_{Z,l}/[1 + \chi(c^w) - \varepsilon_\gamma(l)]$, with $c^w = Al[f(x_{ss}) - f'(x_{ss})x]/h(lf(x_{ss}))$ and l being given by $l(\tau)$. Hence, given that $1 + \chi_{(\tau)} - \varepsilon_{\gamma(\tau)} \neq 0$ for any $\tau \geq 0$, a change in the sign of $(\epsilon_{M,\tau} - \epsilon_{H,\tau})$ is equivalent to a change in the sign of $\epsilon_{Z,l}$, a condition already shown in Proposition 3 to be necessary for steady state multiplicity.

11.4 Local Stability Properties

In this section we analyze the local stability properties of our dynamic system around a steady state solution $k_t = k_{t-1} = k_{ss}$ and $l_{t+1} = l_t = l_{ss}$ of that system.

The log-linearized system (11.8)–(11.9) around that steady state, is represented by:

$$\begin{bmatrix} \widehat{k_t} \\ \widehat{l_{t+1}} \end{bmatrix} = [J] \begin{bmatrix} \widehat{k_{t-1}} \\ \widehat{l_t} \end{bmatrix} \tag{11.18}$$

where hat-variables denote percentage deviation rates from their steady-state values and J is the Jacobian matrix of the system (11.8) and (11.9) evaluated at the steady state. Its trace T, and determinant, D, are given by:

$$T = 1 + \frac{\sigma \varepsilon_\gamma - \theta (1 - s)(1 + \chi)(1 - \eta)}{[\sigma - s - \eta \sigma (1 - s)](1 + \chi)} \tag{11.19}$$

$$D = \frac{\sigma - \theta(1 - s)}{[\sigma - s - \eta \sigma (1 - s)](1 + \chi)} \varepsilon_\gamma \tag{11.20}$$

where $\varepsilon_\gamma > 1$ is the elasticity of the offer curve $\gamma(l)$, $\chi \geq 0$ is the degree of consumption externalities, i.e. the elasticity of function $\varphi(c^w)$, and η denotes the elasticity of $h(y)$, all evaluated at the steady state under analysis.[18]

In the following, as typically done in Woodford economies, we assume that $0 < \theta(1 - s) < s < 1/2$, i.e., we assume that the period is short so that θ is small, and that the capital share of output, s, is smaller than the labor share. Moreover, we also assume that $\sigma > \sigma^* \equiv s / [1 - \eta (1 - s)]$ and $\eta < 1$, so that effective consumption is increasing in labor as in Lloyd-Braga et al. (2014). To guarantee that $\sigma^* \in (\theta(1 - s), 1)$ which allows us to consider the most relevant parameterizations for σ, we further assume that $\eta \in \left(-[s - \theta(1 - s)]/\theta(1 - s)^2, 1\right)$. Recall that we also have $\varepsilon_\gamma > 1$ and $\chi \geq 0$. Moreover, we assume that $\chi < 1$, i.e., consumption externalities are not too strong, in accordance with empirical studies (see Maurer and Meier (2008)). All these assumptions are summarized below in Assumption 4 and we consider them satisfied in the rest of the paper.

---

[18]Remark that our model is nested in the general framework of analysis provided in Lloyd-Braga et al. (2014) with effective consumption per unit of labor given by $\Omega(k, l) = \frac{\omega(k,l)}{h(Alf(k/l))} \varphi \left(\frac{\omega(k,l)l}{h(Alf(k/l))} \right)$, the real interest rate relevant to capitalists given by $R(k, l) = Af'(x_t)$, and the (generalized) offer curve given by $\Gamma(k, l) = \gamma(l)$. Therefore, the trace and the determinant can be obtained using their expressions (11.7)–(11.8), where, from their expression (9) we have: $\varepsilon_{Rk} = -\frac{1-s}{\sigma}$, $\varepsilon_{Rl} = \frac{1-s}{\sigma}$, $\varepsilon_{\Gamma k} = 0$, $\varepsilon_{\Gamma l} = \varepsilon_\gamma$, $\varepsilon_{\Omega k} = (1 + \chi)\left[\frac{s}{\sigma} - \eta s\right]$ and $\varepsilon_{\Omega l} = \chi - (1 + \chi)\left[\frac{s}{\sigma} + \eta(1 - s)\right]$.

Assumption 4

1. $s < 1/2$ and $0 < \theta < s/(1-s)$;
2. $\eta^* < \eta < 1$ with $\eta^* \equiv -[s-\theta(1-s)]/\theta(1-s)^2 < 0$ and $\sigma > \sigma^*$ with $\sigma^* \equiv s/[1-\eta(1-s)] \in (\theta(1-s), 1)$;
3. $\varepsilon_\gamma > 1$ and $1 > \chi \geq 0$.

The local stability properties of the model are determined by the eigenvalues of the Jacobian matrix J or, equivalently, by its trace, T, and determinant, D, which correspond respectively to the product and sum of the two roots (eigenvalues) of the associated characteristic polynomial $Q(\lambda) \equiv \lambda^2 - \lambda T + D$. The steady state is a sink (both eigenvalues with modulus lower than one) when $D > -T - 1, D > T - 1$ and $D < 1$. Since only capital is a predetermined variable, when the steady state is a sink, it is locally indeterminate[19] and there are infinitely many stochastic endogenous fluctuations (sunspots) arbitrarily close to the steady state.[20] The steady state is saddle stable (one eigenvalue with modulus higher than one and one eigenvalue with modulus lower than one) when $|T| > |D + 1|$ and it is an unstable source (both eigenvalues with modulus higher than one) in the remaining cases.

Straightforward computations show that, under Assumption 4, we always have $D > 0$ and $D > -T - 1$. We will have a source when $D > \max\{1, T - 1\}$, a saddle when $D < T - 1$ and a sink when $T - 1 < D < 1$. Using (11.19) and (11.20), it can be seen that $D < 1$ is equivalent to $\varepsilon_\gamma < \varepsilon_{\gamma 1}$ and $D < T - 1$ is equivalent to $\varepsilon_\gamma > \varepsilon_\gamma^*$, with $\varepsilon_{\gamma 1}$ and ε_γ^* defined in Proposition 5 below. Therefore, we can state that:

Proposition 5 *Under Assumptions 1–4, and defining*

$$\varepsilon_{\gamma 1} \equiv \frac{[\sigma - s - \eta\sigma(1-s)](1+\chi)}{\sigma - \theta(1-s)}$$
$$\varepsilon_\gamma^* \equiv (1+\chi)(1-\eta),$$

the following applies:

1. *The steady state under analysis is a saddle if and only if $\varepsilon_\gamma > \varepsilon_\gamma^*$.*
2. *The steady state under analysis is a source (unstable) if and only if $\varepsilon_{\gamma 1} < \varepsilon_\gamma < \varepsilon_\gamma^*$.*
3. *The steady state under analysis is a sink (indeterminate) if and only if $1 < \varepsilon_\gamma < \min\{\varepsilon_{\gamma 1}, \varepsilon_\gamma^*\}$.*

From Proposition 5, $1 < \varepsilon_\gamma < \varepsilon_\gamma^*$ is a necessary condition for indeterminacy. Therefore, in the absence of externalities, $\chi = 0$, indeterminacy requires a sufficiently countercyclical consumption tax rate, $\eta < 1 - \varepsilon_\gamma < 0$. This result was already obtained for a Woodford model, in Lloyd-Braga et al. (2008), and for a Ramsey model, in Nourry et al. (2013). We can also see that in the absence of

---

[19]Indeterminacy occurs when the number of eigenvalues strictly lower than one in absolute value is larger than the number of predetermined variables.

[20]See for instance Grandmont et al. (1998).

government or with constant tax rates, i.e. $\eta = 0$, local indeterminacy requires sufficiently positive consumption externalities, i.e., $\chi > \varepsilon_y - 1 > 0$, as already shown in Lloyd-Braga and al. (2014).[21]

The steady state is a saddle when ε_y exceeds the critical value ε_y^*, which is equivalent to $\eta > \left(1 + \chi - \varepsilon_y\right)/(1 + \chi)$. Therefore, we can immediately conclude that a procyclical consumption tax rate promotes local stability.[22] In particular, when consumption externalities are sufficiently strong, so that indeterminacy would prevail in the absence of government or with constant tax rates, procyclicality of the tax rate is a necessary condition for local saddle path stability. In contrast, a procyclical consumption tax rate becomes a sufficient condition for saddle path stability and uniqueness of the steady state (see Corollary 2.1), in the absence of consumption externalities ($\chi = 0$).

Note that ε_y^* denotes the value of $\varepsilon_y^*(l)$, defined in Proposition 3, at the steady state under analysis. When $\varepsilon_y = \varepsilon_y^*$, i.e., $D = T - 1$, one of the eigenvalues takes the value 1. Therefore, when ε_y crosses the critical value ε_y^*, a transcritical bifurcation generically occurs through which two steady states exchange stability properties.[23] From Proposition 5 one can see that if $\varepsilon_{y1} < \varepsilon_y^*$ the two steady states exchanging stability properties are a saddle and a source. This happens for $\sigma < \sigma^{**} \equiv \frac{s-\theta(1-s)(1-\eta)}{\eta s}$ where, under Assumption 4, $\sigma^{**} > 1$. Otherwise, if $\sigma > \sigma^{**}$ so that $\varepsilon_{y1} > \varepsilon_y^*$, a source never appears, and the two steady states exchanging stability properties are a saddle and a sink.

Combining now Propositions 3 and 5 we conclude that whenever the steady is a saddle, i.e., $\varepsilon_y > \varepsilon_y^*$, $Z(l)$ is decreasing, i.e., $Z'(l) < 0$. On the contrary, when $Z'(l) > 0$ the steady state is either a sink or a source. Since existence of the normalized steady state is assumed, B is always identical to B_{nss} and the function Z must cross B_{nss} at $l_{nss} = 1/Af(x_{ss})$. See Proposition 1. Hence, from Definition 2, if there is another steady state it means that $Z(l)$ crosses again B_{nss} for another value of l. Of course, since $Z(l)$ is a continuous function, if $Z'(l) < 0$ at l_{nss} and there are other steady states, then $Z'(l)$ must take a positive sign at least for one of them, and vice versa. Accordingly, we obtain the following result:

Proposition 6 *Under Assumptions 1–4, consider that under Proposition 1 a normalized steady state, l_{nss}, exists and that under Proposition 3 there is at least another steady state, l_a. Let ε_y and ε_y^* denote $\varepsilon_y(l)$ and $\varepsilon_y^*(l)$ evaluated at the normalized steady state. Then, if*

---

[21] See also Alonso-Carrera et al. (2008) for a similar result.

[22] This result could be immediately obtained from Lloyd-Braga et al. (2014). Indeed our parameterization falls into their configuration (i). Therefore, from their table 1, we can immediately see that the steady state is always a saddle when ε_y exceeds the critical value for which a transcritical bifurcation occurs, i.e., ε_y^* in the current paper.

[23] We disregard the case of a saddle node bifurcation since in Sect. 11.5 we apply our local dynamics analysis to a normalized steady state whose persistence is ensured. We also disregard Pitchfork bifurcations since they are non generic. When ε crosses the value ε_{y1}, we expect that a Hopf bifurcation occurs, through which a pair of complex conjugate eigenvalues cross the unit root.

1. $\varepsilon_\gamma > \varepsilon_\gamma^*$, the normalized steady state is a saddle $(Z'(l_{nss}) < 0)$ and there is another one, l_a, that is a source or a sink $(Z'(l_a) > 0)$;
2. $\varepsilon_\gamma < \varepsilon_\gamma^*$, the normalized steady state is a source or a sink $(Z'(l_{nss}) > 0)$ and there is another one, l_a, that is a saddle $(Z'(l_a) < 0)$.

Moreover, when the normalized steady state is a saddle with $Z'(l_{nss}) < 0$, if condition (i) of Proposition 2 is satisfied, then, by continuity, there must exist another steady state $l_a < l_{nss}$ with a corresponding level of output $y_a = l_a A f(x_{ss}) < y_{nss} = 1$, such that $Z'(l_a) > 0$. Hence, we can state:

Proposition 7 *Under Assumptions 1–4, consider that under Proposition 1 a normalized steady state, l_{nss} with $y_{nss} = 1$, exists. Then, under the conditions of Proposition 2.(i), when the normalized steady state is a saddle, there is at least another one, l_a, with a corresponding lower level of output $y_a < 1$, that is a source or a sink.*

11.5 Stabilization Policy

We now illustrate our results on steady state multiplicity and indeterminacy considering isoelastic specifications for $\varphi(\bar{c}_{t+1}^w)$, $\gamma(l)$ and $\tau_c(y_t)$. Accordingly we have $U(c_{t+1}^w \varphi(\bar{c}_{t+1}^w)/B) = c_{t+1}^w (\bar{c}_{t+1}^w)^\chi / B$ and $V(l_t) = l_t^{\varepsilon_\gamma}$ with $l \in (0, \tilde{l})$, where $\tilde{l} = +\infty$, so that the degree of externalities and the elasticity of the offer curve at the individual level are given by constant parameters, $\chi \geq 0$ and $\varepsilon_\gamma > 1$ respectively. For the consumption tax rate, we feature an isoelastic specification *a la* Lloyd-Braga et al. (2008):

$$\tau_c(y_t) = \alpha y_t^\phi \tag{11.21}$$

with parameters $\alpha \geq 0$ and $\phi \in R$. The parameter ϕ denotes the constant elasticity of the tax rate with respect to income. When $\phi < 0$ the tax rate decreases when output increases, i.e., the tax rate moves countercyclically. The case of $\phi > 0$ corresponds to the cases where the tax rate increases with output, i.e. the tax rate is procyclical. For $\phi = 0$ the tax rate is constant at the level α. Remark that, although we assume an isoelastic specification for the tax rate, the function $h(y) = 1 + \tau_c(y)$, is not isoelastic when $\phi \neq 0$. Indeed, its elasticity, given by $\eta(y) = \phi \frac{\alpha y_t^\phi}{1+\alpha y_t^\phi}$ is not constant, increasing or decreasing with y depending on whether $\phi > 0$ or $\phi < 0$. Also, regarding technology, we consider a constant elasticity of substitution (CES) production function, the intensive production function being given by $f(x) = \left[ax^{\frac{\sigma-1}{\sigma}} + 1 - a\right]^{\frac{\sigma}{\sigma-1}}$ with $\sigma > 0$ and $a \in (0, 1)$.

We assume the conditions of Proposition 1 are satisfied, so that a normalized steady state exists.[24] We depart from a situation where in the absence of govern-

---

[24]Note that with $\tau_c(y_t)$ given as in (11.21), we have that at the normalized steady state $h(1) = 1 + \alpha$, i.e., the tax wedge does not depend on ϕ. Hence, existence of the normalized steady state is persistent for all values of ϕ, including $\phi = \eta = 0$.

ment intervention with stabilization purposes ($\phi = \eta = 0$), the normalized steady state, although unique by Corollary 1.(i), is locally indeterminate (sink) and discuss whether (cyclical) consumption taxes can be used to stabilize the economy. Therefore, according to Proposition 5, we assume that at the normalized steady state $\varepsilon_\gamma < \min \left\{ \frac{(\sigma-s)(1+\chi)}{\sigma-\theta(1-s)}, (1+\chi) \right\}$. This implies that $1 + \chi - \varepsilon_\gamma > 0$, i.e., in the absence of government intervention and since $\varepsilon_\gamma > 1$, the (normalized) steady state would be locally indeterminate due to the presence of sufficiently strong consumption externalities. In this case, a sufficiently procyclical consumption tax rate at the normalized steady state, $\phi > \phi_1 \equiv \frac{1+\alpha}{\alpha} \frac{1+\chi-\varepsilon_\gamma}{1+\chi} > 0$, is required to ensure local saddle path stability of the normalized steady state. See Proposition 5.1. Also, as $1 + \chi - \varepsilon_\gamma > 0$, from Proposition 4, a Laffer curve does not exist. However, when $1 + \chi - \varepsilon_\gamma > 0$, from Corollary 2.2, we know that steady state multiplicity may emerge when tax rates are procyclical. Indeed, with the functional

forms considered, we obtain $Z(l) \equiv \dfrac{l^{1+\chi-\varepsilon_\gamma}\{A(1-a)\}^{1+\chi}\left[ax_{ss}^{\frac{\sigma-1}{\sigma}}+1-a\right]^{\frac{1+\chi}{\sigma-1}}}{\left\{1+\alpha\left(A\left[ax_{ss}^{\frac{\sigma-1}{\sigma}}+1-a\right]^{\frac{\sigma}{\sigma-1}}\right)^{\phi}l^{\phi}\right\}^{1+\chi}}$. It is easy to

check that $\lim_{l \to 0} Z(l) = 0 < B_{nss}$. Also, $\lim_{l \to \infty} Z(l) = 0 < B_{nss}$ if and only if $\phi > \phi_2 \equiv \frac{1+\chi-\varepsilon_\gamma}{1+\chi}$.[25] Therefore, according to Proposition 2.(i), we will have steady state multiplicity whenever the tax rate is sufficiently procyclical, $\phi > \phi_2 > 0$, where $\phi_2 < \phi_1$. This means that whenever the government successfully locally stabilizes the normalized steady state, another steady state appears. We conclude that although a sufficiently strong procyclical tax rate is able to bring local saddle path stability to the normalized steady state, in a situation where this steady state would be indeterminate in the absence of taxation, procyclicality of the tax rate is not, per se, able to stabilize globally the economy.

We consider parameter values consistent with empirical studies and satisfying Assumption 4. We assume that $A = 0.4$, $a = 0.25$, $\beta = 0.9$ and $\theta = 0.2$ (which corresponds to a depreciation rate around 11 % consistent with annual data). We also assume $\sigma = 2.9$. This value is consistent with the estimates obtained by Duffy and Papageorgiou (2000).[26] Then, using Proposition 1, we see that $\left[ax_{ss}^{\frac{\sigma-1}{\sigma}}+1-a\right]^{\frac{1}{\sigma-1}}$ $ax_{ss}^{\frac{-1}{\sigma}} = \frac{\theta}{A\beta}$, so that $x_{ss} = 1$ and the steady state capital share of output

---

[25]To compute the $\lim_{l \to \infty} Z(l)$ we apply l'hopital rule, obtaining $\lim_{l \to \infty} Z(l) =$

$\lim_{l \to \infty} \dfrac{(1+\chi-\varepsilon_\gamma)\left[A(1-a)\left[ax_{ss}^{\frac{\sigma-1}{\sigma}}+1-a\right]^{\frac{1}{\sigma-1}}\right]^{1+\chi}}{(1+\chi)\phi\alpha\left(A\left[ax_{ss}^{\frac{\sigma-1}{\sigma}}+1-a\right]^{\frac{\sigma}{\sigma-1}}\right)^{\phi}\left[l^{\frac{\phi-1-\chi+\varepsilon_\gamma}{\chi}}+\alpha\left(A\left[ax_{ss}^{\frac{\sigma-1}{\sigma}}+1-a\right]^{\frac{\sigma}{\sigma-1}}\right)^{\phi}l^{\frac{\phi(1+\chi)-1-\chi+\varepsilon_\gamma}{\chi}}\right]}$. It is

easy to see that this limit takes the value 0 or ∞ depending on the parameter's values. Therefore $\lim_{l \to \infty} Z(l) < B_{nss}$ can only be satisfied when $\lim_{l \to \infty} Z(l) = 0$. This will happen if and only if $\phi > \frac{1+\chi-\varepsilon_\gamma}{1+\chi}$.

[26]They report robust estimates contained in [1.24, 3.24].

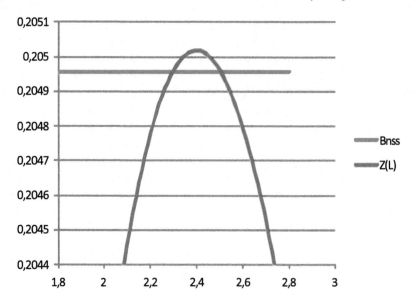

Fig. 11.1 Steady state multiplicity with procyclical tax rates

$s(x_{ss}) = 0.25$. Also $l_{nss} = 1/A = 2.5$.[27] We further assume $\varepsilon_y = 1.1$ and $\chi = 0.35$.[28] We consider that $B = B_{nss}$ so that the normalized steady state ($y_{nss} = 1$) exists. Under the parameterization considered, the normalized steady state is inde-terminate in the absence of taxation, i.e., $\varepsilon_y < \frac{(1-s)(1+\chi)}{1-\theta(1-s)}$, so that $\chi > \varepsilon_y - 1$. Note also that in this case, where ε_y and χ are constants, if the government is absent or fixes a constant tax rate ($\alpha = 0$ or $\phi = 0$ so that $\eta = 0$ for all values of $y > 0$), the normalized steady state, although indeterminate, would be unique. See Corollary 2.2. We set $\alpha = 0.15$ so that, at the normalized steady state the consumption tax rate is in line with the VAT tax rates observed in most industrialized economies. We also set $\phi = 1.5 > \frac{1+\alpha}{\alpha} \frac{1+\chi-\varepsilon_y}{1+\chi}$ so that the normalized steady state is a saddle, i.e., $\varepsilon_y > \varepsilon^*_{y\,(l_{nss})} = (1+\chi)(1-\eta(1)) = 1.086$. Then, from Proposition 3, we know that $\epsilon_{Z,l}(l_{nss}) < 0$, i.e. $Z(l)$ is decreasing at l_{nss}. Also, from Proposition 7, besides the normalized, there is another steady state l_a with a corresponding lower level of output, $y_a = Al_a < y_{nss} = Al_{ss} = 1$. This is illustrated in Fig. 11.1, where $l_a = 2.3$ is the other value of l such that $Z(l)$ crosses $Bnss$. Note that at l_a, $\epsilon_{Z,l} > 0$, i.e. $Z(l)$ is increasing at l_a. Since in our example we have $\sigma = 2.9 > \sigma^{**}$, i.e. $\varepsilon_{y1} > \varepsilon^*_y$, from Proposition 5 this other steady state is a sink.[29]

[27] Since $x_{ss} = 1 = f(x_{ss})$, this implies $k_{nss}/y_{nss} = l_{nss} = 2.5$, i.e., the capital-output ratio takes a value that is empirically plausible at annual data.

[28] Concerning the value of χ Maurer and Meier (2008) estimate significant peer effects, but in any case lower than 0.44.

[29] Remark that σ^{**} must be evaluated at the lower steady state, i.e., $\sigma^{**} \equiv \frac{s-\theta(1-s)(1-\eta(y_a))}{\eta(y_a)s} = 2.88$.

Our numerical example shows that, even if a sufficiently procyclical tax rate is able to eliminate local indeterminacy of the normalized steady state, (which is due to the existence of consumption externalities) it may not be able to stabilize globally the economy. Hence, using fiscal policy with stabilization purposes seems to be a hopeless task when other significant, but empirically plausible distortions are present. Moreover, in this case, government intervention may end up having pernicious effects since, instead of stabilizing the economy around the higher activity steady state, it may introduce fluctuations around a lower activity steady state. Note however that if consumption externalities are small or absent, procyclical consumption tax rates are able to guarantee saddle path stability and uniqueness of the steady state, as in Nourry et al. (2013). See Corollary 2.1.

Before closing this section it is worth referring that our results apply to a wide choice of functional forms for preferences and for the tax rate rule. In the special case where the tax wedge, $h(y)$, the externality function, $\varphi(c^w)$, and the offer curve, $\gamma(l)$, are isoelastic as in Corollary 1.(ii), procyclicality of the tax rates is able to eliminate indeterminacy, rendering the steady state saddle stable and unique. However, with our example, we have shown that when $\eta(y)$, the elasticity of the $h(y)$ function, is not constant, all the other relevant elasticities being constant, this uniqueness result no longer holds. It is easy to see that, even if the consumption tax wedge is an isoelastic function of income, it may be sufficient to have at least one non isoelastic specification, for $\varphi(c^w)$ or for $\gamma(l)$, such that $1 + \chi(c^w) - \varepsilon_\gamma(l) - (1 + \chi(c^w))\eta(y)$ changes sign for some $l > 0$ (see Proposition 3) in order to loose the uniqueness result. This means that our results are quite robust, and that guaranteeing uniqueness of the steady state with procyclical tax rates is almost impossible if other significant distortions are present in the economy.

11.6   Concluding Remarks

This work shows that saddle path stability around a steady state, which, in the absence of government intervention, would be indeterminate due to the existence of sufficiently important consumption externalities, is achievable under balanced budget through a sufficiently procyclical consumption tax rate. However, this procyclicality may be responsible for the appearance of another steady state, that can never be a saddle.

Our conclusions are illustrated using a numerical example for an economy with isoelastic consumption externalities, offer curve and consumption tax rate, but where the tax wedge is not isoelastic. When, under sufficiently strong consumption externalities, the consumption tax rate becomes sufficiently procyclical, the normalized steady state becomes a saddle, but another steady state with lower levels of output appears. Therefore, government intervention with stabilization purposes may end up deteriorating the performance of the economy, instead of improving it, leading to lower employment and output levels which are not on a stable saddle path. Our results were obtained in the context of a Woodford(1986)/Grandmont (1998) set up.

We conjecture that similar results apply in other frameworks, for instance, by considering externalities in the Ramsey model with non separable preferences as is Nourry et al. (2013). We think that further research confirming this result is worthwhile.

References

Alonso-Carrera, J., Caballé, J., & Raurich, X. (2008). Can consumption spillovers be a source of equilibrium indeterminacy? *Journal of Economic Dynamics and Control, 32,* 2883–2902.

Carlsson, F., Johansson-Stenman, O., & Martinsson, P. (2003). Do you enjoy having more than others? Survey evidence of positional goods. *Economica, 74,* 586–598.

Duffy, J., & Papageorgiou, C. (2000). A cross-country empirical investigation of the aggregate production function specification. *Journal of Economic Growth, 5,* 87–120.

Ferrer-i-Carbonell, A. (2005). Income and well-being: An empirical analysis of the comparison income effect. *Journal of Public Economics, 89,* 997–1019.

Gali, J. (1994). Keeping up with the joneses: consumption externalities, portfolio choice and asset prices. *Journal of Money Credit and Banking, 26,* 1–8.

Giannitsarou, C. (2007). Balanced budget rules and aggregate instability: The role of consumption taxes. *Economic Journal, 117,* 1423–1435.

Gokan, Y. (2006). Dynamic effects of government expenditure in a finance constrained economy. *Journal of Economic Theory, 127,* 323–333.

Gokan, Y. (2008). Alternative government financing and aggregate fluctuations driven by self-fulfilling expectations. *Journal of Economic Dynamics & Control, 32,* 1650–1679.

Grandmont, J.-M., Pintus, P., & de Vilder, R. (1998). Capital-labour substitution and competitive nonlinear endogenous business cycles. *Journal of Economic Theory, 80,* 14–59.

Guo, J. T., & Lansing, K. (1998). Indeterminacy and stabilization policy. *Journal of Economic Theory, 82,* 481–490.

Guo, J. T., & Harrison, S. (2008). Useful government spending and macroeconomic (in)stability under balanced-budget rules. *Journal of Public Economic Theory, 10*(3), 383–397.

Kamiguchi, A., & Tamai, T. (2011). Can productive government spending be a source of equilibrium indeterminacy? *Economic Modelling, 28,* 1335–1340.

Ljungqvist, L., & Uhlig, H. (2000). Tax policy and aggregate demand management under catching up with the Joneses. *American Economic Review, 90,* 356–366.

Lloyd-Braga, T., Modesto, L., & Seegmuller, T. (2008). Tax rate variability and public spending as sources of indeterminacy. *Journal of Public Economic Theory, 10*(3), 399–421.

Lloyd-Braga, T., Modesto, L., & Seegmuller, T. (2014). Market distortions and indeterminacy: A general approach. *Journal of Economic Theory, 151,* 216–247.

Maurer, J., & Meier, A. (2008). Smooth it like the 'Joneses'? Estimating peer-group effects in intertemporal consumption choice. *Economic Journal, 118*(527), 454–476.

Nourry, C., Seegmuller, T., & Venditti, A. (2013). Aggregate instability under balanced-budget consumption taxes: A re-examination. *Journal of Economic Theory, 148,* 1977–2006.

Pintus, P. (2004). *Aggregate instability in the fixed-cost approach to public spending.* Mimeo, Aix-Marseille.

Schmitt-Grohé, S., & Uribe, M. (1997). Balanced-budget rules, distortionary taxes, and aggregate instability. *Journal of Political Economy, 105,* 976–1000.

Wendner, R. (2010). Growth and keeping up with the Joneses. *Macroeconomic Dynamics, 14* (Supplement 2), 176–199.

Woodford, M. (1986). Stationary sunspot equilibria in a finance constrained economy. *Journal of Economic Theory, 40,* 128–137.

Part III
Growth

Chapter 12
Uncertainty and Sentiment-Driven Equilibria

Jess Benhabib, Pengfei Wang and Yi Wen

Abstract We construct a simple neoclassical model to capture the Keynesian idea that equilibrium aggregate supply is determined by aggregate demand and thus influenced by consumer sentiments about aggregate income. This result is nontrivial because in a standard neoclassical setting it is the aggregate supply (productive capacity) that determines aggregate demand, instead of the other way around, through factor-income shares and market-clearing mechanisms. However, we show that when firms' production and employment decisions must be based on expectations of aggregate demand and that realized demand follows from firms' production and employment decisions through market-clearing mechanisms, rational expectations about aggregate demand can lead to stochastic sentiment-driven equilibria despite the absence of production externalities, incomplete financial markets, strategic complementarity or any non-convexities in the model. The key is imperfect information arising naturally from the fact that production and employment decisions by firms, and consumption and labor supply decisions by households, are all made prior to goods being produced and exchanged, and before market clearing prices are realized. The sentiment-driven equilibria in our model are not based on randomizations over multiple fundamental equilibria as in many sunspot-driven business cycle models.

J. Benhabib (✉)
Department of Economics, New York University, 19 West 4th Street,
6th Floor, New York 10003, USA
e-mail: jess.benhabib@nyu.edu

P. Wang
Department of Economics, The Hong Kong University of Science
and Technology, Clear Water Bay, Hong Kong
e-mail: pfwang@ust.hk

Y. Wen
School of Economics and Management, Tsinghua University, Federal Reserve
Bank of St. Louis, P.O. Box 442, Beijing 63166, St. Louis, China
e-mail: yi.wen@stls.frb.org

© Springer International Publishing AG 2017
K. Nishimura et al. (eds.), *Sunspots and Non-Linear Dynamics*,
Studies in Economic Theory 31, DOI 10.1007/978-3-319-44076-7_12

Keywords Keynesian self-fulfilling equilibria · Sentiments · Extrinsic uncertainty ·
Sunspots

JEL codes: D8 · D84 · E3 · E32

12.1 Introduction

The Great Depression (1929–1939), the deepest and longest-lasting economic down-
turn in the history of the Western industrialized world featured 46 % plunge in industrial production, 32 % drop in whole sale prices, 25 % decline in aggregate employment, 70 % collapse in foreign trade, and nearly 50 % bank failures in the United
States and similarly in the Great Britain and other parts of Europe. It made John
Maynard Keynes (1936) ponder that such devastating outcome of market economies
could not have been a supply-side problem (or technology shocks to production
frontiers), because all the production capacities were still there during the recession
but simply remained idle. Keynes' explanation for the Great Depression was insuf-
ficient aggregate demand. His arguments for the dramatic failure of supply to create
its own demand during the Great Depression are based on the fact that in market
economies demand and supply are separated from each other, and made by imperfectly informed households and firms. Under various types of economic frictions,
Keynes argued, the invisible hand often fails to coordinate demand and supply. Such
market-coordination failures driven by pesmistic outlooks of the economy were the
root cause of the Great Depression. However, such self-fulfilling sentiment-driven
economic fluctuations are not possible in standard neoclassical general equilibrium
models where aggregate supply determines aggregate demand through factor-income
share and market-clearing mechanisms.

We construct a neoclassical model to capture the Keynesian insight. In the model
production and employment decisions by firms, and consumption and labor supply
decisions by households are made prior to goods being produced and exchanged and
before market clearing prices are realized. We characterize the rational expectations
equilibria of this model. We find that despite the lack of any non-convexities in
technologies and preferences, there can be multiple rational expectations equilibria
and fluctuations driven by consumer sentiments. The sentiment-driven equilibria are
the result of the signal extraction problem faced by firms when production decisions
must be made prior to the realization of demand, and household consumption plans
must be made prior to the realization of their income. Our model thus captures
the Keynesian notion that animal spirits, unconnected to fundamentals, can drive
employment and output fluctuations even under rational expectations and full-fledged
market clearing.[1]

---

[1]For the classical work on extrinsic uncertainty and sunspots, see Cass and Shell (1983).

Our model is inspired by Angeletos and La'O (2013) as well as by the Lucas (1972) island model.[2] Trades take place in our model in centralized markets rather than bilaterally through random matching, and at the end of the period all trading history is public knowledge. Informational asymmetries exist only within the period as firms decide on how much to produce on the basis of the signals they receive at the beginning of the period. Consumers are subject to aggregate preference shocks or equivalently productivity shocks to the final good aggregator. This differs from Benhabib et al. (2015) where preference shocks on each consumption good are idiosyncratic, and where the demand signal received by each firm reflects the idiosyncratic preference shock for its own good. Instead firms here make production and employment decisions based on a signal of aggregate consumer demand, possibly emanating from survey based private or governmental forecasts of aggregate demand. Since aggregate consumer demand reflects fundamental aggregate preference shocks as well as pure sentiments, the firms face a signal extraction problem even without the idiosyncratic preference shocks as in the model of BWW, and even if the aggregate demand signal that they receive is precise and not noisy. So in contrast to BWW, we show that under reasonable conditions there exists not a unique but a continuum sentiment-driven self-fulling equilibria, as well as multiple fundamental equilibria rather than a unique one. The sentiment-driven equilibria are not randomizations over the multiple fundamental equilibria of our model.

We describe the baseline model, the behavior of households, and the equilibria in the sections that follow. In Sect. 12.5 we extend our model to the case where consumer sentiments are heterogenous but correlated, and show that the earlier results continue to hold.

12.2   Baseline Model

We consider a simple Dixit-Stiglitz model where the final consumption good is produced by a representative final-good firm from a continuum of intermediate goods. Each intermediate good is produced by a single monopolistic firm. Producers of intermediate goods must make production and employment decisions before the demand for intermediate goods is realized. The output from each firm is then combined by a representative final-good producer to yield the final consumption good. Producers can perfectly observe the entire history of the economy up to the current decision period. At this stage production has not yet taken place, so households have only expectations or sentiments about their real wage and employment to guide their con-

---

[2]In the island model where agents on each island face a signal extraction problem Lucas (1972) showed that there is a unique fundamental equilibrium as realized prices reveal the ratio of the monetary shock to the the island-specific (population) shock. In a private communication to Lucas, Jean-Michel Grandmont noted a flaw in the uniqueness proof, which is acknowledged and discussed by Lucas (1983). For an extensive follow-up analysis see Chiappori and Guesnerie (1991).

sumption plans, along with aggregate shocks to their preferences—which is the only fundamental shock we consider in this paper. [3]

Households can infer what the real wage is based on their sentiments about aggregate demand and their aggregate preference shock. The firms, through market research, consumer surveys or private or government forecasts, obtain a noisy signal about consumer sentiments of aggregate demand. They try to infer the demand for their own particular intermediate good from this signal, so they face a signal extraction problem. They then hire workers from households by offering a nominal wage. Households, as already noted, as well as firms have an expectation of t prices and the real wage, which are correct in equilibrium.

We show that in equilibrium firms' expectations on the sentiments of the households will be self-fulfilling, in the sense that at realized prices the goods markets and the labor market will clear, household expectations of the prices and real wages will be correct, and firms' forecasts of aggregate consumption demand will be confirmed. Furthermore the actual equilibrium distribution of output will be consistent with the distribution of consumer sentiments in a stochastic self-fulfilling equilibrium. We obtain, therefore, a stochastic rational expectations equilibrium driven by consumer sentiments. In contrast to the indeterminacy literature (e.g., Benhabib and Farmer 1994), this equilibrium is not based on radomizations over multiple fundamental equilibria.[4] In addition to the sentiment-driven equilibria, wealso obtain multiple fundamental rational expectations equilibria driven only by fundamental shocks, despite the lack of any non-convexities in technologies and preferences.

To generate aggregate fluctuations, sentiments in our model must be correlated across households.[5] In the benchmark model, as in Benhabib et al. (2015), the aggregate sentiment is identical for all consumers. In the extension here in Sect. 12.5, the consumer sentiments have a common as well as an i.i.d. idiosyncratic component. In this augmented setup we obtain essentially the same results as in the benchmark case.

To be more explicit, a representative household derives utility from a final good and leisure. The final good is produced by a representative final goods producer using a continuum of intermediate goods indexed by $j \in [0, 1]$. Each intermediate good is produced using labor. We use labor as the numéraire so the wage rate is

---

[3]We can also interpret the preference shock as an aggregate productivity shock to the production of the aggregate final good.

[4]See Cass and Shell (1983) for the classical case of sunspot equilibrium that is not based on randomizations over fundamental equilibria. Peck and Shell (1991) also show the existence of sunspot equilibria in a finite economy with a unique fundamental equilibrium by allowing non-Walrasian trades prior to trading on the post-sunspot spot markets. Spear (1989) shows the possibility of an equilibrium for an OLG model with two islands where sunspot driven prices in one island act as sunspots for the other and vice versa. The equilibrium in each of the two islands depend on prices on both islands with sunspot variables eliminated, and the equilibrium is non-trivially stochastic.

[5]The correlated sentiments induce correlated optimal choices for firms to generate additional stochastic equilibria, similar to correlated equilibria in games where the introduction of correlations in players' strategies can enlarge a game's equilibrium possibilities beyond the set of Nash equilibria. See Aumann (1974, 1987), Maskin and Tirole (1987); Aumann et al. (1988); Peck and Shell (1991); Bergemann and Morris (2013, 2015); Bergemann et al. (2015).

fixed at 1. The real wage (in terms of final goods) can of course fluctuate with the price of the final goods. The households are subject to aggregate preference (fundamental) shocks and sentiment (non-fundamental) shocks in each period. In the equilibrium of the benchmark model the households have perfect foresight. Namely, conditional on the aggregate shock and their sentiments, they can perfectly forecast the price level. Based on the forecasted price, and therefore the real wage, they make their consumption and labor supply decisions. The consumption decisions made by the households are the source of noisy demand signals for the intermediate goods producers. Based on their demand signals, obtained through market research, intermediate goods producers decide how much to produce, and the price of each intermediate good adjusts to equalize demand and supply on each island. These prices then determine the average cost of the final good and hence the price of the final good. In equilibrium this realized price coincides with the price expected by households based on sentiments. The results extend to the case where consumer sentiments are heterogenous but correlated.

The following road-map specifies the sequence of actions by consumers and firms, the information structure, and the rational expectations equilibria of our benchmark model:

1. At the beginning of each period, households form expectations on aggregate output/income based on their sentiments. They also form demand functions for each differentiated good based on their sentiments and the aggregate preference shocks realized in the beginning of the period, contingent on the market prices to be realized when the goods markets open.
2. Like households, firms also believe that aggregate output/demand could be driven by sentiments. Unlike households, firms do not directly observe households' sentiments or preference shocks. Firms instead receive a noisy signal about aggregate demand.
3. Given a nominal wage, households make labor supply decisions based on their sentiments, and firms make employment and production decisions based on their signals, taking the expected real wage and prices as given. [6]
4. Goods markets open and goods are exchanged at market clearing prices, the real wage and actual consumption are realized.

12.2.1 Households

A representative household derives utility from final goods and leisure according to the utility function

---

[6]At this point, since goods markets have not opened and goods prices have not yet been realized, there is no guarantee that labor demand should magically (automatically) equal labor supply. We will show, however, that in equilibrium, where the distribution of sentiments is pinned down, labor supply will always equal labor demand.

$$U_t = A_t \frac{C_t^{1-\gamma}}{1-\gamma} - \frac{N_t^{1+\eta}}{1+\eta}, \tag{12.1}$$

where C_t is consumption of the final good, A_t is the preference shock, and N_t is labor supply. We assume that $\eta = 0$ for convenience. [7] The parameter γ is the inverse of the price elasticity of final good consumption. We normalize the nominal wage to 1 and write the household's budget constraint as $P_t C_t \leq N_t + \Pi_t$, where P_t is the price of the final good and Π_t is the aggregate profit income from all intermediate firms. Define $\frac{1}{P_t}$ as the real wage, then the budget constraint becomes

$$C_t \leq \frac{1}{P_t} N_t + \frac{\Pi_t}{P_t}. \tag{12.2}$$

Note that the real incomes of households fluctuate with P_t. The first-order condition for C_t is

$$A_t C_t^{-\gamma} = P_t. \tag{12.3}$$

A conjectured decrease in the price level P_t will induce the household to consume more. Households observe the aggregate preference shock A_t and an aggregate sentiment ("sunspot") shock Z_t and conjecture that the equilibrium aggregate price is given by $P_t = P(A_t, Z_t)$ and therefore that the real wage is $(P_t)^{-1}$. We assume $z_t \equiv \log(Z_t)$ is normally distributed with zero mean and unit variance. An equilibrium is a "fundamental equilibrium" if it is not affected by z_t. Otherwise we call the equilibrium a "sentiment-driven equilibrium".

12.2.2 Firms

The supply side has a representative final good producer and a continuum of intermediate goods producers indexed by $j \in [0, 1]$. The final good producer serves as an aggregator of all intermediate goods, and it does not play an active role in the model. We assume the final good producer makes decisions after all shocks are realized, so its decisions are not subject to any uncertainty.

The final good firm. The final good firm solves

$$\max_{C_{jt}} P_t C_t - \int P_{jt} C_{jt} dj, \tag{12.4}$$

where C_t is produced by a continuum of intermediate goods according to the Dixit-Stiglitz production function,

---

[7] The quasi-linear utility function is assumed for simplicity without loss of generality.

$$C_t = \left[\int_0^1 C_{jt}^{\frac{\theta-1}{\theta}} \, dj \right]^{\frac{\theta}{\theta-1}}. \tag{12.5}$$

The final goods producer's profit maximization problem yields the inverse demand function for each intermediate good,

$$\frac{P_{jt}}{P_t} = C_{jt}^{-\frac{1}{\theta}} C_t^{\frac{1}{\theta}}, \tag{12.6}$$

and the aggregate price index,

$$P_t = \left[\int_0^1 P_{jt}^{1-\sigma} \right]^{\frac{1}{1-\sigma}}. \tag{12.7}$$

The intermediate goods firms. The intermediate goods firms use labor as the only input to produce output according to

$$C_{jt} = N_{jt}. \tag{12.8}$$

Unlike the households and the final good producer, the intermediate goods producers face uncertainty in making their production decisions: they do not have full information regarding the aggregate demand shock A_t and the aggregate price P_t. We assume that intermediate firm j has to choose its production based on a noisy signal about aggregate demand. Denote the signal as S_{jt}. The intermediate good firm j solves

$$\max_{C_{jt}} E[(P_{jt} C_{jt} - C_{jt})|S_{jt}], \tag{12.9}$$

with the constraint (12.6). Substituting out P_{jt}, the first-order condition for C_{jt} is

$$C_{jt} = \left\{ E[P_t C_t^{\frac{1}{\theta}} | S_{jt}] \left(1 - \frac{1}{\theta} \right) \right\}^{\theta}. \tag{12.10}$$

Using the first-order condition of the household in Eq. (12.3), we then have

$$C_{jt} = \left(1 - \frac{1}{\theta} \right)^{\theta} \left\{ E[A_t C_t^{\frac{1}{\theta}-\gamma} | S_{jt}] \right\}^{\theta}. \tag{12.11}$$

We assume that the signal is a mixture of aggregate demand (C_t) and idiosyncratic noise (v_{jt}) given by

$$s_{jt} \equiv \log S_{jt} = \log C_t + v_{jt} \equiv c_t + v_{jt}, \tag{12.12}$$

where v_{jt} is normally distributed with mean of 0 and variance of σ_v^2. For notational convenience we will re-scale the aggregate preference shock A_t as $A_t = \left(\frac{\theta}{\theta-1}\right)\exp(a_t/\theta)$, where a_t is normally distributed with mean 0 and variance σ_a^2.

We note that in what follows, the noise v_{jt} will not be essential for our results: we could have set $\sigma_v^2 = 0$. In that case the signal s_{jt} would fully reveal aggregate consumption c_t to the intermediate goods firms; but, as we will see in Sect. 12.4, sentiment-driven rational expectations equilibria would still exist.

12.2.3 General Equilibrium

We define the general equilibrium recursively as follows:

- Based on the preference shock A_t and sentiment Z_t, households conjecture that the aggregate price is $P_t = P(A_t, Z_t)$, and real wage is $(P_t)^{-1}$;
- Based on the conjectured price P_t and real wage is $(P_t)^{-1}$, the households choose their consumption plan $C_t = C(A_t, Z_t)$ according to (12.3) to maximize their utility;
- The consumption decisions create signals to firms j as $\log S_{jt} = c_t + v_{jt}$;
- Based on the signal S_{jt}, firm j hires workers and produces C_{jt} according to (12.11) to maximize its expected profit;
- Given the production of C_{jt}, price P_{jt} adjusts to equate demand and supply according to Eq. (12.6);
- The total production of final good C_t, according to (12.5), equals the households' planned consumption. Hence the realized aggregate price is equal to the conjectured price P_t and the realized real wage is the conjectured real wage is $(P_t)^{-1}$.

It turns out that Eqs. (12.5), (12.11), and (12.12) are sufficient to characterize the general equilibrium. We conjecture that the equilibrium production (in logarithm) can be written as $\log C_{jt} = \tilde{c} + c_{jt}$ and $\log C_t = \bar{c} + c_t$ and that c_{jt} and c_t are solutions to the following systems of equations:

$$c_{jt} = E\{[a_t + \beta c_t]|s_{jt}\}, \tag{12.13}$$

$$c_t = \int_0^1 c_{jt}\,dj, \tag{12.14}$$

$$s_{jt} = c_t + v_{jt}, \tag{12.15}$$

where

$$\beta \equiv 1 - \gamma\theta. \tag{12.16}$$

Notice that $\theta > 1$ and $\gamma > 0$; hence, we have $\beta \in (-\infty, 1)$. The intermediate firm's output c_{it} would decrease with aggregate demand c_t if $\beta < 0$, which implies that intermediate goods are strategic substitutes; whereas $\beta > 0$ would correspond to the case of strategic complementarity among intermediate goods. Hence, our model is flexible enough to characterize both strategic complementarity and strategic sub-stitutability in production, and our results hold true even if $\beta < 0$. Equilibrium in the model is then fully characterized by $\{\tilde{c}, \bar{c}\}$ and two mappings, $c_{jt} = c_{jt}(s_{jt})$ and $c_t = c(a_t, z_t)$ that solve Eq. (12.13) for all j and Eq. (12.14). We are now ready to characterize all the possible equilibria.

12.3 Fundamental Equilibria

We first study the equilibria driven only by fundamentals, in particular by preference shocks a_t. In a fundamental equilibrium neither aggregate consumption c_t nor the production of each intermediate good c_{jt} is affected by consumer sentiments. We show that this simple model permits multiple fundamental equilibria.

We use a conjecture-and-verification strategy to find the equilibria. A guess for the solution to the system of Eqs. (12.13)–(12.15) is

$$c_t = \phi a_t, \tag{12.17}$$

where ϕ is an undetermined coefficient. Finding equilibrium is then equivalent to determining the coefficient ϕ.

12.3.1 A Constant Output Equilibrium

Proposition 1 *The allocation with $p_t \equiv \log P_t - \bar{p} = a_t/\theta$ and $c_{jt} = c_t = 0$ is always an equilibrium.*

Proof See Appendix. ∎

In this case consumption does not respond to preference shock a_t; namely, the demand elasticity $\phi = 0$ and the equilibrium aggregate price $P_t = \exp(\bar{p} + a_t/\theta)$. To see the intuition, suppose that there is an increase in a_t, which makes households want to spend more if all else is equal. But whether households actually spend more also depends on their expectation of the aggregate price (or real wage). In the above equilibrium, households conjecture that the price will rise exactly in proportion to preference shocks so their incentive to consume more is completely curbed.

12.3.2 Stochastic Fundamental Equilibria

In this case consumption responds to the preference shock a_t with the demand elas-ticity $\phi \in \left(0, \frac{1}{\gamma\theta}\right)$. We show that there can be two such equilibria in the model under certain conditions. Suppose household consumption in logarithm is given by $\log C_t - \bar{c} = c_t = \phi a_t$, with $\phi > 0$. Households conjecture that price is given by

$$\log P_t - \bar{p} = \phi_p a_t = \left(\frac{1}{\theta} - \gamma\phi\right) a_t. \tag{12.18}$$

Note that Eq. (12.3) is satisfied, implying that the households' consumption is optimal. In what follows we use the method of undetermined coefficients to determine the coefficient ϕ and the constants $\bar{c}$ and $\bar{p}$.

To solve for ϕ, we utilize Eq. (12.13). Using the above conjectured equilibrium for c_t, we express the production of each intermediate goods firm as

$$\log C_{jt} - \bar{c} = c_{jt} = E(a_t + \beta c_t)|(\phi a_t + v_{jt}) = \frac{(\phi + \beta\phi^2)\sigma_a^2}{\phi^2\sigma_a^2 + \sigma_v^2}(\phi a_t + v_{jt}). \tag{12.19}$$

Aggregating all firms' output across j gives aggregate output c_t; then, by matching the coefficient of a_t, we obtain $\frac{(\phi+\beta\phi^2)\phi\sigma_a^2}{\phi^2\sigma_a^2+\sigma_v^2} = \phi$. Rearranging terms leads to a quadratic equation in ϕ:

$$(\phi + \beta\phi^2)\sigma_a^2 = \phi^2\sigma_a^2 + \sigma_v^2. \tag{12.20}$$

Notice that in general, there is no guarantee that the solution to the above equation is unique. Denoting

$$\mu = \frac{\sigma_v^2}{\sigma_a^2} \tag{12.21}$$

as the noise ratio, we have the following Proposition:

Proposition 2 *Suppose $0 < \mu < \frac{1}{4(1-\beta)}$ and let $c_t = \log C_t - \bar{c} = \phi a_t$. In a rational expectations equilibrium the aggregate price is*

$$p_t = \log P_t - \bar{p} = \left(\frac{1}{\theta} - \gamma\phi\right) a_t, \tag{12.22}$$

each firm j produces

$$c_{jt} = \phi a_t + v_{jt}, \tag{12.23}$$

where ϕ is given by

$$\phi = \frac{1}{2(1-\beta)} \pm \sqrt{\frac{1}{4(1-\beta)^2} - \frac{\mu}{1-\beta}} \in \left(0, \frac{1}{\gamma\theta}\right), \tag{12.24}$$

and $\{\bar{c}, \bar{p}, \tilde{c}\}$ are given by

$$\bar{c} = \frac{1}{2}\frac{1}{\theta^2\gamma}[(\theta - 1)\sigma_v^2 + (1 + \beta\phi)(1 - (1 - \beta)\phi\sigma_a^2)], \tag{12.25}$$

$$\bar{p} = \log\left(\frac{\theta}{\theta - 1}\right) - \gamma\bar{c}, \tag{12.26}$$

$$\tilde{c} = (1 - \theta\gamma)\bar{c} + \frac{1}{2}\frac{(1 + \beta\phi)(1 - (1 - \beta)\phi)\sigma_a^2}{\theta} \tag{12.27}$$

Proof See Appendix.   ∎

This proposition shows that for any given value of the noise ratio $\mu \in \left(0, \frac{1}{4(1-\beta)}\right)$, there exist two additional fundamental equilibria: each corresponds to a particular value of ϕ. In the special case where $v_{jt} \equiv 0$ so that $\sigma_v^2 = 0$ and the signal fully reveals aggregate demand c_t to the firms, it is easy to see from Eq. (12.20) that one of the equilibria now coincides with the constant output equilibrium with $\phi = 0$. In the second equilibrium we have $\phi = (1 - \beta)^{-1}$. Since the signal reveals c_t fully, by Eq. (12.23) a_t is also fully revealed to firms in equilibrium, so we may call this type of equilibrium (with $\sigma_v^2 = 0$) the full information equilibrium.

In the two additional fundamental equilibria of Proposition 2, the equilibrium price does not respond fully to preference shocks. If the households think price will respond to the preference shocks less strongly, they will consume more in the aggregate when a_t increases. This then sends a more precise signal to the intermediate goods producers, as consumption volatility would be relatively larger relative to the noise in the signal. As a result, the firms produce more and, indeed, the aggregate market clearing price rises less, confirming the initial belief of the households.[8]

12.4   Sentiment-Driven Equilibria

Now we consider another type of equilibrium in which consumption responds to a pure sentiment variable z_t that is completely unrelated to the fundamental shock a_t. More importantly, the variance (uncertainty) of sentiment is itself a self-fulfilling object. We note that the existence of sentiment-driven equilibria is not based on randomizations over the fundamental equilibria studied above.

Suppose households incur a sentiment shock called z_t. After knowing the sentiment shock and observing aggregate preference shock a_t, households choose their optimal consumption based on their conjecture of the price level. Let us conjecture an equilibrium in which household consumption takes the form

$$c_t = \phi a_t + \sigma_z z_t, \tag{12.28}$$

---

[8]In Benhabib et al. (2015), where firms that produce intermediate goods face idiosyncratic demand shocks as opposed to aggregate demand shocks, there also are sentiment-driven stochastic equilibria, but the fundamental equilibrium is always unique.

along with the conjectured price $p_t = \alpha_a a_t + \alpha_z z_t$. For notational convenience, we have normalize the variance of z_t to unity, so the scaler σ_z represents the standard deviation of the sentiment shock (i.e., $var(\sigma_z z_t) = \sigma_z^2$). Given aggregate consumption demand, the production of the individual firm j is

$$c_{jt} = E(a_t + \beta c_t)|(c_t + v_{jt}) \tag{12.29}$$

$$= E(a_t + \beta \phi a_t + \beta \sigma_z z_t)|(\phi a_t + \sigma_z z_t + v_{jt}) \tag{12.30}$$

$$= \frac{(\phi + \beta \phi^2)\sigma_a^2 + \beta \sigma_z^2}{\phi^2 \sigma_a^2 + \sigma_v^2 + \sigma_z^2}(\phi a_t + \sigma_z z_t + v_{jt}).$$

Aggregating firm-level production across j and comparing coefficients of a_t and z_t between this aggregated equation and Eq. (12.28) gives

$$\phi = \frac{(\phi + \beta \phi^2)\sigma_a^2 + \beta \sigma_z^2}{\phi^2 \sigma_a^2 + \sigma_v^2 + \sigma_z^2}\phi \tag{12.31}$$

and

$$\frac{(\phi + \beta \phi^2)\sigma_a^2 + \beta \sigma_z^2}{\phi^2 \sigma_a^2 + \sigma_v^2 + \sigma_z^2} = 1. \tag{12.32}$$

Notice that Eqs. (12.31) and (12.32) are identical as long as $\phi \neq 0$.

Lemma 1 *If $\phi = 0$, then there is no sentiment-driven equilibrium.*

Proof The proof is straightforward. If $\phi = 0$ and a sentiment-driven equilibrium exists, then (12.32) becomes

$$(\beta - 1)\sigma_z^2 = \sigma_v^2 \geq 0 \tag{12.33}$$

Since $\beta < 1$, we have a contradiction.   ∎

Proposition 3 *Suppose $\phi > 0$ and $\sigma_v^2 < \frac{1}{4(1-\beta)}\sigma_a^2$. There exists a continuum of sentiment-driven equilibria indexed by variance of sentiments in the interval $\sigma_z^2 \in \left(0, \frac{1}{4(1-\beta)^2}\sigma_a^2 - \frac{\sigma_v^2}{1-\beta}\right)$. At each sentiment-driven equilibrium within this interval, the equilibrium price is given by*

$$p_t = \log P_t - \bar{p} = \left(\frac{1}{\theta} - \gamma \phi\right) a_t - \gamma \sigma_z z_t, \tag{12.34}$$

and the optimal consumption level is given by

$$c_t = \log C_t - \bar{c} = \phi a_t + \sigma_z z_t, \tag{12.35}$$

where ϕ is given by

$$\phi = \frac{1}{2(1-\beta)} \pm \sqrt{\frac{1}{4(1-\beta)^2} - \frac{\tilde{\mu}}{1-\beta}} > 0 \qquad (12.36)$$

and $\tilde{\mu} = \frac{\sigma_v^2 + \sigma_z^2(1-\beta)}{\sigma_a^2}$. The constants in the price and consumption rules are given by

$$\bar{c} = \frac{1}{2\gamma\theta^2}[(\theta - 1)\sigma_v^2 + (1 + \beta\phi)(1 - (1 - \beta)\phi)\sigma_a^2 - \beta\sigma_z^2(1 - \beta)]. \qquad (12.37)$$

$$\bar{p} = \log\left(\frac{\theta}{\theta - 1}\right) - \gamma\bar{c}. \qquad (12.38)$$

Proof See Appendix. ∎

Re-arranging Eq. (12.32) yields

$$\sigma_z^2 = \frac{\phi(1 - (1 - \beta)\phi)\sigma_a^2 - \sigma_v^2}{1 - \beta}. \qquad (12.39)$$

Hence, the equilibrium aggregate demand (production) is determined by

$$c_t = \phi a_t + \sqrt{\frac{\phi(1 - (1 - \beta)\phi)\sigma_a^2 - \sigma_v^2}{1 - \beta}} z_t, \qquad (12.40)$$

which shows that not only the level of sentiments z_t matters but the variance of sentiments also matters (since ϕ depends on σ_z by Eq. (12.36)). More importantly, the degree of uncertainty, that is σ_z^2, is itself self-fulfilling in a sentiment-driven equilibrium.

We can use Eq. (12.39) to rewrite Eq. (12.37) as

$$\bar{c} = \frac{1}{2\gamma\theta^2}[(\theta - 1 + \beta)\sigma_v^2 + (1 - (1 - \beta)\phi)\sigma_a^2)], \qquad (12.41)$$

where ϕ is given by (12.36). It is evident that the effect of σ_z^2 on the mean consumption depends on the value of ϕ. For the equilibrium with $\phi = \frac{1}{2(1-\beta)} + \sqrt{\frac{1}{4(1-\beta)^2} - \frac{\tilde{\mu}}{1-\beta}}$, an increase in σ_z^2 reduces mean consumption while for the equilibrium with $\phi = \frac{1}{2(1-\beta)} - \sqrt{\frac{1}{4(1-\beta)^2} - \frac{\tilde{\mu}}{1-\beta}}$, an increase in σ_z^2 will increase mean consumption.

The intuition for the sentiment-driven equilibria is similar to the intuition for multiple fundamental equilibria. Which equilibrium prevails depends on consumer's expectation of the aggregate price level, which depends negatively on the sentiments (Eq. (12.34)). If consumers are optimistic, they would anticipate a lower aggregate price level (cheaper consumption goods, higher real wage), so they choose to consume more. Since firms cannot distinguish the fundamental shock a_t from the sentiment shock z_t, they choose to produce more to meet the higher expected consumption demand indicated by the signal, which fulfills the consumer sentiments. On the other

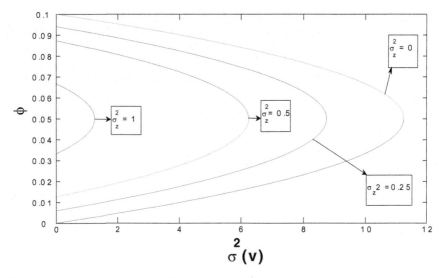

Fig. 12.1 Contour of the relationship between ϕ and σ_v^2 for any given σ_z

hand, if demand is more volatile due to more variable sentiments, firms would opt to attach less weight to fundamentals (preferences) in their signal extraction, rendering production more volatile. Namely at each sentiment-driven equilibrium (indexed by σ_z), for each realization of the sentiment z, the intermediate goods firms produce exactly the amount of goods, aggregated into the final good, that the households want to consume, and markets clear.

The results of Proposition 3 still hold even if we set $\sigma_v^2 = 0$, that is if we allow the signal s_{jt} to fully reveal aggregate demand c_t to intermediate goods firms. Nevertheless, as we see from (12.30), the firms set their optimal outputs under imperfect information using their signal $s_{jt} = c_t$ because they do not directly observe a_t and z_t separately. So in this case even if aggregate consumption c_t is fully observed by intermediate goods firms, we see from (12.35) that sentiments z_t still drive aggregate and firm outputs in rational expectations equilibria, and aggregate output will be equal to the sentiment signal z on aggregate demand for each realization of z.

In Fig. 12.1 we plot the coefficients ϕ for the fundamental stochastic equilibria in Proposition 2, and the corresponding coefficients ϕ for the sentiment-driven equilibria in Proposition 3, against variance of the noise σ_v^2. We calibrate $\theta = 10$,[9] $\gamma = 1$, the variance of log A_t at 4.5, and we plot ϕ against feasible σ_v^2 for various variances of sentiments $\sigma_z^2 = \{0, 0.25, 0.5, 1\}$.

Note that $\sigma_z^2 = 0$ (the outmost contour or hyperbola) yields the pairs of ϕ for the two fundamental stochastic equilibria for each value of σ_v^2. Figure 12.1 thus makes it clear that these fundamental equilibria may be viewed as a special case

---

[9]In typical calibrations $\theta = 10$ implies a markup of about 11 %.

of sentiment-driven equilibria where the variance of sentiments σ_z^2 go to zero. [10]
We can also observe in Fig. 12.1 how changing σ_z^2 generates additional pairs of ϕ
centering sentiment-driven stochastic equilibria for various values of σ_v^2. For each
σ_v^2 we may have up to five types of equilibria: a continuum of pairs of sentiment-
driven equilibria indexed by σ_z^2, a pair of stochastic fundamental equilibria where
aggregate consumption c_t is driven only by fundamental shocks a_t, and a constant
output equilibrium.

Persistence in output can be introduced in a variety of ways. A simple way is to
make the productivity parameter A_t persistent, with its past values only observable
at the beginning of each new period. Alternatively we could easily introduce a simple Markov sunspot process that selects the equilibrium in each period, alternating
between the fundamental and sentiment-driven equilibria, an exercise that we leave
to the interested reader.

12.5 An Extension

In the baseline model we assumed that all households have the same sentiment z_t. In
this section, we show that our results are robust to heterogenous sentiment shocks.
We index individual households by $i \in [0, 1]$. Suppose in the beginning of each
period households receive a noisy sentiment signal z_{it},

$$z_{it} = z_t + e_{it}, \tag{12.42}$$

so that the sentiments are correlated across households because of the common
component z_t. Suppose consumers choose their consumption expenditure C_{it} on the
basis of expected price given their signal z_{it}. As before, suppose each household
conjectures that the aggregate price will be determined by

$$\log P_t - \bar{p} = p_t = \phi_a^p a_t + \sigma_z^p z_t, \tag{12.43}$$

with undetermined coefficients $\{\phi_a^p, \sigma_z^p\}$. In a competitive environment, consumers
have the incentive to figure out the aggregate sentiment z_t because it matters for
the aggregate price level and the real wage. Each consumer therefore faces a signal
extraction problem. [11]

---

[10] So the sentiment equilibria collapse to the fundamental one as σ_z^2 goes to zero. In a different
context Manuelli and Peck (1994) show that in an OLG model with two islands, as in Spear (1989),
that shocks to fundamentals may also act as sunspots thyat amplify fluctuations, even as these shocks
get arbitrarily small.

[11] It is easy to see that the fundamental equilibrium is not affected by sentiments: if the aggregate
price depends only on the aggregate preference shock, the sentiment shocks will not affect the
consumption decision of the households.

The first-order condition for consumers now changes to

$$C_{it} = \left\{ \frac{1}{E(P_t|z_{it})} \exp(a_t/\theta) \left(\frac{\theta}{\theta - 1} \right) \right\}^{\frac{1}{\gamma}}. \tag{12.44}$$

Aggregating across consumers, we obtain the aggregate consumption $c_t = \log C_t = \log(\int_0^1 C_{it} di)$. As before, we assume that each firm receives a noisy signal $\log S_{jt} = c_t + v_{jt}$. The production decision by the firms is given by Eq. (12.10) as before, namely,

$$C_{jt} = \left\{ E[P_t C_t^{\frac{1}{\theta}} | S_{jt}](1 - \frac{1}{\theta}) \right\}^\theta. \tag{12.45}$$

An equilibrium of the economy is defined again as in Sect. 12.2.3. We have the following Proposition:

Proposition 4 *Suppose $\sigma_v^2 < \frac{1}{4(1-\beta)}\sigma_a^2$ and let $\kappa = \frac{1}{1+\sigma_e^2}$. There exists a continuum of sentiment-driven equilibria indexed by $\sigma_z^2 \in \left(0, \frac{\kappa}{4(1-\beta)^2}\sigma_a^2 - \frac{\kappa\sigma_v^2}{1-\beta} \right)$. At each equilibrium the aggregate price is*

$$p_t = \log P_t - \bar{p} = \phi_a^p a_t + \phi_z^p z_t \equiv \left(\frac{1}{\theta} - \gamma\phi \right) a_t - \frac{\gamma}{\kappa} \sigma_z z_t,$$

and the aggregate consumption (output) is

$$\log C_t - \bar{c} = c_t = \phi a_t + \sigma_z z_t, \tag{12.46}$$

where

$$\phi = \frac{1}{2(1 - \beta)} \pm \sqrt{\frac{1}{4(1 - \beta)^2} - \frac{\tilde{\mu}}{1 - \beta}} \tag{12.47}$$

and $\tilde{\mu} = \frac{\sigma_v^2 + \sigma_z^2(1-\beta)/\kappa}{\sigma_a^2}$. Consumers' idiosyncratic consumption demand is

$$\log C_{jt} - \tilde{c} = c_{it} = \phi a_t + \sigma_z(z_t + e_{it}). \tag{12.48}$$

Each individual firm's optimal production is

$$\log C_{it} - \hat{c} = c_{jt} = \phi a_t + \sigma_z z_t + v_{jt}. \tag{12.49}$$

The constant terms are given by

$$\bar{p} = \log(\frac{\theta}{\theta - 1}) - \frac{\theta - 1}{2\theta^2}\sigma_v^2 - \frac{1}{2}\Omega_s \tag{12.50}$$

$$\bar{c} = \frac{1}{\gamma}[\frac{\theta - 1}{2\theta^2}\sigma_v^2 + \frac{1}{2}\Omega_s] - \frac{\gamma}{2}\left(\frac{1}{\kappa}\sigma_z\right)^2 (1 - \kappa) + \frac{1}{2}\sigma_z^2\frac{1 - \kappa}{\kappa} \qquad (12.51)$$

$$\hat{c} = \bar{c} - \frac{1}{2}\sigma_z^2\sigma_e^2, \qquad \tilde{c} = \bar{c} - \frac{1}{2}\frac{\theta - 1}{\theta}\sigma_v^2. \qquad (12.52)$$

$$\tilde{c} = \bar{c} - \frac{1}{2}\frac{\theta - 1}{\theta}\sigma_v^2. \qquad (12.53)$$

Proof See Appendix.   ∎

12.6   Conclusion

We explore the Keynesian idea that sentiments or animal spirits can influence the level of aggregate income and give rise to recurrent boom-bust cycles. We show that in a neoclassical production economy, pure sentiments (completely unrelated to fundamentals) can indeed affect economic performance and the business cycle even though (i) expectations are fully rational and (ii) there are no externalities or non-convexties or even strategic complementarities. In particular, we show that when consumption and production decisions must be made separately by consumers and firms based on mutual forecasts of each other's actions, the equilibrium outcome can indeed be influenced by animal spirits or sentiments, even though all agents are fully rational and all market are cleared by flexible prices. Furthermore the existence of sentiment-driven equilibria is not based on randomizations across the fundamental equilibria. The key to generating our results is a natural friction in information: Even if firms can perfectly observe or forecast aggregate consumption demand (without any noises), they still cannot separately identify the components of aggregate demand stemming from consumer sentiments as opposed to preference shocks (fundamentals). Sentiments matter because they are correlated across households, so they affect aggregate demand and real wages differently than shocks to aggregate productivity (or preferences). Faced with a signal extraction problem, firms make optimal production decisions that depend on the degree of sentiment uncertainty or the variance of sentiment shocks. In our model there exists a continuum of (normal) distributions for sentiment shocks indexed by their variances that give rise to self-fulfilling rational expectations equilibria.

A Appendix

A.1 Proof of Proposition 1

Proof Suppose households conjecture that the aggregate price is given by

$$p_t = \log P_t - \bar{p} = a_t/\theta, \tag{12.54}$$

where P_t satisfies Eq. (12.3). Then aggregate consumption must be a constant $\bar{C}$. This implies that the signal $s_{jt} = \log C_t + v_{jt}$ is nothing but pure noise. Hence, by Eq. (12.10) each firm's production is also a constant given by $C_{jt} = C_t = \bar{C}$. Equation (12.11) can be written as

$$C_{jt} = \left\{ E[\exp(a_t/\theta)C_t^{\frac{1}{\theta}-\gamma}|S_{jt}] \right\}^{\theta} = C_t^{1-\gamma\theta} \left\{ E[\exp(a_t/\theta)|S_{jt}] \right\}^{\theta}, \tag{12.55}$$

which, under the log-normal assumptions, implies

$$\gamma\theta \log C_t = \theta \log E \exp(a_t/\theta) = \frac{1}{2}\theta\frac{\sigma_a^2}{\theta^2}. \tag{12.56}$$

This implies

$$\log C_t = \log C_{jt} = \bar{c} = \tilde{c} = \frac{1}{2\gamma\theta^2}\sigma_a^2. \tag{12.57}$$

Since the conjecture of the aggregate price is self-fulfilling, the total supply is indeed a constant and all markets clear under the conjectured prices.   ∎

A.2 Proof of Proposition 2

Proof From the definition of μ we obtain

$$\phi(1 - (1-\beta)\phi) = \mu. \tag{12.58}$$

Note that there are two solutions for ϕ if $0 < \mu < \max_\phi \phi(1-(1-\beta)\phi) = \frac{1}{4(1-\beta)}$, given by

$$\phi = \frac{1}{2(1-\beta)} \pm \sqrt{\frac{1}{4(1-\beta)^2} - \frac{\mu}{1-\beta}} > 0. \tag{12.59}$$

It is easy to see that for $\mu > 0$

$$0 < \phi < \frac{1}{1 - \beta}. \tag{12.60}$$

Given ϕ, we can calculate the three constants $\tilde{c}, \bar{c}$ and $\bar{p}$ to fully characterize the equilibrium. The fact that aggregate consumption is log-normally distributed implies that we can obtain $\tilde{c}$ from Eq. (12.11),

$$\tilde{c} = (1 - \theta\gamma)\bar{c} + \frac{\theta}{2}\Omega_s, \tag{12.61}$$

where Ω_s is the conditional variance of $a_t/\theta + (\frac{1}{\theta} - \gamma)\phi a_t$ based on the signal. The variance Ω_s is:

$$\begin{aligned}
\Omega_s &= var[(a_t/\theta + (\frac{1}{\theta} - \gamma)\phi a_t)|\phi a_t + v_{jt}] \\
&= \frac{1}{\theta^2} var(a_t + \beta\phi a_t|\phi a_t + v_{jt}) \\
&= \frac{(1 + \beta\phi)^2\sigma_a^2 - (1 + \beta\phi)\phi\sigma_a^2}{\theta^2} \\
&= \frac{(1 + \beta\phi)(1 - (1 - \beta)\phi)\sigma_a^2}{\theta^2}.
\end{aligned} \tag{12.62}$$

Finally, notice that $c_{jt} = \phi a_t + v_{jt}$, so the dispersion in the production of the intermediate goods is purely due to the noisy signal. We then obtain

$$\bar{c} = \frac{1}{2}\frac{\theta - 1}{\theta}\sigma_v^2 + \tilde{c} \tag{12.63}$$

by Eq. (12.5). With the two equations and two unknows $\bar{c}$ and $\tilde{c}$, we obtain

$$\bar{c} = \frac{1}{2}\frac{1}{\theta^2\gamma}[(\theta - 1)\sigma_v^2 + (1 + \beta\phi)(1 - (1 - \beta)\phi)\sigma_a^2]. \tag{12.64}$$

Once we obtain $\bar{c}$, by Eq. (12.3) we can obtain $\bar{p} = \log\left(\frac{\theta}{\theta-1}\right) - \gamma\bar{c}$ and

$$\tilde{c} = (1 - \theta\gamma)\bar{c} + \frac{1}{2}\frac{(1 + \beta\phi)(1 - (1 - \beta)\phi)\sigma_a^2}{\theta}. \tag{12.65}$$

Since both households and the firms' optimization conditions are satisfied and the planned consumption equals the actual consumption, we have a rational expectations equilibrium. ∎

A.3 Proof of Proposition 3

Proof Notice that for $\phi > 0$ Eqs. (12.31) and (12.32) are identical, so we only need to consider Eq. (12.32). After re-arranging terms we obtain

$$\phi(1 - (1 - \beta)\phi)\sigma_a^2 = \sigma_v^2 + \sigma_z^2(1 - \beta). \tag{12.66}$$

Notice that for $\sigma_v^2 < \frac{1}{4(1-\beta)}\sigma_a^2$, we can find a continuum of σ_z^2 to satisfy the above equation. Namely, there exists a continuum of sentiment-driven equilibria indexed by $\sigma_z^2 \in \left(0, \frac{1}{4(1-\beta)^2}\sigma_a^2 - \frac{\sigma_v^2}{1-\beta}\right)$ such that (12.66) is satisfied. Given σ_z^2, we can solve for ϕ as

$$\phi = \frac{1}{2(1 - \beta)} \pm \sqrt{\frac{1}{4(1 - \beta)^2} - \frac{\tilde{\mu}}{1 - \beta}}, \tag{12.67}$$

where $\tilde{\mu} = \frac{\sigma_v^2 + \sigma_z^2(1-\beta)}{\sigma_a^2}$. Once we obtain ϕ, we can then solve for $\tilde{c}$ and $\bar{c}$. The production of each firm is given by

$$c_{jt} = \phi a_t + \sigma_z z_t + v_{jt}. \tag{12.68}$$

To solve the constants we first use expression (12.11) to obtain

$$\tilde{c} = (1 - \theta\gamma)\bar{c} + \frac{\theta}{2}\Omega_s, \tag{12.69}$$

where

$$\begin{aligned}
\Omega_s &= var[a_t/\theta + (\frac{1}{\theta} - \gamma)(\phi a_t + \sigma_z z_t)|\phi a_t + \sigma_z z_t + v_{jt}] \\
&= \frac{1}{\theta^2}var(a_t + \beta\phi a_t + \beta\sigma_z z_t|\phi a_t + \sigma_z z_t + v_{jt}) \\
&= \frac{(1 + \beta\phi)^2\sigma_a^2 + \beta^2\sigma_z^2 - (1 + \beta\phi)\phi\sigma_a^2 - \beta\sigma_z^2}{\theta^2} \\
&= \frac{(1 + \beta\phi)(1 - (1 - \beta)\phi)\sigma_a^2 - \beta\sigma_z^2(1 - \beta)}{\theta^2}.
\end{aligned} \tag{12.70}$$

And again we also have

$$\bar{c} = \frac{1}{2}\frac{\theta - 1}{\theta}\sigma_v^2 + \tilde{c}, \tag{12.71}$$

or

$$\bar{c} = \frac{1}{2\gamma}\left[\frac{\theta - 1}{\theta^2}\sigma_v^2 + \frac{(1 + \beta\phi)(1 - (1 - \beta)\phi)\sigma_a^2 - \beta\sigma_z^2(1 - \beta)}{\theta^2}\right]. \tag{12.72}$$

Finally, we have

$$\bar{p} = \log\left(\frac{\theta}{\theta-1}\right) - \gamma\bar{c},\tag{12.73}$$

so we also have

$$p_t = (\frac{1}{\theta} - \gamma\phi)a_t - \gamma\sigma_z z_t.\tag{12.74}$$

Since all first-order conditions are satisfied and markets clear, we have an equilibrium.   ∎

A.4 Proof of Proposition 5

Proof Denote $\kappa = \frac{1}{1+\sigma_e^2}$. First, taking the log of Eq. (12.44) yields

$$a_t/\theta - \gamma c_{it} = \phi_a^P a_t + \sigma_z^P \kappa (z_t + e_{it}).\tag{12.75}$$

Aggregating across consumers we then obtain

$$a_t/\theta - \gamma c_t = \phi_a^P a_t + \sigma_z^P \kappa z_t.\tag{12.76}$$

Since

$$c_t = \phi a_t + \sigma_z z_t,\tag{12.77}$$

we then have

$$\phi_a^P = \frac{1}{\theta} - \gamma\phi, \ \sigma_z^P = -\frac{\gamma}{\kappa}\sigma_z.\tag{12.78}$$

Hence we obtain

$$c_{it} = \phi a_t + \sigma_z(z_t + e_{it}).\tag{12.79}$$

Taking the log of Eq. (12.45) gives

$$\begin{aligned}c_{jt} &= E(\theta p_t + c_t)|(c_t + v_{jt})\\ &= E[a_t + (1-\gamma\theta)\phi a_t + (1 - \frac{\gamma\theta}{\kappa})\sigma_z z_t]|(\phi a_t + \sigma_z z_t + v_{jt})\\ &= \frac{\phi(1 + (1-\gamma\theta)\phi)\sigma_a^2 + (1 - \frac{\gamma\theta}{\kappa})\sigma_z^2}{\phi^2\sigma_a^2 + \sigma_z^2 + \sigma_v^2}(\phi a_t + \sigma_z z_t + v_{jt}).\end{aligned}\tag{12.80}$$

Aggregating over j yields

$$\frac{\phi(1 + (1-\gamma\theta)\phi)\sigma_a^2 + (1 - \frac{\gamma\theta}{\kappa})\sigma_z^2}{\phi^2\sigma_a^2 + \sigma_z^2 + \sigma_v^2} = 1.\tag{12.81}$$

To be consistent with the production function of final goods (12.5), we must have

$$\phi^2 \sigma_a^2 + \sigma_z^2 + \sigma_v^2 = \phi(1 + \beta\phi)\sigma_a^2 + \left(1 - \frac{1-\beta}{\kappa}\right)\sigma_z^2 \qquad (12.82)$$

or

$$\phi(1 - (1-\beta)\phi)\sigma_a^2 = \sigma_v^2 + \sigma_z^2 \frac{(1-\beta)}{\kappa}. \qquad (12.83)$$

Notice that for $\sigma_v^2 < \frac{1}{4(1-\beta)}\sigma_a^2$, there exists a continuum of sentiment-driven equilibria indexed by $\sigma_z^2 \in \left(0, \frac{\kappa}{4(1-\beta)^2}\sigma_a^2 - \frac{\kappa\sigma_v^2}{1-\beta}\right)$. Given any σ_z^2 we have

$$\phi = \frac{1}{2(1-\beta)} \pm \sqrt{\frac{1}{4(1-\beta)^2} - \frac{\tilde{\mu}}{1-\beta}}, \qquad (12.84)$$

where $\tilde{\mu} = \frac{\sigma_v^2 + \sigma_z^2(1-\beta)/\kappa}{\sigma_a^2}$. The individual production c_{jt} is hence equal to

$$c_{jt} = \phi a_t + \sigma_z z_t + v_{jt}. \qquad (12.85)$$

We still have several remaining constants to be determined. First, by Eq. (12.44), we obtain

$$\hat{c} = \frac{1}{\gamma} \log \frac{\theta}{\theta - 1} - \frac{1}{\gamma}\bar{p} - \frac{1}{\gamma}\frac{1}{2}\left(\frac{\gamma}{\kappa}\sigma_z\right)^2 (1 - \kappa). \qquad (12.86)$$

Denote

$$\Omega_s = var[\frac{1}{\theta}a_t + \left(\frac{1}{\theta} - \gamma\right)\phi a_t + \left(\frac{1}{\theta} - \frac{\gamma}{\kappa}\right)\sigma_z z_t | \phi a_t + \sigma_z z_t + v_{jt}]$$

$$\equiv \frac{1}{\theta^2} var[(a_t + \beta\phi a_t + \tilde{\beta}\sigma_z z_t) | \phi a_t + \sigma_z z_t + v_{jt}]$$

$$= \frac{1}{\theta^2}\left[(1 + \beta\phi)^2\sigma_a^2 + \tilde{\beta}^2\sigma_z^2 - (1 + \beta\phi)\phi\sigma_a^2 - \tilde{\beta}\sigma_z^2\right]$$

$$= \frac{(1 + \beta\phi)(1 - (1-\beta)\phi)\sigma_a^2 - \tilde{\beta}\sigma_z^2(1 - \tilde{\beta})}{\theta^2}. \qquad (12.87)$$

Then by Eq. (12.45) we obtain

$$\tilde{c} = \theta\bar{p} + \bar{c} + \frac{\theta}{2}\Omega_s + \theta\log(1 - \frac{1}{\theta}). \qquad (12.88)$$

Finally, from the aggregate production we obtain

$$\bar{c} = \frac{1}{2}\frac{\theta - 1}{\theta}\sigma_v^2 + \tilde{c}. \qquad (12.89)$$

We then solve

$$\bar{p} = \log(\frac{\theta}{\theta-1}) - \frac{\theta-1}{2\theta^2}\sigma_v^2 - \frac{1}{2}\Omega_s \tag{12.90}$$

and hence

$$\hat{c} = \frac{1}{\gamma}\left[\frac{\theta-1}{2\theta^2}\sigma_v^2 + \frac{1}{2}\Omega_s - \frac{1}{2}\left(\frac{\gamma}{\kappa}\sigma_z\right)^2(1-\kappa)\right]. \tag{12.91}$$

Finally, the relationship between $\hat{c}$ and $\bar{c}$ is

$$\begin{aligned}
\bar{c} &= \hat{c} + \frac{1}{2}\sigma_z^2\sigma_e^2 \\
&= \hat{c} + \frac{1}{2}\sigma_z^2\frac{1-\kappa}{\kappa} \\
&= \frac{1}{\gamma}[\frac{\theta-1}{2\theta^2}\sigma_v^2 + \frac{1}{2}\Omega_s] - \frac{\gamma}{2}\left(\frac{1}{\kappa}\sigma_z\right)^2(1-\kappa) + \frac{1}{2}\sigma_z^2\frac{1-\kappa}{\kappa}.
\end{aligned} \tag{12.92}$$

When $\kappa \to 1$ (or $\sigma_e^2 \to 0$), the above equation reduces to the case with homogenous sentiments. ∎

References

Angeletos, G.-M., & La'O, J. (2013). Sentiments. *Econometrica, 81*, 739–779.

Aumann, R. J. (1974). Subjectivity and correlation in randomized strategies. *Journal of Mathematical Economics, 1*, 67–96.

Aumann, R. J. (1987). Correlated equilibrium as an expression of bayesian rationality. *Econometrica, 55*, 1–18.

Aumann, R. J, Peck, J., & Shell, K. (1988). *Asymmetric information and sunspot equilibria: A family of simple examples*, Working Paper 88–34, CAE.

Benhabib, J., & Farmer, R. (1994). Indeterminacy and increasing returns. *Journal of Economic Theory, 63*, 19–41.

Benhabib, J., Wang, P., & Wen, Y. (2015). Sentiments and aggregate demand fluctuations. *Econometrica, 83*, 549–585.

Bergemann, D., & Morris, S. (2013). Robust predictions in games with incomplete information. *Econometrica, 81*, 1251–1308.

Bergemann, D., Morris, S. (2015). Bayes correlated equilibrium and the comparison of information structures in games (forthcoming), *Theoretical Economics*.

Bergemann, D., Heumann, T., & Morris, S. (2015). Information and volatility. *Journal of Economic Theory, 158*, 427–465.

Cass, D., & Shell, K. (1983). Do sunspots matter? *Journal of Political Economy, 91*, 193–227.

Chiappori, P. A., & Guesnerie, R., (1991). The Lucas equation, indeterminacy and non-neutrality: An example. In P. Dasgupta, D. Gale, O. Hart (Eds.), *Essays in honor of Frank Hahn* (pp. 445–464) MIT Press.

Lucas, R. E, Jr. (1972). Expectations and the neutrality of money. *Journal of Economic Theory, 4*, 103–124.

Lucas, R. E, Jr. (1983). Corrigendum. *Journal of Economic Theory, 31*, 197–99.

Maskin, E., & Tirole, J. (1987). Correlated equilibria and sunspots. *Journal of Economic Theory*, *43*, 364–373.

Peck, J., & Shell, K. (1991). Market uncertainty: Correlated and sunspot equilibria in imperfectly competitive economies. *Review of Economic Studies*, *58*, 1011–1029.

Spear, S. E. (1989). Are sunspots necessary? *Journal of Political Economy*, *97*, 965–973.

Chapter 13
Technological Progress, Employment and the Lifetime of Capital

Raouf Boucekkine, Natali Hritonenko and Yuri Yatsenko

Abstract We study the impact of technological progress on the level of employment in a vintage capital model where: (i) capital and labor are gross complementary; (ii) labor supply is endogenous and indivisible; (iii) there is full employment, and (iv) the rate of labor-saving technological progress is endogenous. We characterize the stationary distributions of vintage capital goods and the corresponding equilibrium values for employment and capital lifetime. It is shown that both variables are non-monotonic functions of technological progress indicators. Technological accelerations are found to increase employment provided innovations are not *too* radical.

Keywords Vintage capital · Technological progress · Employment · Compensation theory

JEL numbers: C61 · D21 · D92 · O33

13.1 Introduction

In the traditional neoclassical growth model, capital goods have an infinite lifetime, the contribution to production of the oldest capital goods get smaller and smaller due to physical depreciation but they are never scrapped. Even more, because technologi-

Special thanks go to a referee whose careful reading and suggestions have markedly improved this work. The usual disclaimer applies.

R. Boucekkine (✉)
Aix-Marseille University (AMSE and IMéRA), CNRS and EHESS,
Institut Universitaire de France, Paris, France
e-mail: Raouf.Boucekkine@univ-amu.fr

N. Hritonenko
Prairie View A&M University, Prairie View, USA
e-mail: NaHritonenko@pvamu.edu

Y. Yatsenko
Houston Baptist University, Houston, USA
e-mail: YYatsenko@hbu.edu

© Springer International Publishing AG 2017
K. Nishimura et al. (eds.), *Sunspots and Non-Linear Dynamics*,
Studies in Economic Theory 31, DOI 10.1007/978-3-319-44076-7_13

cal progress is disembodied (that's it applies to all the stock of capital whatever the age of its components), a technological acceleration fosters productivity of all the capital goods in the same way. The vintage capital approach differentiates the capital goods (say the machines) by their vintage to reflect that every vintage embodies the best technology available at the date of its construction, that's technological progress is embodied (see the seminal paper by Solow et al. 1966). As a consequence, a capital good may be withdrawn at finite time under certain mild assumptions (see Benhabib and Rustichini 1991; Hritonenko and Yatsenko 1996, 2012, and Boucekkine et al. 1997, 1998). Moreover, an acceleration of *embodied* technological progress affects the lifetime of capital goods, therefore inducing a further mechanism linking technological progress to the main macroeconomic variables such like investment, employment and of course GDP.

The traditional vintage capital structure borrowed from Solow et al. (1966) has a fixed labor supply and an exogenous embodied technological progress. Very few papers relax such assumptions (see Boucekkine et al. 2014a, b), and when they do so, they do not systematically investigate the full macroeconomic implications of embodied technological progress. Moreover, even in those considering the traditional vintage capital set-up, such an investigation is not complete. For example, Boucekkine et al. (1998) use a specific utility function and are unable to come out with an accurate enough picture of the relationship between technological progress accelerations and the lifetime of capital goods, as they ultimately restrict their analysis to parametric cases where the latter relationship is monotonic. This paper is intended to supply with a much broader and more accurate analysis of this relationship starting with the traditional frame envisaged by Boucekkine et al. (1998). Extensions to endogenous labor supply and endogenous (embodied) technical progress are also detailed. In contrast to the early studies mentioned above, non-monotonic interactions between technological progress and the lifetime of capital are highlighted at equilibrium, and their implications on other important macroeconomic variables are carried out.

We shall specially examine the implications for employment, typically omitted in the early vintage capital literature. Perhaps one the most debated questions in economics is the impact of technological progress on employment (see the excellent survey by Vivarelli 2007). In particular, labor-saving technical change has by definition a direct employment cost, which raises the question of indirect induced mechanisms which may compensate the direct job losses. Several compensation theories have been proposed. Different supply and demand side mechanisms have been put forward. Perhaps the most elementary is due to Say (1964) who argued that technological progress comes through new machines, jobs have to be created to produce these machines, which may well compensate the direct job losses. Innovations may also lead to create new goods (product innovation), giving birth to new industries and therefore to job creations (see for example Freeman and Soete 1994).

Other very known compensation theories are demand-based: for example, one would argue that as innovations decrease production costs, product prices should in turn drop provided markets are competitive, which would stimulate demand, production and therefore employment (see again Say 1964, or more recently Smolny 1998).

One may also more straightforwardly put forward an increase in wages pushed either by productivity gains and/or by unions claim for a larger share of savings allowed by innovations (Boyer 1988). Finally, other compensation theories target technological unemployment and the subsequent wage adjustment, which typically goes in the opposite direction to the story told just before: provided labor markets are flexible enough, wages should drop in response to unemployment, which favors job recovery (as in Addison and Teixeira 2001).

As mentioned above, the traditional vintage capital theory à la Solow et al. (1966) didn't pay attention to this question, labor supply was taken exogenous and employment was only treated residually as the main research question was the modernization role of investment and its impact on short-term dynamics and long-term growth. Indeed, most of the historical papers in this vein assume constant saving rate (so no preferences at all, and a fortiori no labor supply), the Leontief technology assumption allowing to focus on one of the two production factors (that's, capital). The recent vintage capital literature following Benhabib and Rustichini (1991) does consider full-fledged general equilibrium models with explicit preferences. Still, a vast majority of these papers ignore labor adjustment questions, and do not consider elastic labor supply.

Two significant exceptions are worth mentioning however. First, a few papers (like Cooley et al. 1997) include elastic labor supply and do marginally study labor adjustment to technological shocks in general equilibrium vintage capital models, the approach being fully quantitative. Second, an important literature led by Caballero and Hammour (1996) borrows the vintage capital structure to discuss labor market variables dynamics in frictional economics. Caballero and Hammour's (1996) model builds on bargaining over appropriability surpluses to design inefficient labor markets and to quantitatively study the resulting pace of job creation and destruction. Boucekkine et al. (1999) provide with an analytical investigation of the labor market dynamics for a special version of Caballero and Hammour's model. A few other vintage capital models designed for the quantitative analysis of inefficient labor markets, including wage inequality, have been proposed (for example, the search model proposed by Hornstein et al. 2005).

In this paper, we argue that employment covariation with technological progress is already a nontrivial question in the framework of vintage capital models with elastic labor supply, even though labor markets were efficient. To this end, we proceed in a few steps. First of all, we shed light on the nontrivial relationship between (exogenous) technological accelerations and the (endogenous) lifetime of capital goods. Second, even though one accepts the idea that capital lifetime should be shortened by innovations as new and more efficient capital goods lead to replace the oldest and obsolete ones earlier, the implications for employment (in an efficient labor market) are nontrivial. Replacement of obsolete capital amounts to closing some production units with the subsequent job losses. Whether the jobs created thanks to the new machines are enough to compensate the losses is a general equilibrium question. At equilibrium, employment depends both on the lifetime of capital (and more broadly on the age structure of capital) and on investment in new machines (in efficiency units). Whether the latter will be stimulated enough to compensate the fall in the

capital lifetime for employment to increase in response to technological accelerations is not obvious. Indeed, as wages go up in the latter case, income and substitution
effects depending on preferences will drive households' consumption and labor supply responses. So once again, the general equilibrium outcomes are nontrivial. We
shall investigate this issue in the indivisible labor case (Hansen 1985), which is one of
the preferred specifications in RBC theory. Though such a specification is particular
(as it implies linearity in labor), it's far from making the problem trivial as we will
show.

Third, as a strong departure with respect to the vintage capital contributions cited
above, we shall endogenize technological progress by introducing purposive investment in labor-saving R&D at the firm level as in Boucekkine et al. (2011). As R&D
expenditures are rival to investment in new capital, the general equilibrium outcomes
are even less obvious. It's important to notice that the modeled R&D leads to process
innovation, not to a product innovation. This is an important characteristic, especially
regarding the empirical literature.

13.1.1 Relation to the Literature and Main Findings

As mentioned above, this paper can be seen as an extension of previous work of the
authors on vintage capital models (notably Boucekkine et al. 1997; Hritonenko and
Yatsenko 1996, 2012; Yatsenko et al. 2009, or Jovanovic and Yatsenko 2012). They
can be also related to other more recent contributions of the authors like Boucekkine
et al. (2014a, b). Among several noticeable differences with respect to this literature, the focus on employment and the general equilibrium appraisal of the impact
of technological accelerations are distinctive enough. Compared to Caballero and
Hammour (1996) and Hornstein et al. (2005), we keep labor markets efficient in our
set-up (no unemployment) but allow for endogenous technical progress. Last but
not least, our approach is mainly analytical (complemented with some numerical
simulations) in contrast to the papers à la Cooley et al.

On the empirical ground, our findings can be discussed in the light of the large
empirical literature on employment and innovations (see again the survey of Vivarelli
2007). An interesting reference is Greenan and Guellec (2000) who worked on French
manufacturing sectors over the period 1986–90. While they identified a positive
correlation between innovation and employment at the firm's level with both product
and process innovation, they found that such a net result does not hold at the sectoral
level: compensation only works for product innovation, process innovation does
generate jobs within the innovative firms but it also destroys jobs in the competing
firms through a business stealing effect, leading to an overall negative effect at the
sectoral level.

We may interpret our vintage capital model as a model of a manufacturing sector
subject to process innovation: different machines correspond to different production
units running different technologies, technological accelerations lead to destroy the
jobs matched with the obsolete machines at the same time as they create jobs to

operate the new machines. One of our main findings is that at stationary equilibria, process innovation leads to increase (aggregate) employment in the vintage capital economy provided innovation is not *too* radical (that's if the resulting rate of embodied technical progress is not *too* high). Indeed, we will show through successive extensions of the benchmark vintage capital model that both employment and the lifetime of capital are non-monotonic functions of the rate of (embodied) technological progress (when it is exogenous) or of the efficiency in the R&D sector (in the case of endogenous technological progress). So strictly speaking, employment can either increase or decrease in response to technological progress, and we clarify the demand-based and supply based compensation mechanisms at work. However, using analytical inspection and numerical exercises, we end up finding that employment keep growing with the rate of embodied technological progress unless innovations are *too* radical.

13.2 The Benchmark Vintage Capital Growth Model

As mentioned in the introduction section, we build on the seminal work of Solow et al. (1966). Optimal growth versions of the model developed in the latter have been developed by Boucekkine et al. (1997, 1998) and Hritonenko and Yatsenko (1996, 2005). We present here a decentralized equilibrium counterpart of this model, which will allow us to shed light on some original properties inherent to our decentralized framework, although the general equilibrium outcomes are identical to the optimal allocations in this model. A significant contribution of this section is to deliver a clear picture of the (mostly non-monotonic) long-term relationship between technological progress, detrended investment and the lifetime of capital when the utility function is strictly concave. Such a task has been undertaken by Boucekkine et al. (1998) using a non-standard nonlinear utility function, we shall get much clearer results in the standard case where the utility function is logarithmic. Boucekkine et al. (1997) have a linear utility function.

13.2.1 Firms

The major ingredient of the theory is the vintage capital technology. We shall consider the problem of a price-taking firm seeking to maximize the net profit that produces $Y(t)$ units of output, uses $L(t)$ units of labour and invests $I(t)$ into new capital:

$$\max_{I,L} \int_0^\infty [Y(t) - w(t)L(t) - I(t)]\mu(t)dt, \tag{13.1}$$

$w(t)$ is the unit wage at time t, and $\mu(t)$ is the discounting factor, which depends on the stream of interest rate up to t according to:

$$\mu(t) = e^{-\int_0^t r(s)ds},$$ (13.2)

Before proceeding any further, commenting on capital costs in our competitive equilibrium set-up is worthwhile. In the profit expression given above, the price of capital is equal to 1, just like one unit of output which plays the role of the numeraire. This makes sense in a one-sector model like ours. With the traditional notation inherited from Solow et al. (1966), the price of new capital at time t is $P(t, t)$ (while $P(t, \tau)$ would designate the price at time t of a unit of capital of vintage τ), and the cost of acquiring $I(t)$ is $P(t, t)I(t)$. In competitive equilibrium, $P(t, t)$ should be equal the production cost of one unit of the new vintage t. $P(t, t) = 1$ follows naturally from our one-sector setting.[1] Accordingly, note that firms only buy the new capital goods $I(t)$ at any time t, never the older vintages. Buying older vintages would make sense in a model with learning costs as in Feichtinger et al. (2006) where it is postulated that running a new machine is harder than an old and more familiar one. We abstract away from these aspects in this paper.

The production function is a vintage capital Leontief technology: any unit of capital of vintage τ available at time t produces the same amount of output (say one unit), for any vintage τ, but operating one unit of vintage τ requires $\frac{1}{\beta(\tau)}$ units of labor. Accordingly:

$$Y(t) = \int_{a(t)}^t I(\tau)d\tau,$$ (13.3)

$$L(t) = \int_{a(t)}^t \frac{I(\tau)}{\beta(\tau)}d\tau,$$ (13.4)

where $a(t)$ is the oldest vintage in use at time t, and $L(t)$ is total labor required to operate all the machines used. In the benchmark, we consider exogenous technical progress: $\beta(t) = e^{\gamma t}$. Technological progress is Harrod neutral labor-saving at a given rate $\gamma > 0$.

The *constraints* of the firm optimization problem are given by the positivity condition

$$I(t) \geq 0,$$ (13.5)

---

[1]Needless to say, the price at t of a unit of vintage τ, P(t,τ), is not equal to 1 if $\tau < t$ as this price should reflect the decreasing pattern of quasi-rents associated to the use of these vintages due to obsolescence. See Solow et al. (1966), p. 99.

and the standard requirement that scrapped machines cannot be reused:

$$a'(t) \geq 0, \quad a(t) \leq t. \tag{13.6}$$

We shall also specify the *initial conditions* as follows:

$$a(0) = a_0 < 0, \quad I(\tau) \equiv I_0(\tau), \quad \tau \in [a_0, 0]. \tag{13.7}$$

The optimization problem (*OP*) (13.1)–(13.7) includes four unknown functions I, a, Y, and L connected by equalities (13.3)–(13.4). We choose I and a as the *independent* controls of the OP and consider the rest of the unknown functions Y and L as the *dependent* (*state*) variables. The necessary conditions for **an interior** extremum are:

$$\int_{t}^{a^{-1}(t)} \mu(\tau) [\beta(t) - w(\tau)] d\tau = \mu(t)\beta(t), \tag{13.8}$$

$$w(t) = \beta(a(t)), \tag{13.9}$$

where $a^{-1}(t)$ is the inverse function of $a(t)$.

Equality (13.8) is the optimal investment rule common in vintage capital models. Dividing both sides of the equality by $\mu(t) \beta(t)$, one gets on the right hand side the marginal cost of new capital (equal to 1 in our one-sector model as argued above), and on the left hand side its expected discounted revenue over its lifetime. The flow revenue is $1 - \frac{w(\tau)}{\beta(t)}$ since each unit of capital bought at t yields one unit of output, and requires $\frac{1}{\beta(t)}$ units of labor to be operated.

Equation (13.9) is the optimal scrapping condition: for given wage, it allows to compute the cut-off vintage index beyond which all the machines are scrapped. It simply states that it is optimal to keep on using machines until their labor productivity ($\beta(a(t))$) equals labor cost ($w(t)$).

A quick look at Eqs. (13.8)–(13.9) is sufficient to get that the interior extremum is **not** generally implementable from $t = 0$ for given prices (that is for given wage and interest rate). This point has been already made by Boucekkine et al. (1997) on the optimal growth version of the model. But in our framework, this property stems from different reasons. For a given sequence of wages, the scrapping condition (13.9) allows to determine the optimal trajectory for the vintage index $a(t)$. This in turn determines the inverse function $a^{-1}(t)$. For given interest rate data, the left and right hand sides of the optimal investment rule (13.8) are therefore predetermined and need not be equal. In general so, the interior solution is not implementable from t = 0, and a transitory corner regime will set in.[2] In this paper, we are focusing on interior and permanent regimes, and we will show in a few paragraphs that such regimes exist in general equilibrium.

---

[2] A typical example of these dynamics can be found in Boucekkine et al. (1997).

13.2.2   Consumers

The consumers' block is standard: consumers consume and save out of a total income generated by initial wealth and labor income. Financial markets are perfect, and the return to financial assets is equal to the interest rate $r(t)$. For simplification, we shall consider a constant population size and logarithmic utility throughout this paper. Boucekkine et al. (1998) used instead $u(c) = c^{\sigma}$ with $0 < \sigma < 1$, which is not standard and does not degenerate into the benchmark logarithmic case when σ goes to zero. In the benchmark model, labor supply is inelastic, normalized to 1. So the consumers face the typical program:

$$\max_{c} \int_{0}^{\infty} e^{-\rho t} \ln c(t) dt, \quad \rho > 0, \qquad (13.10)$$

with the budget constraint

$$\dot{A} = r(t)A(t) + w(t) - c(t), \qquad (13.11)$$

where $A(t)$ is the wealth of people, $A(0)$ is given, ρ is the impatience rate. The corresponding first order conditions are as usual

$$\frac{\dot{c}}{c} = r(t) - \rho \quad \text{with} \quad \lim_{t \to \infty} \phi(t)A(t) = 0, \qquad (13.12)$$

where $\phi(t)$ is the co-state variable associated with the wealth accumulation equation (13.11).

13.2.3   Decentralized Equilibrium

We define the equilibrium of this economy in the following traditional way.

Definition An equilibrium for this economy is a trajectory $(I(t), Y(t), L(t), a(t), c(t), A(t), \phi(t), r(t), \mu(t), w(t), t \geq 0)$ which solves the firms optimization problem (13.1)–(13.7) and the consumers optimization problem (13.10)–(13.11) while both the good and labor markets clear.

The latter clearing conditions are straightforward:

$$y(t) = c(t) + I(t), \qquad (13.13)$$

$$L(t) = 1. \qquad (13.14)$$

It's then possible to characterize the equilibrium of the economy from a system of seven simultaneous equations:

$$\frac{\dot{c}}{c} = r(t) - \rho, \tag{B1}$$

$$\mu(t) = e^{-\int_0^t r(s)d\tau}, \tag{B2}$$

$$Y(t) = \int_{a(t)}^{t} \beta(\tau)m(\tau)d\tau, \tag{B3}$$

$$\int_{t}^{a^{-1}(t)} \mu(\tau)\left[\beta(t) - \beta(a(\tau))\right]d\tau = \mu(t)\beta(t), \tag{B4}$$

$$w(t) = \beta(a(t)), \tag{B5}$$

$$Y(t) = c(t) + \beta(t)m(t), \tag{B6}$$

$$1 = \int_{a(t)}^{t} m(\tau)d\tau, \tag{B7}$$

with the corresponding boundary conditions, and where for convenience we change the decision variable $I(t)$ to $m(t) = I(t)/\beta(t)$. Again, the system above only considers interior solutions for the firms' optimization problem. Nonetheless, and in contrast to the partial equilibrium firms' problem, the existence of such interior regimes cannot be immediately discarded. Indeed, prices are no longer given, and the solution to the system is far from obvious. It is easy to understand why: for given past investment profile, the labor market equilibrium condition (B7) allows to compute $a(t)$, which determines wages by (B5), and features the interest rate as a solution to an advanced integral equation from the optimal (interior) investment rule (B4). Whether this equation admits a solution is another very complicated story, which goes much beyond this section. Partial solutions are provided in Boucekkine et al. (1997) who solved the optimal growth version under linear utility or in Boucekkine et al. (1998) who numerically investigated the optimal short term dynamics of the same optimal growth model with nonlinear utility. Here, as mentioned above, we focus on permanent interior regimes, and we show hereafter that balanced growth paths do exist.

13.2.4 Balanced Growth

Let us assume that $t - a(t) = T = \text{const}$, $Y(t) = \bar{y}e^{\gamma t}$, and $c(t) = \bar{c}e^{\gamma t}$. Then $m(t) = \bar{m}$, and the system (B1)–(B7) leads to

$$r(t) = \gamma + \rho = \text{constant}, \tag{BG1}$$

$$\mu(t) = e^{-(\gamma + \rho)t}, \tag{BG2}$$

$$\bar{y} = \bar{m}\frac{1 - e^{-\gamma T}}{\gamma}, \tag{BG3}$$

$$\frac{1 - e^{-(\rho+\gamma)T}}{\rho + \gamma} - e^{-\gamma T}\frac{1 - e^{-\rho T}}{\rho} = 1, \tag{BG4}$$

$$w(t) = e^{\gamma(t-T)}, \tag{BG5}$$

$$\bar{y} = \bar{c} + \bar{m}, \tag{BG6}$$

$$1 = \bar{m}T. \tag{BG7}$$

In contrast to the dynamic system (B1)–(B7), the system characterizing the balanced growth paths (BGP) is straightforward. Indeed, under exogenous growth, the consumption Euler equation (BG1) immediately fixes the BGP interest rate, and Eq. (BG4) only depends on capital lifetime T. If this equation admits a solution, then the BGP values of all the other variables can be determined accordingly following a trivial recursive scheme. For example, (BG7) determines investment (in efficiency units), then output thanks to (BG3), and finally consumption using (BG6). The Eq. (BG4) is actually (BG4) is actually well-behaved.

Lemma 1 *Let $\rho < 1$. For any given $0 < \gamma < 1 - \rho$, the T-equation (BG4) has a unique positive solution $0 < T$.*

Proof Let us denote the left-hand side of equation (BG4) as $F(T) = \frac{1-e^{-(\rho+\gamma)T}}{\rho+\gamma} - e^{-\gamma T}\frac{1-e^{-\rho T}}{\rho}$. Now using the properties $F(0) = 0$, $F(T) \xrightarrow[T\to\infty]{} \frac{1}{\rho+\gamma}$, and observing that $F'(T) = e^{-(\rho+\gamma)T} + \gamma e^{-\gamma T}\frac{1-e^{-\rho T}}{\rho} - e^{-\gamma T}e^{-\rho T} = \gamma e^{-\gamma T}\frac{1-e^{-\rho T}}{\rho} > 0$, one can conclude that the function $F(T)$ increases from 0 to $\frac{1}{\rho+\gamma}$. Therefore, a finite solution T to (BG4) exists only if $\frac{1}{\rho+\gamma} > 1$, or $\rho + \gamma < 1$. The solution is unique because the function $F(T)$ is monotonic. **Q.E.D**

It's important to note here that the conditions of Lemma 1 hold by far (notably $0 < \gamma < 1 - \rho$) under realistic parametrizations of the vintage capital model. We now move to the study of the impact of technological progress along the balanced growth paths.

13.2.5   Technological Progress, Investment and the Lifetime of Capital

We first characterize the impact of technological accelerations on the long-term lifetime of capital.

Proposition 1 *Let T be the solution of equation (BG4) under the conditions of Lemma 1. The value $T \to \infty$ as $\gamma \to 0$ or $\gamma \to 1 - \rho$, so T decreases in γ for small γ and increases for larger γ. The value $\gamma T \to -ln(1 - \rho)$ as $\gamma \to 0$ and γT is larger for a larger γ. If $\rho << 1$ and $\gamma << 1$, then $T \approx \sqrt{2/\gamma}$.*

Proof If both $\rho << 1$ and $\gamma << 1$, then applying the Taylor series up to the second order to equation (BG4), we obtain $\frac{\gamma T^2}{2} \approx 1$ or $T \approx \sqrt{\frac{2}{\gamma}}$.

In order to understand relations between γ, T, and γT for an arbitrary $\rho < 1$, let us introduce the auxiliary $G(T, \gamma) = \frac{1-e^{-(\rho+\gamma)T}}{\rho+\gamma} - e^{-\gamma T}\frac{1-e^{-\rho T}}{\rho} - 1$.

Applying the Theorem on the Implicit Function to the equality $G(T, \gamma) = 0$, we get

$$\frac{dT}{d\gamma} = -\frac{\partial G/\partial \gamma}{\partial G/\partial T} = -\frac{T\left(1 - \gamma e^{-(\rho+\gamma)T}/(\rho+\gamma)\right) - \left(e^{\gamma T} - e^{-\rho T}\right)\rho/(\rho+\gamma)^2}{\gamma\left(1 - e^{-\rho T}\right)}.$$

Now, calculating and estimating the derivative

$$\frac{d(\gamma T)}{d\gamma} = T + \gamma\frac{dT}{d\gamma} = \frac{\rho^2 e^{-\rho T}}{(\rho+\gamma)^2\left(1-e^{-\rho T}\right)}\left(e^{-(\rho+\gamma)T} - 1 - (\rho+\gamma)\right) > 0,$$

we obtain that the value γT monotonically increases in γ for any $0 < \gamma < 1-\rho$. Next, presenting equation (BG4) in the form

$$\frac{1 - e^{-\rho T}}{\rho} - \frac{\gamma\left(1 - e^{-(\rho+\gamma)T}\right)}{\rho(\rho+\gamma)} = 1,$$

we see that the value $\gamma T \to -\ln(1-\rho) > 0$ as $\gamma \to +0$. Correspondingly, $T \to \infty$ as $\gamma \to 0$. Finally, since the value γT remains finite as $\gamma \to +0$, the above derivative $dT/d\gamma \to -\infty$ as $\gamma \to +0$. Therefore, the value T decreases in γ for certain small values γ, but it increases for larger γ because $T \to \infty$ again at $\gamma \to (1-\rho)$. **Q.E.D**

Proposition 1 establishes the non-monotonic nature of the relationship between the rate of technological progress and the lifetime of capital. More precisely, it proves on one hand that the BGP lifetime of capital is a decreasing function of labor-saving technical progress rate, γ, when γ is small enough, which covers most of the real-life technological accelerations. This property is obtained under much less clean conditions in Boucekkine et al. (1998) who found it to hold provided $\gamma T \leq 1$ (Proposition 2, p. 368). On the other hand, and in addition to this improvement, here we show clearly that the correlation between capital lifetime and the rate of labor-saving technical progress turns out to be positive for γ large enough, a property not demonstrated in Boucekkine et al. (1998). The reason is the following: while an increase in γ pushes the firms to shorten the lifetime of machines to take advantage of the better characteristics of the new ones, the discount rate of their profits plays in the opposite direction as it neatly transpires from (BG2). This second effect is dominated for γ small enough but should be predominant for large γ-values. Proposition 1 provides with a simple and clear statement of this non-monotonicity, not uncovered in the previous papers on related models. Nonetheless, from now on, we shall mostly concentrate on the realistic parameterizations of the model, that is when both ρ and γ are small enough.

Once (BG4) has a unique solution T, the whole BGP is uniquely determined as explained above. The following proposition can then be stated. It clarifies, among others, how technological progress affects detrended investment, output and consumption along the BGP.

Proposition 2 *The decentralized equilibrium (BG1)–(BG7) possesses a unique BGP as T is uniquely determined by Eq. (BG4). c(t) is positive, at least for small γ and ρ. For these parameterizations, when γ rises, investment level increases while both output and consumption levels go down.*

The unique pending point in the theorem is the positivity of consumption at the BGP, which is not granted for any parameterization of the model. Yet one can straightforwardly see that consumption is indeed positive when both γ and ρ are small enough. To this end, it is enough to combine (BG3) and (BG6) under Proposition 1, and the approximation $T \approx \sqrt{2/\gamma}$.

The impact of technological progress on investment level (or detrended investment) at the BGP derives immediately from the clearing condition of the labor market (BG7). It implies that $\bar{m} = \frac{1}{T}$, and since T is decreasing in γ when ρ and γ are small enough, one gets that $\bar{m}$ is increasing in γ in this parametric case. That is in this benchmark model with exogenous technical progress and inelastic labor supply, a technological acceleration shortens the lifetime of capital goods and increases the level of investment. Under Proposition 1, the model has a more precise prediction in this respect: since $T \approx \sqrt{2/\gamma}$, an acceleration in γ by one percentage point leads to an increase in the investment level by half a point. Finally observe that as one departs from the case where ρ and γ are small enough, one may get the opposite picture: the discount rate effect of technological progress on profitability of investment may dominate its scrapping time effect leading to firms using machines for a longer time and investing less (in level). We shall tackle the precise quantitative conditions and implications in the last section of this paper on an enlarged model.

Here we conclude our study of the benchmark model by looking at the effect of technological progress on detrended output and consumption when ρ and γ are small enough. The impact on output comes straightforwardly from the combination of (BG3), (BG6) and (BG7) under $T \approx \sqrt{2/\gamma}$. The decrease in the lifetime of machines is so harmful for the level of output that it is not compensated by the rise in investment. It should be noted here that (detrended) output is also a decreasing function of the rate of technological progress in the standard neoclassical growth model, say the Solow model, due to decreasing returns. The result established here is of course much less obvious because the returns are no longer decreasing and specially because there are two conflicting forces in the vintage framework (shorter capital lifetime, bigger investment level) with no obvious trade-off. The same property can be established for consumption level.

We now move to the extensions announced in the introduction section. We start by endogenizing labor supply.

13.3 Introducing Endogenous (Indivisible) Labor Supply

In this section, we introduce endogenous labor supply into the benchmark vintage capital studied above. Traditional vintage capital theory *à la* Solow et al. (1966) does consider inelastic labor supply. The unique departure from the benchmark is the consumption block which incorporates now disutility of work, in a linear manner consistently with the framework of indivisible labor supply put forward by Hansen (1985) and Rogerson (1988):

$$\int_0^\infty e^{-\rho t}[\ln c - \theta N]dt \longrightarrow \max_{C,N}, \quad \theta > 0, \tag{13.15}$$

with the budget constraint

$$\dot{A} = r(t)A(t) + w(t)N(t) - c(t), \tag{13.16}$$

where $A(t)$ is the wealth of people, θ measure disutility of work (or preference for leisure), $A(0)$ is given. The decision variable $N(t)$ satisfies the following constraint:

$$0 \leq N(t) \leq 1, \tag{13.17}$$

This is consistent with the previous attempt to endogenize labor supply in the same general frame due to Boucekkine et al. (2014a). However, these authors didn't analyze the model from the point of view of compensation theory (in the sense given in the introduction). The first order optimality conditions corresponding to the consumer problem for an interior maximum in c, N, A are

$$1/c = e^{\rho t}\lambda, \tag{13.18}$$

$$e^{-\rho t}\theta = \lambda w \quad \text{or} \quad \theta = w/c, \tag{13.19}$$

$$\dot{\lambda} = \lambda r, \tag{13.20}$$

which yields, after some trivial algebra:

$$\frac{\dot{c}}{c} = r(t) - \rho, \quad \text{with} \quad \lim_{t\to\infty} \phi(t)A(t) = 0, \tag{13.21}$$

and

$$w = \theta c. \tag{13.22}$$

(13.22) is the new optimality equation with respect to the benchmark, it characterizes optimal (interior) labor supply by equalizing its marginal cost (θ) and marginal benefit (w/c). We now study whether the endogenization of labor supply changes the economic mechanisms (and/or their relative sizes) disentangled in the benchmark. Since the firms' block is identical, we move directly to the decentralized equilibrium. After characterizing this equilibrium, we develop the compensation mechanisms inherent in this model.

13.3.1 Decentralized Equilibrium and Compensation Mechanisms at Work

With respect to the benchmark model at the decentralized equilibrium (B1)–(B7), we have now one equation more, and the resulting system writes like

$$\frac{\dot{c}}{c} = r(t) - \rho, \tag{C1}$$

$$\mu(t) = e^{-\int_0^t r(s)d\tau}, \tag{C2}$$

$$Y(t) = \int_{a(t)}^t \beta(\tau)m(\tau)d\tau, \tag{C3}$$

$$\int_{t}^{a^{-1}(t)} \mu(\tau)\left[\beta(t) - \beta(a(\tau))\right] d\tau = \mu(t)\beta(t), \qquad (C4)$$

$$w(t) = \beta(a(t)), \qquad (C5)$$

$$Y(t) = c(t) + \beta(t)m(t), \qquad (C6)$$

$$0 < N(t) = \int_{a(t)}^{t} m(\tau)d\tau < 1, \qquad (C7)$$

$$w(t) = \theta c(t), \qquad (C8)$$

with the initial conditions (13.7) and $A(0)$. Equations (C1)–(C6) are identical to (B1)–(B6). (C7) is the new labor market clearing condition and it integrates the fact that labor supply, $N(t)$, is now endogenous. (C8) is the new equation characterizing optimal labor supply. As in the benchmark case, we look for interior solutions.

Before starting the analysis of the corresponding BGPs, it is worth pointing out that the addition of endogenous labor supply strikingly complicates the involved dynamic systems. In the benchmark case, for given past investment profile, the labor market equilibrium condition (B7) allows to compute $a(t)$, which determines wages by (B5), and allows to identify the interest rate as a solution to an advanced integral equation based on the optimal (interior) investment rule (B4). This partial recursive scheme is no longer possible in this extension: the equilibrium condition for the labor market, here (C7), has two unknowns for given past investment profile, labor supply, $N(t)$, and the vintage index, $a(t)$. The full simultaneity of the dynamic system will unfortunately remain at the BGP as it is shown hereafter.

Before getting to this point, it is important to disentangle at this stage the impact of technological progress at equilibrium in the light of the principles of the traditional compensation theory. Since Eqs. (C1)–(C6) are identical to (B1)–(B6), one might use some of the lessons obtained from the benchmark model to comment in this respect. One is that a technological acceleration will lead to higher wages and a shorter capital lifetime (when ρ and γ are small enough). Because capital and labor are gross complementary, the latter property features the negative impact of technological progress on employment through a pure obsolescence effect: obsolete machines are scrapped (in other words $a(t)$ increases in Eq. (C7)), with the subsequent job losses. However, workers fired from obsolete production units to be closed can be reallocated to the newer units to be opened. Again because of capital/labor complementarity, this positive effect plays through the size of investment in new machines as measured by $m(t)$ in (C7).

Whether the latter effect will be strong enough to compensate the job losses caused by obsolescence is largely a general equilibrium question, and one has to look at the demand side of the economy. Indeed the increase in wages following technological acceleration raises consumption (through the optimality condition C8) but have ambiguous effects on investment and output (via Eqs. (C3) and (C6)), and therefore on employment (again thanks to the capital/labor complementarity).

We shall evaluate the overall impact of the latter supply and demand side compensation mechanisms along the balanced growth paths, given that an analytical exploration of this issue on the system of dynamic equations (C1)–(C8) is far intractable.

13.3.2 Balanced Growth

Let us assume that $t - a(t) = T = \text{const}$, $Y(t) = \bar{y}e^{\gamma t}$, and $c(t) = \bar{c}e^{\gamma t}$. Then $N(t) = \bar{N}, m(t) = \bar{m}, w(t) = e^{\gamma(t-T)}$, and the system (C1)–(C8) leads, after some substitutions (to eliminate the discount term $\mu(t)$ and $w(t)$), to the following system of 6 equations

$$r(t) = \gamma + \rho = \text{const}, \tag{CG1}$$

$$\bar{y} = \bar{m}\frac{1 - e^{-\gamma T}}{\gamma}, \tag{CG2}$$

$$\frac{1 - e^{-(\rho+\gamma)T}}{\rho + \gamma} - e^{-\gamma T}\frac{1 - e^{-\rho T}}{\rho} = 1, \tag{CG3}$$

$$\bar{y} = \bar{c} + \bar{m}, \tag{CG4}$$

$$\bar{N} = \bar{m}T, \tag{CG5}$$

$$\theta\bar{c} = e^{-\gamma T} \tag{CG6}$$

As announced above, the BGP system is no longer recursive: while the BGP scrapping time, T, is still the solution of the same equation as in benchmark, that is (CG3) is identical to (BG4), the rest of BGP values cannot be determined recursively from T. One has to combine simultaneously (CG2), (CG4), (CG5) and (CG6) to write employment level $\bar{N}$ in terms of T, that is

$$\bar{N} = \frac{1}{\theta}\frac{\gamma T e^{-\gamma T}}{1 - \gamma - e^{-\gamma T}}. \tag{CG7}$$

It's then easy to characterize the BGPs.

Proposition 3 *The decentralized equilibrium (CG1)–(CG6) possesses a unique BGP:*

$$\bar{m} = \bar{N}/T, \tag{CG8}$$

$$Y(t) = \bar{N} \frac{1 - e^{-\gamma T}}{\gamma T} e^{\gamma t}, \tag{CG9}$$

$$c(t) = \frac{e^{-\gamma T}}{\theta} e^{\gamma t}, \tag{CG10}$$

where T is determined by equation (BG4), and $\bar{N}$ is given by Eq. (CG7). Moreover, if and only if

$$\theta > \frac{\gamma T e^{-\gamma T}}{1 - \gamma - e^{-\gamma T}}, \tag{CG11}$$

then the optimal $\bar{N}$ is interior. The optimal $\bar{N}$ is corner, $\bar{N} = 1$, if (CG11) fails. At $\gamma < \rho << 1$, the condition (CG11) is $\theta > (1 - \gamma)^{-1} + o(\gamma)$.

The proof is obvious. It's useful to verify that at small values of the leisure parameter θ, the optimal $\bar{N} = 1$ is boundary, i.e., people must work maximum possible hours. For larger θ, the optimal $\bar{N} < 1$ is interior, i.e., people can work less (but always $\bar{N} > 0$). The last condition for interior optimal labor supply uses the fact that when $\rho << 1$ and $\gamma << 1$, we have $T \approx \sqrt{2/\gamma}$. We now move to the key economic question, namely the overall impact of technological progress on employment.

13.3.3 Exogenous Technological Progress and Employment

We shall focus on the case where both ρ and γ are small enough to extract analytical results. If $\rho << 1$ and $\gamma << 1$, implying $T \approx \sqrt{2/\gamma}$, optimal $\bar{N}$ is indeed given by the approximate formula $\bar{N} \approx \frac{1+o(\gamma)}{\theta(1-\gamma)}$.

In contrast to the strongly intractable dynamic system (C1)–(C8) which delivers several possible supply and demand side interactions between technical progress and employment but no clear overall correlation, things are largely simplified along the BGPs: taking all these interactions into account, BGP restrictions yield that employment is an unambiguously increasing function of the rate of technological progress when ρ and γ are small enough. Interestingly enough note that if $\gamma \to 0$, then $T \to \infty$ (no scrapping) by Proposition 1, and $\bar{N} \to 1/\theta$. So, if the rate of technological progress is negligible, scrapping machines is not profitable but people still work and operate existing machines. If the leisure parameter θ as smaller than unity, then the people work maximum hours.

An important conclusion of the exercise is that just like in the typical RBC model with indivisible labor, a rise in γ does increase employment in the canonical vintage capital setting we are considering. This said, this property is much less obvious here as explained repeatedly in this section. Ambiguity has a double source in our vintage capital setting. A first ambiguity comes from the fact that technological progress affects detrended investment $\bar{m}$ (in new capital goods) through various supply side and demand side mechanisms (see the decentralized equilibrium above) which do not *a priori* yield a non-ambiguous overall impact. Next, even though one would comfortably conjecture that detrended investment would normally go up with technological accelerations, the impact on employment is still ambiguous since $\bar{N} = \bar{m}T$: in our vintage capital set-up, capital lifetime goes unambiguously down in response to labor-saving innovations, which in turn drives employment down. Our closed form (approximate) solution for $\bar{N}$ indicates that the latter obsolescence destructive effect is indeed dominated by the employment creative effect through investment in new and more efficient equipment $\bar{m}$ (under labor/capital gross complementarity), which is therefore not only positive but also sizeable enough to compensate the job losses due to obsolescence.

The latter comments call for the analysis of the dependence of the optimal investment, output and consumption levels on the rate γ. In the benchmark case, we show that a technological acceleration raises investment in level but decreases the levels of output and consumption.

The following proposition shows that these properties still hold under endogenous labor supply.

Proposition 4 *If $\gamma < \rho <\!\!< 1$, then the levels of optimal output and consumption are smaller for larger values of γ while investment level does increase.*

Proof We focus here on the impact of higher g on output level, which is the less easy part of the proof. Substituting $\bar{N}$ and $T \approx \sqrt{2/\gamma}$ into (CG9), we obtain that

$$Y(t)e^{-\gamma t} = \frac{1}{\theta}e^{-\gamma T}\frac{1 - e^{-\gamma T}}{1 - \gamma - e^{-\gamma T}} = \frac{1}{\theta}\frac{e^{-x} - e^{-2x}}{1 - e^{-x} - x^2/2} = f(x),$$

where $x = \gamma T$. The differentiation of $f(x)$ gives

$$f'(x) = \frac{B}{\theta}e^{-x}\frac{(1 - e^{-x})(e^{-x} - 1 + x + x^2/2) - e^{-x}x^2/2}{(1 - e^{-x} - x^2/2)^2}.$$

Hence, at small γ, $\gamma T \approx \sqrt{2\gamma}$ is also small. Presenting the exponent e^{-x} as the Taylor series, we obtain

$$f'(x) \approx -\frac{B}{2\theta}e^{-x}\frac{x^2(1 - 3x + 4x^2/3)}{(1 - e^{-x} - x^2/2)^2}.$$

Hence, $f'(x) < 0$ at small x and γ. **Q.E.D**

That's to say that the endogenous scrapping mechanism inherent to the vintage capital model is quantitatively strong enough to dominate the impact on output level of a simultaneous rise of employment and investment induced by exogenous technological accelerations. The next section evaluates whether the same conclusion holds if technological progress is endogenous (and of course costly).

13.4   Introducing Endogenous Labor-Saving Technological Progress

In this section, we assume that firms choose the optimal lifetime of their vintage capital and invest in new capital and in adoptive and/or innovative R&D. We shut up at the moment the endogenous labor supply channel to isolate the mechanisms generated by endogenous technical change. Therefore the consumption block is identical to the one of the benchmark model. Only the firms' block is re-designed. Mathematically, this section draws on previous work by Boucekkine et al. (2009, 2011, 2014b), applied to energy-saving technical progress under pollution quota. The 2009 paper explores the properties of a Solow version of the model (constant saving rate), the 2011 article solves a partial equilibrium (firm) model while the most recent one examines the optimal growth counterpart. In this section with inelastic labor supply, we study the decentralized equilibrium version of this class of models where energy is replaced by labor and the pollution quota by the labor market clearing condition. Several results of this section can be easily adapted from Boucekkine et al. (2011, 2014b) but not the general equilibrium outcomes which are new.

A key modelling aspect is R&D investment decision. As outlined in the introduction, innovation may consist in adding a new product or moving to a new production process. Consistently with the traditional vintage capital setting, innovation is labor-saving, and though it is exclusively conveyed by new capital goods, we interpret it as process innovation: this is a one-sector model and the successive capital goods represent the available frontier technologies. As explained in the introduction, the related empirical literature (like Greenan and Guellec 2000) tend to deliver that process innovations are associated with a negative effect on employment at the sectoral level. Though these innovations have a strong negative effect in our model through shutting down the obsolete production units, we show hereafter (notably in Sect. 13.5) that the overall effect on employment is positive unless the innovation is *too* radical. Last but not least, for a matter of simplification, we assume that R&D is undertaken by the same representative firm producing the final good. This is consistent with the observation that the largest part of R&D (private) expenditures are undertaken by the major companies in the automotive, ICT and pharmaceutical sectors.

13.4.1 The New Firm Problem

We shall consider the problem of a firm seeking to maximize the net profit that produces $Y(t)$ units of output, uses $L(t)$ units of labor, invests $R(t)$ into innovative and/or adoptive R&D, and invests $I(t)$ into new capital:

$$\max_{R,I,L} \int_0^\infty [Y(t) - w(t)L(t) - R(t) - I(t)]\mu(t)dt,$$

under the same constraints (13.2)–(13.4) as the benchmark firm. The novelty is the control $R(t)$, which is the amount (in terms of final good) the firm spent in improving its labor use through adoptive or innovative R&D, and reorganization. The associated production function is

$$\frac{\beta'(\tau)}{\beta(\tau)} = \frac{bR^n(\tau)}{\beta^d(\tau)}, \quad 0 < n < 1, \quad d > 0, \quad b > 0, \tag{13.23}$$

Following Boucekkine et al. (2011, 2014b), we postulate that the level $\beta(\tau)$ of the labor-saving technological progress evolves endogenously according to (13.23), where n is the parameter of "R&D efficiency" and d is the parameter of "R&D complexity". The constraints of the problem are identical to those of the benchmark firm problem. Initial conditions include those of the benchmark, plus a data on past $R(t)$ values because the delayed integral labor demand equation (13.4) requires such a data:

$$a(0) = a_0 < 0, \quad \beta(a_0) = \beta_0, \quad I(\tau) \equiv I_0(\tau), \quad R(\tau) \equiv R_0(\tau), \quad \tau \in [a_0, 0].$$

The nonlinear ODE (13.23) has an exact solution of the form:

$$\beta(\tau) = \left(d \int_0^\tau bR^n(v)dv + B^d \right)^{1/d}, \quad \tau \in [0, \infty), \tag{13.24}$$

where $B = \beta(0) = \left(d \int_{a_0}^0 bR_0^n(v)dv + \beta_0^d \right)^{1/d}$ using the initial conditions.

The optimization problem (OP) includes six unknown functions R, β, I, a, Y, and L connected by equalities (13.3), (13.4) and (13.23). Following Hritonenko and Yatsenko (1996) and Yatsenko (2004), we choose R, I, and a as the *independent* controls of the OP and consider the rest of the unknown functions β, y, and L as the *dependent (state)* variables. Using again the variable change $m(t) = I(t)/\beta(t)$, one can find the following *necessary first order conditions*:

Lemma 2 *Let (R, m, a, β, Y, L) be an interior solution of the OP. Then*

$$bn R^{n-1}(t) \int_t^\infty \beta^{1-d}(\tau)m(\tau) \left[\int_\tau^{a^{-1}(\tau)} \mu(z)dz - \mu(\tau) \right] d\tau = \mu(t), \qquad (13.25)$$

$$\int_t^{a^{-1}(t)} \mu(\tau)[\beta(t) - w(\tau)]\,d\tau = \mu(t)\beta(t), \qquad (13.26)$$

$$w(t) = \beta(a(t)), \qquad (13.27)$$

where $\beta(t)$ is determined from (13.24), $Y(t)$ from (13.3) and $L(t)$ from (13.4).

The proof is similar to the one provided in Boucekkine et al. (2011) with some routine modifications. The new optimality condition (13.25) features the optimal R&D investment rule. It equalizes the actualized marginal cost of one unit of good invested in R&D, the right-hand side, and its marginal benefit, the left-hand side: Technological advances made at t benefit to all machines produced from this date during their lifetime, which explains the structure of (13.25). Unlike the benchmark, nothing obvious can be said about the implementability or not of the interior solution for given prices from $t = 0$. We move to the decentralized equilibrium and its balanced growth paths in sake for more clarity.

13.4.2 Decentralized Equilibrium

The equilibrium of this economy is described by the following system of equations in the unknowns $R, \beta, \mu, m, a, y, c, r,$ and w

$$\frac{\dot{c}}{c} = r(t) - \rho, \qquad (D1)$$

$$\mu(t) = e^{-\int_0^t r(s)d\tau}, \qquad (D2)$$

$$y(t) = \int_{a(t)}^t \beta(\tau)m(\tau)d\tau, \qquad (D3)$$

$$\int_t^{a^{-1}(t)} \mu(\tau)\left[\beta(t) - \beta(a(\tau))\right] d\tau = \mu(t)\beta(t),$$ (D4)

$$w(t) = \beta(a(t)),$$ (D5)

$$y(t) = c(t) + R(t) + \beta(t)m(t),$$ (D6)

$$L(t) = \int_{a(t)}^t m(\tau)d\tau = 1,$$ (D7)

$$bnR^{n-1}(t)\int_t^\infty \beta^{1-d}(\tau)m(\tau)\left[\int_\tau^{a^{-1}(\tau)} \mu(z)dz - \mu(\tau)\right] d\tau = \mu(t),$$ (D8)

$$\beta(\tau) = \left(d\int_0^\tau bR^n(v)dv + B^d\right)^{1/d},$$ (D9)

under the given boundary conditions. Equality (D6) is the new clearing condition of the final good market. Equations (D8) and (D9) come from the endogeneity of technological progress, they are specific to the latter. Notice that just like in the benchmark model, given the initial investment profile, one can determine the scrapping time $a(t)$ from the labor market clearing condition (D7). But the similarity stops here: because technological progress is endogenous, the scrapping condition (D5) does not allow to compute wages. That is, while the initial profile $R(t)$ does allow to compute the initial profile in $\beta(t)$, the computation of actual $\beta(t)$ requires a sequence of $R(t)$ values (by Eq. (D9)), which are determined by the forward-looking equation (D8). So we move to balanced growth paths.

13.4.3 Balanced Growth Paths: Indeterminacy in Levels

We will explore the possibility of exponential solutions for $R(t)$, while $m(t)$ and $t - a(t)$ are constant, to the system (D1)–(D9). First of all, we start with the following preliminary result: if $R(t)$ is exponential, then $\beta(t)$ is *almost exponential* and practically undistinguishable from an exponent at large t in the sense of the following lemma:

Lemma 3 (Boucekkine et al. 2011). *If $R(t) = R_0 e^{Ct}$ for some $\gamma > 0$, then[3]*

$$\beta(t) \approx \bar{R}^{n/d} \left(\frac{bd}{\gamma n}\right)^{1/d} e^{\gamma nt/d} \qquad (13.28)$$

at large t. In particular, $\beta(t) = \bar{R}^{n/d} (bd/\gamma n)^{1/d} e^{\gamma nt/d}$ if $bd\bar{R}^n = \gamma n B^d$.

Boucekkine et al. (2011) actually studied a problem similar to the firm problem under consideration here, that is they postulated given prices. In particular, they showed that such a partial equilibrium problem cannot admit balanced growth paths in the cases $n > d$ and $n < d$. This property still holds in general equilibrium as one can guess. We therefore restrict our attention to the case $n = d$. The key result of this section is:

Proposition 5 *At $n = d$, the decentralized equilibrium (D1)–(D9) possesses an interior optimal regime (BGP)*

$$R(t) \approx \bar{R} e^{\gamma t}, \qquad \beta(t) \approx \bar{R} (b/\gamma)^{1/d} e^{\gamma t}, \qquad \text{(DG1)}$$

$$t - a(t) = T = \text{const}, \qquad m(t) = \bar{m} = 1/T, \qquad \text{(DG2)}$$

$$r(t) = \gamma + \rho = \text{const}, \qquad w(t) = \bar{R} (b/\gamma)^{1/d} e^{\gamma(t-T)}, \qquad \text{(DG3)}$$

$$y(t) \approx \bar{R} \left(\frac{b}{\gamma}\right)^{1/d} \frac{1 - e^{-\gamma T}}{\gamma T} e^{\gamma t}, \qquad \text{(DG4)}$$

$$c(t) = \bar{R} \left\{ \frac{1}{T} \left(\frac{b}{\gamma}\right)^{1/d} \left[\frac{1 - e^{-\gamma T}}{\gamma} - 1\right] - 1 \right\} e^{\gamma t}, \qquad \text{(DG5)}$$

where the constants γ and T are determined by the nonlinear equations (BG4) and

$$\gamma^{(1-d)/d} (\rho + \gamma d) e^{\gamma T} = db^{1/d} \frac{1 - e^{-\rho T}}{\rho T}, \qquad \text{(DG6)}$$

which has a positive solution, at least, at small ρ and $b \leq 1$.
Namely, if $\rho << 1$ and $b \leq 1$, then:

• *the rate γ is a unique positive solution $0 < \gamma < b$ of the nonlinear equation*

$$\gamma^{1/d} + \rho \gamma^{(1-d)/d}/d = b^{1/d} e^{-\gamma T} + o(\rho), \qquad (13.29)$$

---

[3]For brevity, we will omit the expression "at large t" when using the notation $f(t) \approx g(t)$.

- *a positive T is uniquely determined from Eq. (BG4) by Lemma 1.*
- *the optimal consumption given by (DG5) is positive.*

The proof is long and tricky, we report it in the appendix. The main result of the proposition is that BGPs still exist once technological progress is endogenized. The uniqueness properties are more subtle than in the benchmark and the first extension. Indeed, all the BGP variables are undetermined as it transpires from (DG4) and (DG5) for example. Strictly speaking, this is not surprising: The BGP systems are undetermined under endogenous technical progress since we add an unknown (the growth rate) to the same set of equations. As in traditional endogenous growth models (see Barro and Sala-i-Martin 1995, Chap. 4), all variables are undetermined in level: the BGP restrictions only determine ratios of variables. This is also the case here. Since T and the growth rate γ are uniquely determined, the ratios consumption to output, investment to output and R&D effort to output are also uniquely determined.

Beside indeterminacy, the endogenization of technological progress leads to the simultaneous and unique determination of capital lifetime and the rate of technological progress, which complicates dramatically the computations. Moreover, the BGP is proved to exist, and notably the scrapping time is finite along the BGP, only if the R&D productivity parameter b is bounded: $b \leq 1$. The condition $b \leq 1$ of Proposition 5 is sufficient but not necessary. Such a condition arises because of the way how the theorem is proved. As below numerical simulation shows, the BGP exists for larger values b as well. An interesting way to investigate why is to study the comparative statics of this BGP when b goes up. In particular, one would like to know how T, γ and the three ratios listed just above react to an increase in b.

It is natural to expect that the endogenous technological progress rate γ is larger for larger values of the R&D efficiency b. It can be proved analytically for a reasonable range of model parameters.

Proposition 6 (comparative statics in b) *If $\rho \ll 1$ and a BGP exists, then the endogenous rate of technological progress γ increases as the R&D efficiency b increases.*

Proof Let us rewrite Eq. (DG6) at $\rho \ll 1$ as

$$F(\gamma, b) = d\gamma^{1/d} + \rho\gamma^{1/d-1} - db^{1/d}e^{-\gamma T} = 0.$$

By Proposition 1, γT increases in γ, so the last term of function F increases in γ. The first two terms also increase in γ, therefore, the function F increases in $\gamma : \partial F/\partial \gamma > 0$. Next, $\partial F/\partial b = -b^{1/d-1}e^{-\sqrt{2\gamma}} < 0$. By the Theorem about Implicit Function, $d\gamma/db = -\partial F/\partial \gamma / \partial F/\partial b > 0$, which proves that γ is larger for larger values b. The proposition is proved. **Q.E.D**

We cannot obtain such a simple monotonic result as Proposition 6 for the lifetime T because, by Proposition 1, T is not monotonic in γ. Namely, T decreases in γ for "small enough" γ, but increases for γ close to $1 - \rho$ (and becomes infinite as $\gamma = 1 - \rho$). The question is when (and whether) the endogenous γ can grow close to

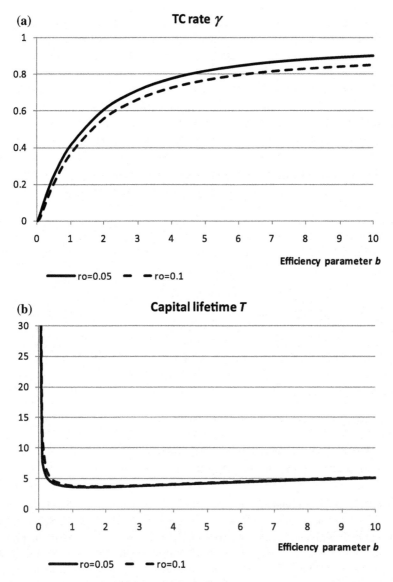

Fig. 13.1 **a** Simulated TC rate γ for $d = 0.5$ and b ranging from 0.01 to 10. **b** Simulated optimal capital lifetime for $d = 0.5$ and b ranging from 0.01 to 10

$1 - \rho$. One can notice that Proposition 6 gives no further details on the character of γ increase (in particular, whether γ increases indefinitely for indefinite b). Finding corresponding analytic conditions in the terms of given model parameters appears to be not possible. Applying various small parameter approximations to the nonlinear system (BG4) and (DG6) is inefficient because of the complex interplay of parameters b, ρ and d in the nonlinear equation (33).

So, to find more subtle comparative statics, we resort to numerical simulation. Namely, we have found an approximate solution of the system of two nonlinear equations (BG4) and (DG6) in γ and T for $d = 0.5$ and multiple combinations of parameters ρ and b. The obtained values of γ and T are of obvious interest and help to identify the reasonable range of the important given parameter—the R&D efficiency b (which is difficult to estimate using statistics or any other means). The corresponding calculations are done in Matlab.

13.4.4 A Numerical Exploration of the Covariations of Technological Progress and Capital Lifetime

To start, let $\rho = 10\%$ and the R&D efficiency b change from 0.02 to 10. The corresponding growth rate γ is shown in Fig. 13.1a. It increases from 0 at $b = 0.01$ to $\approx 40\%$ and slows down later and saturates around the value $\gamma \approx 0.8$. At simulation provided, the rate γ never reaches the above mentioned "critical" value $1 - \rho = 0.9$.

The corresponding capital lifetime T decreases fast at $b = 0.01$ to 9 years at $b = 0.1$ and later stabilizes around 3.6 years (see Fig. 13.1b). As predicted theoretically by Proposition 1, the capital lifetime starts to increase at some large b (and corresponding to "large" γ) but the subsequent path is rather flat (from 3, 5 years to 4.6 years when b changes 1.3 from to 10). A realistic range for b seems to be [0.05, 0.2] as γ would then increase from $\approx 0.5\%$ to $\approx 10\%$ and T decreases from 27 years to 5.6 years. The simulation results are similar for $\rho = 5\%$ and shown on the same Fig. 13.1a, b.

13.5 Endogenous Technical Progress, Employment and the Lifetime of Capital

We now consider the model incorporating the two extensions considered so far, that is putting together the consumer block with endogenous labor supply of Sect. 13.3, and the firm block with endogenous technical progress of Sect. 13.4. As in the previous section, we set $n = d$, and we seek for BGP solutions to the obtained dynamic system checking in particular $t - a(t) = T = \text{const}$ and $R(t) \approx \bar{R}e^{\gamma t}$. After the needed tedious computations, it turns out that the resulting BGP restrictions lead to

$$\beta(t) \approx \bar{R}\,(b/\gamma)^{1/d}\,e^{\gamma t}, \tag{EG1}$$

$$t - a(t) = T = \text{const}, \qquad m(t) = \bar{N}/T, \tag{EG2}$$

$$r(t) = \gamma + \rho = \text{const}, \qquad w(t) = \bar{R}\,(b/\gamma)^{1/d}\,e^{\gamma(t-T)}, \tag{EG3}$$

$$\bar{y} = \bar{c} + \bar{R} + B\bar{m}, \tag{EG4}$$

$$\theta\bar{c} = \bar{R}\left(\frac{b}{\gamma}\right)^{1/d} e^{-\gamma T}, \tag{EG5}$$

$$y(t) \approx \bar{N}\bar{R}\left(\frac{b}{\gamma}\right)^{1/d} \frac{1 - e^{-\gamma T}}{\gamma T} e^{\gamma t}, \tag{EG6}$$

$$\gamma^{1/d} + \gamma^{(1-d)/d}\rho/d = \bar{N}b^{1/d} \frac{1 - e^{-\rho T}}{\rho T} e^{-\gamma T}, \tag{EG7}$$

where T is still given by Eq. (BG4). Needless to say, (EG7) is the counterpart of (DG6) when labor supply is endogenous. Substituting of m, y, and c from (EG2), (EG5), (EG6) into the good market equilibrium condition leads to the following equality:

$$\bar{R}\frac{\bar{N}}{T}\left(\frac{b}{\gamma}\right)^{1/d}\left[\frac{1 - e^{-\gamma T}}{\gamma} - 1\right] = \frac{\bar{R}}{\theta}\left(\frac{b}{\gamma}\right)^{1/d} e^{-\gamma T} + \bar{R}, \tag{EG8}$$

where the left-hand side is $\bar{y} - B\bar{m}$ (which is proportional to $\bar{N}$) and the right-hand side is $\bar{c} + \bar{R}$. After further transformations, we obtain the following system of three nonlinear equations with respect to the unknown $\bar{N}$, γ, and T, namely (BG4), (EG7) and Eq. (EG8) rewritten as:

$$\bar{N}\left[1 - \gamma - e^{-\gamma T}\right] = \frac{1}{\theta}\gamma T e^{-\gamma T} + \gamma T\left(\frac{\gamma}{b}\right)^{1/d}. \tag{EG9}$$

Compared to the previous extension in Sect. 13.2, we move from a set of two simultaneous equations determining $\bar{N}$ and T, to a set of three simultaneous equations that determine $\bar{N}$, γ, and T. Since these equations are highly nonlinear, little can be proved analytically. This said, and before numerical experimentation, a few analytical things can be extracted from the system above. In particular, note that (EG9) gives:

$$\bar{N} = \frac{\gamma T}{1 - \gamma - e^{-\gamma T}}\left[\frac{e^{-\gamma T}}{\theta} + \left(\frac{\gamma}{b}\right)^{1/d}\right].$$

Analytical comparison with the counterpart under exogenous technical progress is possible.

Proposition 7 *At the same rate of technical progress γ, employment is larger under endogenous technological progress.*

The proof is trivial. It's enough to compare the expression above with the corresponding formula under exogenous technological progress, namely Eq. (CG7):

$$\bar{N} = \frac{e^{-\gamma T}}{\theta} \frac{\gamma T}{1 - \gamma - e^{-\gamma T}}.$$

This means that for the same value of the rate of technological progress, γ, inducing the same value for capital lifetime T by Eq. (BG4), the economy with endogenous technical progress is associated with more employment, which also implies, by $m(t) = \bar{N}/T$, that the former economy will experience a higher level of investment. In simple words, the presence of an innovative R&D sector in economy increases the willingness of people to invest and work at stationary equilibrium. It's easy to figure out why: in our model technological advancement is costly, the economy has to devote part of the final good to this activity, which (at the same intended rate of technological progress) means extra-labor supply compared to the case where innovations are exogenous since our technology is Leontief. Of course this nontrivial but partial result does not say much on the covariations of employment and technological progress in the extended economy, which we investigate numerically here below.

13.5.1   Numerical Exploration of the Covariations of Technological Progress and Employment

In summary, a possible interior optimal regime (BGP) in the decentralized equilibrium model (EG1)–(EG7) is completely characterized by the system of three nonlinear equations (BG4), (EG7) and (EG8) with respect to $\bar{N} > 0$, $\gamma > 0$ and $T > 0$. If the BGP exists, then the optimal $\bar{N}$ is always positive and may be interior, $0 < \bar{N} < 1$, or corner: $\bar{N} = 1$. In the case of full employment $\bar{N} = 1$, the BGP is equivalent to the problem of Sect. 13.4. However, the optimal $\bar{N}$ will be interior $0 < \bar{N} < 1$, at least, at small b, which is confirmed by the numerical simulation below.

We assumed $\rho = 5\%$ at $d = 0.5$ and provided a series of simulations for the R&D efficiency b changing from 0.02 to 10 and the leisure parameter θ changing from 0.1 to 20. The results are shown in Figs. 13.2a–c. The figures demonstrate several interesting effects.

Figures 13.2a, b depict the behavior of the endogenous rate of technological progress γ and capital lifetime T when the R&D efficiency b increases. One can see that the basic qualitative picture is the same as in the model with fixed labor supply. Namely, Figs. 13.2a, b resemble Figs. 13.1a, b from Sect. 13.4 in the b ranges where (optimal) employment is interior: $\bar{N} < 1$. As in Fig. 13.1b, the optimal capital lifetime T decreases fast at small b and starts to increase very slowly at larger b. However and in contrast to the previous models studied in Sects. 13.3 and 13.4, both γ and T depend on θ now. In particular, as the preference for leisure increases, technological progress is shifted down. This is a natural outcome: increasing the rate of techno-

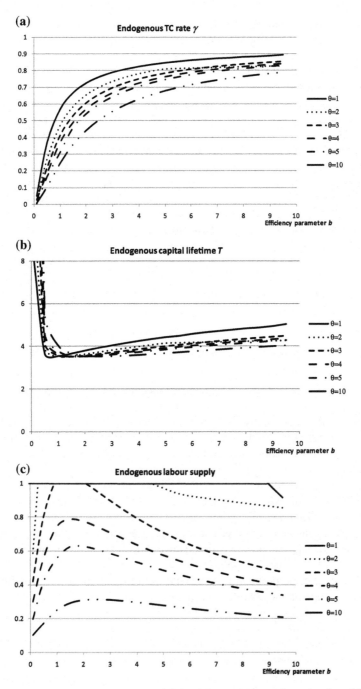

Fig. 13.2 a The endogenous growth rate γ for $d = 0.5$, $\rho = 0.05$, and b ranging from 0.01 to 10.
b The endogenous capital lifetime T for $d = 0.5$, $\rho = 0.05$, and $b = 0.01 \div 10$. **c** Employment N
for $d = 0.5$, $\rho = 0.05$, and $b = 0.01 \div 10$

logical progress requires more final good, which itself requires more investment and more labor given labor/capital complementarity. As households are less willing to work, the resulting technological path goes down.

Fig. 13.2c is new and nontrivial. First of all, it confirms the above theoretical findings that optimal labor is corner (that's $\bar{N}=1$) for not too large leisure parameter θ. The simulation shows that if $\theta = 0.2$, then employment is corner for any $0 < b < 22$. If $\theta = 0.5$, then $\bar{N} = 1$ for any $0 < b < 15$. If $\theta = 1$, then $\bar{N} = 1$ for $0 < b < 9$ and $\bar{N} < 1$ is interior for $9 \leq b$. Starting with $\theta \geq 4$, $\bar{N} < 1$ is always interior for any b, i.e., employment is definitely smaller when the preference for leisure is high enough. Next, the dependence of employment on the R&D efficiency parameter b appears to be non-monotonic with a maximum reached at some value of b. Note that the usual calibration of the RBC models using indivisible labor typically yields $\bar{N} = 1/3$ (people work one third on their time). Non-monotonicity of the employment path is indeed apparent at all values of θ. Nonetheless, it is fair to observe that the turning points (that's the b values at which employment is maximal) correspond to γ values (in Fig. 13.2a) which are close to 40 % for reasonable values of $\bar{N}$. It follows that compensation does not always work in the vintage capital model: process innovation does create employment provided the innovation is not *too* radical.

13.6 Conclusion

In this paper, we have studied the impact of technological progress on the level of employment in a vintage capital model where: (i) capital and labor are gross complementary; (ii) labor supply is elastic; (iii) there is full employment, and (iv) the rate of labor-saving technological progress is endogenous. We restrict our analysis to the indivisible labor case. To this end, we have characterized the stationary distributions of vintage capital goods and the corresponding equilibrium values for employment and the lifetime of capital. We have shown through successive extensions of a benchmark vintage capital model that both employment and the lifetime of capital are indeed non-monotonic functions of the rate of technological progress (when it is exogenous) or of the efficiency in the R&D sector (in the case of endogenous technological progress). In particular, we show that employment keep growing with the rate of technological progress unless innovations are too radical.

Studying the same problem for a general utility function and/or departing from the Leontief technologies (while assuring finite time scrapping for capital goods) is of course much more demanding. Despite our simplifying assumptions, we believe that our analysis makes already clear that the traditional "compensation problem" is definitely much less trivial when taking into account the vintage structure of capital, even though labor markets are taken efficient.

Appendix

Proof of Proposition 4: By Lemma 3, $\beta(t)$ is given by (28) at $n = d$. Then, the substitution of this relationship into Eq. (D8) at the BGP leads to

$$bd(\bar{R}e^{\gamma t})^{d-1} \int\limits_t^\infty \left(\bar{R} \left(\frac{b}{\gamma} \right)^{\frac{1}{d}} e^{\gamma \tau} \right)^{1-d} \frac{1}{T} \left[\frac{e^{-r\tau} - e^{-r(\tau+T)}}{r} - e^{-r\tau} \right] d\tau = e^{-rt},$$

and, after integration, to

$$\frac{db^{\frac{1}{d}} \gamma^{\frac{d-1}{d}}}{T[\gamma(1-d)-r]} \left[\frac{1 - e^{-rT}}{r} - 1 \right] e^{-rt} = e^{-rt}. \qquad (13.30)$$

Substituting $r = \gamma + \rho$ into (13.30) and using Eq. (BG4) leads to (DG6).

The system of Eqs. (BG4) and (DG6) may have a positive solution γ, T at natural assumptions. Namely, let $\rho \ll 1$. Then, presenting the exponent $e^{-\rho T}$ in (DG6) as the Taylor series, we obtain

$$\gamma^{(1-d)/d}(\rho + \gamma d)e^{\gamma T} = db^{1/d} + o(\rho). \qquad (13.31)$$

The function $F(\gamma) = \gamma^{(1-d)/d}(\rho + \gamma d)e^{\gamma T}$ monotonically increases to ∞ in γ and $F(0) = 0$. Therefore, Eq. (13.31) has a unique positive solution γ. By (13.31), $(\gamma/b)^{1/d}(1 + \rho/\gamma d)e^{\gamma T} \approx 1$, therefore, $\gamma < b(1 + \rho/\gamma d)^{-d} \approx b(1 - \rho/\gamma)$. Since $b \leq 1$ in the theorem statement, then $\gamma < 1-\rho$ and, by Lemma 1, Eq. (BG4) has a unique positive solution T at the known γ. At $\rho \ll 1$, presenting two exponents $e^{-\rho T}$ in (BG4) as the Taylor series, we have

$$1 - e^{-\gamma T} - \gamma T e^{-\gamma T} = \rho + \gamma + o(\rho). \qquad (13.32)$$

The left-hand side of (13.32) monotonically increases in γT and equals zero at $\gamma T = 0$ and $\gamma = 0$. Therefore, the equality (13.32) establishes a one-to-one relationship between γT and γ.

If also $b \ll 1$, then the (13.32) solution $\gamma \ll 1$ and $T \approx \sqrt{2/\gamma}$ by Lemma 1.

Knowing $\beta(t)$ and T, we can easily find the rest of the unknowns. In particular, let us look at $c(t)$ and analyze when $c(t) > 0$. We have

$$c(t) = y(t) - \beta(t)m(t)a - R(t)$$

$$\approx \left\{ \bar{R} \left(\frac{b}{\gamma} \right)^{1/d} \left[\frac{1 - e^{-\gamma T}}{\gamma T} \right] - \bar{R} \left(\frac{b}{\gamma} \right)^{1/d} \frac{1}{T} - \bar{R} \right\} e^{\gamma t},$$

which gives (DG5). Now let us rewrite the latter as

$$c(t) = \bar{R} \left\{ \frac{1}{\gamma T} \left(\frac{b}{\gamma} \right)^{1/d} \left(1 - e^{-\gamma T} - \gamma \right) - 1 \right\} e^{\gamma t}, \tag{13.33}$$

Substituting the expression of γ through γT from the formula (13.32) into (13.33), we obtain

$$c(t) \approx \bar{R} \left\{ \left(\frac{\rho}{\gamma T} + e^{-\gamma T} \right) \left(\frac{b}{\gamma} \right)^{1/d} - 1 \right\} e^{\gamma t}.$$

Therefore, $c(t) > 0$ if

$$\left(\frac{\rho}{\gamma T} + e^{-\gamma T} \right) \left(\frac{b}{\gamma} \right)^{1/d} - 1 > 0. \tag{13.34}$$

By (13.29), $\frac{\rho}{\gamma d} + 1 = \left(\frac{b}{\gamma} \right)^{1/d} e^{-\gamma T}$ at $\rho \ll 1$. So, (13.34) becomes

$$\frac{\rho}{\gamma d} + 1 + \left(\frac{b}{\gamma} \right)^{1/d} \frac{\rho}{\gamma T} > 0, \tag{13.35}$$

which holds for any positive ρ, γ, d, and T. The theorem is proved. **Q.E.D**

References

Addison, J., & Teixeira, P. (2001). Technology, employment and wages. *Labour, 15*, 191–219.
Barro, R., & Sala-i-Martin, X. (1995). *Economic growth*. McGraw-Hill.
Benhabib, J., & Rustichini, A. (1991). Vintage capital, investment, and growth. *Journal of Economic Theory, 55*, 323–339.
Boucekkine, R., del Rio, F., & Martinez, B. (2009). Technological progress, obsolescence, and depreciation. *Oxford Economic Papers, 61*, 440–466.
Boucekkine, R., Hritonenko, N., & Yatsenko, Yu. (2014a). Health, work intensity, and technological innovations. *Journal of Biological Systems, 22*, 219–233.
Boucekkine, R., Hritonenko, N., & Yatsenko, Yu. (2014b). Optimal investment in heterogeneous capital and technology under restricted natural resource. *Journal of Optimization Theory and Applications, 163*, 310–331.
Boucekkine, R., Hritonenko, N., & Yatsenko, Yu. (2011). Scarcity, egulation and endogenous technical progress. *Journal of Mathematical Economics, 47*, 186–199.
Boucekkine, R., del Rio, F., & Licandro, O. (1999). Exogenous vs endogenously driven fluctuations in vintage capital growth models. *Journal of Economic Theory, 88*, 161–187.
Boucekkine, R., Germain, M., Licandro, O., & Magnus, A. (1998). Creative destruction, investment volatility and the average age of capital. *Journal of Economic Growth, 3*, 361–384.
Boucekkine, R., Germain, M., & Licandro, O. (1997). Replacement echoes in the vintage capital growth model. *Journal of Economic Theory, 74*, 333–348.

Boyer, R. (1988). New technologies and employment in the 1980s: From science and technology to macroeconomic modelling. In J. A. Kregel, E. Matzner, & A. Roncaglia (Eds.), *Barriers to full employment* (pp. 233–68). London: Macmillan.

Caballero, R., & Hammour, M. (1996). On the timing and efficiency of creative destruction. *The Quarterly Journal of Economics, 111*, 805–851.

Cooley, T., Greenwood, J., & Yorokoglu, M. (1997). The replacement problem, *Journal of Monetary Economics 40*, 457–500.

Feichtinger, G., Hartl, R., Kort, P., & Veliov, V. (2006). Capital accumulation under technological progress and learning: A vintage capital approach. *European Journal of Operation Research, 172*, 293–310.

Freeman, C., & Soete, L. (1994). *Work for all or mass unemployment? Computerised technical change into the twenty-first century*. London-New York: Pinter.

Greenan, N., & Guellec, D. (2000). Technological innovation and employment reallocation. *Labour, 14*, 547–90.

Hansen, G. (1985). Indivisible labor and the business cycle. *Journal of Monetary Economics, 16*, 309–327.

Hornstein, A., Krusell, P., & Violante, G. (2005). The replacement problem in frictional economies: A near-equivalence result. *Journal of the European Economic Association, 3*, 1007–1057.

Hritonenko, N., & Yatsenko, Yu. (2005). Turnpike properties of optimal delay in integral dynamic models. *Journal of Optimization Theory and Applications, 127*, 109–127.

Hritonenko, N., & Yatsenko, Yu. (2012). Technological modernization under resource scarcity. *Optimal Control Applications and Methods, 33*, 249–262.

Hritonenko, N., & Yatsenko, Yu. (1996). *Modeling and Optimization of the Lifetime of Technologies*. Dordrecht: Kluwer Academic Publishers.

Jovanovic, B., & Yatsenko, Yu. (2012). Investment in vintage capital. *Journal of Economic Theory, 147*, 551–569.

Rogerson, R. (1988). Indivisible labor, lotteries and equilibrium. *Journal of Monetary Economics, 21*, 3–16.

Say, J.B. (1964). *A treatise on political economy or the production, distribution and consumption of wealth*, New York (M. Kelley, first edition 1803).

Smolny, W. (1998). Innovations, prices and employment: A theoretical model and an empirical application for West German manufacturing firms. *Journal of Industrial Economics, 46*, 359–81.

Solow, R., Tobin, J., Von Weizsacker, C., & Yaari, M. (1966). Neoclassical growth with fixed factor proportions. *Review of Economic Studies, 33*, 79–115.

Vivarelli, M. (2007). Innovation and employment, a survey. *IZA Discussion Paper, 2621*.

Yatsenko, Yu., Boucekkine, R., & Hritonenko, N. (2009). On explosive dynamics in R&D-based models of endogenous growth. *Nonlinear Analysis: Theory, Methods and Applications, 71*, e693–e700.

Yatsenko, Yu. (2004). Maximum principle for Volterra integral equations with controlled delay time. *Optimization, 53*, 177–187.

Chapter 14
Nonbalanced Growth in a Neoclassical Two-Sector Optimal Growth Model

Harutaka Takahashi

Abstract For a neoclassical two-sector optimal growth model with Cobb-Douglas technologies and sector specific technical progress, we examine three properties: (i) each sector has an optimal path by which it will grow at a constant growth rate (an optimal constant growth path); (ii) the optimal constant growth paths satisfy saddle path stability; (iii) the elasticity of substitution between total labor and total capital is less than one along the optimal constant growth paths. These results, presented by Acemoglu and Guerrieri in their model with two intermediate good sectors and one final good sector, will give a firm theoretical base for establishing the Kaldor and Kuznets facts, and are proven herein for a neoclassical growth model.

JEL Classification: O14 · O21 · O24 · O41

14.1 Introduction

A resurgence of an interest in growth and structural change has swept the field of economics recently. In facts, industry-based empirical studies across countries show clearly that growth in an individual industry's per-capita capital stock and output grow at the industry's own growth rate, which is related closely to its technical progress measured by total factor productivity (TFP) of the industry. As presented in Fig. 14.1, the correlation coefficient between the per-capita GDP average growth rate and TFP average growth rate is very high: 0.65. Furthermore, Fig. 14.2 shows per-capita output of several sectors in log-scale term. One might read that the agriculture grows at about 5 % per annum along its own constant growth path indicated by its trend line, while the total manufacturing industry grows at about 10 % annually, also following the industry's constant growth path.

H. Takahashi (✉)
Department of Economics, Meiji Gakuin University, 1-2-37 Shirokanedai,
Tokyo, Minato-ku 108-8636, Japan
e-mail: haru@eco.meijigakuin.ac.jp

© Springer International Publishing AG 2017
K. Nishimura et al. (eds.), *Sunspots and Non-Linear Dynamics*,
Studies in Economic Theory 31, DOI 10.1007/978-3-319-44076-7_14

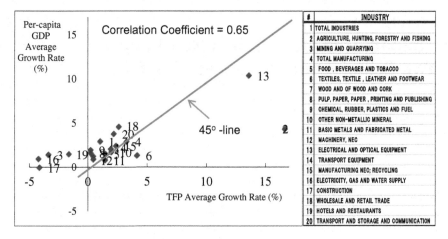

Fig. 14.1 US economy, 1970–2005. *Source* EU-KLEMS database

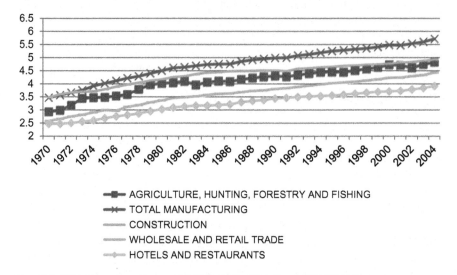

Fig. 14.2 US industry-based per-capita GDP, 1970–2004. *Source* EU-KLEMS database

Syverson (2011) recently examined these observations explained above. This phenomenon has been designated as **"nonbalanced growth among industries"** following Acemoglu and Guerrieri (2008).

Attempts to elucidate this phenomenon have generated the strong theoretical necessity for constructing nonbalanced growth models. One can classify these studies into two categories[1]: preference-driven and technology-driven structural change. In the first category, structural change is derived by differences in the income

---

[1]Recently, Alvarez-Cuadrado and Long (2011) present the third mechanism for a structural change, where they emphasize differences in the elasticity of substitution across sectors.

elasticity of demand across goods, as explained for instance in Kongsamut et al. (2001), Foellmi and Zweimuller (2008), Alonso-Carrera and Raurich (2015), Laitner (2000) and Boppart (2013). In the second category, structural change is derived mainly from technological differences across sectors measured by TFP, for which argument can be traced back to Baumol (1967). Ngai and Pissarides (2007)[2] set up a multi-sector optimal growth model with common technologies and demonstrated structural change, where, in contrast to Baumol's claim, the economy's growth rate is not on an indefinitely declining trend. Other examples include explanations by Echevarria (1997) and Acemoglu and Guerrieri (2008). Setting up an optimal growth model with three sectors: primary, manufacturing, and service, the former has applied numerical analysis to solve it and demonstrate structural change. On the other hand, the latter has studied the model with two physically differentiated intermediate good sectors and single final good sector. In the contrast to the effort by Ngai and Pissarides (2007), both papers emphasize sectoral differences in factor proportions evaluated by the sector's capital intensity and present the following three main results:

(i) Each sector has an optimal path along which it will grow at its own constant growth rate (an optimal constant growth path).
(ii) The optimal constant growth paths satisfy saddle path stability.
(iii) The aggregate elasticity of substitution between labor and capital is less than one along the optimal constant growth paths.

Note that the above results will give a firm theoretical base for establishing the Kaldor and Kuznets facts as discussed in Acemoglu and Guerrieri (2008).

This paper proves the three main results mentioned above in a neoclassical two-sector optimal growth model with Cobb-Douglas technologies and sector specific technical progress. Any models listed above shares a common feature: a single final good. In contrast, a standard two-sector neoclassical growth model will include two physically different goods. A study by Echevarria (1997) is the only exception, with three final goods studied. The present study uses a discrete time structure, while Acemoglu and Guerrieri (2008) set up their model in terms of a continuous time structure. Furthermore, because we assume herein that each good is produced with a different technology, the consumption good and capital good have completely differentiated physical characters and there is no direct relationship between outputs. In contrast, in Acemoglu and Guerrieri (2008), because sector's outputs will be used as inputs for producing the single final good, they have strong correlation. As we will demonstrate later, although there exists this feature of our model, the factor share proportion differences within and across sectors take important roles in demonstrating the nonbalanced growth as emphasized by Acemoglu and Guerrieri (2008) and Acemoglu (2009).

In Sect. 14.2, the model and assumptions are presented. In Sect. 14.3, we derives the optimal steady state (OSS) in the efficiency-unit, which exhibits nonbalanced

---

[2]Since they assume the same production functions among sectors, except sector specific exogenous technical progress, their model would be identified as one-commodity economy. Because of this property, they could aggregate sector's output even in a transition process. Contrast to their model, we will study a two-commodity economy in this paper.

growth in the original-unit. The growth rate of the aggregated GDP defined along the OSS will converge to that of the technologically progressive sector. In Sect. 14.4, the aggregate elasticity of substitution is studied and the local stability in terms of saddle path stability will be proved. Local stability implies that each sector's per-capita capital stock and output eventually grow at the rate of sector's TFP growth. Section 14.5 presents the salient conclusions of these efforts.

14.2   Model and Assumptions

The symbols used in this paper are as follows:

r	subjective rate of discount,
$C(t) \in \mathbb{R}_+$	total good consumed at t,
$c(t) \in \mathbb{R}_+$	$C(t)/L(t) \in \mathbb{R}_+$,
$Y(t) \in \mathbb{R}_+$	tth period capital output of the capital goods sector,
$K(t) \in \mathbb{R}_+$	total capital goods at t,
$K(0) \in \mathbb{R}_+$	initial total capital goods,
$K_i(t) \in \mathbb{R}_+$	tth period capital stock of the ith sector,
$L(t) \in \mathbb{R}_+$	total labor input at t,
$L(0) \in \mathbb{R}_+$	initial total labor input,
$L_i(t) \in \mathbb{R}_+$	tth labor input of the ith sector,
δ	depreciation rate,
$A_i(t)$	Hicks neutral technical-progress of the ith sector.

In that notation, $i = 0$ and $i = 1$ respectively indicate the consumption goods sector and the capital goods sector.

The following two assumptions are made for the model:

Assumption 1

(1) The utility function u($\cdot$) is defined on $\mathbb{R}_{++}$ as the following standard form of,

$$u(c(t)) = u\left(C(t)/L(t)\right) = \frac{c(t)^{1-\sigma}}{1-\sigma} \ for \ t \geq 0 \ and \ \sigma > 0.$$

(2) $L(t) = (1 + g)^t L(0)$, where g is the rate of population growth.

Assumption 2

(1) All goods are produced with the following Cobb-Douglas production functions with Hicks-neutral technical progress:

$$C(t) = A_0(t)K_0(t)^{\alpha_1} L_0(t)^{\alpha_2} \ and \ Y(t) = A_1(t)K_1(t)^{\beta_1} L_1(t)^{\beta_2}$$
$$where \ \alpha_1 + \alpha_2 = 1 \ and \ \beta_1 + \beta_2 = 1,$$

and they satisfy the "Inada-conditions."

(2) $A_i(t) = (1 + a_i)^t A_i(0)$ $(i = 0, 1)$, where a_i is a rate of output-augmented (Hicks-neutral) technical-progress of the ith sector and given as $0 < a_i < 1$.

It is noteworthy that (2) of Assumption 2 implies that the sector-specific TFP is measured by the sector-specific output-augmented technical progress (Hicks-neutral technical progress), which is given externally.

Let us first define the capital intensity of a sector.

Definition When $\alpha_1/\alpha_2 > \beta_1/\beta_2 (\alpha_1/\alpha_2 < \beta_1/\beta_2)$ hold, one may say that the consumption good sector is *capital intensive (labor intensive)* in comparison with the capital goods sector.

Takahashi et al. (2012) present firm evidence that in any OECD country, the consumption good sector is the more capital intensive. Based on this observation and the detailed discussion by Acemoglu and Guerrieri (2008) and Acemoglu (2009),[3] the following technology parameter condition, which exhibits **factor share proportion differences across sectors**, must be imposed.

Assumption 3 (*Factor Share Proportion Differences*) $\alpha_1 > \alpha_2$ and $\beta_2 > \beta_1$.

Make sure that Assumption 3 indicates not only the factor proportional differences within sectors but also the factor share proportion differences across sectors. Clearly Assumption 3 implies that $\alpha_1/\alpha_2 > \beta_1/\beta_2$: the consumption good sector is capital intensive.

Before setting up the model, we will divide all variables by $A_i(t)L(t)$ and will transform the original variables into *per-capita efficiency-unit* variables. Next, we define the following normalized variables as shown below:

$$\widetilde{y}(t) = \frac{Y(t)}{A_1(t)L(t)}, \quad \widetilde{c}(t) = \frac{C(t)}{A_0(t)L(t)}, \quad y(t) = \frac{Y(t)}{L(t)}, \quad c(t) = \frac{C(t)}{L(t)},$$

$$k_1(t) = \frac{K_1(t)}{L_1(t)}, \quad k_0(t) = \frac{K_0(t)}{L_0(t)}, \quad \ell_1(t) = \frac{L_1(t)}{L(t)}, \quad \ell_0(t) = \frac{L_0(t)}{L(t)}.$$

Therein, "$\sim$" indicates the efficiency-unit variables.

Transforming both sector's production functions into the efficiency-unit ones as follows; dividing both sides by $A_1(t)L(t)$, we obtain

$$\widetilde{c}(t) = f^0(k_0(t), \ell_0(t)) = k_0(t)^{\alpha_1} \ell_0(t)^{\alpha_2} \text{ and } \widetilde{y}(t) = f^1(k_1(t), \ell_1(t)) = k_1(t)^{\beta_1} \ell_1(t)^{\beta_2}.$$

The next step is to derive the efficiency-unit production possibility frontier (PPF), which is shown in Lemma 1:

Lemma 1 *The efficiency-unit production possibility frontier (efficiency PPF for short):* $\widetilde{c} = T\ (\widetilde{y}, k)$ *is calculated explicitly as follows:*

$$\widetilde{c} = T\ (\widetilde{y}, k) = \left[\frac{\beta_1(\alpha_1 - \alpha_2)}{(\alpha_2\beta_1 - \alpha_1\beta_2)e\ (k, \widetilde{y})} \right]^{\alpha_2} [k - e\ (k, \widetilde{y})]. \qquad (14.1)$$

---

[3] See Sect. 20.2 of Acemoglu (2009) for detailed discussion.

In that equation, $k_1 = e(k, \tilde{y})$ is the function obtained by solving the following equation with respect to k_1 :

$$(\alpha_1\beta_2 - \alpha_2\beta_1)^{\beta_2} k_1^{\beta_2 - \beta_1} \tilde{y} = [(\alpha_2 - \alpha_1)\beta_1 k - (\beta_2 - \beta_1)\alpha_1 k_1]^{\beta_2} .$$

Proof We apply the analytical method explained in Baierl et al. (1998).[4] Under Assumption 2, let us consider the problem (∗) where the time index is dropped for simplicity:

(∗) $Max \; \tilde{c} = k_0^{\alpha_1} \ell_0^{\alpha_2}$ *s.t.* $\tilde{y} = k_1^{\beta_1} \ell_1^{\beta_2}$, $\ell_0 + \ell_1 = 1$ *and* $\ell_0 k_0 + \ell_1 k_1 = k$.

Because of profit-maximization, the following F.O.C. will be derived:

(i) $\alpha_1 k_0^{\alpha_1 - 1} \ell_0^{\alpha_2} - w\ell_0 = 0$

(ii) $\alpha_2 k_0^{\alpha_1} \ell_0^{\alpha_2 - 1} - w_0 - wk_0 = 0$

(iii) $-p\beta_1 k_0^{\beta_1 - 1} \ell_0^{\beta_2} - w\ell_1 = 0$

(iv) $-p\beta_2 k_1^{\beta_1 - 1} \ell_1^{\beta_2 - 1} - w_0 - wk_1 = 0$

Multiplying k_0 and ℓ_0 to (i) and (ii) respectively, we have

$$\frac{w_0}{w} = \left(\frac{\alpha_2}{\alpha_1} - 1\right) k_0.$$

From (iii) and (iv), we obtain

$$\frac{w_0}{w} = \left(\frac{\beta_2}{\beta_1} - 1\right) k_1.$$

Accordingly it holds that

$$\left(\frac{\alpha_2}{\alpha_1} - 1\right) k_0 = \left(\frac{\beta_2}{\beta_1} - 1\right) k_1.$$

Solving the equation above with respect to k_0,

$$k_0 = \left(\frac{\beta_2 - \beta_1}{\alpha_2 - \alpha_1}\right)\left(\frac{\alpha_1}{\beta_1}\right) k_1 = Dk_1 \; where \; D = \left(\frac{\beta_2 - \beta_1}{\alpha_2 - \alpha_1}\right)\left(\frac{\alpha_1}{\beta_1}\right).$$

From the resource constraints and the result presented above,

$$\ell_1 = \frac{k - k_0}{k_1 - k_0} = \frac{k - Dk_1}{(1 - D)k_1} = \frac{(\alpha_2 - \alpha_1)\beta_1 k - (\beta_2 - \beta_1)\alpha_1 k_1}{(\alpha_2\beta_1 - \alpha_1\beta_2)k_1} \tag{14.2}$$

---

[4]One crucial difference from Baierl et al. (1998) is that the labor force keeps growing at g in our model. Accordingly we have the per-capita capital constraint: $\ell_0 k_0 + \ell_1 k_1 = k$ instead of $k_0 + k_1 = k$.

Furthermore, substituting (14.2) into $\tilde{y}(t) = k_1(t)^{\beta_1} \ell_1(t)^{\beta_2}$ yields

$$(\alpha_2\beta_1 - \alpha_1\beta_2)^{\beta_2} k_1^{\beta_2-\beta_1} \tilde{y} = [(\alpha_2 - \alpha_1)\beta_1 k - (\beta_2 - \beta_1)\alpha_1 k_1]^{\beta_2} = [\Delta(k, \tilde{y})]^{\beta_2}$$
$$where \ \Delta(k, k_1) = (\alpha_2 - \alpha_1)\beta_1 k - (\beta_2 - \beta_1)\alpha_1 k_1.$$
(14.3)

Solving (14.3) with respect to k_1, one obtain

$$k_1 = e(k, \tilde{y}).$$
(14.4)

Similarly substituting (14.2) and (14.4) into $\tilde{c} = k_0^{\alpha_1} \ell_0^{\alpha_2} = (k - k_0)^{\alpha_1} \ell_0^{\alpha_2}$ and after some routine calculations, we have the following efficiency PPF:

$$\tilde{c} = \left(1 - \frac{\Delta(k,\tilde{y})}{(\alpha_2\beta_1 - \beta_2\alpha_1)k_1}\right)^{\alpha_2} (k - k_1)^{\alpha_1} = \left(\frac{\beta_1(\alpha_1 - \alpha_2)(k - k_1)}{(\alpha_2\beta_1 - \beta_2\alpha_1)k_1}\right)^{\alpha_2} (k - k_1)^{\alpha_1}$$
$$= \left(\frac{\beta_1(\alpha_1 - \alpha_2)}{(\alpha_2\beta_1 - \beta_2\alpha_1)k_1}\right)^{\alpha_2} (k - k_1) = \left(\frac{\beta_1(\alpha_1 - \alpha_2)}{(\alpha_2\beta_1 - \beta_2\alpha_1)e(k,\tilde{y})}\right)^{\alpha_2} (k - e(k, \tilde{y})). \quad \blacksquare$$

Remark 1 The derived efficiency-unit PPF is constructed such that the original-unit PPF at each period will be pulled back to the corresponding efficiency-unit PPF by discounting with each sector's rate of TFP growth ($A_i(t)$) as depicted in Fig. 14.3, where four PPF curves of t and (t + 1) periods are drawn. They respectively correspond to the original-unit PPF and its corresponding efficiency-unit PPF curves. Efficiency-unit PPF curves at t period will be constructed by discounting back the original-unit PPF at each sector's TFP growth rate along each axis. The efficiency-

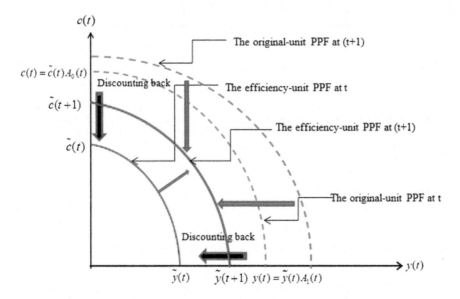

Fig. 14.3 xxx

unit PPF at $(t+1)$ period will be obtained by applying the same procedures to the original-unit PPF at $(t+1)$ period.

Next the following optimal growth model is constructed in terms of the per-capita efficiency-unit:

Per-capita Efficiency-unit Optimal Growth Model

$$Max \sum_{t=0}^{\infty} \rho^t u\left(\widetilde{c}(t)\right) = \sum_{t=0}^{\infty} \rho^t \frac{\widetilde{c}(t)^{1-\sigma}}{1-\sigma} \ where \ \rho = \frac{(1+a_0)^{1-\sigma}}{1+r}$$
$$s.t. \ k(0) = \overline{k},$$

$$\widetilde{c}(t) = T\left(\widetilde{y}(t), k(t)\right) \quad (t = 0, 1, \ldots)$$

$$\widetilde{y}(t) + (1-\delta)k(t) - (1+g)k(t+1) = 0 \quad (t = 0, 1, \ldots). \tag{14.5}$$

The objective function in terms of per-capita term is derived as follows: Let us transform the consumption into the efficiency-unit terms.

$$\widetilde{c}(t) = \frac{C(t)}{A_0(t)L(t)} = \frac{c(t)}{(1+a_0)^t A_0(0)} = \frac{c(t)}{(1+a_0)^t}.$$

Therein, it is assumed that $A_0(0) = 1$ for simplicity.

Then the infinite discounted sum of consumptions will be rewritten in terms of efficiency-unit as

$$\sum_{t=0}^{\infty} \left[\frac{(1+a_0)^{1-\sigma}}{(1+r)}\right]^t \frac{\widetilde{c}(t)^{1-\sigma}}{1-\sigma} = \sum_{t=0}^{\infty} \rho^t \frac{\widetilde{c}(t)^{1-\sigma}}{1-\sigma},$$

where

$$\rho = \frac{(1+a_0)^{1-\sigma}}{1+r}.$$

We will make the following additional assumption here for establishing $0 < \rho < 1$[5]:

Assumption 4

(1) $A_0(0) = 1$
(2) If $0 < \sigma \le 1$, $0 < a_0 < r < 1$ holds. If $1 < \sigma$, a_0 and r can be chosen arbitrarily so that $a_0 \in (0, 1)$ and $r \in (0, 1)$.

Note that (2) of Assumption 4 establishes that $0 < \rho < 1$ for $0 < \sigma$.

---

[5]Since Cobb-Douglas production functions satisfy the "Inada conditions", so all variables are bounded and Assumption 4 guarantees that the objective function of the problem is infinitely summable.

The accumulation equation (14.5) is constructed based on the efficiency-unit PPF as explained in Remark 2, where

$$\frac{K(t)}{L(t)} = k(t) \ and \ \frac{K(t+1)}{L(t)} = \frac{(1+g)K(t+1)}{(1+g)L(t)} = (1+g)k(t+1).$$

An important remark is in order now:

Remark 2 It is noteworthy that the accumulation equation (14.5) will be derived directly from rewriting the following efficiency-unit accumulation equation constructed based on the efficiency-unit PPF by dividing both sides with $L(t)$:

$$\widetilde{Y}(t) + (1-\delta)K(t) - K(t+1) = 0.$$

Observing Fig. 14.3, the TFP growth effect will be counteracted by discounting the original-unit variables with each sector's TFP growth rate, but the efficiency PPF will still expand outward because of capital deepening alone. Indeed, Eq. (14.5) shows this process. This property comes from the fact that each sector's output is independent each other and shows a sharp contrast to that of Acemoglu and Guerrieri (2008), where sector's outputs are used as inputs for producing the final good and are correlated.

If x and z respectively indicate initial and terminal capital stocks, then the reduced form utility function $V(x,z)$ and the feasible set D will be defined as[6]

$$V(x,z) = u(T[(1+g)z - (1-\delta)x, x])$$

and

$$D = \{(x,z) \in \mathbb{R}_+ \times \mathbb{R}_+ : T[(1+g)z - (1-\delta)x, x] \geq 0\},$$

where $x = k(t) \ and \ z = k(t+1)$. We eliminate the time index for simplicity.
Finally, the per-capita efficiency-unit model will be summarized as the following standard reduced form model, which has been studied intensively by Scheinkman (1976) and McKenzie (1983).

Reduced Form Model

$$(**) \ Max \ \sum_{t=0}^{\infty} \rho^t V(k(t), k(t+1))$$
$$s.t. \ (k(t), k(t+1)) \in D \ for \ t \geq 0 \ and \ k(0) = \overline{k}.$$

In addition, it is noteworthy that any interior optimal path must satisfy the following Euler equation, which exhibits an inter-temporal efficiency allocation condition:

---

[6]We will use the function symbol "u(.)" instead of the explicit form defined in Assumption 1 for a notational convenience.

$$V_z(k(t-1), k(t)) + \rho V_x(k(t), k(t+1)) = 0 \; for \; all \; t \geq 0, \quad (14.6)$$

where the partial derivatives mean that

$$V_x(k(t), k(t+1)) = \frac{\partial V(k(t), k(t+1))}{\partial k(t)} \; and \; V_z(k(t-1), k(t)) = \frac{\partial V(k(t-1), k(t))}{\partial k(t)}.$$

Under differentiability assumptions, because of the envelope theorem, all prices will be obtained as the following relations:

$$q = \frac{du(\tilde{c})}{d\tilde{c}} = 1, \; p = -q\frac{\partial T(\tilde{y}, k)}{\partial \tilde{y}}, \; w = q\frac{\partial T(\tilde{y}, k)}{\partial k}, \; and \; w_0 = q\tilde{c} + p\tilde{y} - wk,$$

where the price of the consumption good is normalized as 1.

We define the optimal steady state as explained below.

Definition An optimal steady state path (OSS) k^ρ is an optimal path which solves the reduced-form model (∗∗) and which satisfies $k^\rho = k(t) = k(t+1) \; for \; all \; t \geq 0$.

14.3 Nonbalanced Growth

In this section, based on Cobb-Douglas technologies, we will derive the optimal steady state in the efficiency unit and demonstrate nonbalanced growth in terms of the original unit.

14.3.1 Optimal Steady State

The optimal steady state k^ρ is a solution of the following equation:

$$V_z(k^\rho, k^\rho) + \rho V_x(k^\rho, k^\rho) = 0.$$

Using the accumulation equation, we define

$$V(x, z) = u(T[\tilde{y}, x]) = u(T[(1+g)z - (1-\delta)x, x])$$
$$where \; x = k(t), \; z = k(t+1).$$

Then, the Euler equation evaluated at the OSS can be derived as follows.

$$\frac{\partial T}{\partial \tilde{y}}\Big|_{(k^\rho, k^\rho)} + \left(\frac{\rho}{1+g}\right)\left[-(1-\delta)\frac{\partial T}{\partial \tilde{y}}\Big|_{(k^\rho, k^\rho)} + \frac{\partial T}{\partial k}\Big|_{(k^\rho, k^\rho)}\right] = 0. \quad (14.7)$$

Therein, the utility term $u'(\tilde{c}^\rho) = (\tilde{c}^\rho)^{-\sigma}$ has been eliminated from the Euler equation dividing both sides by $(1 + g)$. Consequently, the utility term evaluated at the OSS will never affect the calculation of the Euler equation.

Simplifying (14.7) yields the expression below.

$$\left(\frac{\rho}{1+g}\right)\left[p^\rho(1-\delta)+w^\rho\right]=p^\rho \tag{14.8}$$

Because $\beta_1 = \frac{w^\rho k_1^\rho}{p^\rho \tilde{y}^\rho}$ holds, it follows that

$$k_1^\rho = \beta_1 \tilde{y}^\rho \left(\frac{w^\rho}{p^\rho}\right). \tag{14.9}$$

From (14.8),

$$\left(\frac{p^\rho}{w^\rho}\right)=\frac{\rho}{(1+g)-\rho(1-\delta)}\equiv m. \tag{14.10}$$

Eliminating $\left(\frac{p^\rho}{w^\rho}\right)$ from (14.9) and (14.10) yields the equation below.

$$k_1^\rho = \beta_1 m \tilde{y}^\rho \tag{14.11}$$

Furthermore, substituting again (14.11) into (14.3) and using the fact that $\tilde{y}^\rho = (g + \delta)k^\rho$, one obtains the following.

$$\begin{aligned}(\alpha_2\beta_1 - \alpha_1\beta_2)^{\beta_2}&[(\beta_1 m)(g+\delta)k^\rho]^{\beta_2-\beta_1}(g+\delta)(k^\rho)\\&=[(\alpha_2-\alpha_1)\beta_1 k^\rho - (\beta_2-\beta_1)\alpha_1\beta_1 m(g+\delta)k^\rho]^{\beta_2}\end{aligned} \tag{14.12}$$

Solving (14.12) with respect to k^ρ, we will show Proposition 1 under the following extra assumption which guarantees viability of the OSS.

Assumption 5 $0 < (1 - \delta) < (g + \delta)$

To understand the assumption, multiplying k^ρ both sides yields

$$\tilde{y}^\rho = (g+\delta)k^\rho > (1-\delta)k^\rho.$$

Then Assumption 5 exhibits that the OSS per-capita output of the capital goods must be greater than the OSS net per-capita capital stock.

Proposition 1 *Under Assumptions 1 through 5, the OSS can be derived as follows, and it is positive:*

$$k^\rho = \left[\frac{(\alpha_2-\alpha_1)\beta_1 - (\beta_2-\beta_1)\alpha_1\beta_1 m(g+\delta)}{(\alpha_2\beta_1-\alpha_1\beta_2)(\beta_1 m)^{\frac{\beta_2-\beta_1}{\beta_2}}(g+\delta)^2}\right] > 0, \quad where\ m = \frac{\rho}{(1+g)-\rho(1-\delta)}.$$

Fig. 14.4 xxx

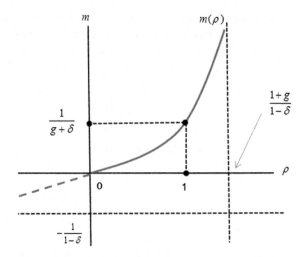

Proof Solving the Eq. (14.12) yields k^ρ. To show that k^ρ is positive *for* $\rho \in (0, 1)$, let us denote $m = m(\rho)$ *for* $\rho \in (0, 1)$. From Assumptions 1 through 5, $m(\rho)$can be depicted as Fig. 14.4. Since m is positive *for* $\rho \in (0, 1)$, the denominator and numerator of $k^\rho$are clearly positive from Assumptions 3 through 5. ∎

14.3.2   Nonbalanced Growth

Because of Proposition 1 and the accumulation equation (14.5), $\tilde{y}^\rho = (g + \delta)k^\rho$ holds. Therefore, the per-capita output of the capital good sector at the original-unit value: $y^\rho(t)$ will be expressed as

$$y^\rho(t) = (1 + a_1)^t A_1(0)\tilde{y}^\rho.$$

This expression will exhibit that the per-capita optimal output of the capital goods sector will grow at its TFP growth rate along the OSS path. Similarly, because $\tilde{c}^\rho = T(\tilde{y}^\rho, k^\rho)$ holds, it follows that $c^\rho(t) = (1 + a_0)^t A_0(0)\tilde{c}^\rho$, and that the per-capita output of consumption good sector will also grow at its TFP growth rate. These constant growing paths are designated as **optimal constant growth paths** (OCG for short).

The result presented above is summarized as the following corollary:

Corollary *The consumption good and the capital good sectors have their own optimal constant growth paths denoted by $c(t)^\rho$ and $y(t)^\rho$, which are growing respectively at their own TFP rate: a_0 and a_1.*

Instead of using those expressions to investigate the motions of prices, it is more convenient to rewrite the prices in terms of the conventional wage-rental rate used by Uzawa (1964). The following relations concerned with the wage-rental rate denoted by ω will be obtained:

$$\omega(t) = \frac{w_0(t)}{w(t)} = \frac{A_0(t)\left[f_0(k_0(t),1)-k_0(t)\frac{\partial f_0}{\partial k_0(t)}\right]}{A_0(t)\frac{\partial f_0}{\partial k_0(t)}} = \frac{A_1(t)\left[f_1(k_1(t),1)-k_1(t)\frac{\partial f_1}{\partial k_1(t)}\right]}{A_1(t)\frac{\partial f_1}{\partial k_1(t)}},$$

where $k_0(t) = \frac{K_0(t)}{L_0(t)}$ and $k_1(t) = \frac{K_1(t)}{L_1(t)}$.

Using these relations and due to the fact that k^ρ is constant along the OSS, ω^ρ is also constant. The relative price of capital good $p(t)$ is a function of $\omega(t)$. Therefore, it will also be constant along the OSS. To clarify the description, let us illustrate Fig. 14.5, where each sector's production function: $\tilde{c}(t) = f_0(k_0(t), \ell_0(t))$ and $\tilde{y}(t) = f_1(k_1(t), \ell_1(t))$ shifts at their own TFP growth rates. To avoid complexity, each axis represents a two-dimensional real space.

Because the wage-rental rate is constant along the optimal constant growth paths, two straight lines, respectively tangent to the sector's production function, will pivot simultaneously upward on the fixed point M, which indicates the constant wage-rental rate along OSS. An upward shift of each production function from period t to period (t + 1) attributable to the sector's Hicks neutral technical progress illustrated in Fig. 14.5 satisfies the following relation:

$$k_0^\rho = k_0^\rho(t) = k_0^\rho(t+1), \quad \ell_0^\rho = \ell_0^\rho(t) = \ell_0^\rho(t+1),$$

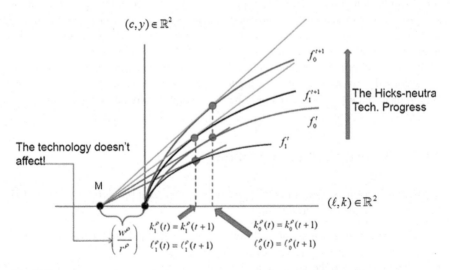

Fig. 14.5 xxx

and

$$k_1^\rho = k_1^\rho(t) = k_1^\rho(t+1), \ \ell_1^\rho = \ell_1^\rho(t) = \ell_1^\rho(t+1).$$

Based on this description, the per-capita gross domestic products along the optimal constant growth path (OCG-GDP) can be defined as shown below:

$$c^\rho(t) + p^\rho y^\rho(t) = A_0(t)\widetilde{c}^\rho + A_1(t)p^\rho \widetilde{y}^\rho.$$

Because the growth rate of the OCG-GDP will be defined as the weighted average of the growth rate of both sectors, we have the expression below:

$$The \ growth \ rate \ of \ the \ OCG\text{-}GDP = \left(\frac{A_0(t)\widetilde{c}^\rho}{A_0(t)\widetilde{c}^\rho + A_1(t)p^\rho \widetilde{y}^\rho}\right) a_0$$
$$+ \left(\frac{A_1(t)p^\rho \widetilde{y}^\rho}{A_0(t)\widetilde{c}^\rho + A_1(t)p^\rho \widetilde{y}^\rho}\right) a_1$$

Results show that the OCG-GDP grows at a certain constant rate calculated as the weighted sum of the growth rates of both sectors. Based on the relation above, we can easily demonstrate the following proposition:

Proposition 2 *If, $a_0 > a_1 (a_0 < a_1)$, then the per-capita OCG-GDP growth rate will converge to $a_0(a_1)$.*

If $a_0 > a_1 (a_0 < a_1)$, then the ***asymptotically dominant sector*** turns out to be the consumption goods sector (the capital goods sector). Note that this proposition presents a sharp contrast to the properties discussed in Section D of Acemoglu and Guerrieri (2008),[7] where the asymptotically dominant sector depends on the augmented rates of technological progress and the elasticity of substitution of the final good CES production function. Contrastingly, Proposition 2 is free from those conditions. In our model, giving a higher rate of technological progress to either sector, then that sector will turn out to be the asymptotically dominant sector.

14.4   Elasticity and Local Stability

This section presents proof of the following two important properties demonstrated by Acemoglu and Guerrieri (2008):

(i) The aggregated elasticity of substitution among labor and capital evaluated along the optimal constant growth paths is less than one.
(ii) Local stability in the sense of saddle-point stability.

---

[7]Acemoglu (2009) also discusses on the factor proportion differences in Sect. 20.2.

We will see later that factor proportion differences across and within sectors expressed by factor share parameters of sector's Cobb-Douglas production functions will take important roles to establish the both propositions.

14.4.1 Elasticity of Substitution

The elasticity of substitution between total labor (L) and total capital (K) is defined using our notation as shown below.

Definition $\sigma = -\dfrac{dk/k}{d(w_0/w)/(w_0/w)} = -\dfrac{(w_0/w)}{k} \times \dfrac{dk}{d(w_0/w)}$

Acemoglu and Guerrieri (2008) have demonstrated as an important result that the elasticity of substitution between L for K (denoted by σ) is less than one, which is also a stylized fact examined in many studies. For instance, Chirinko (2008) presents an excellent summary of the empirical literature, listing estimates from numerous and diverse sources and empirical methods. Chirinko concludes that "the weight of the evidence suggests that σ lies in the range between 0.40 and 0.60." We prove that this property is satisfied also in our model under the following assumption, which is obtained by restricting Assumption 3.

Assumption 3'. $\frac{\alpha_2}{\alpha_1} < \frac{4}{3} < \frac{\beta_2}{\beta_1}$.

Because of Assumption 3, Assumption 3' implies $\frac{\alpha_2}{\alpha_2} < 1 < \frac{4}{3} < \frac{\beta_2}{\beta_1}$. Namely, it makes the factor share proportion differences across sectors wider.

Proposition 3 *Under Assumptions 1, 2, 3', 4 and 5, the elasticity of substitution between aggregated labor and capital along OSS denoted by $\sigma^\rho = \sigma(\rho)$ for $\rho \in [0, 1]$ is less than one.*

Proof Equation (14.3) yields

$$\tilde{y} = \frac{\Delta^{\beta_2}}{(\alpha_2\beta_1 - \alpha_1\beta_2)^{\beta_2} k_1^{\beta_2-\beta_1}}.$$

Totally differentiating the above produces the following expression:

$$d\tilde{y} = \left\{ \frac{(\beta_2-\beta_1)(\alpha_2\beta_1-\alpha_1\beta_2)^{\beta_2}k_1^{\beta_2-\beta_1-1}\Delta^{\beta_2} - (\alpha_2\beta_1-\alpha_1\beta_2)^{\beta_2}k_1^{\beta_2-\beta_1}\cdot\frac{\partial\Delta}{\partial k_1}}{[(\alpha_2\beta_1-\alpha_1\beta_2)^{\beta_2}k_1^{\beta_2-\beta_1}]^2} \right\} dk_1$$

$$+ \left\{ \frac{\beta_2\Delta^{\beta_2}}{(\alpha_2\beta_1-\alpha_1\beta_2)^{\beta_2}k_1^{\beta_2-\beta_1}} \cdot \frac{\partial\Delta}{\partial k} \right\} dk,$$

where $\frac{\partial\Delta}{\partial k_1} = -(\beta_2-\beta_1)\alpha_1$ *and* $\frac{\partial\Delta}{\partial k} = -(\alpha_2-\alpha_1)\beta_1$.
Some additional calculation yields the following.

$$\frac{dk}{dk_1} = -\frac{(\beta_2 - \beta_1)(\Delta + \alpha_1\beta_1 k_1)}{\beta_1\beta_2(\alpha_2 - \alpha_1)k_1}.$$

From $\left(\frac{w_0}{w}\right) = \left(\frac{\beta_2-\beta_1}{\beta_1}\right)k_1, d\left(\frac{w_0}{w}\right) = \left(\frac{\beta_2-\beta_1}{\beta_1}\right)dk_1$ can be established. Combining these results and evaluating them along the OSS yields the following.

$$\sigma(\rho) = -\frac{\left(\frac{w_0^\rho}{w^\rho}\right)}{k^\rho} \cdot \frac{dk}{d\left(\frac{w_0}{w}\right)}\bigg|_\rho = \frac{(\beta_2-\beta_1)(\Delta^\rho+\alpha_1\beta_1 k_1)}{\beta_1\beta_2(\alpha_2-\alpha_1)k_1} = \frac{(\beta_2-\beta_1)[\Delta^\rho+\alpha_1\beta_1^2 m(\rho)(g+\delta)]}{\beta_1\beta_2(\alpha_2-\alpha_1)}$$

$$= \left(\frac{\beta_2-\beta_1}{\beta_1}\right)\left[1 + \frac{(2\beta_1-\beta_2)\alpha_1}{(\alpha_2-\alpha_1)}m(\rho)(g+\delta)\right].$$

Therein, $k_1^\rho = \beta_1 m(\rho)(g+\delta)k^\rho$ and $m(\rho) = \frac{\rho}{(1+g)-\rho(1-\delta)}$.
Note that $0 \le m(\rho) \le \frac{1}{g+\delta}$ for $\rho \in [0, 1]$.
   Using Assumption 3', we establish that

(i) $\sigma(0) = \frac{\beta_2-\beta_1}{\beta_1} = 1 - \frac{\beta_1}{\beta_2}$

and

(ii) $\sigma(1) = \left(1 - \frac{\beta_1}{\beta_2}\right)\left[1 + \frac{(2\beta_1-\beta_2)\alpha_1}{(\alpha_2-\alpha_1)}\right] = \left(1 - \frac{\beta_1}{\beta_2}\right)\left[1 - \frac{(2\beta_1-\beta_2)\alpha_1}{(\alpha_1-\alpha_2)}\right].$

It is straightforward to show that $0 < \sigma(0) < 1$ from Assumption 3'. In contrast, showing $0 < \sigma(\rho) < 1$ necessitates rewriting of the second term in the second square bracket of $\sigma(1)$ as

$$D = \frac{(2\beta_1 - \beta_2)\alpha_1}{(\alpha_1 - \alpha_2)} = \frac{(2\beta_1 - \beta_2)}{\left[1 - \left(\frac{\alpha_2}{\alpha_1}\right)\right]},$$

and it is necessary to show that $0 < D < 1$. The denominator of D is greater than $\frac{1}{2}$ and positive due to Assumption 3. The numerator of D will be rewritten as $\beta_2\left[2\left(\frac{\beta_1}{\beta_2}\right) - 1\right]$. Assumption 3' gives that $\frac{1}{2} > 2\left(\frac{\beta_1}{\beta_2}\right) - 1 > 0$. Using this result and the fact that $0 < \beta_2 < 1$ yield $\beta_2\left[2\left(\frac{\beta_1}{\beta_2}\right) - 1\right] < \frac{1}{2}$. Thus the numerator of D is less than that of the denominator. It follows that $0 < D < 1$. Because $\sigma(\rho)$ is a monotone-decreasing linear function of ρ for $0 \le \rho \le 1$. Consequently, $0 < \sigma(\rho) < 1$ holds for $\rho \in [0, 1]$.∎

We derive the similar result reported by Acemoglu and Guerrieri (2008). Contrastingly, we needed to restrict the factor share proportion difference within the capital goods sector.

14.4.2 Local Stability

Expanding the Euler equation (14.7) around k^ρ, and redefining

$$V_x(x, z) = T_{\tilde{y}} \text{ and } V_z(x, z) = -(1 - \delta)T_{\tilde{y}} + T_x,$$

where the utility terms are eliminated in advance.
Differentiating around OSS yields the following simplified characteristic equation:

$$\rho^* V^\rho_{xz}\lambda^2 + (V^\rho_{zz} + \rho^* V^\rho_{xx})\lambda + V^\rho_{zx} = 0 \quad where \; \rho^* \equiv \frac{\rho}{1+g},$$

If $V^\rho_{xz} \neq 0$, then

$$\rho^*\lambda^2 + \frac{(V^\rho_{zz} + \rho^* V^\rho_{xx})}{V^\rho_{xz}}\lambda + \frac{V^\rho_{zx}}{V^\rho_{xz}} = 0. \tag{14.13}$$

Local stability will be analyzed using this equation. Solving (14.13) yields Lemma 2.

Lemma 2 [8] *Characteristic equation (14.13) has the following two roots:*

$$\begin{cases} \lambda_1 = \frac{m(\alpha_2 - \beta_2)}{\alpha_2[1 - m(1-\delta)] + m(1-\delta)(\alpha_2 - \beta_2)} \\ and \\ \lambda_2 = \frac{1}{\rho^*}\left\{\frac{\alpha_2[1 - m(1-\delta)] + m(1-\delta)(\alpha_2 - \beta_2)}{m(\alpha_2 - \beta_2)}\right\} \end{cases},$$

In other words if λ is a root, then $\frac{1}{\rho^*\lambda}$ is also the root of the characteristic equation (14.13).

Proof Linearizing the Euler equation (14.6) around the OSS yields the well-known characteristic equation (14.13) above, making sure that the utility terms evaluated at the OSS disappear from the Euler equation. To simplify our calculations below, one uses the simplified terms V_x and V_z in advance.

Expanding this simplified terms around the OSS and by the facts that $V_x = -(1-\delta)T_{\tilde{y}} + T_x$ and $V_z = T_z$ where $\tilde{y} = (1+g)z - (1-\delta)x$, will eventually yield the following results:

$$\begin{aligned} V^\rho_{xx} &= (1-\delta)^2 T^\rho_{\tilde{y}} - (1-\delta)T^\rho_{\tilde{y}\tilde{y}} - (1-\delta)T^\rho_{x\tilde{y}} + T^\rho_{xx} \\ &= (1-\delta)\left[(1-\delta)T^\rho_{\tilde{y}\tilde{y}} - T^\rho_{\tilde{y}x}\right] + \left[\frac{\Delta^\rho_x}{\Delta^\rho_{\tilde{y}}} - (1-\delta)\right]T^\rho_{x\tilde{y}}, \end{aligned} \tag{14.14}$$

$$V^\rho_{xz} = V^\rho_{zx} = -T^\rho_{\tilde{y}\tilde{y}}(1-\delta) + T^\rho_{x\tilde{y}}, \tag{14.15}$$

$$V^\rho_{zz} = T^\rho_{\tilde{y}\tilde{y}}, \tag{14.16}$$

where $\Delta^\rho_x = \alpha_2\beta_1 + (\alpha_1\beta_2 - \alpha_2\beta_1)\dfrac{\partial k^\rho_1}{\partial x}$ *and* $\Delta^\rho_z = (\alpha_1\beta_2 - \alpha_2\beta_1)\dfrac{\partial k^\rho_1}{\partial \tilde{y}}$.

From (14.3),

$$\frac{\partial k^\rho_1}{\partial x} = \frac{\tilde{y}^\rho \beta_2(\Delta^\rho)^{\beta_2-1}\Delta^\rho_x}{(\alpha_1\beta_2)^{\beta_2}} \quad and \quad \frac{\partial k^\rho_1}{\partial \tilde{y}} = \frac{(\Delta^\rho)^{\beta_2} + \tilde{y}^\rho\beta_2(\Delta^\rho)^{\beta_2-1}\Delta^\rho_y}{(\alpha_1\beta_2)^{\beta_2}}.$$

---

[8]Levhari and Liviatan (1972) has proved this property in a general optimal growth model.

It is then possible to show that

$$T^\rho_{x\tilde{y}} = -\frac{\alpha_1}{\beta_1}\left(\frac{\Delta^\rho}{\alpha_2\beta_1}\right)^{-\alpha_2-1}\Delta^\rho_{\tilde{y}}, \quad T^\rho_{xx} = T^\rho_{x\tilde{y}}\frac{\Delta^\rho_x}{\Delta^\rho_{\tilde{y}}}, \quad T^\rho_{\tilde{y}\tilde{y}} = -T^\rho_{x\tilde{y}}\frac{(\alpha_2-\beta_2)m}{\alpha_2}.$$

(14.17)

Some manipulations yield the following expression.

$$\frac{\Delta^\rho_x}{\Delta^\rho_{\tilde{y}}} = \frac{\alpha_2}{(\alpha_1\beta_2-\alpha_2\beta_1)m} = \frac{\alpha_2}{-(\alpha_2-\beta_2)m}$$

(14.18)

Substituting (14.17) and (14.18) into Eqs. (14.14) through (14.16) produces

$$\frac{\rho^* V^\rho_{xx}}{V^\rho_{xz}} = -\rho^*\left[\frac{(1-\delta)m(\alpha_2-\beta_2)+\alpha_2[1-m(1-\delta)]}{m(\alpha_2-\beta_2)}\right],$$

and

$$\frac{V^\rho_{zz}}{V^\rho_{xz}} = -\frac{m(\alpha_2-\beta_2)}{(1-\delta)m(\alpha_2-\beta_2)+\alpha_2[1-m(1-\delta)]}.$$

The Euler equation will eventually be rewritten as

$$\rho^*\lambda^2 - \left\{\left[\frac{(1-\delta)m(\alpha_2-\beta_2)+\alpha_2[1-m(1-\delta)]}{m(\alpha_2-\beta_2)}\right] + \rho^*\left[\frac{m(\alpha_2-\beta_2)}{(1-\delta)m(\alpha_2-\beta_2)+\alpha_2[1-m(1-\delta)]}\right]\right\}\lambda + 1 = 0.$$

Thereby, one obtains the two roots of Eq. (14.13). ∎

We strengthen Assumption 5, called Assumption 5′, in order to demonstrate local stability, where the right hand side inequality is added to the viability assumption, Assumption 5.

Assumption 5′. $(1-\delta) < (g+\delta) < \left(\frac{\beta_2}{\alpha_2}\right)(1-\delta)$

Multiplying k^ρ to the RHS inequality yields $y^\rho < \left(\frac{\beta_2}{\alpha_2}\right)(1-\delta)k^\rho$, which implies that not only the OSS per capita output must satisfy viability, but also it must have the upper bound related closely to the factor share proportion difference across sectors indicated by $\left(\frac{\beta_2}{\alpha_2}\right) > 1$.

Based on Lemma 2, we can present the following proposition under Assumption 5′:

Proposition 4 *Under Assumptions 1, 2, 3, 4, and 5′, there exists a $\rho\prime > 0$ such that $\rho \in [\rho\prime, 1)$, which implies that $1 < \lambda$ and $1 < \rho\lambda$ hold simultaneously.*

Proof Remember that $m(\rho)$ is depicted as Fig. 14.4 and $m(0) = 0$ and $m(1) = \frac{1}{g+\delta}$. Let us define one of two roots as

$$\lambda(\rho) = \frac{m(\rho)(\alpha_2-\beta_2)}{(1-\delta)m(\rho)(\alpha_2-\beta_2)+\alpha_2[1-m(\rho)(1-\delta)]}.$$

First, it can be shown that $\lambda(1)$ is greater than one. Using $m(1) = \frac{1}{g+\delta}$, one obtains the expression shown below.

$$\lambda(1) = \frac{(\alpha_2 - \beta_2)}{(1-\delta)(\alpha_2 - \beta_2) + \alpha_2[(g+\delta) - (1-\delta)]}.$$

Subtracting one from both sides of the above equation will yield $\lambda(1) - 1 = \frac{A}{B}$, where

$$A = \delta(\alpha_2 - \beta_2) - \alpha_2[(g+\delta) - (1-\delta)] < 0$$

and

$$B = \alpha_2[(g+\delta) - (1-\delta)] + (\alpha_2 - \beta_2)(1-\delta) < 0.$$

From Assumption 4, the signs of terms A and B are both negative. Consequently, $\lambda(1) > 1$. When $\rho = 1$, $\rho^* = \frac{1}{1+g}$ holds. Therefore, it follows that

$$\frac{1}{1+g}\lambda(1) - 1 = \frac{C}{B} \; where \; C = \alpha_2 \left\{ [1 - (1+g)(g+\delta)] - \left(\frac{\beta_2}{\alpha_2}\right)[1 - (1+g)(1-\delta)] \right\}.$$

The term B is negative. To show that the term C is also negative, we need to rewrite Assumption 5′ as shown below.

$$(g+\delta) > (1-\delta) \Rightarrow (1+g)(g+\delta) > (1+g)(1-\delta) \Rightarrow 1 - (1+g)(1-\delta) > 1 - (1+g)(g+\delta)$$
$$\Rightarrow \left(\frac{\beta_2}{\alpha_2}\right)[1 - (1+g)(1-\delta)] > [1 - (1+g)(g+\delta)]$$

Consequently, the term C is negative. Figure 14.6 plots the graph of $\lambda(\rho)$. Choosing $\rho\prime > 0$ close enough to 1, it follows that $1 < \lambda(\rho)$ and $1 < \rho^*\lambda(\rho)$ for $[\rho\prime, 1).$[9]■

From Proposition 4, it is straightforward to show that the characteristic equation (14.13) has $\lambda(\rho) > 1$ and $0 < \frac{1}{\rho^*\lambda(\rho)} < 1$ as its roots. Therefore, we have proved the following.

Proposition 5 *Under Assumptions 1, 2, 3, 4, and 5′, the optimal steady state k^ρ for $\rho \in [\rho\prime, 1)$ is locally stable in terms of the saddle-point.*

Thereby, one finally obtains the desired result.

Corollary *Under Assumptions 1, 2, 3, 4, and 5′, each sector's optimal path in the neighborhood of the optimal steady state k^ρ for $\rho \in [\rho\prime, 1)$ will converge to its own optimal constant growth path:*

$$(\widetilde{c}^\rho(t), \widetilde{y}^\rho(t)) \to (\widetilde{c}^\rho, \widetilde{y}^\rho) \; as \; t \to \infty.$$

---

[9] Note that the case of $\rho = 1$ is excluded because of Assumption 3.

Fig. 14.6 xxx

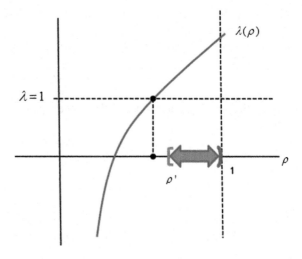

It follows that, in terms of the original-unit variables, a sector's per-capita capital stock and output eventually converge to the optimal constant growth paths, which grow at the rate of a sector's TFP growth:

$$c^\rho(t) = (1+a_0)^t \tilde{c}^\rho \ and \ y^\rho(t) = (1+a_1)^t \tilde{y}^\rho.$$

14.5 Conclution

We have demonstrated the existence of nonbalanced optimal constant growth paths and, by demonstrating saddle-point stability, the optimal path of each sector will converge to its own optimal constant growth path with its TFP growth rate, where the factor share proportion differences within and across sectors took important roles. Therefore, in the end, the sector with the higher TFP growth rate will dominate asymptotically at the aggregated GDP growth rate. This result presents a sharp contrast to that reported by Baumol (1967). As discussed in Acemoglu and Guerrieri (2008), it is noteworthy that the local stability asymptotically features both nonbalanced growth at the sectoral level and aggregate growth consistent with the Kaldor facts.

The existence and the global stability of the efficiency-unit OSS under the assumptions of general production functions among sectors will be left as subject for further research paper. It will be done by applying the turnpike theory developed by McKenzie (1983) and Scheinkman (1976).

Acknowledgments The paper was presented at the following meetings; The IV CICSE Conference on Structural Change, Dynamics, and Economic Growth, September, 2013 in Livorno, Italy, The International Conference on Macrodynamics—Financial and Real Interdependencies, 2015 in Lisbon and Society for the Advancement of Economic Theory (SAET) Meeting, 2015 in Cambridge. I thank Alain Venditti, Carine Nourry, Arrigo Opocher and Rachel Ngai for their useful comments to the earlier version of the paper. The research was supported by Grant-in-Aid for Scientific Research #22530187 and #25380238. I would like to dedicate the paper to Jean-Michel Grandmont. At the end, the author would like to express my sincere gratitude to an anonymous referee.

References

Acemoglu, D., & Guerrieri, V. (2008). Capital deepening and nonbalanced economic growth. *Journal of Political Economy, 116*, 467–498.

Acemoglu, D. (2009). Introduction to Modern Economic Growth. Princeton UP.

Alonso-Carrera, J., & Raurich, X. (2015). Demand-based structural change and balanced economic growth. *Journal of Macroeconomics, 46*, 356–374.

Alvarez-Cuadrado, F., & Long, N. (2011). Capital-labor substitution, structural change and growth. *CIRANO Scientific Series* 2011s-68.

Baierl, G., Nishimura, K., & Yano, M. (1998). The role of capital depreciation in multi-sectoral models. *Journal of Economic Behavior and Organization, 33*, 467–479.

Baumol, W. J. (1967). Macroeconomics of nonbalanced growth: The anatomy of urban crisis. *American Economic Review, 57*(3), 415–426.

Boppart, T. (2013). Structural change and the Kaldor facts in a growth model with relative price effects and non-Gorman preferences. *Working Paper.* http://hdl.handle.net/10419/79777.

Chirinko, R. S. (2008). σ: The long and short of it. *Journal of Macroeconomics, 30*(2), 671–686.

Foellmi, R., & Zweimuller, J. (2008). Structural change Engel's consumption cycles and Kaldor's facts of economic growth. *Journal of Monetary Economics, 55*, 1317–1328.

Echevarria, C. (1997). Changes in sectoral composition associated with economic growth. *International Economic Review, 38*, 431–452.

Kongsamut, P., Rebelo, S., & Xie, D. (2001). Beyond balanced growth. *Review of Economic Studies, 68*, 869–882.

Laitner, J. (2000). Structural change and economic growth. *Review of Economic Studies, 67*, 545–561.

Levhari, D., & Liviatan, N. (1972). On stability in the saddle-point sense. *Journal of Economic Theory, 4*, 88–93.

McKenzie, L. (1983). Turnpike theory, discounted utility, and the von Neumann facet. *Journal of Economic Theory, 30*, 330–352.

Ngai, R., & Pissarides, C. (2007). Structural change in a multisector model of growth. *American Economic Review, 97*(1), 429–443.

Scheinkman, J. (1976). An optimal steady state of n-sector growth model when utility is discounted. *Journal of Economic Theory, 12*, 11–20.

Syverson, C. (2011). What determines productivity? *Journal of Economic Literature, 49*(2), 326–365.

Takahashi, H., Mashiyama, K., & Sakagami, T. (2012). Does the capital intensity matter?: Evidence from the postwar Japanese economy and other OECD countries. *Macroeconomic Dynamics, 16*(Supplement 1), 103–116.

Uzawa, H. (1964). Optimal Growth in a Two-Sector Model of Capital Accumulation. *Review of Economic Studies, 31*, 1–24.

Part IV
General Equilibrium

Chapter 15
An Argument for Positive Nominal Interest

Gaetano Bloise and Herakles Polemarchakis

Abstract In a dynamic economy, money provides liquidity as a medium of exchange. A central bank that sets the nominal rate of interest and distributes its profit to shareholders as dividends is traded in the asset market. A nominal rates of interest that tend to zero, but do not vanish, eliminate equilibrium allocations that do not converge to a Pareto optimal allocation.

Keywords Nominal rate of interest · Dynamic efficiency

JEL Classification D-60 · E-10

15.1 Introduction

The Pareto optimality of competitive equilibrium allocations is a major tenet of classical welfare economics and the main argument in favor of competitive markets for the allocation of resources. Deviations from the classical paradigm sever the link between Pareto optimal and competitive equilibrium allocations, with repercussions both for the theory and practice of economic policy.

Competitive equilibrium allocations may fall short of Pareto optimality in two distinct, if related, situations: (i) in economies that extend over an infinite horizon with a demographic structure of overlapping generations (Gale 1973; Samuelson

---
We want to thank Jean-Jacques Herings, Felix Kubler and Yannis Vailakis and the Hotel of Gianicolo for hospitality.

G. Bloise (✉)
Department of Economics, Yeshiva University, New York, USA
e-mail: gaetano.bloise@yu.edu

G. Bloise
Department of Economics, University of Rome III, Rome, Italy

H. Polemarchakis
Department of Economics, University of Warwick, Coventry, England
e-mail: h.polemarchakis@warwick.ac.uk

K. Nishimura et al. (eds.), *Sunspots and Non-Linear Dynamics*,
Studies in Economic Theory 31, DOI 10.1007/978-3-319-44076-7_15

1958) and (ii) in economies with an operative transactions technology with money that provides liquidity services as a medium of exchange (Clower 1967; Lucas and Stokey 1987).[1]

The dynamic failure of optimality in economies of overlapping generations is well understood: competitive prices that attain market clearing may fail to provide consistent accounting over infinite streams of output. Long-lived productive assets, with streams of output that extend to the infinite future, when traded in asset markets, guarantee that equilibrium prices provide consistent intertemporal valuation and restore the optimality of competitive allocations (Wilson 1981; Santos and Woodford 1997).

When money serves as a medium of exchange, the nominal rate of interest does not allow competitive prices to exhaust the static gains from trade. Vanishing nominal rates of interest or, equivalently, the payment of interest on money balances on par with the rate of return on alternative stores of value eliminates the suboptimality of monetary equilibria (Friedman 1969).

The argument here is that low, but not vanishing, nominal rates of interest shield the economy from intertemporal suboptimality at the cost of some static inefficiency. Differently from other arguments for a positive nominal interest, the argument does not appeal to nominal rigidities, imperfect competition or any other imperfection or incompleteness of financial markets.

In an economy of overlapping generations with cash-in-advance constraints, a central bank issues balances in exchange for bonds and distributes its profits, seignorage, as dividends to shareholders Bloise et al. (2005), Nakajima and Polemarchakis (2005). Importantly, shares to the bank are traded in the asset market and the bank is, initially, owned by a finite number of individuals, most simply among those active at the starting date of economic activity. Even if not common practice, trades in shares of the centrak bank is not without precedent: shares of the Bank of England were traded until 1946. And it is correct and explicit accounting for the market in shares (Drèze and Polemarchakis 2000) that has allowed recent formulations to resolve a conundrum (Hahn 1965) and establish the existence of monetary equilibria even over a finite horizon.

At equilibrium, the market value of the bank is at least equal to the present value of seignorage. Seignorage corresponds to the intertemporal value of net transactions, which is, thus, finite. A condition of intra-generational heterogeneity ensures gains to trade even at intergenerational autarky, which guarantees that, provided that the nominal interest is small enough, some commodities are non-negligibly traded over the entire infinite horizon. As net transactions are finitely valued, so is the aggregate endowment of non-negligibly traded commodities. And, as a consequence, the aggregate endowment is finitely valued at equilibrium, for, otherwise, the relative prices

---

[1] Search theoretic models of monetary economies (Diamond 1984) or (Kiyotaki and Wright 1969) are, evidently, more satisfactory, but the simple cash-in-advance formulation here, as in much of the literature, offers analytical tractability and does not play an otherwise important role in the argument argument.

of negligibly traded to non-negligibly traded commodities would explode across periods of trade.

As long as the nominal rate of interest is arbitrarily low, but bounded away from zero, the static inefficiency associated with non-vanishing nominal rates remains but is essentially negligible; more importantly, with the stream of seignorage bounded away from zero, the bank substitutes for the long-lived productive assets that guarantee intertemporal optimality.

In Bloise and Polemarchakis (2006) we gave an argument for the very special case of a simple economy of overlapping generations.

The connection between costly transactions and intertemporal efficiency was recognized in Weiss (1980); the argument there, however, was restricted to steady-state allocations and relied on real balances entering directly the utility functions of individuals with a positive marginal utility everywhere. The argument identified debt with money balances and, more importantly, it did not ensure dynamic efficiency of non-stationary equilibrium allocation.

We organize the development of the argument as follows: In Sect. 15.2, we give simple examples that illustrate the argument. In Sect. 15.3, we present the argument in abstract terms, at a level of generality that is comparable to that of Wilson (1981). This only requires the modification of budget constraints of individuals that is inherited from a primitive description of sequential trades through cash-in-advance constraints. We prove the result under a hypothesis of gains to trade that we show (in Sect. 15.4) to be generically satisfied in standard stationary economies of overlapping generations with intra-generational heterogeneity. In Sect. 15.4, we describe a monetary economy of overlapping generations with cash-in-advance constraints where a central bank, whose ownership is sequentially traded in the stock market, pegs the nominal rate of interest, accommodates the demand for balances and distributes the seignorage to shareholders as dividends. Not surprisingly, a canonical intertemporal consolidation of sequential budget constraints reveals that relevant equilibrium restrictions of this sequential economy are exactly those in the abstract analysis. We conclude with some remarks.[2]

15.2 Examples

Simple, stationary economies of overlapping generations illustrate the argument.[3] Dates or periods of trade are $\mathcal{T} = \{0, 1, 2, \ldots, t, \ldots\}$. Each non-initial generation has a life span of two periods and consists of two individuals, $i \in \mathcal{J} = \{a, b\}$. An initially old generation is active at $t = 0$.

---

[2]A reader might prefer to reverse the order of presentation we chose by reading Sect. 15.4 before Sect. 15.3. This creates no difficulty, after a preliminary reading of the beginning of Sect. 15.3 for the notation we use.

[3]Minor changes of notation from the abstract argument that follows facilitate the exposition.

2.1

One commodity is exchanged and consumed at each date; the commodity is perishable.

The intertemporal utility function of an individual is

$$u^i(x^i, z^i) = x^i + \ln z^i,$$

where x^i is the excess consumptions of the individual when young, while z^i is the consumption when old.

The endowment of an individual when old is $e^i > 0$—with quasi linear preferences, it is not necessary to specify the endowment when the individual is young, when a sufficiently large endowment guarantees positive consumption.

The spot price of the commodity is p_t.

Nominal bonds, b_t, of one period maturity, serve to transfer revenue across dates. The nominal rate of interest is $r_t \geq 0$.

Balances, m_t, provide liquidity services; they also serve as a store of value, but they are dominated as such by bonds.

At each date, a central bank or monetary authority issues bonds in exchange of balances, with

$$\frac{1}{1+r_t}b_t + m_t = 0,$$

that it redeems at the following date, earning seignorage

$$b_t + m_t = r_t m_t$$

at date $t + 1$, that it distributes as dividend to shareholders.

Shares to the bank, their number normalised to one, are traded at each date and serve as a store of value. In the absence of uncertainty, no-arbitrage requires that the returns to bonds and shares coincide, and as a consequence, the *cum dividend* price of shares, v_t, satisfies

$$v_t = \frac{r_t}{1+r_t}m_t + \frac{1}{1+r_t}v_{t+1};$$

if it is finite,

$$v_0 = \frac{r_0}{1+r_0}m_0 + \sum_{t=1}^{\infty}\frac{1}{1+r^{t-1}}(\frac{r_t}{1+r_t}m_t),$$

where

$$(1+r^t) = (1+r_0) \times \ldots \times (1+r_t), \quad t = 1, \ldots.$$

The rate of inflation is $\pi_{t+1} = (p_{t+1}/p_t) - 1$, and the real rate of interest is $\rho_{t+1} = [(1+r_{t+1})/(1+\pi_{t+1})] - 1$; real balances are $\mu_t = m_t/p_t$.

An individual, young at t, faces the budget constraints

$$p_t x_t^i + \frac{1}{1+r_t} b_t^i + m_t^i \leq 0,$$

$$p_{t+1} z_{t+1}^i \leq b_t^i + m_t^i + p_{t+1} e^i$$

and the cash in advance constraint[4]

$$m_t^i \geq p_t x_t^{i-}, \quad m_t^i \geq 0.$$

Equivalently, an individual faces the intertemporal budget constraint

$$x_t^i + \frac{r_t}{1+r_t} x_t^{i-} + \frac{1}{1+\rho_t}(z_{t+1}^i - e^i) \leq 0,$$

with

$$\mu_t^i \geq x_t^{i-}, \quad \mu_t^i \geq 0$$

the associated holdings of real balances.

Similarly, the cum dividend price of shares in real terms φ_t, satisfies

$$\varphi_t = \frac{r_t}{1+r_t} \mu_t + \frac{1}{1+\rho_t} \varphi_{t+1};$$

if it is finite,

$$\varphi_0 = \frac{r_0}{1+r_0} \mu_0 + \sum_{t=1}^{\infty} \frac{1}{1+\rho^{t-1}} (\frac{r_t}{1+r_t} \mu_t),$$

where

$$(1+\rho^t) = (1+\rho_0) \times \ldots \times (1+\rho_t), \quad t = 1, \ldots.$$

Since shares and bonds are perfect substitutes, it is not necessary either to introduce shares explicitly in the intertemporal optimization of individuals or to distinguish between the initial value of the bank, v_0, and debt held by the initially old.

With $e^a \ll e^b$, along any equilibrium path, $x_t^a < 0$, while $x_t^b > 0$.

The solutions to the optimization problems of individuals are

$$x_t^a(\rho_t) = \frac{1}{(1+\rho_t)(1-\theta_t)} e^a - 1 \leq 0, \quad z_{t+1}^a(\rho_t) = (1+\rho_t)(1-\theta_t),$$
$$\mu_t^a(\rho_t) = -x_t^a,$$

---

[4] $x^{-}i$ is the negative part of x.

and

$$x_t^b(\rho_t) = \tfrac{1}{(1+\rho_t)}e^b - 1 \geq 0, \quad z_{t+1}^b(\rho_t) = (1+\rho_t),$$
$$\mu_t^b(\rho_t) = 0,$$

where, $\theta_t = [r_t/(1+r_t)] < 1$.

Along an equilibrium path,

$$x_t^a + x_t^b + z_t^a + z_t^b = e,$$

where $e = e^a + e^b$ is the aggregate endowment of individuals when old.

With $r_t = r \geq 0$, and, as a consequence, $\theta_t = \theta$, an equilibrium path of real rates of interest satisfies

$$\rho_{t+1} = \frac{e(1-\theta) + \theta e^a}{(1-\theta)(e - (2-\theta)\rho_t + \theta)} - 1,$$

where

$$e = e^a + e^b < 2$$

is the aggregate endowment of individuals at the second date in their life spans.

If $r = 0$, there exist two steady-states, one with $\rho^* = 0$ and another with $\bar{\rho} = (e/2) - 1 < 0$; in addition, there is a continuum of non-stationary paths indexed by the initial real rate of interest, $\rho_0 \in (\bar{\rho}, \rho^*)$. The steady-state path with $\rho^* = 0$, the golden rule, supports a Pareto optimal allocation, while all other equilibrium paths support suboptimal and Pareto ranked allocations; $\bar{\rho} = (e/2) - 1 < 0$ support intergenerational autarky. Note that, at the autarkic equilibrium, $\bar{x}^a = (4e^a - e^2)/(2e) < 0$, and, as a consequence, the associated real balances that support the equilibrium are $\bar{\mu} = -\bar{x}^a > 0$.

If $r > 0$, but sufficiently small, there is a steady-state equilibrium path with

$$\rho^*(r) = \frac{(2+e) + \sqrt{(2+e)^2 - 4(2-\theta)(e + \frac{\theta}{1-\theta}e^a)}}{2(2-\theta)} - 1 > 0.$$

By a standard argument, there is no equilibrium path with $\rho_0 \notin [\bar{\rho}(r), \rho^*(r)]$, where $\bar{\rho}(r) = [(2+e) - \sqrt{(2+e)^2 - 4(2-\theta)(e + \theta/(1-\theta)e^a)}]/[2(2-\theta)] - 1 < 0$.

For $\rho(t) \in [\bar{\rho}(r), \rho^*(r)]$, real balances are bounded below by $\mu^a(\bar{\rho}(r)) > 0$ and, as a consequence, the value of the bank is well defined and, in particular finite, only if $\rho(t) > 0$.

Since $\rho(t) \to \bar{\rho}(r) < 0$ if $\rho(t) \in [\bar{\rho}(r), \rho^*(r))$, the steady-state at $\rho^*(r)$ is the unique equilibrium path.

Importantly,

$$\lim_{r \to 0} \rho^*(r) = \rho^*;$$

as the nominal rate of interest tends to 0, the unique, steady-state real rate of interest tends to the golden rule and the associated allocation to a Pareto optimum.

The argument fails in the absence of intragenerational heterogeneity, when real balances need not be bounded away from zero as the economy tends to autarky.

Alternatively, Weiss (1980) allows real balances to enter directly the intertemporal utility function of a representative individual, $u(x, z, \mu)$, and he writes the intertemporal budget constraint as

$$x_t + \frac{1}{1+\rho}(z_{t+1} - e) + \frac{r_t - \pi_t}{1+r_t}\mu_t \leq 0,$$

which follows from the hypothesis that changes in the supply of balances are distributed as lump-sum transfers to individuals when old.

At a steady-state, optimization requires that

$$\frac{u_\mu}{u_x} = \frac{r}{1+r},$$

while market clearing requires that

$$\frac{r - \pi}{1+r}\mu^i = \frac{r - \pi}{1+r}(z - e);$$

the outstanding debt is

$$b = (z - e) + \mu.$$

At equilibria with debt, $r = \pi$ and $\rho = 0$. As a consequence, the liquidity services that balances, distinct from debt, provide, do not shield the economy from intertemporal inefficiency.

Alternatively, without debt (or, equivalently, if debt provides liquidity services and is not distinguishable from money), $\mu = z - e$ and, with $\pi = 0$ the real rate of interest is necessarily positive, $\rho = r > 0$, which, indeed, guarantees intertemporal efficiency. The hypothesis of non-vanishing marginal utility for money balances plays a role similar to that of intragenerational heterogeneity in our construction, but the logic of the arguments is different.

2.2

Two perishable commodities, $l \in \mathcal{N} = \{a, b\}$, are exchanged and consumed at each date. Individual i only consumes commodity i, but is endowed with one unit of the other commodity, $-i$, when young and nothing when old. The intertemporal utility function of an individual i is

$$u^i\left(x^i, z^i\right) = x^i + 2z^i,$$

where x^i and z^i are the consumptions of the individual in commodity i, respectively, when young and when old.

The price of commodity i at date t in present value terms is p_t^i. The constant nominal rate of interest is $r \geq 0$. An individual faces the single budget constraint

$$p_t^i x_t^i + p_{t+1}^i z_{t+1}^i \leq \left(\frac{1}{1+r}\right) p_t^{-i},$$

which reflects an underlying cash-in-advance constraint. In addition, the budget constraint of an initially old individual is

$$p_0^i z_0^i \leq \mu^i \left(\frac{r}{1+r}\right) \sum_t \left(p_t^a + p_t^b\right),$$

which reflects the hypothesis that the individual is entitled to a share $\mu^i \geq 0$ in intertemporal seignorage, so that $\mu^a + \mu^b = 1$.

Market clearing simply requires that

$$x_t^i + z_t^i = 1.$$

At equilibrium, sequential Walras Law implies

$$\left(\frac{r}{1+r}\right) \sum_i p_t^i + \sum_i p_{t+1}^i z_{t+1}^i = \sum_i p_t^i z_t^i.$$

This completes the description of the economy.

Let $\theta = [r/(1+r)] \leq 1$, for $r \geq 0$. Reinterpreting terms, one might suppose that every individual i with only $1 - \theta$ units of commodity $-i$ when young and nothing when old. In addition, a real productive asset i, initially owned by old individuals, deliver ϵ units of commodity i at every date.

We consider equilibria in two distinct cases.

First, $r = 0$. From the budget constraints of initially old individuals, it follows that $z_0^i = 0$ and, so, exploiting sequential Walras Law, that $x_t^i = 1$ and $z_t^i = 0$ for every t. This requires $p_{t+1}^i \geq 2 p_t^i$ for every t. The equilibrium allocation clearly fails Pareto optimality.

Alternatively, $r > 0$. From the budget constraints of initially old individuals, it follows that

$$p_0^i z_0^i = \mu^i \left(\frac{r}{1+r}\right) \sum_t \left(p_t^a + p_t^b\right),$$

and, as a consequence, that $\sum_t \left(p_t^a + p_t^b\right)$ is finite. By a canonical argument, the equilibrium allocation achieves Pareto optimality. We show that a steady state equilibrium exists under an equal distribution of seignorage, $\mu^a = \mu^b$.

Assume that $x_t^i = 0$ and $z_t^i = 1$ for every t. To obtain equilibrium prices, observe that, from the budget constraints of young individuals,

$$p_{t+1}^i = \left(\frac{1}{1+r}\right) p_t^{-i},$$

while, from the budget constraints of initially old individuals, $p_0 = p_0^a = p_0^b$; it follows that

$$p_t = p_t^a = p_t^b = \left(\frac{1}{1+r}\right)^t p_0,$$

at every date t.

For an arbitrary distribution of seignorage, a steady state equilibrium might not exist. To verify this, observe that, at a stationary equilibrium, $x_t^i = x^i$ and $z_t^i = z^i$ for every t, with $x^i + z^i = 1$. If $x^i > 0$, by utility maximization, $p_{t+1}^i \geq 2p_t^i$, which would violate the fact that $\sum_t \left(p_t^a + p_t^b\right)$ is finite. Hence, $x^i = 0$, which implies, by the budget constraint of a young individual and utility maximization,

$$2p_t^i \geq p_{t+1}^i = \left(\frac{1}{1+r}\right) p_t^{-i}.$$

In addition, an initial condition requires

$$p_0^i = \mu^i \left(\frac{r}{1+r}\right) \sum_t \left(p_t^a + p_t^b\right).$$

From both conditions, it follows that

$$2\mu^b \geq \left(\frac{1}{1+r}\right) \mu^a$$

and

$$2\mu^a \geq \left(\frac{1}{1+r}\right) \mu^b.$$

Hence, a stationary equilibrium might not exist for an arbitrary distribution of shares—well known for (stationary) economies of overlapping generations with multiple individuals in each generation and multiple commodities.

This example is designed to deliver an extremely clear conclusion about efficiency at equilibrium. In particular, a simplifying assumption, that each individual is endowed only with the commodity that he does not consume, eliminates price distortions due to cash-in-advance constraints, which only operates through pure wealth effects. Within each generation, there are actually infinite gains to trade, as young individuals are clearly better off by exchanging their endowments. Intergenerational trade allows for a further increase in welfare.

15.3   The Abstract Argument

There is a countable set of individuals, $\mathcal{I} = \{\ldots, i, \ldots\}$, a countable set of periods of trade, $\mathcal{T} = \{0, 1, 2, \ldots, t, \ldots\}$, and a finite set of physical commodities in every period of trade, $\mathcal{N}$. The commodity space is $\boldsymbol{L} = \mathbb{R}^{\mathcal{L}}$, where $\mathcal{L} = \mathcal{T} \times \mathcal{N}$.[5]

The consumption space of an individual is $\boldsymbol{L}^+$, the positive cone of the commodity space, and an element, $\boldsymbol{x}^i$, of the consumption space is a consumption plan. An individual is characterized by a preference relation, $\succeq^i$, on the consumption space and an endowment, $\boldsymbol{e}^i$, of commodities, an element of the consumption space itself. He is also entitled to a share $\mu^i \geq 0$ of aggregate revenue, so that, across individuals, $\sum_i \mu^i = 1$.

Fundamentals, $(\ldots, (\succeq^i, \boldsymbol{e}^i, \mu^i), \ldots)$, are restricted by canonical assumptions, so that every single individual is negligible. The aggregate endowment is $\sum_i \boldsymbol{e}^i$, which is understood to be a limit in the product topology.

Assumption 1 (*Preferences*) The preference relations of individuals are convex, continuous, weakly monotone and locally non-satiated.

Assumption 2 (*Endowments*) The endowments of individuals are positive, negligible elements of the commodity space.

Assumption 3 (*Aggregate Endowment*) The aggregate endowment is a positive element of the commodity space.

An allocation, $\boldsymbol{x} = (\ldots, \boldsymbol{x}^i, \ldots)$, is a collection of consumption plans. It is balanced whenever $\sum_i \boldsymbol{x}^i = \sum_i \boldsymbol{e}^i$. It is feasible whenever $\sum_i \boldsymbol{x}^i \leq \sum_i \boldsymbol{e}^i$. It is individually rational whenever, for every individual i, $\boldsymbol{x}^i \succeq^i \boldsymbol{e}^i$. For a feasible allocation $\boldsymbol{x}$, aggregate consumption, $\sum_i \boldsymbol{x}^i$, is an element of $\boldsymbol{L}(\boldsymbol{e})$, where $\boldsymbol{e} = \sum_i \boldsymbol{e}^i$ is the aggregate endowment.

Trade occurs intertemporally subject to transaction costs. In an abstract formulation, it simplifies matters to assume that individuals can only trade if they deliver a value that is proportional to the value of their net transactions. Such revenues from transactions accrue to a central authority that redistributes them to individuals as lump-sum transfers, according to given shares. This abstraction corresponds to the description of a sequential monetary economy under a complete asset market and a central bank that pegs a constant nominal rate of interest and accommodates

---

[5]The set of all real valued maps on $\mathcal{L}$ is $\boldsymbol{L} = \mathbb{R}^{\mathcal{L}}$. An element $\boldsymbol{x}$ of $\boldsymbol{L}$ is said to be positive if $\boldsymbol{x}(l) \geq 0$ for every l in $\mathcal{L}$; negligible if $\boldsymbol{x}(l) = 0$ for all but finitely many l in $\mathcal{L}$. For an element $\boldsymbol{x}$ of $\boldsymbol{L}$, $\boldsymbol{x}^+$ and $\boldsymbol{x}^-$ are, respectively, its positive part and its negative part, so that $\boldsymbol{x} = \boldsymbol{x}^+ - \boldsymbol{x}^-$ and $|\boldsymbol{x}| = \boldsymbol{x}^+ + \boldsymbol{x}^-$. The positive cone, $\boldsymbol{L}^+$, of $\boldsymbol{L}$ consists of all positive elements of $\boldsymbol{L}$. Also, $\boldsymbol{L}_0$ is the vector space consisting of all negligible elements of $\boldsymbol{L}$. Finally, for every element $\boldsymbol{x}$ of $\boldsymbol{L}$,

$$\boldsymbol{L}(\boldsymbol{x}) = \{\boldsymbol{v} \in \boldsymbol{L} : |\boldsymbol{v}| \leq \lambda |\boldsymbol{x}|, \text{ for some } \lambda > 0\}$$

is a principal ideal of $\boldsymbol{L}$. Unless otherwise stated, every topological property on $\boldsymbol{L}$ refers to the traditional product topology. We remark that, throughout the paper, the term 'positive' is used to mean 'greater than or equal to zero'.

the demand for balances. In addition, the central bank, whose ownership is sequentially trade on the asset market, redistributes its profit (seignorage) as dividends to shareholders.

Prices of commodities p are also an element of L^+. These are, in a sense, discounted or Arrow-Debreu prices. The duality operation on $L^+ \times L^+$ is defined by

$$p \cdot v = \sup \{p \cdot v_0 : v_0 \in [0, v] \cap L_0\},$$

that may be infinite.

The budget constraint of an individual is

$$\left(\frac{r}{1+r}\right) p \cdot \left(x^i - e^i\right)^- + p \cdot \left(x^i - e^i\right) \leq \mu^i w,$$

where $\mu^i \geq 0$ is the share of the individual i in the aggregate positive transfer w.

For given a positive nominal rate of interest, r, an *(abstract) r-equilibrium* consists of a balanced allocation, x, prices, p, and a aggregate positive transfer, w, such that

$$\left(\frac{r}{1+r}\right) p \cdot \sum_i \left(x^i - e^i\right)^- \leq w$$

and, for every individual,

$$\left(\frac{r}{1+r}\right) p \cdot \left(x^i - e^i\right)^- + p \cdot \left(x^i - e^i\right) \leq \mu^i w$$

and

$$z^i \succ^i x^i \implies \left(\frac{r}{1+r}\right) p \cdot \left(z^i - e^i\right)^- + p \cdot \left(z^i - e^i\right) > \mu^i w.$$

An (abstract) r-equilibrium involves a *speculative bubble* if

$$b = w - \left(\frac{r}{1+r}\right) p \cdot \sum_i \left(x^i - e^i\right)^- > 0.$$

Notice that an (abstract) 0-equilibrium is what the literature traditionally refers to as an equilibrium with (possibly) positive outside money, or with (possibly) a positive speculative bubble.

Lemma 1 *The value of net transaction, $p \cdot \sum_i \left(x^i - e^i\right)^-$, is finite at every r-equilibrium with $r > 0$.*

Proof Obvious.   **Q.E.D.**

Allocation z Pareto dominates allocation x if, for every individual, $z^i \succeq^i x^i$ with $z^i \succ^i x^i$ for some. Allocation z Malinvaud dominates allocation x if z Pareto dominates x, while $z^i = x^i$ for all but finitely many individuals (Malinvaud 1953).

For a given positive nominal rate of interest, r, an allocation, x, is *Pareto (Malinvaud) r-undominated* if it is not Pareto (Malinvaud) dominated by an alternative allocation, z, that satisfies

$$\left(\frac{r}{1+r}\right)\sum_i \left(z^i - e^i\right)^- + \sum_i z^i \leq \left(\frac{r}{1+r}\right)\sum_i \left(x^i - e^i\right)^- + \sum_i x^i.$$

Evidently, a Pareto (Malinvaud) 0-undominated allocation coincides with a standard Pareto (Malinvaud) efficient allocation.

Lemma 2 *Every r-equilibrium allocation is a Malinvaud r-undominated allocation.*

Proof If not, there is an allocation z that Malinvaud dominates allocation x and satisfies

$$\left(\frac{r}{1+r}\right)\sum_i \left(z^i - e^i\right)^- + \sum_i z^i \leq \left(\frac{r}{1+r}\right)\sum_i \left(x^i - e^i\right)^- + \sum_i x^i.$$

Thus, for every individual,

$$\left(\frac{r}{1+r}\right)p \cdot \left(z^i - e^i\right)^- + p \cdot z^i \geq \left(\frac{r}{1+r}\right)p \cdot \left(x^i - e^i\right)^- + p \cdot x^i,$$

with at least one strict inequality. Since the allocation z coincides with the allocation x for all but finitely many individuals, aggregation across individuals yields a contradiction.   **Q.E.D.**

An allocation, x, involves *uniform trade* if there is a decomposition $L_f \oplus L_b$ of the (reduced) commodity space $L(e)$, with

$$L_f \subseteq \left\{ v \in L : |v| \leq \lambda \sum_i \left(x^i - e^i\right)^-, \text{ for some } \lambda > 0 \right\},$$

and an allocation v such that $\sum_i v^i$ belongs to $L(e)$ and, for some $\lambda > 0$ small enough, $x^i - \lambda x_b^i + v_f^i \succeq^i x^i$ for every individual. This requires that the set of commodities can be partitioned into commodities that are traded in some uniformly strictly positive amount and commodities that are not, in such a way that all individuals can increase their welfare by a large enough increase in consumption in the former set of commodities, even when consumption in the latter set of commodities is slightly reduced.

Assumption 4 (*Gains to Trade*) Every individually rational balanced Malinvaud r-undominated allocation, with $r > 0$ sufficiently small, involves uniform trade.

The gains to trade hypothesis extends the condition Bloise, Drèze and Polemarchakis (2004) and Dubey and Geanakoplos (2005). It has as consequence that the value of the aggregate endowment is finite at equilibrium.

Lemma 3 *The value of the aggregate endowment, $p \cdot \sum_i e^i$, is finite at every r-equilibrium with $r > 0$ sufficiently small.*

Proof Consider the decomposition of the (reduced) commodity space $L_f \oplus L_b = L(e) \subseteq L$ in the hypothesis of a uniform trade. Clearly, p defines a positive σ-additive linear functional on L_f. Thus, $p \cdot \sum_i e^i$ is unbounded only if $p \cdot \sum_i e^i_b$ is unbounded and, hence, only if $p \cdot \sum_i x^i_b$ is unbounded. Also, $p \cdot \sum_i v^i_f$ is finite, where v is the allocation mentioned in the definition of uniform trade.

For every individual, $z^i \succeq^i x^i$ implies

$$\left(\frac{r}{1+r}\right) p \cdot \left(z^i - e^i\right)^- + p \cdot z^i \geq \left(\frac{r}{1+r}\right) p \cdot \left(x^i - e^i\right)^- + p \cdot x^i.$$

Since $\left(z^i - x^i\right)^- + \left(x^i - e^i\right)^- \geq \left(z^i - e^i\right)^-$, it follows that

$$p \cdot \left(z^i - x^i\right)^+ \geq \left(\frac{1}{1+r}\right) p \cdot \left(z^i - x^i\right)^-.$$

As $x^i - \lambda x^i_b + v^i_f \succeq^i x^i$ for some $\lambda > 0$ sufficiently small, using the previous argument, with $z^i = x^i - \lambda x^i_b + v^i_f$ implies that

$$p \cdot v^i_f \geq \left(\frac{1}{1+r}\right) \lambda p \cdot x^i_b.$$

Aggregation across individuals yields a contradiction. **Q.E.D.**

As the aggregate endowment is finitely valued at equilibrium, canonical conclusions about efficiency and the absence of speculative bubbles can be drawn.

Proposition 1 (Almost Pareto Optimality)
No r-equilibrium allocation, x, with $r > 0$ sufficiently small, is Pareto dominated by an alternative allocation, z, that satisfies

$$\sum_i z^i \leq \left(\frac{1}{1+r}\right) \sum_i e^i.$$

Proof As the aggregate endowment is finitely valued, it is clear that every r-equilibrium allocation, with $r > 0$ small enough, is a Pareto r-undominated allocation (the proof is just an adaptation of the proof of Lemma 2). So, in order to prove that the statement in the proposition holds true, suppose not. It follows that x is Pareto dominated by an alternative allocation z that satisfies

$$\left(\frac{r}{1+r}\right) \sum_i (z^i - e^i)^- + \sum_i z^i \leq$$

$$\left(\frac{r}{1+r}\right) \sum_i e^i + \sum_i z^i \leq \sum_i e^i$$

$$\leq \left(\frac{r}{1+r}\right) \sum_i (x^i - e^i)^- + \sum_i x^i.$$

This contradicts Pareto r-undomination.   **Q.E.D.**

Proposition 2 (No Speculative Bubbles)
  No r-equilibrium, with $r > 0$ sufficiently small, involves a speculative bubble.

Proof As the aggregate endowment is finitely valued, the result follows from the aggregation of budget constraints across individuals.   **Q.E.D.**

It remains to understand the restrictions implied by the gains to trade hypothesis (Assumption 4).

15.4   Gains to Trade in a Stationary Economy

The hypothesis on gains to trade (Assumption 4) is generically satisfied in a standard stationary economy of identical overlapping generations of heterogenous individuals. We shall simply provide the core argument, as details are straightforward but heavy in terms of notation.

The set of individuals is $\mathcal{I} = \mathcal{J} \times \mathcal{T}$, where $\mathcal{T} = \{0, 1, 2, \ldots, t, \ldots\}$ are dates or periods of trade and $\mathcal{J}$ is a finite set of individuals within a generation: for every t in $\mathcal{T}$, $\mathcal{I}^t = \{(j, t) : j \in \mathcal{J}\}$ is generation t. All generations $\mathcal{I}^{t+1}$ are identical and have life spans $\mathcal{T}^{t+1} = \{t, t+1\} \subseteq \mathcal{T}$. The initial generation $\mathcal{I}^0$ has life span $\mathcal{T}^0 = \{0\} \subseteq \mathcal{T}$.

Preferences are strictly monotone over the life span of an individual: for an individual in generation t in $\mathcal{T}$, preferences are strictly monotone on the positive cone of $\boldsymbol{L}^t = \mathbb{R}^{\mathcal{L}^t} \subseteq \mathbb{R}^{\mathcal{L}} = \boldsymbol{L}$, where $\mathcal{L}^t = \mathcal{T}^t \times \mathcal{N}$.

Endow the (reduced) commodity space $\boldsymbol{L}$ ($\boldsymbol{e}$) with the supremum norm

$$\|\boldsymbol{v}\|_\infty = \sup\{\lambda > 0 : |\boldsymbol{v}| \leq \lambda \boldsymbol{e}\}.$$

As the economy is stationary, this involves no loss of generality. Suppose that there is $\epsilon > 0$ such that, for every individually rational, balanced, Malinvaud efficient allocation, x, the aggregate net trade of every generation t in $\mathcal{T}$ is ϵ-bounded away from autarky, that is,

$$\left\| \sum_{i \in \mathcal{I}^t} \left(x^i - e^i \right)^+ \right\|_\infty > \epsilon.$$

In a stationary economy of identical overlapping generations, this is a rather weak requirement when there are at least two individuals in each generation.[6] It follows that there is $\epsilon > 0$ such that, provided that $r > 0$ is small enough, for every individually rational, balanced, Malinvaud r-undominated allocation, x, the aggregate net trade of every generation t in $\mathcal{T}$ is ϵ-bounded away from autarky.

This is evident.

Let e_l be the aggregate endowment of commodity l in $\mathcal{L}$ (regarded as an element of the commodity space L). For a generation t in $\mathcal{T}$, let $g(t)$ in $\mathcal{L}$ be a commodity such that

$$e_{g(t)} \leq \frac{1}{\epsilon} \sum_{i \in \mathcal{I}^t} \left(x^i - e^i \right)^+.$$

Such a commodity exists because net trades are uniformly bounded away (in the sup norm) from zero by $\epsilon > 0$. Decompose the aggregate endowment as $e = e_f + e_b$, where

$$e_f = \sum_{l \in g(\mathcal{T})} e_l,$$

and

$$e_b = \sum_{l \notin g(\mathcal{T})} e_l.$$

Clearly, $L_f = L\left(e_f\right)$ and $L_b = L(e_b)$ are such that $L(e) = L_f \oplus L_b$. In addition,

$$e_f \leq$$

$$\sum_{t \in \mathcal{T}} e_{g(t)} \leq \tfrac{1}{\epsilon} \sum_{t \in \mathcal{T}} \sum_{i \in \mathcal{I}^t} \left(x^i - e^i \right)^+ =$$

$$\tfrac{1}{\epsilon} \sum_i \left(x^i - e^i \right)^+ = \tfrac{1}{\epsilon} \sum_i \left(x^i - e^i \right)^-,$$

---

[6] As the allocation is Malinvaud efficient, it is Pareto efficient within every generation. As the allocation is individually rational and preferences are strictly monotone, if positive net trades vanish within a generation, so do negative net trades. Thus, using the fact that all generations are identical, $\epsilon > 0$ above does not exist only if no-trade is a Pareto efficient allocation within a typical generation. This does not occur generically in preferences and endowments.

so that

$$L_f \subseteq \left\{ v \in L : |v| \le \lambda \sum_i \left(x^i - e^i \right)^-, \text{ for some } \lambda > 0 \right\}.$$

For an individual i in generation t in $\mathcal{T}$, let $v^i = v^i_f = e_{g(t)}$. Taking into account multiplicities and using the fact that generations overlap for at most two periods, it is easily verified that

$$\sum_i v^i = \sum_{t \in \mathcal{T}} \sum_{i \in \mathcal{I}^t} e_{g(t)} \le (\#\mathcal{J}) \sum_{t \in \mathcal{T}} e_{g(t)} \le 2 (\#\mathcal{J}) \sum_{l \in g(\mathcal{T})} e_l = 2 (\#\mathcal{J}) e_f.$$

Using stationarity hypotheses and the strict monotonicity of preferences over relevant consumption spaces, it is simple to show that there is $1 > \lambda > 0$ such that, for every individual,

$$x^i - \lambda x^i_b + v^i_f \succeq^i x^i.$$

The gains to trade hypothesis (Assumption 4) is satisfied.

15.5   Sequential Trade

The abstract framework accommodates a sequential economy of overlapping generations. We here present the classical arguments for the consolidation of budget constraints that are implied by a sequentially complete asset market.

15.5.1   Prices and Markets

In every period of trade, there are markets for commodities, balances and assets. Balances are the numéraire at every date. A constant positive nominal rate of interest, r, is pegged by the monetary authority.[7]

The asset structure consists of a one-period nominally risk-free bond and an infinitely-lived security that pays off nominal dividends in every period. Short sales are allowed on bonds, but not on the security. Prices of the security q are a positive

---

[7] As far as individuals and commodities are concerned, notation is as in Sect. 15.3. In particular, an element x of $L = \mathbb{R}^{\mathcal{T} \times \mathcal{N}}$ decomposes, across periods of trade, as

$$x = (x_0, \ldots, x_{t-1}, x_t, x_{t+1}, \ldots),$$

where each x_t is an element of $\mathbb{R}^{\mathcal{N}}$; an element x of $E = \mathbb{R}^{\mathcal{T}}$ decomposes, across periods of trade, as

$$x = (x_0, \ldots, x_{t-1}, x_t, x_{t+1}, \ldots),$$

where each x_t is an element of $\mathbb{R}$.

element of $E = \mathbb{R}^T$. These are spot prices. Dividends of the security y are a positive element of E. This security is in positive net supply and, to simplify, the supply is normalized to the unity.

Discount factors, a, a positive element of E, are obtained by setting

$$a_t = \left(\frac{1}{1+r}\right)^t.$$

No arbitrage, jointly with the fact that the security cannot be dominated by bonds at equilibrium, implies that, in every period of trade,

$$a_t q_t = a_t y_t + a_{t+1} q_{t+1}.$$

This condition reflects the innocuous assumption that the security is priced *cum dividend*. As far as the intertemporal transfer of wealth is concerned, bonds and the security are perfect substitutes under this no-arbitrage pricing.

A standard argument implies that, in every period of trade,

$$q_t \geq \frac{1}{a_t} \sum_{s \geq t} a_s y_s.$$

That is, the price of the security is at equal to or greater than its fundamental value. The displacement of the market value of the security from its fundamental value is the speculative bubble.

Prices of commodities p are a positive element of L. To avoid an excess of notation, we interpret p as present value prices of commodities. So, current (or spot) prices of commodities are

$$\left(\frac{1}{a_0} p_0, \ldots, \frac{1}{a_{t-1}} p_{t-1}, \frac{1}{a_t} p_t, \frac{1}{a_{t+1}} p_{t+1}, \ldots\right).$$

15.5.2 Sequential Budget Constraints

Sequential constraints are canonical. Individual i formulates a consumption plan, x^i, a positive element of L, and a financial plan, $\left(m^i, z^i, v^i\right)$, consisting of holdings of balances, m^i, a positive element of E, of the security, z^i, a positive element of E, and of short-term bonds, v^i, an element of L. Individual i enters period of trade t with some accumulated nominal wealth, w_t^i; he trades in assets and balances according to the budget constraint

$$m_t^i + (q_t - y_t) z_t^i + \left(\frac{1}{1+r}\right) v_t^i \leq w_t^i;$$

he uses balances for the purchase of commodities, as prescribed by a cash-in-advance constraint,

$$\frac{1}{a_t} p_t \cdot (x_t^i - e_t^i)^- \leq m_t^i;$$

receives balances from the sale of commodities and he enters the following period of trade $t + 1$ with nominal wealth

$$w_{t+1}^i = m_t^i + q_{t+1} z_t^i + v_t^i - \frac{1}{a_t} p_t \cdot (x_t^i - e_t^i).$$

In addition, a wealth constraint of the form

$$-\frac{1}{a_{t+1}} \sum_{s \geq t+1} p_s \cdot e_s^i \leq w_{t+1}^i$$

is imposed in order to avoid Ponzi schemes. Finally, the initial nominal wealth is given by the initial price of the security, $w_0^i = \mu^i q_0$, where $\mu^i \geq 0$ is the initial share of individual i into the security.

If a consumption plan, x^i, and a financial plan, (m^i, z^i, v^i), satisfy all the above described restrictions at all periods of trade, we say that financial plan (m^i, z^i, v^i) finances consumption plan x^i (equivalently, consumption plan x^i is financed by financial plan (m^i, z^i, v^i). The sequential budget constraint of individual i is the set of all consumption plans, x^i, that are financed by some financial plan.

Literally interpreted, our sequential budget constraint might appear contradicting the hypothesis of overlapping generations of individuals. Indeed, it can be argued that an individual might not be active at some date and, so, it is meaningless to assume that consumptions and wealth accumulation of such an individual are restricted by the entire sequence of constrains. Observe, however, that an individual should be regarded as not being active at some date only if he has no endowment of commodities and his utility is unaffected by the consumption of commodities at that date. These are joint assumptions of preferences and endowments. Letting the individual trade when he should be regarded as not being active adds redundant constraints without altering the substance. A skeptical reader might assume that an individual i is characterized by a time horizon $T^i \subseteq T$ of consecutive periods of trade. Both the consumption plan and the financial plan can be assumed to be zero out of the given time horizon. In the same spirit of the above observation, one might be willing to assume that the initial share into the security is strictly positive only for individuals that are active in the initial period of trade.

15.5.3 Intertemporal Budget Constraints

By a canonical consolidation, provided that there are no arbitrage opportunities, sequential budget constraint reduces to a single intertemporal budget constraint of the form

$$\left(\frac{r}{1+r}\right) \sum_t p_t \cdot \left(x_t^i - e_t^i\right)^- + \sum_t p_t \cdot \left(x_t^i - e_t^i\right) \leq \mu^i q_0.$$

The underlying demand of balances satisfies, in every period of trade t,

$$m_t^i \geq \frac{1}{a_t} p_t \cdot \left(x_t^i - e_t^i\right)^-,$$

with the equality whenever $r > 0$. Also, the holding of bonds and of the security, witch are perfect substitutes as far as intertemporal transfers of wealth are concerned, can be assumed to satisfy, in every period of trade t,

$$m_t^i + q_{t+1} z_t^i + v_t^i = \left(\frac{r}{1+r}\right) \frac{1}{a_t} \sum_{s \geq t+1} \left(x_s^i - e_s^i\right)^- + \frac{1}{a_t} \sum_{s \geq t} p_s \cdot \left(x_s^i - e_s^i\right).$$

As a matter of mere fact, using a more compact notation, a consumption plan, x^i, is restrict by a single intertemporal budget constraint of the form

$$\left(\frac{r}{1+r}\right) p \cdot \left(x^i - e^i\right)^- + p \cdot \left(x^i - e^i\right) \leq \mu^i q_0.$$

The financial plan, $\left(m^i, z^i, v^i\right)$, that finances an intertemporally budget feasible consumption plan, x^i, can be recovered, up to an intrinsic multiplicity due to redundant assets.

15.5.4 The Monetary Authority

The security is backed by the ownership of a central bank, which issues balances against bonds and distributes its profit as a divided to shareholders. A plan, (m, v, y), of the monetary authority consists of a supply of balances, m, a positive element of E, a demand of short-term bonds, v, an element of E, and dividends to shareholders, y, a positive element of E. A sequential budget constraint imposes

$$m - \left(\frac{1}{1+r}\right) v = y.$$

The monetary authority accommodates the demand for balances (that is, $m = \sum_i m^i$) and runs balanced accounts (that is, $m = v$), so that

$$y = \left(\frac{r}{1+r}\right) \sum_i m^i.$$

15.5.5 Sequential Equilibrium

Equilibrium requires market clearing only for commodities and assets, as the demand of balances is accommodated by the monetary authority. Given a positive nominal rate of interest, r, a *sequential r-equilibrium* consists of a collection of plans for individuals,

$$\left(\ldots, \left(x^i, \left(m^i, z^i, v^i\right)\right), \ldots\right),$$

a plan for the monetary authority, (m, v, y), prices, p, and security prices, q, such that the following conditions are satisfied.

(a) For every individual i, consumption plan x^i is $\succeq^i$-optimal, subject to sequential budget constraint, and is financed by financial plan $\left(m^i, z^i, v^i\right)$.

(b) The monetary authority accommodates the demand for balances and runs a balanced budget or

$$m = \sum_i m^i,$$

$$v = m,$$

$$y = \left(\frac{r}{1+r}\right) m.$$

(c) Markets for commodities and assets clear or

$$\sum_i x^i = \sum_i e^i,$$

$$\sum_i z^i = 1,$$

$$\sum_i v^i = v.$$

Clearly, at a sequential equilibrium, security prices involve no arbitrage opportunities and, in addition, the security is not dominated by bonds.

15.5.6 Abstraction

At equilibrium,

$$\left(\frac{r}{1+r}\right)m_t = \left(\frac{r}{1+r}\right)\frac{1}{a_t}p_t \cdot \sum_i \left(x_t^i - e_t^i\right)^+ = \left(\frac{r}{1+r}\right)\frac{1}{a_t}p_t \cdot \sum_i \left(x_t^i - e_t^i\right)^-$$

and, as a consequence,

$$\left(\frac{r}{1+r}\right)\sum_t p_t \cdot \sum_i \left(x_t^i - e_t^i\right)^- = \sum_t a_t y_t \le q_0.$$

Thus, using consolidation of sequential budget constraints, at a (sequential) r-equilibrium, it follows that

$$\left(\frac{r}{1+r}\right)p \cdot \sum_i \left(x^i - e^i\right)^- \le q_0$$

and, for every individual i,

$$\left(\frac{r}{1+r}\right)p \cdot \left(x^i - e^i\right)^- + p \cdot \left(x^i - e^i\right) \le \mu^i q_0$$

and

$$z^i \succ^i x^i \text{ implies } \left(\frac{r}{1+r}\right)p \cdot \left(z^i - e^i\right)^- + p \cdot \left(z^i - e^i\right) > \mu^i q_0.$$

These are the only substantial equilibrium restrictions, as market clearing for bonds and the security can be verified to hold. As a conclusion, a sequential r-equilibrium coincides with an abstract r-equilibrium.

15.6 Concluding Remarks

In the abstract formulation, every individual i is subject to a single budget constraint of the form

$$\left(\frac{r}{1+r}\right)p \cdot \left(x^i - e^i\right)^- + p \cdot \left(x^i - e^i\right) \le w^i,$$

where w^i would be interpreted, depending on the particular institutional framework, as the value of initial asset holdings plus possibly transfers in present value terms. Thus, Walras Law imposes

$$f + b = \left(\frac{r}{1+r}\right) \sum_i p \cdot \left(x^i - e^i\right)^- + \sum_i p \cdot \left(x^i - e^i\right) = \sum_i w^i = w,$$

where w, f and b are understood to be (possibly non-finite) limits.[8] The argument for almost Pareto optimality moves from the observation that the value of net transactions is finite at equilibrium. As long as nominal rate of interest is strictly positive, $r > 0$, this occurs whenever f is finite. In addition, by local non-satiation of preferences, w is finite if at least one individual is entitled to a positive share of it (that is, $w^i = \alpha^i w$, with $\alpha^i > 0$, for some individual i).

If w is finite, then

$$w \geq \left(\frac{r}{1+r}\right) p \cdot \sum_i \left(x^i - e^i\right)^- = f$$

suffices to argue that f is finite. Incidentally, the above inequality rules out a negative speculative bubble, $w - f = b \geq 0$, but the crucial point is only that it guarantees a finite value of f. Sequential trades and, in particular, a central bank quoted on the stock market serve to interpret w as the initial market value of the central bank and f as the initial fundamental value of the central bank. Thus, $w \geq f$, with w finite, is inherited by a primitive description of sequential trades under the assumption of free disposal on long-term securities, so as to rule out a negative market value of the central bank. Could the same conclusion be drawn in other institutional frameworks?

In Bloise, Drèze and Polemarchakis (2004), a central bank trades balances for bonds and runs a balanced account by redistributing its profit to shareholders. This basically requires $f = w$, which by itself does not ensure a finite value of f. However, if this redistribution of the profit is interpreted as occurring intertemporally (that is, shares are into the intertemporal value of seignorage w), w would be finite and conclusions would be equivalent.

Alternatively, in the spirit of the fiscal theory of price determination (Woodford (1994)), one interprets w as a given stock on public debt, which is, thus, finite. A priori, it does not follow that $w \geq f$, which incidentally shows that the price level might still be indeterminate (in that context, $f = w$ implies an intertemporally balanced public budget). However, if one assumes that public debt cannot be negative, with ambiguous implications for sequential public budget constrains, then $b \geq 0$ and, so, $w \geq f$, thus leading to analogous conclusions.

---

[8]The discussion here is only suggestive, so that we avoid details on conditions for well-defined, though not finite, limits.

References

Bloise, G., & Polemarchakis, H. (2006). Monetary policy and dynamic efficiency in economies of overlapping generations. *International Journal of Economic Theory, 2*, 319–330. http://www.polemarchakis.org/a70-ogr.pdf.

Bloise, G., Drèze, J. H., & Polemarchakis, H. (2005). Monetary equilibria over and infinite horizon. *Economic Theory, 25*, 51–74. http://www.polemarchakis.org/a65-mei.pdf.

Clower, R. (1967). A reconsideration of the microfoundations of monetary theory. *Western Economic Journal, 6*, 1–8.

Diamond, P. A. (1984). Money in search equilibrium. *Econometrica, 52*, 1–20.

Drèze, J. H., & Polemarchakis, H. (2000). Monetary equilibria. In G. Debreu, W. Neufeind, & W. Trockel (Eds.), *Economic essays: A Festschrift in honor of W. Hildenbrand* (pp. 83–108). Springer. http://www.polemarchakis.org/o15-meq.pdf.

Friedman, M. (1969). The optimum quantity of money. In M. Friedman (Ed.), *The optimum quantity of money and other essays* (pp. 1–50). Aldine.

Gale, D. (1973). Pure exchange equilibrium of dynamic economic models. *Journal of Economic Theory, 6*, 12–36.

Hahn, F. H. (1965). On some problems in proving the existence of an equilibrium in a monetary economy. In F. H. Hahn, & F. P. R. Brechling (Eds.), *The theory of interest rates* (pp. 126–135). Macmillan.

Kiyotaki, N., & Wright, R. (1989). On money as a medium of exchange. *Journal of Political Economy, 97*, 927–954.

Lucas, R. E., & Stokey, N. L. (1987). Money and rates of interest in a cash-in-advance economy. *Econometrica, 55*, 491–513.

Nakajima, T. & Polemarchakis, H. (2005). Money and prices under uncertainty. *Review of Economic Studies, 72*, 223–246. http://www.polemarchakis.org/a64-mpu.pdf.

Samuelson, P. A. (1958). An exact consumption-loan model of interest with or without the contrivance of money. *Journal of Economic Theory, 66*, 467–482.

Santos, M. S., & Woodford, M. (1997). Asset pricing bubbles. *Econometrica, 65*, 19–57.

Weiss, L. (1980). The effects of money supply on economic welfare in the steady state. *Econometrica, 48*, 565–576.

Wilson, C. (1981). Equilibrium in dynamic models with an infinity of agents. *Journal of Economic Theory, 24*, 95–111.

Chapter 16
Winners and Losers from Price-Level Volatility: Money Taxation and Information Frictions

Guido Cozzi, Aditya Goenka, Minwook Kang and Karl Shell

Abstract We analyze an economy with taxes and transfers denominated in dollars and an information friction. It is the information friction that allows for volatility in equilibrium prices and allocations. When the price level is expected to be stable, the competitive equilibrium allocation is Pareto optimal. When the price level is volatile, it is not Pareto optimal, but the stable equilibrium allocations do not necessarily dominate the volatile ones. There can be winners and losers from volatility. We identify winners and losers and describe the effect on them of increases in volatility. Our analysis is an application of the weak axiom of revealed preference in the tax-adjusted Edgeworth box.

16.1 Introduction

Finance is an important source of efficiency in modern economies, but it is also a source (perhaps *the* major source) of excess economic volatility, i.e., the potential for volatility of economic outcomes beyond the volatility of the economic fundamentals. Securities and contracts that pay off in dollars or taxes and transfers fixed in dollars can be sources of proper sunspot equilibrium outcomes.

G. Cozzi
University of St.Gallen, St.Gallen, Switzerland

A. Goenka
University of Birmingham, Birmingham, England

M. Kang
Nanyang Technological University, Singapore, Singapore

K. Shell (✉)
Cornell University, 402 D Uris Hall, Ithaca, NY 14853-7601, USA
e-mail: karl.shell@cornell.edu

© Springer International Publishing AG 2017
K. Nishimura et al. (eds.), *Sunspots and Non-Linear Dynamics*,
Studies in Economic Theory 31, DOI 10.1007/978-3-319-44076-7_16

In our model, lump-sum money taxes are set before the price level is known and expectations are formed.[1] The taxes are exogenous. The policy maker sets money taxes and the agents form expectations. Given these, there is an equilibrium outcome. In equilibrium, the price expectations of the agents must be consistent with the outcomes: rational expectations obtain. The price level is sunspot-driven. The set of instruments is incomplete: sunspot-dependent money taxation is assumed to be unavailable to the government. Nominal taxes do not depend on the realization of sunspots, but real taxes do.[2]

There are 3 consumers. The 2 full-information consumers can "see" sunspots and hedge on the securities market against the effects of sunspot-driven price-level volatility. The third consumer is the restricted-information consumer. He cannot see sunspots. He cannot hedge against the effects of price-level volatility: his participation on the securities market is restricted by the information friction.[3] He must raise money in the spot market for paying his dollar tax by selling some of his commodity endowment or, if he is subsidized, use his money subsidy to buy the consumption good in the spot market. He is always hurt by volatility. The full-information consumers trade *ex ante* in the state-contingent Edgeworth box defined by their tax-adjusted endowments.

When the price level is stable, the competitive equilibrium allocation is Pareto optimal. When the price level is volatile, it is not Pareto optimal, but the stable equilibrium allocation does not always dominate the volatile equilibrium allocations. There can be winners as well as losers from volatility. The full-information consumers hedge by trading securities. One of them (but not both of them) can gain enough to be better off than he would have been without volatility.

Our basic tool is the tax-adjusted Edgeworth box in which the full-information agents hedge against price-level volatility. As a group taken together, the full-information agents are harmed by sunspots. Their aggregate tax-adjusted endowment is negatively correlated with the price-level shocks. A simple condition on taxes and transfers ensures that the tax-adjusted endowment of one of the full information agents is positively correlated with the price-level shocks. He can afford to consume his non-sunspot equilibrium consumption, but he chooses another allocation. By the weak axiom of revealed preference, he is better off. He benefits from volatility. He does so by taking on risk from the other full-information agent. Since the total

[1] Our present interpretation is that the government sets money taxes. Hence we have outside money. Another interpretation (due to Neil Wallace) is that what we call taxes and transfers actually represent past private money borrowing and lending, a case of inside money. Either interpretation is okay. The tax interpretation is the better one for our 2 companion papers on endogenous money taxation.

[2] One might think that, in practice, all observed taxes are real taxes. We disagree. Even income taxes are due in dollars this year but based on last year's dollar income. The money taxes in this paper are meant to be suggestive of general issues arising in modern economies, ones with dollar-denominated financial instruments.

[3] Our model is an extension of the exogenous taxation model of Bhattacharya et al. (1998). We are currently working on volatility and *endogenous* taxation. We are preparing 2 papers on endogenous money taxation, one on optimal taxation–the other on voting. See Cozzi et al. (2015, 2016).

endowment of the full-information agents is negatively correlated with price-level shocks, the other consumer is necessarily worse off.

We are not the first to observe that there can be winners from sunspot volatility. Goenka and Préchac (2006) address the same issues but in another economy, the incomplete financial-markets (GEI) economy of Cass (1992). They provide a condition on the utility function ensuring that there are winners and losers from volatility. They require a sufficiently high precautionary motive. Kajii (2007) extends their results to more general utility functions. In our paper, we display similar results but in an economy with information frictions (Aumann 1987) or alternatively with some consumers who are restricted from participating in financial markets (Cass and Shell 1983).

We provide in Proposition 2 conditions on taxes and transfers for one of the full-information consumers to be better off with price-level volatility while the other full-information consumer is worse off. In the proposition, we allow for (1) heterogeneous preferences and (2) utility functions that merely possess positive first derivatives and negative second derivatives. We do not show that expected utilities are monotone in volatility for this general case. We conjecture that monotonicity does not apply generally. Our intuition for this conjecture is based on the possibility in the general case of multiple sunspot equilibria. However, with identical homothetic preferences (in Sect. 4), there is a representative agent (for the full-information agents) and thus, a unique equilibrium is guaranteed.[4] For the special case of identical CRRA preferences, we show in Proposition 5 that the expected utility of the winner is indeed strictly increasing in volatility while the expected utilities of the losers are strictly decreasing in volatility.

16.2 The Model

We analyze a simple exchange economy with lump-sum taxes-and-transfers denominated in money units (say dollars), a single commodity (say chocolate), 3 consumers $h = 1, 2, 3$, and 2 sunspots states $s = \alpha, \beta$. The consumption of Mr. h in state s is $x_h(s) > 0$ (measured in chocolate). His endowment of chocolate is independent of s, $\omega_h(\alpha) = \omega_h(\beta) = \omega_h > 0$. His lump-sum dollar tax is also independent of s, $\tau_h(\alpha) = \tau_h(\beta) = \tau_h$. If τ_h is negative, he is subsidized. If τ_h is zero, then he is neither taxed nor subsidized. Mr. h's expected utility is given by

$$V_h = \pi(\alpha) u_h(x_h(\alpha)) + \pi(\beta) u_h(x_h(\beta)),$$

where $\pi(s)$ is the probability of realization $s = \alpha, \beta$. We assume that $u'_h > 0, u''_h < 0$, and that indifference curves in $(x_h(\alpha), x_h(\beta))$ space do not intersect the axes, thus ensuring interior solutions to the consumer problems.

---

[4]See Chipman (1974) Theorem 3, p. 32.

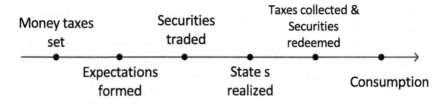

Fig. 16.1 The time line

We assume that the government sets τ_h before expectations are formed and s is realized. The timing is the source of incomplete instruments: $\tau_h(\alpha) = \tau_h(\beta) = \tau_h$. See our time line, Fig. 16.1.

We restrict attention to the case of balanced taxation,

$$\tau_1 + \tau_2 + \tau_3 = 0.$$

Otherwise the chocolate price of money must be zero[5] and autarky is the only equilibrium.

Let $p(s)$ be the *ex-ante* (accounting) price of chocolate delivered in state s and $p^m(s)$ be *ex-ante* (accounting) price of money delivered in state s. Then $P^m(s) = p^m(s)/p(s)$ is the chocolate price of money in s, while $1/P^m(s)$ is the money price of chocolate in s, or the general price level in s. We assume that consumer 3 is restricted from participation in the securities market because he is blind to sunspots (or for any of many possible other reasons including that he is not born in time to hedge his bets), but consumers 1 and 2 are unrestricted; they see sunspots perfectly. This is a special example of "information frictions" (or correlated, or asymmetric, information).[6]

Consumer 3's problem is simple. He chooses $x_3(s) > 0$ to

$$\text{maximize } u_3(x_3(s))$$

subject to

$$p(s)x_3(s) = p(s)\omega_3 - p^m(s)\tau_3$$

for $s = \alpha, \beta$.

Define the tax-adjusted endowment $\widetilde{\omega}_h(s) = \omega_h - P^m(s)\tau_h$. Then, Mr. 3's budget constraints reduces to

$$x_3(s) = \widetilde{\omega}_3(s)$$

for $s = \alpha, \beta$. Mr. 3 is passive: he consumes his tax-adjusted endowment in state s.

---

[5]See Balasko-Shell (1993) on balancedness and bonafidelity.

[6]See Aumann (1987) for the definition of correlated equilibrium in games. See Peck and Shell (1991) for correlated sunspots in imperfectly competitive market economies. See also Aumann et al. (1985).

Mr. 1 and Mr. 2 trade in the securities market and the spot market. Each faces a single budget constraint. Mr. h's problem is to choose $(x_h(\alpha), x_h(\beta)) > 0$ to

$$\text{maximize } V_h$$

subject to

$$p(\alpha)x_h(\alpha) + p(\beta)x_h(\beta) = (p(\alpha) + p(\beta)) \omega_h - (p^m(\alpha) + p^m(\beta)) \tau_h$$

for $h = 1, 2$. From the first-order conditions, we have

$$\frac{p(\beta)}{p(\alpha)} = \frac{\pi(\beta) u_1'(x_1(\beta))}{\pi(\alpha) u_1'(x_1(\alpha))} = \frac{\pi(\beta) u_2'(x_2(\beta))}{\pi(\alpha) u_2'(x_2(\alpha))}. \tag{16.1}$$

Market clearing implies

$$x_1(s) + x_2(s) + x_3(s) = \omega_1(s) + \omega_2(s) + \omega_3(s)$$

or simply

$$x_1(s) + x_2(s) + x_3(s) = \tilde{\omega}_1(s) + \tilde{\omega}_2(s) + \tilde{\omega}_3(s) \tag{16.2}$$

for $s = \alpha, \beta$. But $x_3(s) = \tilde{\omega}_3(s)$, so we have

$$x_1(s) + x_2(s) = \tilde{\omega}_1(s) + \tilde{\omega}_2(s) \quad \text{for } s = \alpha, \beta. \tag{16.3}$$

Equation (16.3) defines the relevant tax-adjusted Edgeworth box (typically a proper rectangular).

In this financial economy, there is a wide range of possible rational beliefs about the price level, generating in turn a wide range of rational, sunspot equilibria. Our goal is to focus on the effects of increased volatility on the behavior of the agents. Hence we focus on economies that can be ranked on volatility. We therefore focus on rational beliefs that are generated as mean-preserving spreads about some non-volatile price level, $P^m(\alpha) = P^m(\beta) = P^m \geq 0$. We measure volatility by the non-negative mean-preserving spread parameter σ defined by

$$P^m(\alpha) = P^m - \frac{\sigma}{\pi(\alpha)}$$

and

$$P^m(\beta) = P^m + \frac{\sigma}{\pi(\beta)},$$

where P^m is the non-sunspot equilibrium chocolate price of dollars and $\sigma \in [0, \pi(\alpha) P^m)$. When $\sigma = 0$, the equilibrium allocations are not affected by sunspots (a non-sunspots equilibrium). When $\sigma > 0$, the economy is a proper sunspots economy. State α is the inflationary state: a dollar buys less chocolate in state α than in state β. State β is the deflationary state: a dollar buys more chocolate in state β than in state α.

Proposition 1 *The non-sunspot-equilibrium ($\sigma = 0$) allocation is Pareto optimal. The proper sunspot-equilibrium allocation ($\sigma > 0$ and $\tau_3 \neq 0$) is not Pareto optimal.*

Proof When $\sigma = 0$, we have $\tilde{\omega}_h(\alpha) = \tilde{\omega}_h(\beta) = \omega_h$ for $h = 1, 2, 3$. The tax-adjusted endowments are Pareto optimal because we have

$$\frac{\pi(\beta)\,u_1'(\omega_h)}{\pi(\alpha)\,u_1'(\omega_h)} = \frac{\pi(\beta)\,u_2'(\omega_h)}{\pi(\alpha)\,u_2'(\omega_h)} = \frac{\pi(\beta)\,u_3'(\omega_h)}{\pi(\alpha)\,u_3'(\omega_h)} = \frac{\pi(\beta)}{\pi(\alpha)}.$$

Each consumer consumes his tax-adjusted endowments, i.e., $x_h(s) = \tilde{\omega}_h(s)$ where $h = 1, 2, 3$ and $s = \alpha, \beta$, and the equilibrium allocations are Pareto optimal.

For $\sigma > 0$ and $\tau_3 > 0$, we assume (for purposes of contradiction) that the equilibrium allocations are Pareto optimal, which would imply

$$\frac{\pi(\beta)\,u_1'(x_1(\beta))}{\pi(\alpha)\,u_1'(x_1(\alpha))} = \frac{\pi(\beta)\,u_2'(x_2(\beta))}{\pi(\alpha)\,u_2'(x_2(\alpha))} = \frac{\pi(\beta)\,u_3'(x_3(\beta))}{\pi(\alpha)\,u_3'(x_3(\alpha))}. \tag{16.4}$$

Because $\tau_3 > 0$, we have $\tilde{\omega}_3(\alpha) > \tilde{\omega}_3(\beta)$ and therefore $x_3(\alpha) > x_3(\beta)$. Because u_h is strictly concave, we have

$$\frac{\pi(\beta)\,u_3'(x_3(\beta))}{\pi(\alpha)\,u_3'(x_3(\alpha))} > \frac{\pi(\beta)}{\pi(\alpha)}. \tag{16.5}$$

Because $\tilde{\omega}_3(\alpha) > \tilde{\omega}_3(\beta)$, from the market clearing condition (see Eq. 16.2) we have

$$\tilde{\omega}_1(\alpha) + \tilde{\omega}_2(\alpha) < \tilde{\omega}_1(\beta) + \tilde{\omega}_2(\beta). \tag{16.6}$$

Inequality (16.6) and the market-clearing condition (see Eq. 16.3) imply that one of the two following inequalities obtains:

$$x_1(\alpha) < x_1(\beta), \tag{16.7}$$
$$x_2(\alpha) < x_2(\beta). \tag{16.8}$$

Inequalities (16.7) and (16.8) imply

$$\frac{\pi(\beta)\,u_1'(x_1(\beta))}{\pi(\alpha)\,u_1'(x_1(\alpha))} < \frac{\pi(\beta)}{\pi(\alpha)} \quad \text{and} \quad \frac{\pi(\beta)\,u_2'(x_2(\beta))}{\pi(\alpha)\,u_2'(x_2(\alpha))} < \frac{\pi(\beta)}{\pi(\alpha)} \tag{16.9}$$

respectively. Either inequality in (16.9) with inequality (16.5) violates Eq. (16.4). The case of $\sigma > 0$ and $\tau_3 < 0$ can be established in like manner. $\square$

Proposition 1 is in the spirit of Cass-Shell (1983). Although our model is different from Cass-Shell, the proof is similar. Another similarity with Cass-Shell (1983) is that if everyone has full information, sunspots cannot matter. A dis-similarity with Cass-Shell (1983) is that in the money taxation model when τ is not equal to 0 and everyone is blind to sunspots, there is typically a continuum of sunspot equilibria. In Cass-Shell, when every individual is restricted, the sunspot equilibria are randomizations over a *finite* number of certainty equilibria.

16.3 The Price Level

See Fig. 16.2. Consider the tax-adjusted Edgeworth box for Mr. 1 and Mr. 2 in the case in which volatility $\sigma > 0$. The dimensions of the box are $(\tilde{\omega}_1(\alpha) + \tilde{\omega}_2(\alpha)) \times (\tilde{\omega}_1(\beta) + \tilde{\omega}_2(\beta))$. If $\tau_1 + \tau_2 \neq 0$, the Edgeworth box is a proper rectangle with height different from width, so that $p(\alpha)/\pi(\alpha) \neq p(\beta)/\pi(\beta)$. If $\tau_1 + \tau_2 > 0$, then the α-dimension is larger than the β-dimension, $\tilde{\omega}_1(\alpha) + \tilde{\omega}_2(\alpha) > \tilde{\omega}_1(\beta) + \tilde{\omega}_2(\beta)$, which implies that we have $p(\alpha)/\pi(\alpha) < p(\beta)/\pi(\beta)$ so the total tax-adjusted-endowment of the 2 unrestricted consumers is negatively correlated with the price level.

Lemma 1 *If $\tilde{\omega}_1(\alpha) + \tilde{\omega}_2(\alpha) > \tilde{\omega}_1(\beta) + \tilde{\omega}_2(\beta)$, then $p(\alpha)/\pi(\alpha) < p(\beta)/\pi(\beta)$.*

Proof (by contradiction) From the first-order conditions, we have

$$\frac{\pi(\alpha) u_1'(x_1(\alpha))}{\pi(\beta) u_1'(x_1(\beta))} = \frac{\pi(\alpha) u_2'(x_2(\alpha))}{\pi(\beta) u_2'(x_2(\beta))} = \frac{p(\alpha)}{p(\beta)}. \tag{16.10}$$

Assume that $p(\alpha)/\pi(\alpha) \geq p(\beta)/\pi(\beta)$. This implies that $u_1'(x_1(\alpha)) \geq u_1'(x_1(\beta))$ and $u_2'(x_2(\alpha)) \geq u_2'(x_2(\beta))$ by Eq. (16.10). Because u_h is strictly concave, we know that $x_1(\alpha) \leq x_1(\beta)$ and $x_2(\alpha) \leq x_2(\beta)$. This implies that

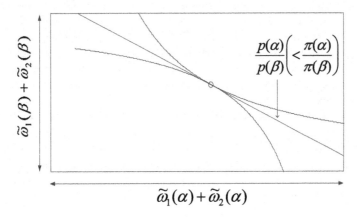

$$\frac{p(\alpha)}{p(\beta)}\left(<\frac{\pi(\alpha)}{\pi(\beta)}\right)$$

$\tilde{\omega}_1(\beta) + \tilde{\omega}_2(\beta)$

$\tilde{\omega}_1(\alpha) + \tilde{\omega}_2(\alpha)$

Fig. 16.2 Tax-adjusted edgeworth box

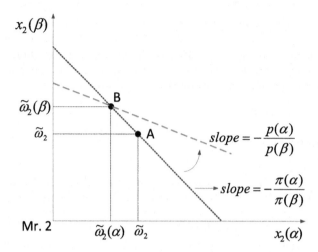

Fig. 16.3 The case of $\tau_2 < 0$

$$x_1(\alpha) + x_2(\alpha) \leq x_1(\beta) + x_2(\beta). \tag{16.11}$$

By the market clearing conditions, $x_1(\alpha) + x_2(\alpha) = \tilde{\omega}_1(\alpha) + \tilde{\omega}_2(\alpha)$ and $x_1(\beta) + x_2(\beta) = \tilde{\omega}_1(\beta) + \tilde{\omega}_2(\beta)$. Because $\tilde{\omega}_1(\alpha) + \tilde{\omega}_2(\alpha) > \tilde{\omega}_1(\beta) + \tilde{\omega}_2(\beta)$, the inequality (16.11) violates the market clearing conditions. $\qquad \square$

When is there a winner in the sunspots economy? Since the total tax-adjusted endowment for Mr. 1 and Mr. 2 is negatively correlated with the price level, a larger tax-adjusted endowment in state s decreases the price $p(s)$ in that state. Even though the total tax-adjusted endowment of the full-information consumers is negatively correlated with the price-level, some consumers' tax-adjusted endowment can be positively correlated with the price level. One possible case is that Mr. 1's nominal tax is larger than Mr. 2's nominal subsidy. In this situation, Mr. 2 can increase his wealth and his expected utility due to volatility by taking on some of Mr. 1's risk. This can be established by the weak axiom of revealed preference; see Fig. 16.3. As σ increases, the tax-adjusted endowment moves from A to B along the dotted line. The dotted line, whose slope is given by the ratio of the probabilities, can be interpreted as (1) the budget line in the non-sunspots economy and also as (2) the set of mean-preserving spreads about the certainty endowment. A is the (unadjusted) endowment. (A is also the equilibrium allocation in the certainty economy.) B is the tax-adjusted endowment and the dashed line represents the budget line for the sunspots economy. In Fig. 16.3, the certainty equilibrium allocation A is affordable in the budget set of the sunspots economy. Therefore, by WARP, Mr. 2's expected utility in the sunspots economy is higher than it is in the certainty economy (because he can afford A, but he chooses B).

Proposition 2 *If $\tau_1 + \tau_2 > 0 \, (< 0)$ and $\tau_2 \leq 0 \, (\geq 0)$, Mr. 2. is better off with price volatility and Mr. 1 and Mr. 3 are worse off with price volatility.*

Proof Case 1: $\tau_1 + \tau_2 > 0$ and $\tau_2 \leq 0$

Utility functions are strictly concave and hence Mr. 3 is obviously worse off from price volatility because his equilibrium allocations are the same as his tax-adjusted endowments, which are (by construction) mean-preserving spreads of the non-sunspots allocation.

Mr. 2's non-sunspot equilibrium allocation is $(x_2(\alpha), x_2(\beta)) = (\tilde{\omega}_2, \tilde{\omega}_2)$ where $\tilde{\omega}_2 = \omega_2 - P^m \tau_h$. We need to show that $(\tilde{\omega}_2, \tilde{\omega}_2)$ is affordable in the proper sunspots economy. Then, by the WARP, Mr. 2 would be better off with the sunspots allocation.

The condition that $(\tilde{\omega}_2, \tilde{\omega}_2)$ is affordable in the sunspots economy is

$$p(\alpha)\,\tilde{\omega}_2 + p(\beta)\,\tilde{\omega}_2 \leq p(\alpha)\,\tilde{\omega}_2(\alpha) + p(\beta)\,\tilde{\omega}_2(\beta), \qquad (16.12)$$

where $p(s)$ is ex-ante price of commodity in state s.

In the case where $\tau_2 < 0$, we have $\tilde{\omega}_2(\alpha) < \tilde{\omega}_2(\beta)$. By $\tilde{\omega}_2(\alpha) < \tilde{\omega}_2(\beta)$ and $\pi(\alpha)\,\tilde{\omega}_2(\alpha) + \pi(\beta)\,\tilde{\omega}_2(\beta) = \tilde{\omega}_2$, inequality (16.12) is equivalent to

$$\frac{p(\alpha)}{\pi(\alpha)} \leq \frac{p(\beta)}{\pi(\beta)}. \qquad (16.13)$$

In the case where $\tau_2 = 0$, inequality (16.13) is not sufficient to make Mr. 2 better off with volatility because Mr. 2's non-sunspot-equilibrium allocation $(\tilde{\omega}_2, \tilde{\omega}_2)$ still lies on the budget line in the sunspots economy. See Fig. 16.4. Therefore, we need another condition, namely that the slope of indifference curve at $(\tilde{\omega}_2, \tilde{\omega}_2)$ is different from the slope of the budget line in the sunspots economy. The slope of the indifference curve is $-\pi(\alpha)/\pi(\beta)$ and the slope of the sunspots budget line is $-p(\alpha)/p(\beta)$. Therefore, the condition is

$$\frac{p(\alpha)}{\pi(\alpha)} \neq \frac{p(\beta)}{\pi(\beta)}. \qquad (16.14)$$

Merging inequalities (16.13) and (16.14), we have

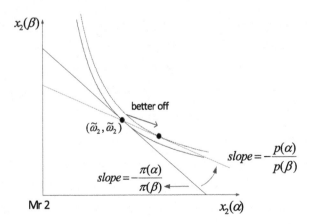

Fig. 16.4 The case of $\tau_2 = 0$

$$\frac{p\,(\alpha)}{\pi\,(\alpha)} < \frac{p\,(\beta)}{\pi\,(\beta)}, \tag{16.15}$$

which is the sufficient condition for Mr. 2 being better off with volatility. Inequality (16.15) is proven in Lemma 1.

Mr. 1: Given strictly positive prices, $p(\alpha)$ and $p(\beta)$, there are two cases;

$$(a)\ \ x_1(\alpha) > \tilde{\omega}_1\,(\alpha)\ \text{ and } x_1(\beta) < \tilde{\omega}_1\,(\beta)\,, \tag{16.16}$$

$$(b)\ \ x_1(\alpha) < \tilde{\omega}_1\,(\alpha)\ \text{ and } x_1(\beta) > \tilde{\omega}_1\,(\beta)\,. \tag{16.17}$$

In case (b), Mr. 1 will necessarily be worse off with volatility because of WARP: The equilibrium allocation $(x_1(\alpha), x_1(\beta))$ is affordable with the prices in the non-sunspots economy. (See Fig. 16.5) Assume by contradiction that case (a) is correct. Then, by the market-clearing conditions, we have

$$x_2(\alpha) < \tilde{\omega}_2\,(\alpha)\ \text{ and } x_2(\beta) > \tilde{\omega}_2\,(\beta)\,. \tag{16.18}$$

Because $\tilde{\omega}_2\,(\alpha) < \tilde{\omega}_2\,(\beta)$, inequality (16.18) implies that $x_2(\alpha) < x_2(\beta)$. Therefore, we have

$$\frac{u_2'\,(x_2(\alpha))}{u_2'\,(x_2(\beta))} > 1, \tag{16.19}$$

Fig. 16.5 Mr. 1 is worse off with volatility

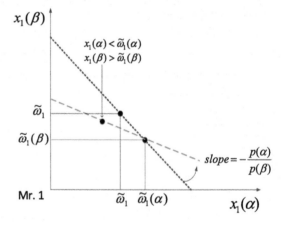

By Eq. (16.10), inequality (16.19) implies that

$$\frac{p(\alpha)}{\pi(\alpha)} > \frac{p(\beta)}{\pi(\beta)},$$

which violates inequality (16.15).

Case 2: $\tau_1 + \tau_2 < 0$ and $\tau_2 \geq 0$: This can be established as in Case 1. □

Proposition 2 shows that in the case where the sign of $\tau_1 + \tau_2$ is different from the sign of τ_2, Mr. 2 is better off with the volatile allocation while Mr. 1 and Mr. 3 are worse off. With the same logic, we can also show that if the sign of $\tau_1 + \tau_2$ is different in sign from τ_1, Mr. 1 is better off with volatility, while Mr. 2 and Mr. 3 are worse off.

Because of balancedness of the tax-transfer plans, the sign of $\tau_1 + \tau_2$ is always opposite to that of τ_3, if τ_3 is not zero. Therefore, both (1) "$\tau_1 + \tau_2 > 0$ and $\tau_2 < 0$" or (2) "$\tau_1 + \tau_2 < 0$ and $\tau_2 > 0$" imply that $sign(\tau_2) = sign(\tau_3)$. The following corollary summarizes this.

Corollary 1 *If Mr. h and Mr. 3 are both taxed (or both subsidized) where $h \neq 3$, Mr. h is better off with volatility and the other two consumers are worse off.*

Proof Directly from Proposition 2. □

Remark 1 Note that one of the full-information consumers, say Mr. 2. without any loss of generality, who is receiving a subsidy is still worse off. There are four different effects: a price effect (related to magnitude of $\tau_1 + \tau_2$), a direct loss of expected utility from increased volatility from risk averseness, a trade effect as the post-tax endowment moves further away from the minor diagonal of the post-tax Edgeworth box, and the gain from the subsidy (since $\tau_2 < 0$). Corollary 1 says that the first three effects can outweigh the fourth effect. This is reminiscent of the transfer paradox (see the formulation in Balasko 1978) where the welfare reversal depends on both the change in prices and the size of the net trade. However, our result is different from the classical transfer paradox as we hold the nominal taxes and transfers constant, and the change in price volatility induces the change in the real taxes and transfers. If there were no price volatility, then $p(\alpha) = p(\beta)$ and Mr. 2 would be unambiguously better off.

The following corollary summarizes how the 3 consumers' expected utilities change with price volatility.

Corollary 2 *The following table summarizes the pattern of winners and losers from price volatility:*

	Full-information consumers		Restricted-information consumer	Full-information consumers		Restricted-information consumer
	Mr. 1	Mr. 2	Mr. 3	Mr. 1	Mr. 2	Mr. 3
Case 1	S	T or 0	T	L	W	L
Case 2	T	S or 0	S	L	W	L
Case 3	T or 0	S	T	W	L	L
Case 4	S or 0	T	S	W	L	L

S denotes subsidized ($\tau_h < 0$), T denotes taxed ($\tau_h > 0$), 0 denotes neither subsidized nor taxed ($\tau_h = 0$), W denotes winner from volatility, and L denotes loser from volatility.

Proof Cases 1–4 follows directly from the proof of Proposition 2.   □

16.4 CRRA Preferences and Global Analysis

We assume in this section that preferences are identical CRRA. We provide the analysis of individual expected utilities as functions of volatility. The main questions are:

(1) Does increasing σ increase the ratio $p(\beta)/p(\alpha)$? (Proposition 3)

(2) Does a higher CRRA risk aversion parameter ρ make the inter-state price ratio more sensitive to money price volatility? (Proposition 4)

(3) Does increasing volatility σ increase the welfare of winners and decrease the welfare of losers? (Proposition 5)

For identical CRRA preferences, we establish that the answer for each of these 3 questions is "yes". Assume that each of the 3 consumers has CRRA preferences given by

$$u(x) = \frac{x^{1-\rho}}{1-\rho} \quad \text{when } \rho \neq 1$$
$$= \log x \quad \text{when } \rho = 1,$$

where ρ is the relative-risk-aversion parameter, i.e., $\rho = -xu''/u' > 0$.

Proposition 3 *Since the 3 consumers have identical CRRA preferences, as σ increases, we have that*
$$\frac{p(\beta)/\pi(\beta)}{p(\alpha)/\pi(\alpha)}$$
increases (decreases) when $\tau_1 + \tau_2 > 0$ (< 0).

Proof Case 1: $\tau_1 + \tau_2 > 0$

From Eqs. (16.1) to (16.3), we have

$$\left(\frac{x_1(\beta)}{x_2(\alpha)}\right)^{-\rho} = \left(\frac{\widetilde{\omega}_1(\beta) + \widetilde{\omega}_2(\beta) - x_1(\beta)}{\widetilde{\omega}_1(\alpha) + \widetilde{\omega}_2(\alpha) - x_2(\alpha)}\right)^{-\rho}. \tag{16.20}$$

Equation (16.20) implies that

$$\frac{x_1(\beta)}{x_2(\alpha)} = \frac{\widetilde{\omega}_1(\beta) + \widetilde{\omega}_2(\beta)}{\widetilde{\omega}_1(\alpha) + \widetilde{\omega}_2(\alpha)}. \tag{16.21}$$

From Eqs. (16.21) to (16.1), we have

$$\frac{p(\beta)\,\pi(\alpha)}{p(\alpha)\,\pi(\beta)} = \left(\frac{\widetilde{\omega}_1(\beta) + \widetilde{\omega}_2(\beta)}{\widetilde{\omega}_1(\alpha) + \widetilde{\omega}_2(\alpha)}\right)^{-\rho} \tag{16.22}$$

Equation (16.22) is equivalent to

$$\frac{p(\beta)\,\pi(\alpha)}{p(\alpha)\,\pi(\beta)} = \left(\frac{\omega_1 + \omega_2 - P^m(\beta)\,(\tau_1 + \tau_2)}{\omega_1 + \omega_2 - P^m(\alpha)\,(\tau_1 + \tau_2)}\right)^{-\rho},$$

which in turn is equivalent to

$$\begin{aligned}
\log\left(\frac{p(\beta)\,\pi(\alpha)}{p(\alpha)\,\pi(\beta)}\right) &= -\rho \log\left(\omega_1 + \omega_2 - P^m(\beta)\,(\tau_1 + \tau_2)\right) \\
&\quad + \rho \log\left(\omega_1 + \omega_2 - P^m(\alpha)\,(\tau_1 + \tau_2)\right) \\
&= -\rho \log\left(\omega_1 + \omega_2 - \left(P^m + \frac{\sigma}{\pi(\beta)}\right)(\tau_1 + \tau_2)\right) \\
&\quad + \rho \log\left(\omega_1 + \omega_2 - \left(P^m - \frac{\sigma}{\pi(\alpha)}\right)(\tau_1 + \tau_2)\right). \tag{16.23}
\end{aligned}$$

Implicitly differentiating Eq. (16.23) with respect to σ, we have

$$\frac{d \log \frac{p(\beta)}{p(\alpha)}}{d\sigma} = \rho \times \underbrace{\left\{\frac{1/\pi(\beta)}{\frac{\omega_1 + \omega_2}{\tau_1 + \tau_2} - \left(P^m + \frac{\sigma}{\pi(\beta)}\right)} + \frac{1/\pi(\alpha)}{\frac{\omega_1 + \omega_2}{\tau_1 + \tau_2} - \left(P^m - \frac{\sigma}{\pi(\alpha)}\right)}\right\}}_{\text{Positive}} > 0. \tag{16.24}$$

□

Case 2: $\tau_1 + \tau_2 < 0$: We establish this as for Case 1.

Equation (16.24) shows that as σ is increased, the price ratio increases. The higher is risk-aversion ρ, the higher is the rate of increase in the interstate price ratio $p(\beta)/p(\alpha)$. The inter-state commodity price ratio deviates more from its benchmark certainty equilibrium price when either σ, or ρ, or both is increased.

Proposition 4 *If the 3 consumers have identical CRRA preferences, the greater the risk-aversion parameter ρ, the greater is the rate of increase (decrease) of the price ratio $p(\beta)/p(\alpha)$ for $\tau_1 + \tau_2 > 0$ (< 0).*

Proof Directly from Eq. (16.24). $\qquad\qquad\qquad\qquad\qquad\qquad\qquad\qquad\qquad$ □

Proposition 5 *If the consumers have identical CRRA preferences, the expected utility of the winner is strictly increasing in σ and the expected utilities of the losers are strictly decreasing in σ. The winner and the loser are determined by the conditions in Proposition 2 or Corollaries 1 and 2.*

Proof Case 1: $\tau_1 + \tau_2 > 0$ and $\tau_2 \leq 0$

The Lagrangian is

$$L = \pi\,(\alpha)\,u(x_h(\alpha)) + \pi\,(\beta)\,u(x_h(\beta))$$
$$+\lambda\left\{\widetilde{\omega}_h\,(\alpha) + \frac{p(\beta)}{p(\alpha)}\widetilde{\omega}_h\,(\beta) - x_h\,(\alpha) - \frac{p(\beta)}{p(\alpha)}x_h\,(\beta)\right\}.$$

By the envelope theorem, $dV_h/d\sigma$ is

$$\frac{dV_h}{d\sigma} = \lambda\left\{\frac{\partial\widetilde{\omega}_h\,(\alpha)}{\partial\sigma} + \frac{p(\beta)}{p(\alpha)}\frac{\partial\widetilde{\omega}_h\,(\beta)}{\partial\sigma} + \frac{d\left(\frac{p(\beta)}{p(\alpha)}\right)}{d\sigma}(\widetilde{\omega}_h\,(\beta) - x_h\,(\beta))\right\}. \quad (16.25)$$

We have $d\,(p(\beta)/p(\alpha))\,/d\sigma > 0$ from Proposition 4.

For Mr. 1, we have

$$\frac{\partial\widetilde{\omega}_1\,(\alpha)}{\partial\sigma} + \frac{p(\beta)}{p(\alpha)}\frac{\partial\widetilde{\omega}_1\,(\beta)}{\partial\sigma} = \frac{\tau_1}{\pi\,(\alpha)} - \frac{p(\beta)}{p(\alpha)}\frac{\tau_1}{\pi\,(\beta)} < 0,$$

because $p(\beta)/p(\alpha) > \pi\,(\beta)\,/\pi\,(\alpha)$ from the proof of Proposition 2 and $\tau_1 > 0$. We know that $\widetilde{\omega}_1\,(\beta) - x_1\,(\beta) < 0$ from the proof of Proposition 2. Therefore, we have $dV_1/d\sigma < 0$ from Eq. (16.25).

For Mr. 2, we have

$$\frac{\partial\widetilde{\omega}_2\,(\alpha)}{\partial\sigma} + \frac{p(\beta)}{p(\alpha)}\frac{\partial\widetilde{\omega}_2\,(\beta)}{\partial\sigma} \geq 0,$$

because $p(\beta)/p(\alpha) > \pi\,(\beta)\,/\pi\,(\alpha)$ and $\tau_2 \leq 0$. We know that $\widetilde{\omega}_2\,(\beta) - x_2\,(\beta) > 0$ from the proof of Proposition 2. Therefore, we have $dV_2/d\sigma > 0$ from Eq. (16.25).

Case 2: $\tau_1 + \tau_2 < 0$ and $\tau_2 \geq 0$: We establish this as in Case 1. $\qquad\qquad$ □

16.5 Numerical Example

In this section we compute a family of numerical examples. Mr. 1 is rich. Mr. 2 and Mr. 3 each have middle class endowments, but only Mr. 3 suffers from the information friction.

$$\omega = (\omega_1, \omega_2, \omega_3) = (116, 100, 100)$$

$$\tau = (\tau_1, \tau_2, \tau_3) = (1, -0.5, -0.5)$$

This is an example of Case 1 taxation since $\tau_1 + \tau_2 = 1 - 0.5 = 0.5 > 0$ and $\tau_2 = -0.5 < 0$.

Utilities are identical CRRA with risk aversion $\rho > 0$.

$$u = \frac{c^{1-\rho}}{1-\rho} \quad \text{for } \rho \neq 1$$
$$= \log c \quad \text{for } \rho = 1$$

We assume that the 2 sunspot states are assumed to be equally probable, i.e.,

$$\pi(\alpha) = \pi(\beta) = 0.5.$$

The family of mean-preserving spreads is defined by

$$P^m(\alpha) = P^m - \frac{\sigma}{\pi(\alpha)}$$

$$P^m(\beta) = P^m + \frac{\sigma}{\pi(\beta)},$$

where $P^m = 10$ and $\sigma \in [0, 5)$.

Mr. 1 is rich and heavily taxed. He has full information. His expected utility V_1 is strictly declining in volatility σ. Mr. 2 and Mr. 3 have the same endowments, but Mr. 2 has full information while Mr. 3 receives no sunspot information. Mr. 2's expected utility V_2 is strictly increasing in σ. V_3 is strictly decreasing in σ. See Fig. 16.6, which illustrates Proposition 5. Given risk aversion ρ, the inter-state commodity price ratio

Fig. 16.6 Expected utilities as functions of volatility for the case of $\rho = 4$

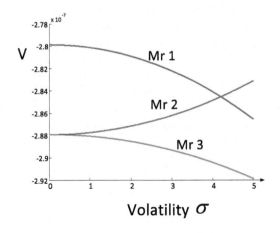

Fig. 16.7 The inter-state
price ratio as a function of
volatility σ for different
values of risk aversion ρ

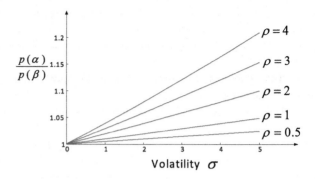

is linear in volatility σ. The effect of volatility is amplified as ρ is increased. See
Fig. 16.7, which illustrates Propositions 3 and 4.

References

Aumann, R. J. (1987). Correlated equilibrium as an expression of bayesian rationality. *Economet-rica*, *55*(1), 1–18.

Aumann, R. J., Peck, J., & Shell, K. (1985, Revised 1988). *Asymmetric information and sunspot equilibria: A family of simple examples*. Working Paper 88–34, Center for Analytic Economics, Cornell University.

Balasko, Y. (1978). The transfer problem and the theory of regular economies. *International Eco-nomic Review*, *19*(3), 687–694.

Balasko, Y. & Shell, K. (1993). Lump-sum taxation: The static economy. In R. Becker, M. Boldrin, R. Jones, & W. Thomson (Eds.), *General equilibrium, growth and trade. Essays in honor of Lionel McKenzie* (Vol. 2, pp. 168–180). San Diego, CA: Academic Press.

Bhattacharya, J., Guzman, M. G., & Shell, K. (1998). Price level volatility: A simple model of money taxes and sunspots. *Journal of Economic Theory*, *81*(2), 401–430.

Cass, D. (1992). Sunspots and incomplete financial markets: The general case. *Economic Theory*, *2*(3), 341–358.

Cass, D., & Shell, K. (1983). Do sunspots matter? *Journal of Political Economy*, *91*(2), 193–227.

Chipman, J. S. (1974). Homothetic preferences and aggregation. *Journal of Economic Theory*, *8*(1), 26–38.

Cozzi, G., Goenka, A., Kang, M., & Shell. K. (2015). Price-level volatility and optimal nominal taxes: Aggregate welfare. In *Working paper*. Cornell University.

Cozzi, G., Goenka, A., Kang, M., & Shell, K. (2016). Price-level volatility and optimal nominal taxes: Individual welfare. In *Working paper*. Cornell University.

Goenka, A., & Préchac, C. (2006). Stabilizing sunspots. *Journal of Mathematical Economics*, *42*(4–5), 544–555.

Kajii, A. (2007). Welfare gains and losses in sunspot equilibria. *Japanese Economic Review*, *58*(3), 329–344.

Peck, J., & Shell, K. (1991). Market uncertainty: Correlated and sunspot equilibria in imperfectly competitive economies. *Review of Economic Studies*, *58*(5), 1011–1029.

Chapter 17
A Note on Information, Trade and Common Knowledge

Leonidas C. Koutsougeras and Nicholas C. Yannelis

Abstract We recast the well known no trade result in Milgrom and Stokey (1982) using the appropriate definition of efficiency among several available in the asymmetric information framework.

Keywords Information · Efficiency · Common knowledge

JEL Classification Number D5 · D81 · D82 · D86

17.1 Introduction

Milgrom and Stokey (1982) developed a well known result in an economy with uncertainty which is characterized by asymmetric information. It stated that, once an efficient allocation is reached in the *ex ante* stage, it cannot be common knowledge that it can be improved upon, even after individuals receive signals and update their knowledge about the state of nature (i.e., in the *interim*). A nice interpretation suggested in that article is that if an individual is offered a trade, then with the understanding that it is feasible and improving for the proposer(s), would infer that the proposer(s) know something that the individual does not. Using this inference to refine private information, the individual would discover that this trade cannot be beneficial and would turn it down. In conclusion there can be no further trade even after individuals update their beliefs.

Notwithstanding the validity of the result, there is a loose end in its development. In particular, the meaning of Pareto efficiency is unclear in that paper. There are several notions of efficiency that can be defined in the asymmetric information framework, depending on the extent of information that is diffused across individuals. In absence of a suitably defined efficiency notion it is not easy to grasp the result. For instance,

L.C. Koutsougeras (✉)
School of Social Sciences, University of Manchester, Manchester, England
e-mail: leonidas@manchester.ac.uk

N.C. Yannelis
Department of Economics, University of Iowa, Iowa, USA

© Springer International Publishing AG 2017   403
K. Nishimura et al. (eds.), *Sunspots and Non-Linear Dynamics*,
Studies in Economic Theory 31, DOI 10.1007/978-3-319-44076-7_17

there are notions of efficiency in the context of asymmetric information for which the proof of the result does not apply. Notably, if *ex ante* Pareto efficiency is understood relative to trades which are adapted to common knowledge events (i.e., only publicly verifiable trades) then in the interim stage mutually improving re trading may be possible.

Using tools that were developed in the subsequent literature, we set out to reconstruct the Milgrom and Stokey (1982) result, by providing formal definitions of efficiency under asymmetric information, which confirm its validity. Specifically, drawing on Yannelis (1991) we provide definitions for efficiency, based on trades compatible with three possible information sharing regimes: common knowledge (coarse), private and pooled (fine) information efficiency.[1] We show that if Pareto efficiency is understood in the private information sense, i.e., allocations and possible reallocations are adapted to the private information of each individual, then the no trade result is valid.

We proceed by developing the context and definitions in Sect. 17.2. Section 17.3 features our results and some concluding remarks follow in Sect. 17.4.

17.2   The Model

Consider an economy with uncertainty which is described as follows. Let Ω denote the set of all possible states of nature. The uncertainty in the model is described by the triple $(\Omega, \mathcal{F}, \mu)$ where $\mathcal{F}$ is the family of all possible events (a σ-field of subsets of Ω) and μ is a probability distribution on the events in $\mathcal{F}$. There is a finite set of agents, denoted by H and a finite set of physical commodities denoted by L. We will take the commodity space in each state to be $\Re_+^L$. Let L_1 be the space of equivalence classes of integrable functions $x : \Omega \to \Re_+^L$.

An *economy* is defined as $\mathcal{E} = \{(X_i, u_i, e_i, \mu) : i \in H\}$ where for each $i \in H$:

(i) $X_i : \Omega \to 2^{\Re_+^L}$ is the random commodity set
(ii) $u_i : \Omega \times \Re_+^L \to \Re$ is the utility function
(iii) $e_i : \Omega \to \Re_+^L$, $e_i(\omega) \in X_i(\omega)$, *a.e* and e_i is integrable, is the random initial endowment
(iv) μ is the (common) prior.[2]

Define for each $i \in H$, the set of (contingent) consumption plans of agent i as:

$$L_{X_i} = \{x \in L_1 : x(\omega) \in X_i(\omega) \ a.e.\}.$$

---

[1] Koutsougeras and Yannelis (1995) introduce refinements such as the strong-coarse and weak-fine information efficiency.

[2] We have considered for simplicity the same prior for all agents, but different priors can be easily accommodated.

We assume that for each $i \in H$, the utility function $u_i(\cdot, x)$ is measurable, so for each $x \in L_{X_i}$ the *ex ante* expected utility is well defined and given by:

$$v_i(x) = \int_{\omega \in \Omega} u_i(\omega, x(\omega)) d\mu.$$

The *private information* of each individual $i \in H$ is represented by sub σ-field $\mathcal{F}_i \subseteq \mathcal{F}$, generated by some partition of Ω, which captures all the events which are verifiable by an individual. The interpretation is that upon occurrence of a state of nature $\omega \in \Omega$ each individual $i \in H$ perceives that the event $\mathcal{F}_i(\omega) \in \mathcal{F}_i$ has occurred.[3] When $\mathcal{F}_i(\omega) = \mathcal{F}_i(\omega')$ for some pair $\omega \neq \omega'$ information is *incomplete* and when $\mathcal{F}_i(\omega) \neq \mathcal{F}_j(\omega)$ for some pair $i \neq j$ it is *asymmetric* among individuals. We assume that $\sigma(e_i, u_i) \subset \mathcal{F}_i$, $\forall i \in H$, i.e., personal characteristics (preferences and endowment) are not informative signals for each individual.[4]

The effect of private information in an economy with uncertainty is that it limits the contingent consumption sets of individuals to those which are compatible with their information, i.e., those which are compatible with the events they can distinguish. Formally the contingent consumption sets of individuals are:

$$L_{X_i}^P = L_{X_i} \cap \{x \in L_1 : x \text{ is } \mathcal{F}_i - measurable\}$$

We can now define an **ex ante economy with private information** as $\mathcal{E}^a = \{(L_{X_i}^P, v_i, e_i, \mathcal{F}_i) : i \in H\}$.

In the sequel we will need the concept of *interim* expected utility, which encapsulates the way that individuals evaluate contingent plans after they become aware of the occurrence of an event. Given $\omega \in \Omega$ the *interim* expected utility of an individual $i \in H$ is given by:

$$v_i^\omega(x) = \int_{\omega' \in \Omega} u_i(\omega', x(\omega')) dq_i.$$

where

$$q_i(\omega') = \begin{cases} \frac{\mu(\omega')}{\int_{\omega'' \in \mathcal{F}_i(\omega)} \mu(\omega'') d\mu} & if \ \omega' \in \mathcal{F}_i(\omega) \\ 0 & otherwise \end{cases} \tag{17.1}$$

The interpretation is that once individuals receive a signal that a state of nature $\omega \in \Omega$ has occurred, they perform a Bayesian update of their priors according to the event $\mathcal{F}_i(\omega)$ that they have perceived.

---

[3] As a matter of notation, the event $\mathcal{F}_i(\omega)$ is understood to be the smallest one in $\mathcal{F}_i$ containing the state ω. In fact, $\mathcal{F}_i(\omega)$ is the element in the partition that generates $\mathcal{F}_i$ containing ω.

[4] A possible interpretation of this formulation is that contracted allocations are executed after preferences and endowments have materialized.

Definition 1 For a given $T \subseteq H$, an element $x \in \prod_{i \in T} L_{X_i}$ such that $\sum_{i \in T} x_i(\omega) = \sum_{i \in S} e_i(\omega)$, ae is called a **feasible allocation** (or simply an allocation) for T. The set of all possible allocations for T is denoted by $\mathcal{C}_T$.

Various definitions of Pareto efficiency can be developed depending on the information that individuals have at their disposal. In particular, following Yannelis (1991) we can define the following notions of efficiency.

Definition 2 An allocation $x \in \mathcal{C}_H$ is Pareto **private information efficient** (efficient relative to private information) if and only if

(i) x_i is $\mathcal{F}_i$-measurable $\forall i \in H$
(ii) $\nexists\, y \in \mathcal{C}_H$ such that y_i is $\mathcal{F}_i$-measurable and $v_i(y_i) > v_i(x_i)$, $\forall i \in H$.

Definition 3 An allocation $x \in \mathcal{C}_H$ is Pareto **coarse efficient** (efficient relative to common knowledge information) if and only if

(i) x_i is $\mathcal{F}_i$-measurable $\forall i \in H$
(ii) $\nexists\, y \in \mathcal{C}_H$ such that y_i is $\bigwedge_{i \in H} \mathcal{F}_i$-measurable and $v_i(y_i) > v_i(x_i)$, $\forall i \in H$.

If individuals are allowed to use pooled information, i.e., trades verifiable with all available information in the economy, in order to improve over a proposed allocation then we have the following definition

Definition 4 An allocation $x \in \mathcal{C}_H$ is Pareto **fine efficient** (efficient relative to pooled information) if and only if

(i) x_i is $\mathcal{F}_i$-measurable $\forall i \in H$
(ii) $\nexists\, y \in \mathcal{C}_H$ such that y_i is $\bigvee_{i \in H} \mathcal{F}_i$-measurable and $v_i(y_i) > v_i(x_i)$, $\forall i \in H$.

Remark 1 In the preceding definitions we have used as private information for each individual their respective initial information $\mathcal{F}_i$. However, the same definitions can be casted, using an information structure $\mathcal{F}_i' \supseteq \mathcal{F}_i$, which captures information that individuals might have inferred during a negotiation or trading process. The key matter distinguishing the alternative notions above, is how much of this information individuals may exchange in their effort to improve over a given allocation.

17.3 Results

Let $x \in \mathcal{C}_H$ be a Pareto private information efficient allocation. Suppose a state of nature $\omega \in \Omega$ occurs and all individuals receive the corresponding signals which indicate that the event $\mathcal{F}_i(\omega) \in \mathcal{F}_i$ has occurred. Consider a net trade $t : \Omega \to \Re^{LH}$, which is $\mathcal{F}_i$-measurable for each $i \in H$. The event that this net trade is feasible and improving for all parties involved is:

$$S = \{\omega' \in \Omega : \sum_{i \in H} t_i(\omega') = 0 \text{ and } \forall i \in H, \ v_i^{\omega'}(x_i + t_i) > v_i^{\omega'}(x_i)\}$$

The result in Milgrom and Stokey (1982) (Theorem 1, p. 21) can be reconstructed as follows

Theorem 1 $S \in \bigwedge_{i \in H} \mathcal{F}_i \Rightarrow \mu(S) = 0$

Equivalently the theorem asserts that if $\mu(S) > 0$ then $S \notin \bigwedge_{i \in H} \mathcal{F}_i$. Thus, the meaning of the theorem is that a non trivial event that a net trade is feasible and *interim* improving over x for all parties, cannot be common knowledge.

Proof Suppose that $S \in \bigwedge_{i \in H} \mathcal{F}_i$ and $\mu(S) > 0$. It follows that $S \in \mathcal{F}_i$ for each $i \in H$. In particular, $S = \bigcup_{\omega \in S} \mathcal{F}_i(\omega)$.
Consider now $z \in \prod_{i \in H} L_{X_i}$ defined as follows for each $i \in H$:

$$z_i(\omega') = \begin{cases} x_i(\omega') + t_i(\omega') & \text{if } \omega' \in S \\ x_i(\omega') & \text{if } \omega' \notin S \end{cases} \tag{17.2}$$

Clearly $z \in \mathcal{C}_H$ by construction and z_i is $\mathcal{F}_i$-measurable $\forall i \in H$. Furthermore,

$$
\begin{aligned}
v_i(z_i) &= \int_{\omega' \in \Omega} u_i(\omega', z_i(\omega'))d\mu \\
&= \int_{\omega' \in S} u_i(\omega', z_i(\omega'))d\mu + \int_{\omega' \notin S} u_i(\omega', z_i(\omega'))d\mu \\
&= \int_{\omega' \in S} u_i(\omega', x_i(\omega') + t_i(\omega'))d\mu + \int_{\omega' \notin S} u_i(\omega', x_i(\omega'))d\mu \\
&= \int_{\omega'' \in \bigcup_{\omega' \in S} \mathcal{F}_i(\omega')} u_i(\omega'', x_i(\omega'') + t_i(\omega''))d\mu + \int_{\omega' \notin S} u_i(\omega', x_i(\omega'))d\mu \\
&= \int_{\omega' \in S} \int_{\omega'' \in \mathcal{F}_i(\omega')} u_i(\omega'', x_i(\omega'') + t_i(\omega''))d\mu + \int_{\omega' \notin S} u_i(\omega', x_i(\omega'))d\mu \\
&= \int_{\omega' \in S} \mu(\mathcal{F}_i(\omega')) \int_{\omega'' \in \mathcal{F}_i(\omega')} u_i(\omega'', x_i(\omega'') + t_i(\omega''))dq_i + \int_{\omega' \notin S} u_i(\omega', x_i(\omega'))d\mu \\
&= \int_{\omega' \in S} \mu(\mathcal{F}_i(\omega'))v_i^{\omega'}(x_i + t_i) + \int_{\omega' \notin S} u_i(\omega', x_i(\omega'))d\mu \\
&> \int_{\omega' \in S} \mu(\mathcal{F}_i(\omega'))v_i^{\omega'}(x_i) + \int_{\omega' \notin S} u_i(\omega', x_i(\omega'))d\mu \\
&= \int_{\omega' \in S} u_i(\omega', x_i(\omega'))d\mu + \int_{\omega' \notin S} u_i(\omega', x_i(\omega'))d\mu \\
&= v_i(x_i) \tag{17.3}
\end{aligned}
$$

which contradicts the hypothesis that x is a Pareto private information efficient allocation $\qquad \square$

The interpretation of this result is that when an efficient allocation has been reached, even after individuals update their beliefs not everyone agrees that a further trade is feasible and individually rational. An interesting way to restate this is the following: an individual who is offered a trade, with the understanding that the offer is feasible and improving for the other party, should infer that the other party knows something about the state of nature that this individual doesn't. By using this inference to refine information the individual would conclude that the proposed trade cannot be beneficial and so turn it down.

Remark 2 A careful look suggests that the proof of the Theorem (1) does not cover the Definition (3) of Pareto efficiency (coarse efficiency). Indeed, coarse efficiency requires that the allocation z in the proof of Theorem (1) is $\bigwedge_{i\in H} \mathcal{F}_i$-measurable, which is not generally true.

On the other hand that proof suggests that there is no need to insist at common knowledge events. A somewhat stronger result can be proved in a similar way. Specifically, let

$$S_i = \{\omega \in \Omega : \sum_{i\in H} t_i(\omega) = 0 \text{ and } v_i^\omega(x_i + t_i) > v_i^\omega(x_i)\}$$

The following result is shown in the same way as Theorem (1).

Theorem 2 $S_i \in \mathcal{F}_i, \forall i \in H \Rightarrow \prod_{i\in H} \mu(S_i) = 0.$

Similarly as in Theorem (1), this theorem asserts that the nontrivial events that a net trade is feasible and improving over x for each party, cannot be private knowledge for all parties, i.e., if $\prod_{i\in H} \mu(S_i) > 0$ then $\exists i \in H$, s.t. $S_i \notin \mathcal{F}_i$. The interpretation of this theorem is similarly that if a net trade is offered to an individual $i \in H$, with the understanding that $S_j \in \mathcal{F}_j, \forall j \neq i$, it must be that $S_i \notin \mathcal{F}_i$, i.e., the proposed trade is either infeasible or not improving for i. Either way the individual would turn down the offer.

17.4 Concluding Remarks

We have reconstructed the well known result in Milgrom and Stokey (1982), based on an appropriate definition of efficiency in the context of asymmetric information. We also provided a slight extension showing that improving trades cannot be consistent with the private information of all parties.

It is noteworthy that the expected utility hypothesis is crucial for this result. A careful review of its proof (also in its original version) makes clear that it is based on the linearly additive structure of the expected utility and the way it is related to its conditional expectation versions through Bayesian updating. This observation suggests that a failure of this result in non expected utility formulations of asymmetric information is possible. The validity of the no trade conclusion of Milgrom and Stokey (1982) in a non expected utility formulation of asymmetric information is an open question, which we intend to study in subsequent work.

References

Koutsougeras, L., & Yannelis, N. C. (1995). Incentive compatibility and information superiority of the core of an economy with differential information. *Economic Theory, 3*, 195–216.

Milgrom, P., & Stokey, N. (1982). Information, trade and common knowledge. *Journal of Economic Theory, 26*(1), 17–27.

Yannelis, N. C. (1991). The core of an economy with differential information. *Economic Theory, 1*, 183–198.

CPSIA information can be obtained
at www.ICGtesting.com
Printed in the USA
BVOW07*0216130117
473416BV00002B/4/P

9 783319 440743